The Commons in a Glocal World

This volume focuses on how, in Europe, the debate on the commons is discussed in regard to historical and contemporary dimensions, critically referencing the work of Elinor Ostrom. It also explores from the perspective of new institutional political ecology (NIPE) how Europe directly and indirectly affected and affects the commons globally.

Most of the research on the management of commons pool resources is limited to dealing with one of two topics: either the interaction between local participatory governance and development of institutions for commons management, or a political-economy approach that focuses on global change as it is related to the increasingly globalised expansion of capitalist modes of production, consumption and societal reproduction. This volume bridges the two, addressing how global players affect the commons worldwide and how they relate to responses emerging from within the commons in a global-local (glocal) world. Authors from a range of academic disciplines present research findings on recent developments on the commons, including: historical insights; new innovations for participatory institutions building in Europe or several types of commons grabbing, especially in Africa related to European investments; and restrictions on the management of commons at the international level. European case studies are included, providing interesting examples of local participation in commons resource management, while simultaneously showing Europe as a centre for globalized capitalism and its norms and values, affecting the rest of the world, particularly developing countries.

This book will be of interest to students and researchers from a wide range of disciplines including natural resource management, environmental governance, political geography and environmental history.

Tobias Haller is Professor at the Institute of Social Anthropology, University of Bern, Switzerland.

Thomas Breu is Director, Centre for Development and Environment, and Executive Director, International Graduate School North-South, University of Bern, Switzerland.

Tine De Moor is a Professor in the Department of Social and Economic History, Utrecht University, the Netherlands.

Christian Rohr is Professor of Environmental and Climate History, University of Bern, Switzerland.

Heinzpeter Znoj is Professor at the Institute of Social Anthropology and director of the Board of the Center for Development and Environment at the University of Bern, Switzerland.

Earthscan Studies in Natural Resource Management

Southern African Landscapes and Environmental Change
Edited by Peter Holmes and John Boardman

Large Carnivore Conservation and Management
Human Dimensions
Edited by Tasos Hovardas

Forest Management Auditing
Certification of Forest Products and Services
Edited by Lucio Brotto and Davide Pettenella

Agricultural Land Use and Natural Gas Extraction Conflicts
A Global Socio-Legal Perspective
Madeline Taylor and Tina Hunter

Tropical Bioproductivity
Origins and Distribution in a Globalized World
David Hammond

The Commons in a Glocal World
Global Connections and Local Responses
Edited by Tobias Haller, Thomas Breu, Tine De Moor, Christian Rohr and Heinzpeter Znoj

Natural Resource Conflicts and Sustainable Development
Edited by E. Gunilla Almered Olsson and Pernille Gooch

For more information on books in the Earthscan Studies in Natural Resource Management series, please visit the series page on the Routledge website: www.routledge.com/books/series/ECNRM/

The Commons in a Glocal World

Global Connections and Local Responses

Edited by Tobias Haller, Thomas Breu, Tine De Moor, Christian Rohr and Heinzpeter Znoj

Schweizerische Akademie der Geistes- und Sozialwissenschaften
Académie suisse des sciences humaines et sociales
Accademia svizzera di scienze umane e sociali
Academia svizra da scienzas umanas e socialas
Swiss Academy of Humanities and Social Sciences

ssas
ssea
sgas

Swiss Society for African Studies
Société suisse d'études africaines
Schweizerische Gesellschaft
für Afrikastudien

BURGERGEMEINDE BERN

Routledge
Taylor & Francis Group
LONDON AND NEW YORK

earthscan
from Routledge

First published 2019 by Routledge

2 Park Square, Milton Park, Abingdon, Oxon, OX14 4RN
605 Third Avenue, New York, NY 10017

Routledge is an imprint of the Taylor & Francis Group, an informal business

First issued in paperback 2020

British Library Cataloguing-in-Publication Data
A catalogue record for this book is available from the British Library

Library of Congress Cataloging-in-Publication Data
Names: Haller, Tobias, editor. | Breu, Thomas, editor. | Moor, Tine De, editor. | Rohr, Christian, editor. | Znoj, Heinzpeter, editor.
Title: The commons in a glocal world : global connections and local responses / edited by Tobias Haller, Thomas Breu, Tine De Moor, Christian Rohr, Heinzpeter Znoj.
Description: Abingdon, Oxon ; New York, NY : Routledge, 2019. | Includes bibliographical references.
Identifiers: LCCN 2018058038 (print) | LCCN 2019006288 (ebook) | ISBN 9781351050982 (eBook) | ISBN 9781138484818 (hbk) | ISBN 9781351050982 (ebk)
Subjects: LCSH: Commons. | Glocalization. | International business enterprises. | International organizations.
Classification: LCC HD1286 (ebook) | LCC HD1286 .C638 2019 (print) | DDC 333.2–dc23
LC record available at https://lccn.loc.gov/2018058038

ISBN: 978-1-138-48481-8 (hbk)
ISBN: 978-0-367-72963-9 (pbk)

Typeset in Goudy
by Newgen Publishing UK

Contents

Illustrations

Figures

Part I

Part II

Part III

Part IV

Tables

Contributors

Editors

Tobias Haller is Professor at the Institute of Social Anthropology at the University of Bern, Switzerland. He studied social anthropology, geography and sociology at the University of Zurich, Switzerland and also graduated there. He did research on institutional change in agriculture and common-pool resources management in Cameroon and Zambia, led several comparative research projects on the management of the commons in floodplains in Mali, Cameroon, Tanzania, Zambia and Botswana, on land, water and green grabbing with impact on gender relations in Kenya, Sierra Leone, Morocco, Ghana, Tanzania, Malawi, on food systems in Kenya and Bolivia, on social and environmental impacts of oil and mining companies worldwide, on the management of the commons in Switzerland and on constitutionality (participatory bottom-up institution building processes).

Thomas Breu is Professor and Director of the Centre for Development and Environment, and Executive Director of the International Graduate School North-South at the University of Bern, Switzerland. He holds degrees in geography and economics and has over 20 years of experience in conducting research and policy dialogue in developing countries, particularly in Southeast Asia, East Africa and Central Asia. Among other topics, his research focuses on the effects of globalization on natural resources and on the livelihoods of rural populations in developing countries.

Tine De Moor is Professor at the Department of Social and Economic History at Utrecht University, the Netherlands. She studied History at the University of Ghent and Queen Mary's and Westfield College in London, and Environmental Sciences at the University of Antwerp, Belgium. Her work focuses on institutions for collective action, in particular in historical perspective. She is currently involved in various projects focusing on the long-term dynamics of such institutions, amongst others commons. De Moor co-founded The International Journal of the Commons and is past-president of the International Association for the Study of the Commons (IASC). Further information about her work and research team can be found via www.collective-action.info.

Christian Rohr is Professor of Environmental and Climate History at the University of Bern, Switzerland, working both at the Institute of History and the interdisciplinary Oeschger Center for Climate Change Research. He studied history and antique and medieval Latin at the University of Vienna, where he obtained his PhD in 1995. During his time as Assistant Professor and later on as Associate Professor of Medieval History at the Institute of History, University of Salzburg (1996–2010) his research started to concentrate on nature-induced disasters, climatic fluctuations and conflicts over natural resources in the medieval and early modern period. Since being in Bern from 2010 onwards, he widened his scope also on modern and contemporary environmental history. He is currently one of the leaders of the interdisciplinary project *Sustainable Commons Adaptations to Landscape Ecosystems in Switzerland (SCALES): Institutional Change, Constitutional Innovations and Public Policies in Swiss Resource Management* funded by the Swiss National Science Foundation.

Heinzpeter Znoj is Professor at the Institute of Social Anthropology and director of the Board of the Center for Development and Environment at the University of Bern, Switzerland. He is an economic anthropologist specializing on theories of exchange and the anthropology of money. He has done research on the societal impact of accelerated rural development in Sumatra and West Papua, Indonesia. A related research interest is governance under conditions of systemic corruption.

Authors

Ward Anseeuw is a development economist and policy analyst, a research fellow at the Agricultural Research Centre for International Development (CIRAD). He is presently seconded to the International Land Coalition as a Senior Technical Specialist responsible for "Knowledge, Learning, Innovation and data". For the last 12 years, he was seconded to the University of Pretoria, as a senior research fellow to the Post-Graduate School of Agriculture and Rural Development and as the co-director of the Center for the Study of Governance Innovations (GovInn) – which he founded in 2012. His work focuses mainly on issues of agricultural and land policies, agrarian and land reforms, large-scale land acquisitions as well as to participatory approaches of data generation, governance and advocacy regarding land. He has published extensively on these issues in scientific journals and with renowned publishers; including *Land, Transition and Compromise* (with Chris Alden, Palgrave, 2009), *The Struggle Over Land in Africa – Conflicts, Politics and Change* (with Chris Alden, HSRC Press, 2010), *South Africa's Agrarian Reform* (in French, Editions Universitaires Européennes, 2011), *South Africa's Agrarian Question* (HSRC Press, 2016) and *Inclusive Businesses in Agriculture* (SunMedia Press, 2017).

Yudi Bachrioktora is currently a PhD student at the Institute of Social Anthropology at the University of Bern, Switzerland. He is also lecturer at the History Department at the University of Indonesia and a researcher at the Agrarian Resources Center in Bandung, Indonesia. His research mainly focuses on agrarian, human rights and environmental issues in Indonesia. He is also interested in other fields of studies, such as youth and cultural studies.

Azernoosh Bazrafkan is a doctoral researcher at Maastricht University, the Netherlands. Her doctoral research focuses on international investment law from the perspective of development. She obtained her L.L.M. degree in International and European Law from the Vrije Universiteit Amsterdam and her Bachelor degree with a major in Law and a minor in Economics from Leiden University, the Netherlands. Azernoosh's research interests are in the field of globalization, the development of international trade and investment law and human rights.

Sarah Baumgartner has recently obtained her PhD at the Institute of History at the University of Bern, Switzerland. Her PhD research focuses on societies of natural sciences in the eighteenth century. Further research interests are the history of agriculture in the Early Modern Age and the history of commons.

Elisabetta Cangelosi is a gender advisor for the International Land Coalition Secretariat and has been working on Women's Land Rights, Land and Environmental Rights Defenders and Gender Justice for ILC Secretariat since 2013. She holds a PhD in Social Sciences from the Istituto Italiano di Scienze Umane (Florence and Siena), an LLM in Human Rights Law (BSIS-University of Kent, UK), and an MA in Classics (University of Siena, Italy). Since 2012, she has been conducting research and consulted with a variety of NGOs and organisations, on a diverse range of topics such as land and water, human rights, gender justice, commons and social movements. She teaches at SciencesPo in Paris, France ('Water: a contemporary issue' and 'Gender and Development') and at BSIS-University of Kent, UK ('Gender in Conflicts, Development and Migration').

Wytske Chamberlain is a PhD candidate in Rural Development at the University of Pretoria, South Africa, where she conducts research into inclusive business models in the agricultural sector in South Africa. She holds a Master in Human Geography from the University of the Witwatersrand, South Africa, and an Honours Bachelor of Arts in International Economics and Economic Geography from Utrecht University, the Netherlands. In addition, she has worked as Coordinator Regional Focal Point Africa for the Land Matrix Initiative, covering large-scale land acquisitions across Africa. She co-authored the second analytical report 'International Land Deals for Agriculture. Fresh insights from the Land Matrix'.

Maïka De Keyzer is a historian from the University of Leuven (KUL), Belgium, connected to the medieval history research unit. She is an associate professor in the field of socio-economic and ecological pre-industrial history. Her research focuses on the ability of pre-modern societies to manage natural resources sustainably and redistribute benefits equitably in the long run. Her past work has centered on historical commons and natural disasters. Her current research project is titled 'The impact of "degrowth" and market economies on welfare and sustainability: a historical exploration'.

Ramez Eid is a social anthropologist and lecturer at Kinneret Academic College and Sakhnin College, Israel. He studied social anthropology at the University of Texas at Arlington, USA and at the University of Bern, Switzerland. His main research interests include bottom-up common-pool resources management, local democratic practices and the commons, and the social and environmental dynamics in protected areas and world heritage sites. He did his main research projects in different sites in the Mediterranean. His last research project was titled 'Burning Forests, Rising Power: Towards a Constitutionality Process in Mount Carmel Biosphere Reserve'.

Markus Giger is an agricultural economist and presently co-leader of the Sustainability Governance Cluster at the Centre for Development and Environment (CDE) of the University of Bern, Switzerland. Before joining the CDE he has been working for the International Labour Office in Indonesia and the FAO in Rome. Markus Giger is involved in several projects that relate to international land acquisitions, amongst them the Land Matrix Initiative, and the AFGOLAND Project, an interdisciplinary project that investigates large agricultural investments in Kenya, Mozambique and Madagascar. He is the co-author of several scientific papers dealing with land acquisitions, and their impacts on livelihoods, poverty and sustainability. He co-authored the second analytical report 'International Land Deals for Agriculture. Fresh insights from the Land Matrix'.

Désirée Gmür holds a Master's Degree in Social Anthropology, Geography and Social and Economic History from the University of Zurich, Switzerland, and is a PhD student at the Institute of Social Anthropology, University of Bern, Switzerland in the research project 'Large-Scale Land Aquisitions and Gender in Africa' funded by the Swiss National Science Foundation. She did social anthropological research among the Wahehe ethnic group in two village settings affected by a forestry investment in Iringa region in Tanzania in terms of food security, access to common-pool resources and gender relations. Her detailed research data sheds light on the strategies of companies involved in such deals, as well as on local reactions and strategies. It shows that women's already difficult position in society and economy is getting more vulnerable and less resilient due to such investments.

Anne-Lise Head-König is Professor Emerita of Economic History at the University of Geneva, Switzerland. Her main interests and subjects of research are, among others, comparative approaches to rural history, economic changes in alpine regions, the common and collective use of resources, farm transfers, inheritance practices and citizenship. These research activities were also developed within a network of European projects funded by the European

Union. She is currently working on the problems of the necessary adjustment of mountain economies to increasing competition from lowland areas over time (seventeenth to twentieth century).

Rahel Jud is a PhD student at the Institute of Social Anthropology at the University of Bern, Switzerland. She studied social anthropology, political science and psychology at the Universities of Zurich and Bern, Switzerland. For her master thesis she has conducted research on the implementation of the *Reducing Emissions from Deforestation and Forest Degradation* (REDD+) policy in Indonesia and its impact on existing tenure status regarding forest land of the local population.

Girma Kelboro is a senior researcher at the Center for Development Research (ZEF), University of Bonn, Germany. His scientific field of expertise is agro-sociology with a PhD degree in agriculture specializing in the social dimension of natural resource management and farming. His research interests are agricultural and rural development governance, institutions, innovation and knowledge in smallholder agriculture, biodiversity conservation, livelihoods and sustainability. He currently coordinates agricultural innovation governance research and science-policy dialogues in the Department of Political and Cultural Change at ZEF.

Madlen Kobi is a postdoctoral researcher at the Academy of Architecture in Mendrisio, Switzerland (Università della Svizzera Italiana). She obtained a PhD in Social Anthropology from the University of Bern, Switzerland, and worked subsequently as a postdoctoral researcher and lecturer at the Department of Social Anthropology at the University of Zurich, Switzerland. Her research and teaching interests include urban anthropology, architectural anthropology and the anthropology of waste use with a regional focus on China. In previous projects, she has done research on natural resource exploration and the impacts of mining and oil exploration on indigenous/local populations.

Angelika Lätsch is a PhD student at the Institute of Social Anthropology at the University of Bern, Switzerland. She studied social anthropology, ecology and sociology at the University of Bern, Switzerland and also graduated there. She did research on the implementation of indigenous policies on a local level and research on institutional change and common-pool resources management in coastal areas of northern Norway and on constitutionality (participatory bottom-up institution building processes).

Gabriela Landolt is a PhD student at the Institute of Social Anthropology, University of Bern, Switzerland. Her main research interests are the analysis of institutional change in the context of common-pool resource management under special consideration of power relations and ideology. This research project is associated to the following research programs: (1) The Swiss National Centre of Competence in Research (NCCR) North-South, Research Partnerships for Sustainable Development, co-funded by the Swiss

National Science Foundation (SNF) and the Swiss Agency for Development and Cooperation (SDC); and (2) AlpFutur (www.alpfutur.ch). The project was made possible by the financial support of the NCCR North-South, the Department of Agriculture of the Canton of Grisons, and the Centre for Development and Environment CDE (University of Bern).

Kristina Lanz holds a PhD in Social Anthropology from the University of Bern, Switzerland. Her doctoral research focused on the negotiation and implementation and on the manifold gendered impacts of a large-scale land acqusitions in Ghana. In recent years, she has also conducted extensive research on the gendered impacts of mining operations on pastoralist communities in Tanzania and Mongolia. She did her MA in International Studies and Diplomacy at the School of African and Oriental Studies (SOAS), UK and her BA in Development Studies at the University of East Anglia in Norwich, UK. She is interested in the meanings and trajectories of development under globalization and in particular in the role and regulation of global multinational corporations, as well as in local resistance and counter-reactions.

Franziska Marfurt is a PhD student at the Institute of Geography at the University of Bern, Switzerland. She studied Social Anthropology and Middle Eastern Studies at the Universities of Bern and Fribourg, Switerzland. She worked on land rights, gender and changing power relations in the context of large-scale land acquisition projects in Sierra Leone. Currently she is researching the political ecology of work and social-ecological transitions in Senegal.

Peter Messerli is Professor of Sustainable Development at the University of Bern, Switzerland, and the director of the Centre for Development and Environment (CDE). As a land system scientist and human geographer his research interests lie in the sustainable development of social-ecological systems in Africa and Asia. He focuses on increasingly globalized and competing claims on land, rural transformation processes, and spatial manifestations of their outcomes. He has lived and worked more than ten years in Madagascar and Laos directing large-scale research projects focusing on inter- and transdisciplinarity. He has extensive experience in science-policy interactions from the local to the global level. Peter Messerli is the co-chair of Future Earth's Global Land Programme (GLP) and has been appointed in 2016 as the co-chair of the group of 15 independent scientists drafting the UN Global Sustainable Development Report (GSDR).

Thomas Niederberger is currently finishing his PhD thesis at the Institute of Social Anthropology at the University of Bern, Switzerland. His research topic is about resource conflicts and indigenous autonomy in the northern Peruvian Amazon, where he collaborates with the Gobierno Territorial Autónomo de la Nacion Wampis. Previously, he was the coordinator and main editor for the collective research project 'The Open Cut – Mining, Transnational

Corporations and Local Populations' (LIT Verlag, 2016), carried out with the Institute for Ecology and Action Anthropology at the University of Bern.

Kerstin Nolte is an economic geographer holding a PhD in Economics from the University of Göttingen, Germany. She is Assistant Professor at the Institute of Economic and Cultural Geography at the Leibniz University Hannover, Germany. Her research focuses on rural transformation with a focus on large-scale land acquisitions. Her research on the processes in which land acquisitions take place, on the determinants and impacts builds on field research in Mali, Zambia and Kenya. She co-authored the second analytical report 'International Land Deals for Agriculture. Fresh insights from the Land Matrix'.

Christoph Oberlack is a senior researcher at the Centre for Development and Environment (CDE), University of Bern, and a post-doctoral researcher at the Institute for Geography, University of Bern, Switzerland. His research and teaching interests lie in sustainability governance, telecoupled land systems, climate change adaptation, and archetype analysis drawing on ecological economics, human geography, and institutional analysis.

Birsha Ohdedar is a PhD Candidate at the School of Law at SOAS, University of London, UK. His research examines the human right to water in the context of climate change in India, analysing the operation of laws and policies in climate and water stressed districts in West Bengal and Rajasthan. Alongside his doctoral research, Birsha is also a practising lawyer working in areas of climate change, renewable energy and environmental law. For a number of years he has been attending UN climate change negotiations with Legal Response International, providing legal assistance and advice to climate vulnerable developing countries. Birsha holds an LLM (Environmental Law) from SOAS, University of London, UK and a BA/LLB from University of Auckland, New Zealand.

Rodrigo Polanco Lazo is a senior lecturer and researcher at the World Trade Institute (WTI) at the University of Bern, Switzerland. He is also a post-doctoral researcher at the University of Luzern and a visiting professor at the University of Chile, where he also served as an assistant professor and director of international relations of the Faculty of Law. He holds a Bachelor and a Master of Laws from University of Chile, an L.L.M. in International Legal Studies from New York University (NYU), USA, and a PhD from the University of Bern, Switzerland, specialized in international investment law. Rodrigo is a visiting professor at Universidad Externado of Colombia and of the Pontifical Catholic University of Peru (PUCP). He is also a co-founder of Fiscalía del Medio Ambiente (FIMA) a Chilean non-profit environmental organisation where he serves as director of its environmental law journal (Justicia Ambiental). Rodrigo's main research areas are investment law, trade law, international environmental law, and international economic law.

Jesse Ribot is a professor of environmental politics in the School of International Service at American University in Washington DC, USA, since 2008. He was a professor of Geography and Anthropology at the University of Illinois, USA, a Senior Associate at the World Resources Institute, visiting professor in Urban Studies and Planning department at MIT, and has been a fellow at the Stanford Center for Advanced Studies in Behavioral Sciences, the Department of Politics of The New School for Social Research, Agrarian Studies at Yale University, the Center for the Critical Analysis of Contemporary Culture at Rutgers, Max Planck Institute for Social Anthropology, Woodrow Wilson Center for International Scholars and Harvard Center for Population and Development Studies. Ribot is an Africanist studying local democracy, resource access and social vulnerability. He is a 2018 Guggenheim Fellow based at the Wagner School at NYU and CUNY Graduate Center Anthropology Program.

Sarah Ryser is a PhD candidate at the Institute of Social Anthropology at the University of Bern, Switzerland and the Graduate School Gender Studies Switzerland. She completed her masters in social anthropology and German studies as well as a Master of Arts in secondary education. She did fieldwork in Paris, France about riots in the outskirts and in Morocco on impacts of climate change in water irrigation systems as well as on a large-scale land acquisition for solar energy and its effects on local institutional change and common-pool resource management from a gender lens. In that capacity, she is writing various papers with a special focus on gender, land, sustainable energy, politics of development and an ethnography for her dissertation.

Daniel Schläppi holds a PhD in Swiss History and is an associate researcher at the Institute of History at the University of Bern, Switzerland. Besides working as a freelance historian, archivist and university lecturer he has done a lot of research on historical commons over the last 20 years. His doctoral thesis from 2000 dealt with the so-called 'Zunftgesellschaft zu Schmieden' in Bern dating from the late middle ages. In 2006 he wrote a book on the 'Zunftgesellschaft zu Metzgern' in Bern. In 2009 he initiated a SNF research project on 'Common Property, Collective Resources and the Political Culture of the Old Confederation (17th and 18th centuries)'. In addition to numerous articles on collective resources and historical commons he published the volume 'The Economy of Social Relations' (Die Ökonomie sozialer Beziehungen, together with G. Jancke in 2015) and most recently 'From Commons to Share Economy' (Von der Allmende zur Share Economy, together with M. Gruber in 2018) and the highly regarded essay 'Die Eid-Genossenschaft'. In addition, he also has a career as an internationally known musician.

David Schlosberg is Professor of Environmental Politics in the Department of Government and International Relations, Payne-Scott Professor, and Director of the Sydney Environment Institute at the University of Sydney, Australia. His work focuses on environmental political theory, environmental and climate justice, climate adaptation planning and policy, and contemporary environmental movements. He is the author of *Defining Environmental Justice* (Oxford, 2007); co-author of *Climate-Challenged Society* (Oxford, 2013); and co-editor of both *The Oxford Handbook of Climate Change and Society* (Oxford 2011); and *The Oxford Handbook of Environmental Political Theory* (Oxford 2016). His latest book project is on *Sustainable Materialism: Environmental Movements and the Politics of Everyday Life* (Oxford 2019).

Till Stellmacher is a senior researcher and program coordinator at the Center for Development Research (ZEF), University of Bonn, Germany. He studied agricultural sciences, geography and development sociology. He holds a 15 years track record in interdisciplinary research projects on rural transformation and environmental governance in the Global South. The geographical focus of his work is Ethiopia; however, he also conducted research in Burkina Faso, Côte d'Ivoire, Ghana, Tanzania, Bangladesh, India and Nicaragua. Between 2010–2014, he coordinated the PhD Program 'Environmental Planning' at Addis Ababa University, Ethiopia. He published extensively on local governance, protected areas, smallholder farming and natural resource management.

Martin Stuber is senior scientist at the Institute of History at the University of Bern, Switzerland. He published widely on the history of natural resources (forest, cultivated plants, wetlands, sustainability) and on the history of knowledge in the eighteenth century (Albrecht von Haller, Scientific Networks, Economic Enlightenment). In addition, in recent years he explored the history of the commons in Switzerland in the long term. He is particularly interested in how the common-pool resources changed fundamentally in the course of transformations concerning political circumstances and energetic systems.

Insa Theesfeld is Professor of Agricultural, Environmental and Food Policy at the Martin Luther University Halle-Wittenberg, Germany. She is an agricultural and institutional economist, who developed an interest for governance questions of various shared natural resources. A significant strand of her work explored the interdependencies between property rights on land and water resources. Her geographical focus is on post-socialist countries. She is likewise a known 'Commons' scholar, leading the European branch of the International Association for the Study of the Commons since 2018.

Samuel Weissman holds an MA in Social Anthropology from the University of Bern, Switzerland. He conducted his field research in Kenya surrounding the institutional settings of the protected areas under the Northern Rangelands Trust and the Lewa Wildlife Conservancy. His research focused on conservation and the historical to present-day transformation of landscapes and institutions from common-pool resource systems to semi-private arrangements in a global, transnational stakeholder arena, specifically looking at effects of green grabbing and the re-shaping of biodiversity through a current understanding of species valuation. He is currently researching the transnational economy of the coffee supply-and-value chain in a multi-sited ethnography from Kenya to Switzerland.

Michael Zeheter is a historian working as lecturer and research associate at the University of Trier, Germany. He studied history, philosophy, art history and media studies at the University of Konstanz, Germany, where he also obtained his PhD. He is currently studying the cultural history of mineral water consumption in Europe from the early nineteenth century to the present. His previous research explored the sustainability of the Lake Constance fisheries from a *longue durée* perspective and the environmental history of cholera epidemics in the colonial cities of Madras and Quebec during the nineteenth century.

Acknowledgements

The editors would like to thank the following supporters of the book, who contributed ideally, in kind or by funding to the publication. Some of them have also been involved in supporting the conference 'Commons in a Glocal World: Global Connectionos and Local Responses' which took place at the University of Bern in 2016: Institute of Social Anthropology of the University of Bern, Centre for Development and Environment (CDE) of the University of Bern, Institute of History of the University of Bern, Centre for Global Studies (CGS) of the University of Bern, World Trade Institute of the University of Bern, Swiss Society for African Studies (SSAS), Burgergemeinde Bern, Swiss Academy of Humanities and Social Sciences (SAGW), Swiss Network for International Studies (SNIS), The International Association for the Study of the Commons (IASC).

Very big thanks goes to the two coordinators of the book project, Anna von Sury and Fabian Käser. The editors also would like to thank the following peer reviewers for their great comments, which improved the quality of the contributions. These are in alphabetic order Darci Alexandra, Jill Belsky, Julia Eckert, Robert Fletcher, Richard Hoffmann, Hans Hurni, Stuart Kirsch, Patrick Kupper, Peter Larsen, Esther Leemann, Karina Liechti, Jon Mathieu, Iain McKinnon, Ernst-Ulrich Petersmann, Niklaus Schareika, Stefanie Steinebach.

Introduction

Commons in a 'glocal' world

Tobias Haller, Thomas Breu, Christian Rohr, Tine De Moor and Heinzpeter Znoj

Rationale of the edited volume

This edited volume compiles a thematic selection of papers dealing with the commons in a global and local context. Most of them were presented in a first version at a conference of the International Association for the Study of the Commons (IASC), held at the University of Bern, Switzerland, in May 2016. However, the concept of the book was reworked independently of the conference. The volume focuses on how, in Europe, the debate on the commons is discussed in regard to their historical and contemporary dimensions, thereby referring to the work of Elinor Ostrom. At the same time and from the perspective of a new institutional political ecology (NIPE), it explores how Europe directly affected, and still affects, the commons globally in different localities via the economic involvement of transnational companies, private investments and NGOs, as well as indirectly by means of its policies on climate, the environment and trade. The term 'glocal' in the title of the edited volume refers to this interconnectedness of global dynamics and local realities. The book fills a gap between institutional approaches and approaches related to issues of power and post-colonial studies.

The relevance of this edited volume lies in the observation that much research on the management of common-pool resources (CPRs), such as water, pasture, wildlife, fisheries, and veld products, is limited to dealing with just one of two topics: either the interaction between local participatory governance and the development of institutions for commons management, or a political-economy approach that focuses on global change as it relates to the increasingly globalised expansion of capitalist modes of production, consumption, and societal reproduction. This edited volume aims to bridge both of these topics, by investigating how global players such as European multinational companies and organisations affect the commons worldwide and how they relate to responses emerging from within the commons in a global-local ('glocal') world. Various authors from a range of academic disciplines such as geography, social anthropology, history, development studies, economics, political science, and legal studies present their research findings on the latest development in studying the commons, be these historical insights, new innovations for participatory institution-building in Europe, or several forms of commons grabbing and restrictions on the management of commons at the international level.

The Swiss commons and their theoretical legacy

The volume's focus on Europe (and, in particular, on Switzerland) derives from the importance of these cases for debate on the commons. Here, Netting's work on Törbel, Switzerland, has been particularly significant. Netting conducted fieldwork on the peasantry in Nigeria during his early career in the 1960s, and the limitations he encountered in terms of available historical data led him to decide to do his following project in Törbel (see Netting 1993 for an overview of his research). He searched for a site in which to combine participant observation with a historical perspective, and proceeded to analyse the complete set of demographic data stretching back over centuries which he found in the archives of the Catholic church in Törbel, a small village in the German-speaking part of the Canton of Valais. 'Balancing on an Alp' (Netting 1981) was the title of his book on how the villagers managed CPRs based on common property institutions in this remote location in the mountains. His work was later used by Elinor Ostrom in her Nobel Prize-winning book 'Governing the Commons' (1990) as one of the prime examples from which she distilled the famous so-called design principles for robust institutions. Netting's work still shapes the way in which Swiss common property institutions are perceived in the global debate on the sustainable management of CPRs.

The Swiss case shows two crucial elements that are central to this volume: First, in the early modern period local actors received a degree of political independence from feudal forces and, thus, increased their bargaining power for self-governance. The members of the community were able to craft flexible solutions for resource management problems in an environment characterised by extreme seasonal variation and fragile soil conditions. Second, the basic institutional design lay in the peculiar mix between private and common property. Common property and its management institutions for the pastures made sense because hay needed to be produced on the privately-owned meadows for fodder in the winter. This necessitated feeding and collectively herding cattle on a communal pasture in more elevated, yet less productive, areas during the summer. However, in order to ensure sufficient hay production on private land in this arid Alpine landscape, the meadows required irrigation, which could only be achieved through a collective effort to transport water to the village via communal irrigation channels. This link between differing institutional designs – both common *and* private property – as well as the embedding of neighbourhood, reciprocity and trust into local religious norms and values, in this case made the commons management 'robust'. This was further enabled by the political and power-specific conditions, which were already at that time a 'glocal' outcome of negotiations between villagers and the feudal system. The villagers bought their freedom from these powers by providing mercenaries to their feudal lords, who in turn proceeded to rent them out to European powers (see Netting 1981). However, the claim here is not that such a process was only able to develop in Switzerland: global research on the commons shows similar systems throughout Europe and elsewhere (see Feeney at al. 1990; Ostrom 1990, 2005), as well as in the Global South under pre-colonial conditions (see also Oerebech et al. 2005; Haller 2010).

Issues of 'glocality' in commons studies

What do we mean by speaking of the commons in a 'glocal' world? Robertson's (1995) influential work argues that this means the bringing together of global and local processes in interrelation, emphasising that so-called global processes of unification and homogenisation lead both to local fragmentations as well as to differentiation and divisions (Robertson 1995; Roudometof 2005, 2016; Pulcini 2012). Pulcini further suggests that glocalisation unfolds in a pathological process of 'unlimited individualism' on the one hand, and 'endogamous communitarianism' on the other hand. While the former is argued to foster an individualisation of maximising ego's gains, the latter is a reaction to this process yet also linked to the creation of closed identity groups with hostile consequences in processes of 'Otherings' (Pulcini 2012). However, for the commons in a 'glocal' world, a focus on unification, individualisation and the closure of community is misleading. Glocality instead emphasises the interconnectedness of transnational spaces (see Beck 1999) and the forces of transnationalism in a capitalist world where institutions are seen as 'rules of the game'. The concept of glocality highlights the way in which major global actors, who have decision-making power and who are concentrated in the centres of the Global North, shape global institutional trends, which have an impact on local contexts such as common property institutions for the management of CPRs. Since colonial and post-colonial times, but in particular and increasingly since the 1990s, these forces shape neo-liberal rules, which affect not only the South and the East but also the hinterlands of the Global North and its centres. Erik Swyngedouw's (2004) definition of glocalisation elegantly summarises this:

> 'Glocalisation' refers to the twin process whereby, firstly, institutional/regulatory arrangements shift from the national scale both upwards to supra-national or global scales and downwards to the scale of the individual body or to local, urban or regional configurations and, secondly, economic activities and inter-firm networks are becoming simultaneously more localised/regionalised and transnational.
>
> (Swyngedouw 2004: 24)

Glocalisation, then, shows the way how powerful actors in the global capitalist system move up and down on multiple scales and transform the rules of the game mostly flows of capital, yet also for other (cultural, social, symbolic) flows and networks. In this way, areas are de- or re-territorialised (ibid.), meaning that they are of, or out of, interest to the elite in control of economic flows and the ways in which they are institutionalised and legitimised. This process is central to the majority of the contributions in this volume. One of the basic elements of this twin process is that common property institutions have been changed into state property since the colonial and early post-colonial period and, following the logic of the neo-liberal market, have now been converted into private property. Nevertheless, this must not be understood as a uniform process (see also Ensminger 1992; Haller 2010). New responses have been triggered which re-emphasise the local context and which, in turn, have led to responses at the national and transnational levels. Of central importance here is how a local context and its resources are valued on several scales: in Europe CPRs appear to be of little interest and under less, or less direct, pressure; they face a loss of value and, thus, their maintenance as communities and the public sector struggle to meet these costs (see, for example, the costs necessary to maintain an alpine pasture or communal forestry and its infrastructure). Conversely, in the Global South land and its related CPRs were transformed from common into state property and have attracted even more interest following the financial, food and fuel crisis of 2008.

European investments and their impact on the commons in the Global South

This leads us to a further scalar topic of 'glocality' referred to by Swyngedouw: European countries, and especially Switzerland, host major transnational companies that have an important impact on the global commons, especially in the Global South. These companies are also the major forces driving 'large-scale land acquisitions' (LSLAs) or 'land grabbing' processes as re-territorialisation of capital. They search for land for mining or industrial agricultural production; and they provide and install technology for energy production (from dams to solar energy) which consumes the commons in the Global South. They demand open borders for their products and the protection of the privatisation of the commons; or they open up areas for tourism. And, as these processes have an impact on cultural landscape ecosystems which are labelled as pristine nature, conservation organisations (sometimes in alliance with companies and tourism enterprises) promote green enclosures that affect local livelihoods and reduce access to the commons for vulnerable and marginal households. Such developments have repercussions on the way; men and women of all generations (but especially the youth), as well as indigenous peoples and other marginalised groups and classes, are variously excluded from accessing what formerly formed the base for their shared resources. The issue at hand here is that processes of LSLAs are said to bring development. But these take away not only 'land' but always also land-related CPRs such as water, forest, pasture, and veld products. Compensation may be given to more powerful groups, and to men (who may suddenly become individual landowners), yet not to marginal groups or women. Furthermore, as land prices rise in areas of investment, marginal groups and women holding specific land rights in traditional tenure systems might lose such rights due to an increase in the commercial value of the land. The term 'commons grabbing' is highly appropriate for these processes. However, we suggest that this process should also be labelled as 'resilience grabbing', as it reduces the ability of local resource-users to recover from climate or other shocks such as crop failure, animal and human health problems etc (see the chapters by Sarah Ryser, Desirée Gmür, Kristina Lanz and Franziska Marfurt in this volume).

Local responses to glocalisation

The volume also inquires into responses to such 'glocal' connections. There are cases in Switzerland and Europe, as well as in the Global South, where resistance to commons and resilience grabbing lead to locally developed responses and solutions to global challenges. Several of the chapters contained here illustrate how local actors and groups are able to overcome political and power gaps in order to craft new solutions through processes of bottom-up institution building. These allow local actors to acquire a sense of ownership in the building of institutions for the governance of the commons, in a process that has been labelled as 'constitutionality' (see Haller et al. 2016, 2018). This approach was developed based on case studies that showed locally developed, participatory rules and regulations as a response to 'glocally' induced disturbances. However, this demands that states return basic communal property rights and assist in protecting these rights at a local level. Several chapters on Europe and the Global South in this volume provide persuasive reasons to argue that such solutions are of central importance and deserve our research attention.

Outline of the book

This volume bridges a gap in studies on the commons and contributes to a renewed debate on the politics of the commons in 'glocal' contexts. Its focus is not only on institutional interrelations and the power-dynamics they include, but also lies in addressing gender issues and local responses that go beyond a Northern, hegemonic view of resource governance. Further, it includes debates on issues related to structural, scalar and environmental justice and on how processes and fashions in environmental matters impact upon the commons and the commoners.

The book is outlined as follows: The first part includes new theoretical issues in the study of the commons and their transformations. The second part focuses on historical evidence on commons management in Europe, as well as on contemporary adaptation strategies of commons systems in Europe. A historical part concerns itself with an analysis of various actors and triggering environmental factors which lead to specific types of commons management. This is followed by a contemporary part on European cases, which shows how current commons systems have adapted to external economic and neo-liberal political change, including a focus on innovations in the context of the relative political stability of these European countries. The third part deals with the impact of European investors and NGO actors on the commons of the Global South. Here we discuss commons grabbing processes in LSLAs ('land and water grabbing'), the impact of mining and infrastructure as well as the impact from conservation policies driven by the North ('green grabbing'). To conclude, the final part explores international policies on water and the impact on the commons of Sustainable Development Goals (SDGs) and human rights issues. Information on the content of the chapters can be found below.

Theoretical approaches: historical, causal-justice and institutional-political perspectives

The first part comprises three keynotes and two chapters that were presented in different panels during the conference. The two chapters have a broader theoretical thrust in terms of institutional and economic, as well as power-specific, aspects of the commons, and are well-suited to aid analytical reflections on the case studies that follow in the chapters contained in the remaining sections of the book. The three keynotes with which this book opens are edited transcripts of the talks, which have been revised by the authors, yet retain their oral tone: first, Daniel Schläppi (University of Bern) presents a number of founding elements in thoughts on the commons in Switzerland – a place that, since the work of Netting and Ostrom, is seen as a model for the sustainable management of CPRs. In *Shared ownership as a key issue of Swiss history: common-pool resources, common property institutions and their impact on the political culture of Switzerland from the beginnings to our days*, the author argues that the notion of shared ownership is a key element, and shows that the practice of communing has been present from its very beginnings in the Middle Ages and is, hence, a model of the 'co-production of statehood'.

Looking back is also what the geographer and political ecologist Jesse Ribot (School of International Service at American University, New York University Wagner School, and The Anthropology Program of the City of New York Graduate Center) does in his keynote *Social causality of our common climate crisis: towards a sociodicy for the Anthropocene*. By moving to a more global scale, he argues that the analysis of climate vulnerability of poor groups necessitates a look back at the beginnings of the production of precarity. There is a need to understand causes before anything approaching transformative solutions might be found in regard to adaptation to climate change. Ribot argues that the capacity to adapt to changes is often not available to local actors for social and political reasons. This should lead us to questions of local capacity and resilience.

A link is thus made to David Schlosberg's (University of Sydney) talk on *Disruption, community, and resilient governance: environmental justice in the Anthropocene*, which outlines that environmental justice has rarely been considered in debates on the Anthropocene. This is highly problematic for reasons already emphasised by Ribot, yet here Schlosberg refers to two additional causes, leading to what he terms 'disruptions to communities' – and community relationships with the environment and their commons. First, the occurrence of the 'slow violence' of environmental deterioration (Nixon) reproduces inequality and undermines the ability of communities with their cultures to produce food, access land and provide themselves with the means for good health. Second, and often occurring simultaneously, these disruptions intersect with singular 'quick' events like hurricanes, wildfires and heat waves. These two elements of disruptions caused by the Anthropocene must be addressed if social justice and successful adaptation strategies to disruptions are to take place.

These three pre-reflections – on: a) how well working commons systems are rooted in political and power-specific balanced communal property and actions; b) how in the Global South these are disrupted by lack of power and the creation of vulnerability in contexts of climate change; and c) how the political disruptions in the Anthropocene make quick adaptation strategies useless – are also part of the reflections contained in the two chapters that follow. These address the issues of human rights and social change, as well as reflecting on the value of institutional and political ecology-driven analyses of the interlinkages and power dynamics related to the commons. Elisabetta Cangelosi (SciencePo, Paris) provides *A definition of the commons, between human rights, resistance, and social change*. From a purely traditional and customary form of managing natural resources, the commons now represent a key topic for the debate on economic, social and environmental justice. The debate about the commons plays a relevant role at the global level, especially by connecting experiences of resistance around the world. However, the definition of the commons tends to remain open and Cangelosi's analysis builds upon an approach that combines human rights, tradition and resilience. It also focuses on different types of commons (urban, natural, digital) combining theoretical considerations.

This in turn links to the institutional and political ecology approach of Tobias Haller (University of Bern), who proposes a new theoretical approach with his chapter *Towards a new institutional political ecology: how to marry external effects, institutional change and the role of power and ideology in commons studies*. He shows that previously proposed arguments need to be contextualised in a more structural-scalar and interrelated way in order to address issues of institutional change as developed in the New Institutionalism approach in social anthropology; and he combines this with the study and refinement of the definition of power. In political ecology, the analysis of power, its emergence and its function, are well-studied in frameworks that focus on a Marxist political economy, a Foucauldian constructivist, and a de-constructivist feminism, and the post-human turn. In combination, this helps to conceptualise power dynamics at all scales but often lacks the focus on interdependencies. The marriage of these two approaches – that is, Political Ecology and New Institutionalism, so as to form the New Institutional Political Ecology (NIPE) approach – is argued to be helpful in understanding how mechanisms of institutional change unfold in the interrelation of external and internal factors to a local community. A crucial element here is that external factors change the value of contexts and resources (i.e. a change in relative prices) and affect a local community's bargaining power in selecting institutions and legitimating their choice – which in turn affects the distribution and access to CPRs. This may serve as an additional theoretical framework for looking at the case studies in the subsequent three parts.

Following the discussions that emerge from the introduction, important general reflections can be made for the case studies that form the remainder of the book. The historical cases in Switzerland indicate that the robustness of these commons systems results from local people's increased power to secure common property rights. These can manifest what is labelled to be constitutionality in form of common property and bottom-up institutions that are communally agreed upon and developed by local actors, thereby creating a sense of ownership in the institution-building process (see Haller et al. 2016, 2018). We argue that the empowering of local actors and help in establishing power-specific neutral platforms in order to develop local innovations could be the road forward in successfully tackling the problems faced by commoners in a 'glocal' world.

European examples from the past and present: levelled power, balanced bottom-up solutions and innovations in times of crisis

The second part of the edited volume addresses the issue that Europe can be seen as a lab for sustainable CPR management under state-secured common property institutions. It is comprised of two subsections: in section one we present two further cases from Switzerland (Anne-Lise Head-König; Martin Stuber and Sarah Baumgartner) as well as one on the Benelux area and England (Maïka De Keyzer), and one on the German-Swiss borderlands (Michael Zeheter). The four chapters explain how common property institutions developed and how these were transformed in the pre-industrial period and beyond. These processes took place in situations of asymmetric power relations, when local actors jointly carved out spaces of relative freedom from the mediaeval powers of the church and nobles, both in cities and as well in more remote localities.

Section two contains three case studies (Ramez Eid; Angelika Lätsch; Gabriela Landolt) on contemporary Europe following the end of the age of fuel-wood energy and at a time of structural adjustments brought about by industrial and agro-industrial production after the 1950s, when the relative prices for timber, fish, and pastures as well as related goods such as milk, meat, and other agricultural products had declined drastically in comparison to industrial products and salary work. This is our first encounter with the modern 'glocal' world, whose victims, just as in the Global South, have been local agrarian producers who cannot compete with the lower prices of agrarian products stemming from capitalist production. All three areas presented here show local responses. However, these often go further than just aiming to rescue the commons: responses have been increasingly mixed with issues of local identity, changing power relations and alternative ways of life. It is striking that, in all the local responses witnessed in these three cases, processes of institution shopping and innovations for building local, bottom-up institutions have occurred.

European historical cases

In the first chapter of the history subsection, Maïka De Keyzer (KU Leuven) deals with *Common challenges, different fates. The causal factors of failure or success in the commons: the pre-modern Brecklands (England) and the Campine (Southern Low Countries) compared*. She shows that common property institutions and formal collective action developed simultaneously throughout north-western Europe from the Middle Ages onwards. This can be exemplified by two low-land areas – the Brecklands in South-East England and the Campine area in the Dutch-Belgian borderlands – that faced similar conditions of threat to CPRs (pastures and marginal areas with veld products) which were to be covered by sand. Communities in the Campine area were far more successful than communities in the Brecklands in acting collectively and preventing the sand dunes from covering the pasture, fields and veld and producing so-called 'wastelands'. The author argues that higher power asymmetries and higher inequalities in the Brecklands appear to have undermined collective action, while the buffering of such inequalities and relative political independence led to success in fighting the sand dunes in the Campine area.

A similar stance is taken by Michael Zeheter (Trier University) in his contribution on fisheries in Germany, *For the common good: regulating the Lake Constance fisheries from 1350 to 1800*. He shows that the fisheries on Lake Constance were a common property in medieval and early modern times. Although private property rights pertained close to the shore, all of the 'deep' lake and most of the littoral were open to all inhabitants living in the region who were under the jurisdiction of the neighbouring principality, while the 'deep' lake was beyond anyone's jurisdiction. However, this did not lead to overuse of the fishery because, in practice, specialised skills and equipment were needed to exploit the resource and make a living from selling catches on the local markets. In addition, fishermen were also linked to local municipalities through communal ties. The key institutions for this successful resource management were developed by guilds and were based on balanced power relations.

The next chapter takes us to an overview of Swiss commons institutions: Anne-Lise Head-König (University of Geneva) indicates in *The commons in highland and lowland Switzerland over time: transformations in their organisation and survival strategies (seventeenth to twentieth century)* that there was variability in common property institutions that regulated the CPRs. This was due to environmental and topographic heterogeneity, but also because of differences in territorial organisations within the area of what today is Switzerland. A unifying principle, however, was that all communities were collective owners of land and land-related CPRs, at least until the end of the eighteenth century, in both rural and urban areas. By the end of the Old Regime, this communal order had come under pressure due to an increase in population size and intra-corporational conflicts over the right of usage. This presented significant challenges for commons management over the course of the nineteenth century. Nevertheless, numerous civic institutions (both urban and rural) have persisted, with only a slight change in their objectives.

The city of Bern is a case in point of how external changes affecting the commons can be institutionally adapted in a sustainable way. In their chapter *From natural supply to financial yields: the common fields of the Bernese Civic Corporation since the seventeenth century*, Martin Stuber and Sarah Baumgartner (University of Bern) use the example of the civic corporation of Bern ('Burgergemeinde') to show how the political and economic value of its CPRs have changed fundamentally in the course of transformations that have followed changes in the political and energy systems. The commoners of the city of Bern owned fields and forests from the time of the town's origins in the Middle Ages; each member household was granted the right to a certain amount of timber from the forests and a plot in the fields for their subsistence needs. After the revolution brought about by Napoleon, this privilege ended: while all the inhabitants of Bern were now permitted to rent land, the newly established Bernese Civic Corporation remained the owner of the land and distributed annual revenue among its members – in effect a move back to the initial distributional justice of the commons. Today, the rising income from the commons as real estate is widely distributed as a general investment in 'culture and science' projects.

Contemporary European cases

The last three cases in the historical section bring us to contemporary issues of the commons in Europe, which include coping strategies in the constellation of a 'glocal' world and innovative strategies to defend and maintain the centuries-old commons in a modern context where CPRs are of much lower value. Ramez Eid (Sakhnin College, Israel) opens this sub-section with *Universal values and the protection of commons: fighting corruption with bottom-up process in Mallorca*. Eid uses the constitutionality approach to analyse how local inhabitants in central Mallorca perceive the bottom-up process for crafting institutions for the management of the forest commons in a protected area in the western Mediterranean (Serra de Tramuntana in Mallorca, Spain) threatened by mass tourism that affects the whole island. Eid explains how CPRs are losing value in comparison to investments in tourism, which is a sector in the hands of a few powerful owners of capital who profit from neo-liberal policies. In addition, local people have lost trust in the government, which is believed to be corrupt. This perception of the state in crisis led to the narrative of an ineffective and incapable government and contributed to the creation of a new social and political bottom-up dynamic: local actors used the institutional framework of the UNESCO Biosphere Reserve initiative as an opportunity for their participative strategy to regain the commons of the area.

A similar case is provided by Angelika Lätsch (University of Bern) with her chapter *Constitutionality and identity: bottom-up institution building and identity among Coastal Sami in Northern Norway*. Similarly to the case of Mallorca, the process of regaining the commons started with a crisis related to the failure of state governance of fisheries due to an over-capitalisation of commercial fishing fleets. Despite the fact that this leads to overfishing, the state favours the commercial fisheries industry by permitting higher fishing quotas in fjord areas, and at the expense of traditionally equipped local fishermen of indigenous Sami origin. These locals became engaged in a bottom-up institution-building process, which initially presented the classic elements of constitutionality yet later developed into something more: Lätsch shows with her research that the local small-scale fjord fishermen of Sami origin based their claim to access fishing grounds not only on arguments of equal rights but, increasingly, on their Sami identity as an indigenous people. In doing so they claimed collective resource governance rights in their traditional territory, based on the International Labor Organization (ILO) Convention 169 on indigenous rights, which the Norwegian state has ratified. Adding to previous insights, the case illustrates how identity matters in the process of constitutionality for new CPR institution building.

This element of identity can also be found in Gabriela Landolt's (University of Bern) chapter on institutional changes of the alpine pasture commons in the Canton of Graubünden (Grisons), Switzerland. She provides us with *Swiss alpine pastures as common property: a success story of bottom-up institution-building in Sumvitg, Canton of Grisons, Switzerland* by showing how 80 per cent of alpine pastures in the Canton are still managed as common property. Communes are the lowest governmental entity in Switzerland; and as owners of alpine summer pastures they are responsible for maintaining and allocating the CPRs to its farming population, which is a right backed by cantonal law. Despite the structural pressures of reduced agricultural prices and value of the commons, and despite the decreasing number of farmers – at the same time as their diversity (dairy and meat-producing farmers) was increasing – the case of the mountain village of Sumvitg shows that a community can succeed in upholding its common property institutions by adapting them: first, by crafting new organisational structures with new institutions, which made them independent from non-farmers in the community; second, by creating new rules that integrated both types of farmer groups along with their differing interests. The key to success in this case was good leadership among the farmers and the effort that was expended to enable the full participation of all farming actors involved.

Features and effects of European investments in the commons in the Global South

In this part we move on to the more obvious aspects of the 'glocalisation' of the commons. This part of the book is divided into three sub-parts: the first deals with several notions of *commons grabbing* as an effect stemming from European investments in agro-industrial production. The second looks at *impacts of mining and large infrastructure projects* which involve European investors and companies. The final sub-part deals with green grabbing issues related to *conservation of European provenance*, be this in protected areas in several forms, conservation work for European (Swiss) zoological gardens, or climate protection projects such as REDD+. All chapters begin with a historical overview on the commons institutions before showing how these were dismantled through colonial and post-colonial processes and, finally, outline the impact of relative price changes on the commons as these areas become increasingly valuable and accessible to powerful actors and their investments. In conclusion, these chapters also address local reactions to these investments and the losses that have been brought about – losses that often differ once regarded through the lens of gender.

Land grabbing and the commons

The first of four chapters in this section provides a quantitative overview of the state of the commons in the times of LSLAs. Markus Giger, Kerstin Nolte, Ward Anseeuw, Thomas Breu, Wytske Chamberlain, Peter Messerli, Christoph Oberlack, and Tobias Haller (University of Bern, Leibniz University, CIRAD, ILC, University of Pretoria) show in their contribution *Impacts of large-scale land acquisitions on common-pool resources: evidence from the Land Matrix* how land grabbing or LSLAs impact the commons globally. Large-scale agriculture often affects areas containing CPRs used by smallholders and pastoralists as members of former common property systems. The authors describe several mechanisms of impacts on these commons and indicate that the Land Matrix (LM) data on former usage, vegetation cover and ownership of land acquired through LSLAs show that almost 60 per cent of the lands in which investments are made were, in one form or another, previously governed under common property before they became state property. However, the importance of this finding becomes apparent when we consider that in most regions of the Global South, but especially in Africa, the CPRs are still needed by local users of the land.

In the chapters that follow, authors deal with concrete case studies of LSLAs, in which European investors and European capital are involved. Kristina Lanz (University of Bern) takes us to Ghana to show how local elites and an agro-industrial company remove land and resources previously part of the commons. In *"They said they were bringing a development project": 'best-practice' large-scale land acquisition or 'commons grabbing' in Ghana's Volta Region?* her social anthropological study illustrates the impacts of a large-scale rice plantation in Ghana's Volta Region. The investment reduced both individually owned farmland and communally owned lands (for grazing, collection of fuel wood, fish ponds, etc.) via enclosures. Only rich and politically influential local elites (chiefs) were compensated, while former commoners lost important resources for their livelihoods in the floodplain and other areas for which no compensation was paid. However, both the company as well as the local chiefs as 'customary' authorities argued that the loss of CPRs was more than offset by the various opportunities created by the company's operations and *Corporate Social Responsibility* (CSR) programmes. Beyond this, the chapter also presents local reactions of the area's youth to this company and elite commons grabbing, who challenged elites during official meetings.

The next case study deals with the investment of a forestry company from the UK and South Africa on common land in Tanzania. In her chapter *Grabbing the female commons: large-scale land acquisitions for forest plantations and impacts on gender relations in Kilolo district, Iringa Region, Tanzania*, Désirée Gmür (University of Bern) shows that – as seen already in the previous chapter – the loss of the commons affects mostly women in these communities. Since the socialist period, which lasted until the mid-1980s, and the subsequent move towards a neo-liberal order, Tanzanian governments have increasingly invited foreign investments, attracting them by means of new laws. Gmür describes the deal for forest plantations made by a UK-based investor called *New Forests Company* (NFC) in the Kilolo district of the Iringa region. She shows how the company strategically activates and transforms institutions (so-called institution shopping) in order to access common and private land; it also makes use of laws on the protection of forest, as well as water conservation discourses, in order to legitimise land alienations. These, then, affect women differently than men, as they reduce the formers' ability to fulfil their care work due to their loss of access to the best land and its related CPRs (e.g., water, fruit trees, grasses). The chapter shows that commons enclosure has negative effects on women, who face difficulties in organising resistance and coping with their losses.

In contrast to the Tanzanian case, Franziska Marfurt (University of Bern) shows the opposite process in local reactions by women. In *Gendered impacts and coping strategies in the case of a Swiss bioenergy project in Sierra Leone*, the author illustrates how a land deal implemented by a Swiss-based international biofuels company in Sierra Leone was perceived by local actors, especially women. Because of LSLAs locals lost their status as former commoners and were transformed into mere land users lacking property rights in regard to CPRs such as a wetland's water. This water had been used for irrigation and vegetable production for sale at the markets, thereby forming the basis for many women's livelihoods. However, affected women and others organised themselves, using several institutions (so-called multiple institution-shopping) to regain access to water by, first, using customary land rights providing them with traditional water rights; second, using the modern private land-rights system that places a land owner under pressure to withdraw the land from the company; and third, involving local and international human rights organisations struggling to strengthen women's rights to land and its related CPRs.

Mining and infrastructure

Sub-section two deals with case studies of mining and infrastructure, which involve European countries as investors or hosts of mining companies and providers of wind and solar energy technology. The first chapter in this section provides an overview of new mining activities, illustrated here through concrete examples of the impacts that mining has on the commons. With *The open cut: mining, transnational corporations and the commons*, Thomas Niederberger, Tobias Haller (both University of Bern) and Madlen Kobi (University of Zurich) present how mining activities have been expanding since the early 2000s, leading to confrontations with locals over the loss of their water, land, and forest commons. They analyse the multiple impacts that transnational large-scale mining corporations from Europe and other Northern regions have on the commons, and include findings from an extensive comparison of 13 case studies across five continents (based on Niederberger et al. 2016). In order to illustrate important trends, the Mopani Copper Mines in Zambia and the Tampakan copper-gold project in the Philippines are discussed. These two cases illustrate that large-scale mining not only leads to 'commons grabbing' and contamination of CPRs, but also undermines local institutions (such as customary property rights and other regulations) essential for the sustainable governance of commons. This triggers a variety of local responses, which range from negotiation with the investor to resistance.

While mining activities are generally accused of having negative environmental as well as social impacts, the opposite applies to so-called sustainable energy production such as solar energy. It is hard to see problematic aspects for the commons when desert-like areas are transformed into sustainable energy. Nevertheless, on a large scale such projects can be highly problematic and reveal similar processes of the disregard of local actors and grabs as pertain to their unsustainable cousins. Sarah Ryser (University of Bern) indicates in *Are green energy investments levelled by the 'new commons'? Compensations, CSR measures and gendered impacts of a solar energy project in Morocco* that green investments must pass the test of social sustainability. In her case study a local Berber clan was provided with compensation payments for their land, which had been labelled as wasteland, in order to establish one of the world's largest solar projects. This took place under the guidance of the King of Morocco and with the support of the European Union. Via a state-private partnership company called MASEN (Moroccan Agency for Sustainable Energy), 3,000 hectares of communal land were acquired that had previously been owned by the Moroccan *Amazigh* Clan in the Ghessate rural council area near Ouarzazate. Ryser's analysis shows that the discourse of a fair deal in transforming old to new commons does not hold: not all groups – and no women whatsoever – were involved in the deal, which was based on a non-negotiable and low price per land unit that was paid into a fund for local development. Furthermore, the study shows that local people living in proximity to the site do not have access to this fund and that, thus, the CSR projects are not adapted to local needs.

Green enclosures

The previous case links well to the final sub-section in part three on Europe's impact on the commons in the Global South. While Morocco, too, is a case of green grabbing it does not share the same features as those presented in the following three chapters. These deal with the classic green grabbing processes highlighted by Fairhead, Leach and Scoones (2012) as a way of enclosure based on the discourse of conserving pristine nature. However, this process is clearly related to other types of grabs: agro-industrial investments as well as mining and infrastructure development are directly related to conservation issues. While these activities destroy 'pure nature', the relative price for 'protecting' nature rises and thereby creates a conservation value to be tapped (see also Haller and Galvin 2011) illustrated in this third sub-section. In *Global changes in local governance of the commons: the case of the African Parks Foundation engagement in Nech Sar National Park, Ethiopia*, Girma Kelboro and Till Stellmacher (both University of Bonn) show how global conservation trends affect local people in their usage of their cultural landscape ecosystem in the context of the international NGO African Parks Foundation (APF) in Nech Sar National Park (Nech Sar NP). The authors discuss how pastoralists and farmers living within and around this park are greatly affected by attempts of state authorities and the involved NGO to redefine conservation and usage of the park's 'natural resources'. This has led to the reinforcing of exclusionary conservation approaches that are in contradiction to the interests and needs of most local users.

Participatory conservation can take a more fortress-like shape in a neo-liberal context, as shown by Samuel Weissman (University of Bern) in *Discourse and entanglement in a transnational conservation arena: deciphering the ideologies and narratives behind conservation discourse in the 'glocal' commons in Kenya*. His contribution reveals that conservation in this country has, on paper at least, been inclusive for local people and local conservation organisations. This should also apply for national programmes and international donors who embark on establishing a global network of environmental- and development-friendly communities. This is part of a legitimising discourse that includes local participation in the protection of rhinos, which, however, is not defined by local actors themselves but by a hegemonic, external definition of local people that takes place through defining them as either related to tradition or to modernity. This discourse obscures the discriminations against stakeholders that are taking place. In addition, heavily armed forces control the area, and their presence is legitimised due to the activities of so-called poaching groups seen to be criminal in nature.

Striking similarities exist between Weissman's case and the final case in this sub-section by Heinzpeter Znoj, Rahel Jud and Yudi Bachrioktora (all University of Bern). Their chapter *Rain forest anomy: national parks, REDD+ implementation and the run to the forest in Jambi, Indonesia* shows how the implementation of REDD+ schemes should help safeguard the remaining forests in the province of Jambi in central Sumatra, which have experienced a fast rate of forest conversion for oil palm plantations over the past 40 years. But the enforcement of Jambi's forest reserves, in form of national parks and REDD+ project areas, have an indirect impact on the far larger forest areas beyond their perimeter. Pressure on these forests has increased due to shifting formal and informal modes of access, and because of massive population movements. The authors present three ethnographic cases that shed light on how, in the context of the weak institutional framework of the Indonesian state, the formal implementation of national and global norms of forest preservation goes hand-in-hand with the practice of their systematic circumvention. This process furthermore also undermines traditional, institutionalised CPR-norms regulating access to forests. Although REDD+

schemes provide for local indigenous people to collectively use the forest as a CPR in non-destructive ways, this does not result in robust institutions. On the contrary, settlers successfully claim indigenous status to gain access to the forest and convert it into oil palm plantations.

Commons and international law

As power over resources increasingly shifts towards nationally and globally active investors attempting to institute private property as a precondition for maximal value extraction, bottom-up initiatives to defend community-based control over resources are systematically thwarted. This suggests that legal support at an international level is needed to strengthen the commons in the 'glocal' world.

This is precisely what Insa Theesfeld (Martin Luther University Halle-Wittenberg, Germany) proposes in her chapter *A structured checklist to identify connections between land and water grabbing*. She argues that the increase in the amount and frequency of international land investments should also alert us to the threat of so-called water grabbing. To date there is no conceptual legal framework designed to address the issue of grabbed water or to regulate compensations in the context of negative impacts on locals. Theesfeld therefore proposes a checklist based on property rights theory, which should enable us to examine the relationship between land and water. Key to such a list is the obtaining of knowledge on the individual bundles of property rights, which are affected and change when land is transferred from one owner to another – or when it is simply grabbed. Eight analytical patterns of interrelation are proposed that aim to strengthen the systematic comparison of case studies. The author illustrates this with cases of LSLA that have had an impact on the water sector in Ethiopia and Tajikistan. Her analysis shows that these land grabbing cases can also be termed 'water grabbing'.

Similarly, the issue of water rights is the focus of *International investment agreements and mega-regionals: promoting or undermining the right to water?* by Rodrigo Polanco Lazo (University of Bern) and Azernoosh Bazrafkan (Maastricht University). Looking at the state's obligation to provide drinking water, the authors check on how investment in the water sector within the framework of a commodification of these services affects this obligation. Investment protection rules, as established in International Investment Agreements (IIAs) may have an impact on the human right to water. Polanco and Bazrafkan thus focus on recent treaty-making developments in bilateral investment treaties and preferential trade agreements (termed 'mega-regionals'), including investment rules. They argue that companies whose investments affecting water come into conflict with states that must protect the provision of water. Therefore, laws on water-related issues developed for investor-state dispute settlement (ISDS) mechanisms should be included in the IIAs.

The final chapter in this section takes up the issue of the human right to water – which is also alluded to in Schlosberg's keynote on environmental justice. With her chapter *The human right to water in India: in search of an alternative commons-based approach in the context of climate change*, Birsha Ohdedar (School of

African and Oriental Studies, London) provides evidence that access to water as a human right faces challenges especially in the context of prior marginalisation of actors, which also coincides with them being most affected by climate change (see keynote by Jesse Ribot). Contestation over water in places such as India necessitates an analysis of multiple social, political, ecological, and climatic processes that mediate access to water. Thus, a larger 'hydro-social' lens is needed. While the human right to water has been recognised at both an international and domestic level in India, the dominant interpretation of the right both in legal and political discourse has been centred upon a fixed relationship between the rights-bearing subject and the quantity and quality of water. Ohdedar argues that in this way the multiple contestations in the hydro-social cycle that produce rights issues are not analysed. In order to remedy this problem, she proposes that water has to be conceived as a CPR. This view would provide the legal integration of the human right to water by taking into account the hydro-social conditions related to access and distribution of water.

Concluding remarks

The volume provides examples of the commons in a 'glocal' world on several scales, as well as discussing their interrelations. It bridges the gap in the literature between the focus on the sustainable governance of CPRs and external and internal power relations between actors. The case studies show how institutional and political ecology approaches in combination help to analyse agencies of different actors on different scales and the challenges for the commons that result from this. An important lesson lies in the insight that common property institutions for specific resources may have been more widespread in the past than hitherto acknowledged, and that early on these were already linked to private property in the context of specific other resources. In particular resources like water and marginal, less intensely used land and related CPRs were often governed as common property because this institution provided collective solutions for a sustainable use. This also included adaptive mechanisms to react to external changes. Furthermore, it contributed to social sustainability and higher resilience of local groups as well as of individual households. In this way, historical common property institutions were part of the production and maintenance of dynamic cultural landscape ecosystems. State property and, later, neo-liberal private property arrangements in a 'glocal' world are destabilising and dismantling these common property institutions on a large scale and, as a result, also contribute to the undermining of these ecosystems. This led – and continues to lead – to different forms of exclusion of local actors from CPRs on several scalar levels and contributes to higher vulnerability and lowered resilience of former commoners. However, the cases contained in this book also show successful responses that secure common property and the possibility to defend, newly adapt to and justify collective ownership in a 'glocal' world. The successful cases also indicate that processes of fair and participatory bottom-up institutional adaptation strategies are of crucial importance, as is discussed in the constitutionality approach.

More research is needed on how commons institutions have evolved, how they have been challenged by the impacts of 'glocalisation', and under which conditions local actors with varying bargaining power can regain or lose access and control. These experiences provide the key to reacting to problems and disturbances in the 'glocal' world and its powerful neo-liberal actors – theirs is a power that may be challenged by a collective re-appropriation of the commons in the long run. Yet, the ability to locally reshape the institutional design while still being exposed to 'glocalisation' and depends on the bargaining power of local actors. This includes especially the power to define what so-called 'nature' means: for most of the groups discussed in this volume, their environment is not pristine or 'pure nature' but instead a cultural landscape. They have contributed to this landscape with their institutionalised usage, and they have a rightful share in its CPRs. However, the right to this share and to having the power to decide its institutional design does not stop at the local level. Rather, this needs to include an analysis of higher-level, interrelated power-driven processes, the anticipation of its impacts, and international support of local responses.

References

Beck, U. (1999) *Schöne neue Arbeitswelt: Vision: Weltbürgergesellschaft*. Frankfurt am Main: Campus.

Ensminger, J. (1992) *Making a Market. The Institutional Transformation of an African Society*. Cambridge: Cambridge University Press.

Fairhead, J., Leach, M., and Scoones, I. (2012) 'Green grabbing: a new appropriation of nature?', *Journal of Peasant Studies*, vol 39, no 2, pp 237–261.

Feeney, D., Berkes, F., McCay, B.J., and Acheson, J.M. (1990) 'The Tragedy of the Commons: Twenty-Two Years Later', *Human Ecology*, vol 18, no 1, pp 1–19.

Haller, T. (ed) (2010) *Disputing the Floodplains: Institutional Change and the Politics of Resource Management in African Floodplains*. Leiden: Brill.

Haller, T., Acciaioli, G., and Rist, S. (2016) 'Constitutionality: conditions for crafting local ownership of institution-building processes', *Society & Natural Resources*, vol 29, no, pp 68–87.

Haller, T., Belsky, J.M., and Rist, S. (2018) 'The constitutionality approach: conditions, opportunities, and challenges for bottom-up institution building', *Human Ecology*, vol 46, no 1, pp 1–2.

Haller, T. and Galvin, M. (2011) 'Challenges for participatory conservation in times of global change: lessons from a comparative analysis and new developments', in Wiesmann, U., and Hurni, H. (eds), *Research for Sustainable Development: Foundations, Experiences, and Perspectives. Perspectives of the Swiss National Centre of Competence in Research (NCCR) North-South* (pp 467–503, vol 6). Bern: University of Bern, Geographica Bernensia.

Netting, R.M. (1993) *Smallholders, Householders: Farm Families and the Ecology of Intensive, Sustainable Agriculture*. Stanford: Stanford University Press.

Netting, R.M. (1981) *Balancing on an Alp: Ecological Change and Continuity in a Swiss Mountain Community*. Cambridge: Cambridge University Press.

Orebech, P., Bosselman, F., Bjarup, J., Callies, D., Chanock, M., and Petersen, H. (2005) *The Role of Customary Law in Sustainable Development*. Cambridge: Cambridge University Press.

Ostrom, E. (1990) *Governing the Commons. The Evolution of Institutions for Collective Action*. Cambridge: Cambridge University Press.

Ostrom, E. (2005) *Understanding Institutional Diversity*. Princeton, NJ: Princeton University Press.

Pulcini, E. (2012) *Care of the World: Fear, Responsibility and Justice in the Global Age*. Berlin: Springer.

Robertson, R. (1995) 'Glocalization: Time-Space and Homogeneity–Heterogeneity', in Featherstone, M., Lash, S., and Robertson, R. (eds), *Global Modernities* (pp 25–44). London: Sage.

Roudometof, V. (2005) 'Transnationalism, Cosmopolitanism and Glocalization', *Current Sociology*, vol 53, no 1, pp 113–135.

Roudometof, V. (2016) *Glocalization: A Critical Introduction*. New York: Routledge.

Swyngedouw, E. (2004) 'Globalisation or 'glocalisation'? Networks, territories and rescaling', *Cambridge Review of International Affairs*, vol 17, no 1, pp 25–48.

Part I

Key reflections

New theoretical issues on the commons and their transformations

Top left: The communal soup kitchen of Ins (Switzerland), oil on canvas, painting by Albert Anker 1893. Source: Kunstmuseum Bern. More on the impact of common property institutions and their impact on the political culture of Switzerland through the time, see Chapter 1.

Top right: Institutions for the management of the commons are often contested. Picture by Tobias Haller. Pastures in Kafue Flats, Zambia, illustrating the NIPE approach, see Chapter 5.

Bottom left: Discussing poverty, access to the commons and climate change in Senegal. Picture by Tobias Haller. See Chapter 2, Jesse Ribot.

Bottom right: A women's group that is part of the SPP (farmers union) in Tasikmalaya district, Indonesia, September 2018. Picture by Elisabetta Cangelosi. For the role of commons, between human rights, resistance, and social change see Chapter 4.

1 Shared ownership as a key issue of Swiss history

Common-pool resources, common property institutions and their impact on the political culture of Switzerland from the beginnings to our days

Daniel Schläppi

These days we tend to advertise "commons" as counter projects to capitalism and state power. Switzerland has one of the highest concentrations of globally acting banks and corporations. It has a pole position in the competition for locational advantages. All the same, many very ancient "common property institutions" with partially medieval origins exist. It is astonishing and in need of explanation that so many cooperatives and an extreme political federalism have survived in Switzerland. Most of the ancient commons institutions not only outlasted the great epochal upheavals in the transition to the modern era but also the so-called "Dekorporierung" as well as the long-term change of society during the nineteenth and twentieth century (Stuber 2018). In contrast to the situation in other European countries, corporate organizations and common ownership had a very strong tradition in Switzerland. In addition they were firmly anchored in the Swiss collective consciousness. For these reasons cooperatives and similar institutions were able to defend their political autonomy. Up to our days, even the smallest communities with just a few hundred inhabitants dispose of a wide scope of decision-making.

The "Alte Eidgenossenschaft", the "Old Swiss Confederacy"

In historical comparison, Switzerland has developed differently from European monarchies and princely states. The so-called "Old Swiss Confederacy", die "Alte EidGENOSSENSCHAFT" – "Genossenschaft" means "cooperative" – was composed of 13 politically independent states. These so-called cantons were connected by an intricate system of alliances. Each single canton failed to centralize the administration of its territory, to install a tax regime and to build a standing army. There was no standardized jurisdiction or monarchic concentration of power.

How was it possible that such a mellow institution remained politically stable for centuries in spite of confessional schism and manifold external threats?

This paper shows how a strong tradition of "commoning" in politics, economics and everyday life affected the political culture of this small country as a whole. In my view, common property is responsible for the centuries of Swiss institutional constancy (Schläppi 2017).

In the beginning, there was cooperation for peace

The Old Swiss Confederacy started as a peace project. So-called "Landfriedensbündnisse" like the famous "Federal Charter" of 1291 – the German term "Landfriede" is also used in English – were very common at that time. Manifold partners made arrangements to further security and peace within their home territory and to stabilize the relationships with their neighbours.

In the territory of present-day Switzerland, agrarian and civil elites were behind this process. They aimed to defend themselves against incursions from feudal lords, lower nobility, monasteries, and against blood feuds of hostile clans.

The single cantons did not have enough combat power or money to hire mercenaries. Their weakness made them stand to mutual agreements on how they would help each other in the event of war. Nobody should start a war against third parties on his or her own. The obligation of military assistance was only valid under the premise of codetermination concerning the decision for or against a war. The treaties provided procedures and institutions like arbitral tribunals to settle quarrels.

The practical advantages of these alliances were obvious for all involved parties. The network of coalitions grew bigger and denser over time, even though the participants always acted egoistically and autonomously. However, some of their selfish goals could only be achieved by collective action.

A key role in the institutional consolidation of the Old Swiss Confederacy was played by the subject territories, which were conquered since 1415 and governed by the cantons together. These so-called "Gemeine Herrschaften" (common subject territories) were administrated by the congress of the confederate cantons, the so-called "Tagsatzung", the "Federal Diet".

It is obvious that common property had a massive influence on the political activities of the Federal Diet. The administration of the common subject territories and the sharing of the profits accounted for one third of all agenda items of the Federal Diet.

The so-called "Jahrrechnung" – the get-together for the annual financial statement of the subject territories – was even assembled in times of war between the confessional parties of Switzerland and it was drawn up every year from 1470 up to 1797. This makes the Federal Diet a unique example of institutional continuity in Europe and elsewhere (Würgler 2013).

Figure 1.1 The Federal Diet of Baden (not before 1597).
Source: Chronik von Andreas Ryff, Circkell der Eidtgnoschaft, fol. 173v–174r, Musée Historique de Mulhouse, coll. SIM, © photo Ville de Mulhouse (Christian Kempf).

The commons-state – cooperative structures as foundations of politics and society

On the one hand, the willingness to split power surprises in relation to the early modern period. On the other hand, various types of common property institutions were managed cooperatively in every of the 13 cantons. Here you can see a list of some of the associations that provided the institutional foundation of politics and society:

Bürger-, Nachbar- und Bruderschaften, Land-, Tal-, Allmend-, Wald-, Alp-, Säumer-, Arbeits-, See- and Brunnengenossenschaften, Korporationen, Gemeinwerke, Uerten, Tagwen, Roden, Bäuerten, Genossamen, Viertel, Zünfte, Handwerke, Stiftungen, Gesellschaften, Gerichts-, Schwellen-, Kirch-, Abendmahl-, Territorial-, Personalgemeinden, Fideikommisse und Familienkisten.

The power elites learned as aldermen or simple officials in their local communities that decisions always had to be made and supported by groups. There were certain limits to individual displays of power, the expectations and interests of regular members had to be considered, and the profits made from public goods had to be distributed squarely.

Since all decision makers were socialized this way, we should not be astonished that the Federal Diet decided in 1551, in view of a tapering poverty problem, that in the future the communities in all the cantons should take care of their needy members. It took several generations until the communities had saved up the necessary means. Many among them were never able to cope with the additional burden in their own right.

But still this fundamental change brought a huge empowerment for the communities in the long term: During the seventeenth and eighteenth centuries, communities and cooperatives took more and more charge of state functions at their own expense (such as market regulation, fire brigade, maintenance of infrastructure, residence registration office and welfare institutions). In consequence, important parts of the state administration and public services were based on collective resources. Taxes were only levied in case of emergency.

One could see the Swiss model as a perfect example of "co-production of statehood", as a "commons state" rather built bottom-up than top-down (Schläppi 2007).

Collective resources as a means of existence for private households

The subsistence of private households depended on collective resources such as grazing rights, donations of crop and firewood, social welfare and also building materials as well as financial contributions for real estate, building loans and raw materials for commercial production. Fundamental to daily provisioning were the rights to gather food in the common forest, to fish in ponds and creeks plus the permission to pick fruit and nuts from trees on the common land. Some cooperatives took precautions against war, tried to keep stable prices in times of inflation, or helped providing for their old consociates. Viewed in this light, it was existential for many people to have access to collective resources.

Alongside primary care, communities rendered further services. Some kept a breeding bull, or held properties (for instance school buildings, mills, washhouses, smithies). Cooperatives built and maintained flood protection works, bridges and roads, and acquired fire buckets and fire engines. They paid for mousers, herdsmen or foresters. They recorded and imparted knowledge to future generations in the form of apprenticeships or the recurring organizing of collective work.

Cooperatives also provided their members with immaterial goods and created a symbolic world of emblems, flags, valuables, houses and collective memories. They practiced their own rituals and a living sociability financed by common assets. Reputation, honour and prestige of individuals were directly connected with the good reputation of their peer group. Cultural achievements like a new chapel, church extensions, a modern organ, an attractive reliquary, the clubhouse at the gun range, the music culture, church singing, pilgrimage, pastoral care and commemoration of the dead were financed and allocated by pooling and collective action (Schläppi 2018b).

The economic logic behind cooperative housekeeping: husbandry and austerity

Cooperative strategies were necessary and successful when essential necessities could not be satisfied by individuals or single households because of their limited opportunities of action. Good examples are irrigation or drainage of agricultural land. Often it was just cheaper to cover elemental necessities by common effort. That is, for example, why communities sent their poor to eat from house to house or run a soup kitchen.

Sometimes precariously situated families were allowed to lay out small gardens or potato fields on the common land or they were even allocated a building plot. Of course, it was taken into consideration that their households should no longer depend on financial support from the communal poor relief fund. Communities also sold illegitimate children or orphans to wealthy farmers as cheap labour.

One might call this kind of charity 'slavery'. Of course, it may have reduced the worst hardships. However, at the very end well-off classes profited most from it, because they would have had to pay for welfare costs. Anyway, providence was the highest precept. Taxes, which could only be paid by the rich, were to be avoided. Therefore, communal welfare oriented itself more to the revenue of the poor relief fund than to actual misery. The current costs should be financed by investment income only. Money reserves were not to be touched: The ancestry's legacy had to be taken care of (Schläppi 2018a: 37–41).

Marginalization and exclusion

Less well-off households were forced to keep their spending restraint. If families slid below subsistence level and the community considered their housekeeping not to have been sparing enough, their rights of participation were cancelled. That is, the regular contributions were suspended, and the head of the household was no longer allowed to participate in political assemblies and convivial events.

Troublesome members of the community could easily be expelled from political and economic communal life by labelling them as so-called "Übelhauser", which means "bad housekeeper". The accusation of somebody damaging the "gemeinen Nutzen" (the "common good") because of his or her economic behaviour – the most important justification for communal punishment of ordinary people – stigmatized the accused person permanently. In addition to that, they were threatened with withdrawal of the minimal basics of existence as a penalty for needed support.

In the welfare system, the propelling concern of the commoners for maximum proper benefit manifested itself in the suppression of those who harmed the public interest. Permanent peer pressure against poor people even justified brutal interference with their biographies. Prohibition of marriage because of lacking financial resources, revocation of citizenship because of petty misconducts or prolonged job-related absence, criminalization of illegitimate maternity, removal of children born to unmarried mothers, horrendous fees for marriage with women from other communities: manifold harassments reduced the opportunities in life for small means members. People who were considered to be risks of welfare, and therefore represented potential expenses for the community funds, would better not expect any tolerance from their commoners. Swiss communities actively sponsored the overseas emigration of poor members. By signing standardized contracts, emigrants renounced poor relief not only for themselves but also for their descendants even if they would return just a few years later. Collective mistrust towards deviant concepts of life and patronization were the flipside of the positive achievements that the communities rendered for their well-incorporated members – that is, the settled and saturated centre of society.

The marginalization of unpopular people within the group corresponded with the strict dissociation with the exterior. To keep the circle of members as small as possible, the communities closed up hermetically since the seventeenth century and hardly accepted any new members. On rare occasions, a community member might have sold its citizenship to a rich immigrant because of monetary tightness.

Therefore, in every community there were more and more second-class citizens, the so-called "Hintersassen". This led to social tensions and conflicts about resources. Communities and cooperatives usually struggled resolutely against the claims of outsiders and defended their privileges successfully.

Inequality among equals

In spite of common interests, latent conflicts also festered among the privileged. The seventeenth and eighteenth centuries are considered to be the era of aristocratization: This means that the social elites grew apart from common people. They endeavoured to distinct themselves culturally by wardrobe, furniture and architecture. They also tried to increase their power and private gaining by political functions. Thanks to a cooperative political economy, effective checks and balances were an obstacle to autocratic display of power:

- Even powerful cooperatives like the citizenries of Zurich or Bern stressed the horizontal, the egalitarian aspect of their self-representation. The aristocratic elites had to submit themselves to a modest and uniform official attire for public appearance.
- Important offices and profitable charges were assigned temporarily or in rotation. Close relatives were not allowed to be members of the same board. Some elections were decided by drawing lots or ballotage.
- Fundamental decisions had to be presented to the so-called sovereign, this means: to the full members of the peer group. They had the final say, no matter whether they were the community assembly in a small village or the Council of Two Hundred of a big city ruling a large territory.
- Control procedures by commissions were introduced on suspicion of private enrichment at the cost of the commonwealth.

- There was no domination-free discourse among "equals" because of social hierarchies. All the same, those in power had to at least listen to the discussions in conventions. In addition, they were not able to elude the uncontrollable dynamics of communal assemblies.
- Municipal authorities regularly had contact with common people for practical reasons and had to sit down at tables in taverns on many occasions.

These examples show that cooperative institutions were intrinsically formed by dominance and hierarchies. However, they also had to counterbalance to extreme asymmetries. A smart management of inequality was essential for the long-term political stability of commons (Schläppi 2010).

Legitimation by way of redistribution

The most important factor for the legitimation of power was the redistribution of wealth. This took place by means of material and ritualized participation. Wide sections of the population were formally involved in political procedures. The reciprocal logic of this society is especially obvious looking at the distribution of the most beneficial privileges of rule.

In everyday life, full community members enjoyed now and then a shared meal, a Christmas allowance, a heraldic window, a cartload of firewood or timber and other contributions. Such benefits were not to be questioned nor overlooked.

Conflicts and revolts were sparked off when the handling of common goods by the power elites was in discrepancy with the general public's moral economy. During the eighteenth century, revolts and system-threatening crises erupted. Ideas of Enlightenment like equality and liberty were immaterial to this. The inherent inequality between people with or without citizenship was never questioned. Social clashes broke out, when privileges due to the full community members were usurped by only the holders of power. In rebellions, nobody would ask for a new constitution but for a return to customary law, to the old order and to restore traditional access to resources, positions and opportunities in life. This shows that a strong consent to the system existed even in the light of great social injustice (Schläppi 2015).

Citizen-oriented administration

If nothing else, this was owed to an administration that was in touch with the citizens. It is generally known that there was hardly any professional bureaucracy in the Old Swiss Confederacy. Thus, many responsibilities were in the hands of communities and cooperatives. Many decisions could be taken at a lower level. The municipalities settled their own agenda. The hierarchies of power remained flat. Regular formal and informal meetings and daily business created the basis for a face-to-face administration.

Still, communal resource management brought out administrative routines: Community work and collective cultivation had to be planned, trees and fencing

(2.)

Anno.	Lauwis.	Luggarus.	Maythal.	Mondris.	Baden.	Rinthal.	Turgeüw.	Freyämpt.	Sargans.
1690.	Bärn.	Lucern.	Zürich.	Ury.	Underwalden.	Underwalden.	Ury.	Zürich.	Lucern.
1691.	Bärn.	Lucern.	Zürich.	Ury.	Statt. Zug.	Underwalden.	Ury.	Lucern.	Ury.
1692.	Schweitz.	Underwalden.	Ury.	Schweitz.	Statt. Zug.	Egerin. Zug.	Schweitz.	Lucern.	Ury.
1693.	Schweitz.	Underwalden.	Ury.	Schweitz.	Glarus.	Egerin. Zug.	Schweitz.	Ury.	Schweitz.
1694.	Glarus.	Basel.	Mentzigē. Zug.	Underwalden.	Glarus.	Glarus.	Underwalden.	Ury.	Schweitz.
1695.	Glarus.	Basel.	Mentzigē. Zug.	Underwalden.	Bärn.	Glarus.	Underwalden.	Schweitz.	Underwalden.
1696.	Solothurn.	Schaffhausen.	Freyburg.	Baar. Zug.	Bärn.	Appenzell.	Statt. Zug.	Schweitz.	Underwalden.
1697.	Solothurn.	Schaffhausen.	Freyburg.	Baar. Zug.	Zürich.	Appenzell.	Statt. Zug.	Underwalden.	Egerin. Zug.
1698.	Lucern.	Zürich.	Bärn.	Glarus.	Zürich.	Zürich.	Glarus.	Underwalden.	Egerin. Zug.
1699.	Lucern.	Zürich.	Bärn.	Glarus.	Lucern.	Zürich.	Glarus.	Statt. Zug.	Glarus.
1700.	Underwalden.	Ury.	Schweitz.	Basel.	Lucern.	Lucern.	Zürich.	Statt. Zug.	Glarus.
1701.	Underwalden.	Ury.	Schweitz.	Basel.	Ury.	Lucern.	Zürich.	Glarus.	Zürich.
1702.	Basel.	Mentzigē. Zug.	Glarus.	Freyburg.	Ury.	Ury.	Lucern.	Glarus.	Zürich.
1703.	Basel.	Mentzigē. Zug.	Glarus.	Freyburg.	Schweitz.	Ury.	Lucern.	Zürich.	Lucern.
1704.	Schaffhausen.	Freyburg.	Solothurn.	Solothurn.	Schweitz.	Schweitz.	Ury.	Zürich.	Lucern.
1705.	Schaffhausen.	Freyburg.	Solothurn.	Solothurn.	Underwalden.	Schweitz.	Ury.	Lucern.	Ury.
1706.	Zürich.	Bärn.	Lucern.	Schaffhausen.	Underwalden.	Underwalden.	Schweitz.	Lucern.	Ury.
1707.	Zürich.	Bärn.	Lucern.	Schaffhausen.	Baar. Zug.	Underwalden.	Schweitz.	Ury.	Schweitz.
1708.	Ury.	Schweitz.	Underwalden.	Zürich.	Baar. Zug.	Statt. Zug.	Underwalden.	Ury.	Schweitz.
1709.	Ury.	Schweitz.	Underwalden.	Zürich.	Glarus.	Statt. Zug.	Underwalden.	Schweitz.	Underwalden.

34.

(16.)

Anno.	Lauwis.	Luggarus.	Maythal.	Mondris.	Baden.	Rinthal.	Turgeüw.	Freyämpt.	Sargans.
1996.	Vry.	Schweitz.	Underwalden.	Zürich.	Mentzigē. Zug.	Statt. Zug.	Lucern.	Glarus.	Zürich.
1997.	Vry.	Schweitz.	Underwalden.	Zürich.	Glarus.	Statt. Zug.	Lucern.	Zürich.	Lucern.
1998.	[illegible]aar. Zug.	Glarus.	Basel.	Bärn.	Glarus.	Glarus.	Vry.	Zürich.	Lucern.
1999.	Baar. Zug.	Glarus.	Basel.	Bärn.	Bärn.	Glarus.	Vry.	Lucern.	Vry.
2000.	Freyburg.	Solothurn.	Schaffhausen.	Lucern.	Bärn.	Appenzell.	Schweitz.	Lucern.	Vry.

ENDE.

Figure 1.2 Carl Franz Kreuel, Distribvtio Oder Deütlich- und ordentliche Außtheilung der neun Landvogteyen etc. Einsiedeln 1691.

Source: Universitätsbibliothek Basel, Sign. Falk 2947: 33. There are nine vertical columns, one for each subject territory. The list shows that they are governed in rotation. However, most interesting is the fact that the list starts in 1690 and ends in 2000 as one can see on the last page. This shows that people in the seventeenth century were convinced of the stability of this complex style of government.

to be mended, superintendents to be elected and assignments to be made. Finally, yet importantly, the annual account had to be rendered.

Recurring duties became small commissions. Like a network of local public servants established far into the capillaries of society and territory. These servants did not need theoretical skills of administration. Knowledge of local circumstances and practical skills were more important because daily business involved managing buildings, infrastructure or farm- and forestland.

If a river burst its banks or had destroyed a bridge, the representatives of the power elite met with local farmers and community delegates on the spot and discussed further ways of proceeding. Countless visual and on-site inspections furthered a face-to-face administration. Whoever had to cut timber and transport it, had to re-erect a bridge or levee and then maintain it or just finance it, was negotiated to the end there was an all-around accepted distribution of costs and benefits. On-site inspections were often followed by collective meals. Given that all concerned parties were present, chances were high that the proposals of affected people were considered in the measures decreed (Schläppi 2011b).

Persistence beyond historical upheavals and revolutions

In 1798, Napoleon's army occupied the Old Swiss Confederacy. The subject territories were liberated, and the magistracies of the 13 cantons were disempowered and replaced by a centralized government. Due to the newly introduced freedom of trade, guilds and craft guilds were banned. However, the new political leaders did not dare to touch the countless assets and privileges of communities and cooperatives although they ruled in the name of "equality". They were just scared of spontaneous popular tumults.

The venerable commons proved to be the stronghold of the prosperous people. That is why common property institutions were so widely spread and resistant. This failure of the revolution showed that historical commons enjoyed almost unconditional legitimacy. For the always underprivileged "Hintersassen", the revolutionaries invented the so-called "Einwohnergemeinden", which would remain second-class political institutions for a long time.

By contrast, the old "Bürgergemeinden" were allowed to keep their assets and to take care of the issues of their members as before the revolution. It was obviously easier to decapitate a king than to dispossess respected groups of privileged persons (Schläppi 2011a).

The survival of old common property institutions facilitated the return to power of the former elite at the end of the French occupation (Stuber and Baumgartner, this volume; Schläppi 2001, 2006). However, we also have to consider that the persistence of historical common property institutions provided the cultural background and a perfect starting point for the rise of modern types of cooperatives that were formed in the nineteenth and twentieth centuries.

Housing, consumer, insurance and agricultural cooperatives are very common and omnipresent since the nineteenth century. Many of them have established themselves as important independent players despite potent private competitors in the free market economy. The successful "Schweizerische Mobiliar" (founded in 1825) is one of the largest insurance companies. The two cooperatives "Coop" (1890) and "Migros" (1925) are the dominant retail chains in Switzerland. The "Raiffeisenbank" (1902) and the famous "Wir-Bank" (founded in 1934) are important banking institutions for middle-class people up to our days.

Contrary to popular belief, cooperatives are not just conservative and focused on securing value for their members. Some of them are at the forefront of socio-political innovations and exciting socio-cultural experiments. A nice example is the so-called "Genossenschaft Kalkbreite" (2007) in Zurich. Based on the model of historical neighbourhoods this housing estate implements a new style of urban life that pursues the goals of ecological and social sustainability.

Modern Swiss society and politics are optimally predisposed for such advancements thanks to a strong tradition of "commoning" and collective action whose origins date from the Middle Ages.

References

Schläppi, D. 2001. *Die Zunftgesellschaft zu Schmieden in Bern zwischen Tradition und Moderne. Sozial-, struktur- und kulturgeschichtliche Aspekte von der Helvetik bis ins ausgehende 20. Jahrhundert*. Bern: Stämpfli.

Schläppi, D. 2006. Der Lauf der Geschichte der Zunftgesellschaft zu Metzgern seit der Gründung. In *Der volle Zunftbecher. Menschen, Bräuche und Geschichten aus der Zunftgesellschaft zu Metzgern*, 15–199, 307–309. Bern: Lanius Verlag.

Schläppi, D. 2007. Das Staatswesen als kollektives Gut. Gemeinbesitz als Grundlage der politischen Kultur in der frühneuzeitlichen Eidgenossenschaft. In *Neue politische Ökonomie in der Geschichte*, edited by J. Marx and A. Frings, 169–202. Köln: Zentrum für Historische Sozialforschung.

Schläppi, D. 2010. Differenzmaschinen – Kommunen und Korporationen der Vormoderne als Instanzen postmoderner Ungleichheit. In *Die Produktion von Ungleichheiten*, edited by T. David, V. Groebner, J.M. Schaufelbuehl and B. Studer, 23–33. Zürich: Chronos.

Schläppi, D. 2011a. Grenzen der Gleichheit. Wie und warum die helvetischen Regenten vor dem Gemeinbesitz von Korporationen kapitulierten. In *Grenzen des Zumutbaren. Erfahrungen mit der französischen Okkupation und der Helvetischen Republik (1798–1803)*, edited by A. Würgler, 46–65. Basel: Schwabe.

Schläppi, D. 2011b. Verwalten statt regieren. Management kollektiver Ressourcen als Kerngeschäft von Verwaltung in der alten Eidgenossenschaft. *Traverse. Zeitschrift für Geschichte* 18 (2): 42–56.

Schläppi, D. 2015. "Umverteilung" in korporativer Logik am Beispiel der alten Eidgenossenschaft. *Traverse. Zeitschrift für Geschichte* 22 (1): 51–64.

Schläppi, D. 2017. Die Eid-Genossenschaft. *NZZ Geschichte* 9: 92–102.

Schläppi, D. 2018a. Einleitung. In *Von der Allmende zur Share Economy. Gemeinbesitz und kollektive Ressourcen in historischer und rechtlicher Perspektive*, edited by D. Schläppi and M.-C. Gruber, 9–70. Berlin: Berliner Wissenschafts-Verlag.

Schläppi, D. 2018b. Die Ökonomie des Gemeinwesens. Auskömmliche Haushalte als ökonomisches Fundament und sozialer Kern der Gemeinwirtschaft. In *Universum*

Kleinstadt. Die Stadt Zug und ihre Untertanen im Spiegel der Protokolle von Stadtrat und Gemeinde (1471–1798), 61–85. Zürich: Chronos.

Stuber, M. 2018. Kollektive Ressourcen und Nachhaltigkeit in der Longue durée. Die burgerlichen Wälder der Stadt Bern vom Ancien Régime bis in die Gegenwart. In *Von der Allmende zur Share Economy. Gemeinbesitz und kollektive Ressourcen in historischer und rechtlicher Perspektive*, edited by D. Schläppi and M.-C. Gruber, 173–188. Berlin: Berliner Wissenschafts-Verlag.

Würgler, A. 2013. *Die Tagsatzung der Eidgenossen. Politik, Kommunikation und Symbolik einer repräsentativen Institution im europäischen Kontext (1470–1798)*. Epfendorf: Bibliotheca Academica.

2 Social causality of our common climate crisis

Towards a sociodicy for the Anthropocene

Jesse Ribot[1]

The full range of causes of climate-related risk are rarely treated by climate impact modelers or analysts. Similarly, when people look at the causes of climate-related crises in commons, they do not look at a full range of causes. Rather, analysts seem to get caught up in proximate causality or in the singular causal link between the hazard (storm or drought) and disaster. While all causes of crisis could never be fully accounted for, in this chapter I reflect on broader ways of approaching cause.

When I write about cause of climate-related crisis I am definitely not writing about the causes of climate change. I am not writing about smokestacks or drivers in New Jersey or Beijing or anything like that. Rather, I mean the causes of the crises themselves. The causes of hunger, famine, dislocation, economic loss; that is, the outcomes that happen when climate trends or events hit the ground. Indeed, there is something happening on the ground that is called vulnerability. By definition, without vulnerability there is no crisis. Crisis is when hazard and vulnerability come together – you need them both (Blaikie et al. 1994). Indeed, if people are not vulnerable, you would not even call these stressors hazards – since vulnerability and hazard are mutually defined. Without vulnerability, hazards are demoted to mere nuisance. Vulnerability here is the predisposition, in some way or another, to damage. So, hazard and vulnerability together constitute the climate-risk equation. And climate-related crises therefore do not merely fall from the sky when there is a climate event. They are socially produced via conditions on the ground.

The title of this chapter, *Social causality of our common climate crisis*, was inspired by none other than the Pope Francis, who has written the lovely Encyclical *On care for our common home* (Pope Francis 2015). The Pope uses the term *common* basically to mean shared. Of course, the boundaries between what is common and what is not common – that is what is private or public – are discursively produced in a global, national and local grammar of political claim making. The ways in which these things we called *commons* are defined and limited are products of multi-scale social, political and economic struggles. Indeed, the boundaries of the commons are also totally permeable by many higher-scale events – the price of grain or timber, for example, can change the boundaries of commons, a law allowing the export of timber reconfigures those boundaries by transforming the value of and claims on resources.

In the discursive political battle that Pope Francis (2015) is waging, he wants to make sure that we know that most of what '*is*' '*should be*' common. He is a common-ist. He is with us. He writes:

> … the earth is essentially a shared inheritance, whose fruits are meant to benefit everyone … Hence every ecological approach needs to incorporate a social perspective which takes into account the fundamental rights of the poor and the underprivileged. The principle of the subordination of private property to the universal destination of goods [by which, by the ways, he means social service], and thus the right of everyone to their use, is a golden rule of social conduct and "the first principle of the whole ethical and social order."

Wow. He is evoking the golden rule: "do unto others as you would have them do unto you" (Bible, Matt. 7:12) – implying that our human relations are mediated by this material world we call nature or environment. Indeed, property itself is the relation among people vis-à-vis things, and he seems to be calling for an equitable set of relations – commons.

He argues that social mutuality is needed to guard and sustain humans and to sustain our common home. The world is part of our common heritage and our future, and the world's wellbeing (that is, environmental sustainability) depends on a necessary social interdependence. In this sense, the Pope is channeling Judith Butler (2009), who talks about security and precarity and the fundamental social interdependence of people at all stages of life. There is no time when your life is not entirely conditioned by its location within society. Basically, the commons are about caring for each other. And because we care for each other we care about nature because we depend on it, not only individually but together as well. Society is fundamentally interconnected; no person or commons exists independent of society and we are all secure to the degree that society makes us secure. This is true in or out of commons.

In short, Pope Francis (2015) and I (Ribot 2014), among others (Sen 1981; Watts 1983; Blaikie et al. 1994; Ribot 1995), locate what are called 'environmental crises' in multi-scale political-economic cause. He even goes as far as to explain the developing world, environmental and social crises as being about greed, commodification of people and nature, and the structures of the global financial and economic systems. What I think is amazing is that he does this without citing any political ecology authors. Indeed, he cites the likes of Saint Augustin, he cites former Popes, among other religious figures, and he makes the case beautifully. So, he shows that we can draw in a lot more sources than the gods of our own discipline, and I think we should.

The Pope is providing us with some reflections by which we can chose to ignore or deny God but with which we can create a new humanism; one in which the humans do not dominate nature and nature is not just there to serve humans, but one in which humans must manage and care for nature – because they can and because they need to for their own good. Certainly, elephants and scallops and microbes and rocks (despite what Latour [1988] may intimate), cannot manage humans. We are conscious, have foresight (whether we use it or not), we have the capacity to see the recursive relations between what we do to the world and what that doing does to us. Humanism does not need to mean that humans have rights to nature or that nature is here for us. This is a limited and outdated humanism. Humanism also does not have to maintain its narrow definition of humans – that excluded women, slaves and non-European races. My humanism, when inclusive as its title implies, is merely the valuing and care by people for people – all people and all people equally.

Today, we live in an economy that produces precarious people, it produces a class that Pierre Bourdieu called *the precariat* (Bourdieu 1998; see also Standing 2014). These are the vulnerable, these are the people who are pushed over the all-too-low threshold into crisis when expected climate stresses, storms or droughts happen to hit. So damages are caused not by climate or climate change, but rather by pre-existing social, institutional and political-economic marginalization processes, such as undermining rules of collective ownership and access to resources, that cause precarity (Ribot 2014; see also Haller 2015 for a comparable approach). Part of this chapter is about how people are pushed into positions close to that threshold. I really want to examine why the causes of vulnerability are rarely treated in policy response, and what social sciences and humanities have to say about that. The reason to explore causality is so that we humans can identify causes and thus care for those who are vulnerable. The failure to do so is part of what makes people vulnerable and thus is a failure of the humanist project.

The new term that we all know (it probably does not seem so new to most people), *adaptation*, describes public response to climate change risks. We need to adapt to reduce people's risk in the face of new climate stressors. But to get at how to adapt we need to first understand why people are vulnerable, causality. First you understand why they have a problem and then you do something about it: we need to understand the causes of their vulnerability in order to treat those causes in one way or another.

But the major problem is that the causes of vulnerability are rarely treated within scientific struggles that inform climate adaptation policy. Most climate impact analyses remain obsessed with hazards, they are focused just on the things that are coming out of the sky. Most of those who do look at the vulnerability side of the risk equation, unfortunately focus on proximate variables and do not look into causality very deeply. They are obsessed with things like 'adaptive capacity' (whatever that might be) and with some of its proximate determinants; perhaps they even attend to the role of local institutions. They seem to have little interest, however, in looking at the larger context that creates institutions and assets and social protections and the more proximate determinants of vulnerability. Care, however, must go beyond preventing damages at the last moment. It needs to start as far upstream as possible.

The lack of adaptive capacity or of assets, or of social protections, is the product of a larger stratified institutional environment. This formation needs to be explained. Institutions and actions do not just emerge out of some luminiferous ether of Elinor Ostrom's (2010) policentricity – one that does not address issues of power. Rather, these lacks are about hierarchy and exploitation of institutions related to each other and to the 'rules of the game' through unequal material and discursive powers, exercised, as Nancy Peluso and I have emphasised in our work on access (Ribot and Peluso 2003), through law, custom and convention, but also through stealth, force, coercion and direct or indirect structural, silent or slow violence. This refers to Rob Nixon's (2011) concept of slow violence, but one also needs to be reading *Silent Violence* by Michael Watts (1983), on which Nixon builds his slow violence ideas; both contribute to the question of how we understand and treat Galtung's[2] (1969) 'structural violence'.

So, I will briefly illustrate what I mean by causality through a case study. Then I will examine why causality might be so problematic in vulnerability studies by exploring how causes of risk are entangled in social order, and with blame, liability and responsibility, and therefore why they are highly contentious and often avoided. I will then conclude with a note on emancipatory approaches to causality – evoking a humanism that leads us to a productive and just sociodicy.

Multiple causes of climate change and vulnerability in the commons

The climate-risk equation basically states that damages occur at the confluence of climate hazard and vulnerability. Both climate hazard and vulnerability are needed. Without one or the other, you do not have damages – such as hunger, famine and dislocation. Figure 2.1 shows the typical model that the IPCC and others have used for years, and it makes sense. You have the hazard, it comes along, it is filtered through some static figments of society (assets, capacities, sensitivity, etc.) and you have a list of outcomes (reduced wellbeing, etc.). Among these static figments is vulnerability, which is defined in the 4th assessment report of the IPCC (4th IPCC Report; Appendix II:89) as:

> ... the degree to which a *system* is susceptible to, and unable to cope with, adverse effects of *climate change*, including *climate variability* and extremes. Vulnerability is a function of the character, magnitude, and rate of climate change and variation to which a system is exposed, its *sensitivity*, and its *adaptive capacity*.

But are the outcomes outlined above really products of a climate hazard and the lack of adaptive capacity? No, they cannot be reduced to this. They are an outcome of much more.

In my own work, I invert the equation – following Sen (1981). Rather than looking at the multiple outcomes of some single event, I look at the multiple

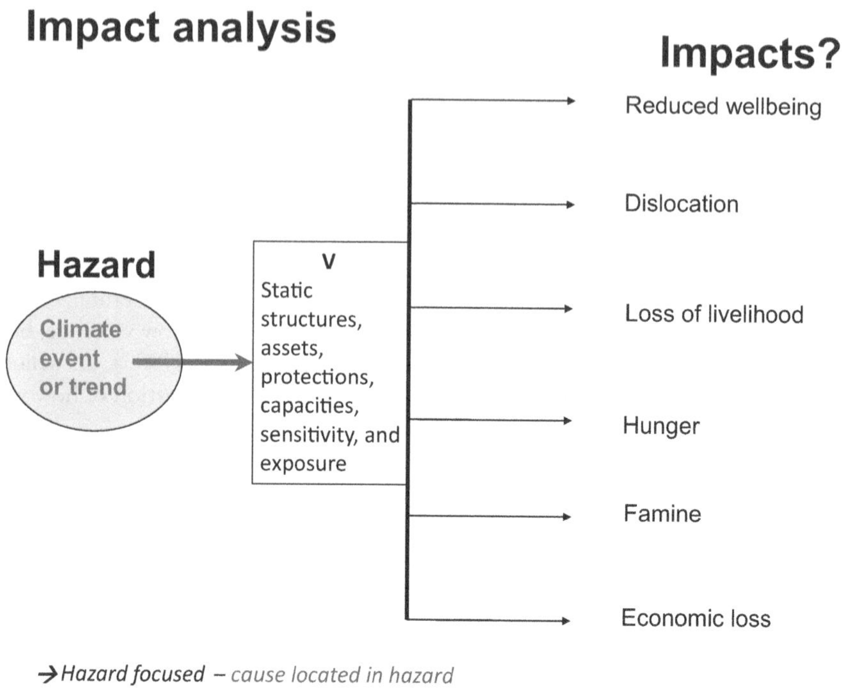

Figure 2.1 Hazards model.

causes of a single outcome, because climate events and trends tend to be one set of stressors amongst a whole host of variables that we must understand (Ribot 1995, 2014) (see Figure 2.2). My representation starts with entitlement failures (*a la* Sen 1981) and explores why there is a lack of assets and a lack of social protections that lead to problems of hunger and famine. You take one value and you find out why it got lost; one damage and explain how it happened. In other words, it is a case study approach. Sen (1984, 1997, 1999) has a secondary link from capacity and democracy back to capabilities to the shaping of assets and social protection. But, like any good economist, Sen takes assets and lack of social protection to be initial conditions. He does not ask where these 'initial' conditions come from, other than saying that they are influenced by democracy, which he defines very broadly as voice and established political processes.

In a political-economic approach, however, you need to go a little further and ask 'assets', 'lack of social protection', where do they come from? They come from exploitation, expropriation, unequal access relations, political exclusions, that themselves are related to other things happening in other parts of society that have to do with the lack of so-called 'capacity' and stratified access to education, lack of representation and democracy. These feed into shaping the broader political economy that shapes one's assets and social protections. Sen

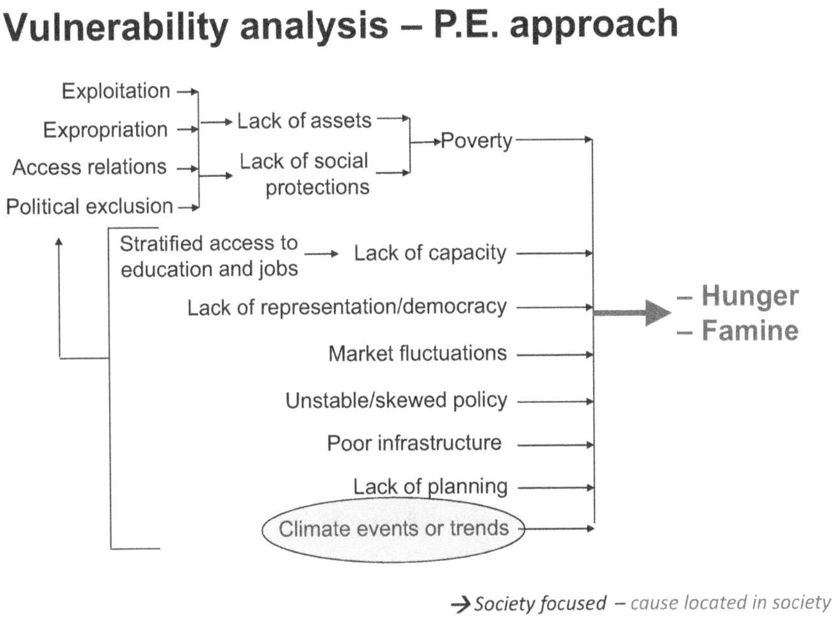

Figure 2.2 Vulnerability model.

calls those assets and protections 'entitlements', and a failure of the entitlements is what leads to hunger, famine and other such crises. Climate in this framing is not at the centre of the universe, but it plays its role. Climate is one of several triggering variables or stressors launching crisis but not the main 'cause'. Indeed, it has its role in the causal chain. Yet, climate could not produce damages were the other causal chains behind each factor of vulnerability – such as enclosures or the breakdown of common property regimes – not already at work.

I will use a case I know well to illustrate this. Dr. Papa Faye[3] – a Senegalese social anthropologist/geographer – and I have been working together in Senegal since 2003. This collaboration, along with a tragedy on 19 April 2015, motivated our current research. That day a boat sank off of Libya, carrying 728 migrants, most of them from Sub-Saharan Africa. There were only 28 survivors, 24 bodies recovered, and the rest, as their parents said in interviews, are at the 'bottom of the sea.'[4] These families did not know where their children were. The newspapers call the Sahelian migrants 'climate refugees' or 'climate migrants' (Foote 2016; Friedman 2016). What nonsense! Refugees are, first of all, defined by law as people running away from oppression or abuse by their governments. Climate is not their government. So, the term refugee does not apply.

A 2017 National Geographic episode, 'Out of Africa,'[5] Thomas Friedman, went to Senegal, interviewed people and basically presents the narrative that out-migration is climate driven. Nonsense. Nobody who lives and works in rural Senegal – unless they work for government, a development agency or an NGO – will tell you that their biggest problem is climate or climate change. That narrative is hard to stomach, because when you get out there and you do that place-based analytic of why and how people have access to the resources and income that they need to eat, to produce, to reproduce, to invest in their own households, you will see that they are producing a lot but keeping almost nothing. Any good analysis of why they are vulnerable (that is, why they lack assets, income and social security), has to explain the core of the problem; it has to explain why they have nothing or feel precarious; it has to explain why they have no surplus to buffer themselves against expected, though unpredictable, periodic stressors.

The case I will discuss is in West Africa in the Tambacounda Region of Senegal. It is a dryland millet, peanut, cotton and corn farming area. There are some woodcutters, and the forests do not look very rich, but they are lucrative, and people make considerable income from them. It is a case like Nancy Peluso (1992) has said, of 'rich forests and poor people'. It is a case of low prices for farm products, expensive inputs and costly credit. It is a case of a commons, the market (not the forest itself), to which farmers who try to supplement their income from woodcutting do not have access. Why?

Crises have histories

The crisis in Tambacounda goes back a long way. It goes back to different moments where West Africa was embedded in different social relations like the slave trade and colonialism (Franke and Chasin 1980; Watts 1983). More recently we see an annual hungry season which is as deep as many years during last 30. We see out migration, including the many from Tambacounda who perished on April 2015. They left for Europe but died at sea. Indeed, more are dying in the desert than die in the Medeterranian (Danish Refugee Council 2016: 2). Youth who depart are aware of some of these dangers, yet they go.

Peasants in Tamba are poor and are vulnerable to hunger and dislocation; indeed they are in a continuous crisis. It is not only a crisis of dislocation or hunger, but the anxiety leading up to that crisis, the insecurity is in itself damaging. These people cannot sleep at night thinking about the fact that they are not going to be able to feed their children (Ribot, Faye and Turner Forthcoming [2020]). Anxiety in Tambacounda was at a level that I had not seen. The pain of anxiety is visible and itself needs to be considered a damage of vulnerability – i.e. vulnerability is damaging even if acute crisis does not arrive; even if it does not lead to the hunger, famine, dislocation or emigration that we are focusing on.

So the rural economy has many elements: farming peanuts, millet, cotton, corn and rice, or herding, labouring in the cities, hunting, selling timber, selling honey, and charcoal production. Each is lucrative, however none are profitable for forest villagers. There are commons and there are private sector opportunities, but farmers find themselves excluded from profits. Here I am going to look at charcoal production. Charcoal provides a case study of why people do not

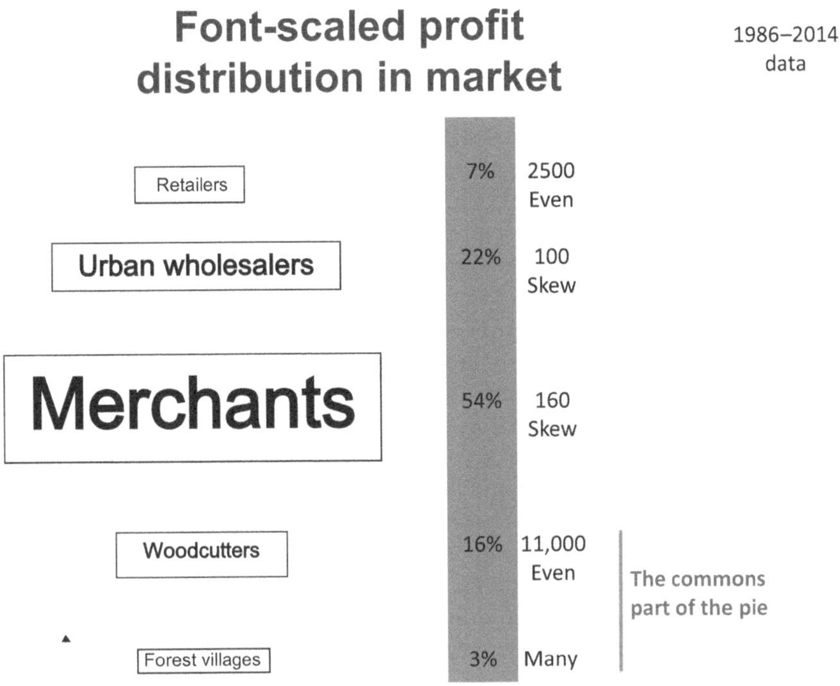

Figure 2.3 Font-scaled distribution.

profit – from forests or markets. It is basically an access analysis that Faye and I have done several times over different periods, starting in the late 1980s, and we have repeated it to see trends and patterns. But, a similar phenomenon is found in all sectors: most producers are wratcheted down to subsistence level or below – different mechanisms in each sector, but all to the same effect of enriching a small elite (usually urban) while leaving the peasants destitute.

The actors in Senegal's charcoal market include forest villagers, migrant woodcutters (some locals are beginning to work there), urban merchants, wholesalers and retailers. When you calculate the profits (a function of margins, expenses and market shares) at each node in the market, it becomes clear very quickly who is getting rich, and it is not the rural people. Figure 2.3 is a font-scaled distribution of profit along the charcoal commodity chain: the font size corresponding to the percentage of profit which goes to each class is applied to that group. So, 54 per cent of the profit (gross income minus the costs for the ensemble of actors at that level in the market) is assigned a 54 point font. What you see is that that the 54 per cent of the whole market's profit is divided among 160 merchants, whereas the 16 per cent of the total market profit profit is divided among 11,000 woodcutters. This last 16 per cent sliver is the commons part of the pie. In short, the merchants are wealthy, while the woodcutters live at subsistence level.

Basically, what you see, using some fairly rough figures (over time this distribution has evolved, although the overall pattern maintains) is that a merchant makes in a few days what a woodcutter makes in a year. I would estimate that today the biggest merchant is making over a million US dollars per year. But since I do not have the exact figures for the recent incomes, earlier ones, 30 years ago, made even as much as 500,000 USD (Ribot 1998). We explain this pattern of distribution by analyzing its components. Basically, you go to each group and you ask: 'How do these people maintain the share of the market that they have, the price that they get, and the cost of their inputs. How do they control it, can they?' And you begin to find out very quickly that their control over these things is embedded in a very complex policy-structured social environment (see Ribot 1998; Larson and Ribot 2007; Faye and Ribot 2017).

But to make a very brief caricature of this, we can, of course, say that exchange always has all kinds of 'regulations' – controls over access to resources, labor, markets, political arenas (Ribot and Peluso 2003) – these are what make markets and shape distributions. Markets are not free, they do not exist without some kind of regulation, at the very minimum a provision of currency. In the charcoal market there are licences and quotas (relabeled 'contracts' in recent neo-liberal tainted years), determined by the forest service, and then the regional forest service offices allocate them to merchants who have the licences and pay taxes (municipal and forestry). With taxes paid, they can then get transport permits. The producers, who have to get producer ID cards and production permits from the merchants, who have the quotas or 'contracts' and the licences, have relations with and get these papers from the foresters. You also have under the Ministry of Interior, an elected Regional Council and an elected Rural Council, who formally have powers since 1998 over the allocation of rights to forest, but that power never came to be in their hands, as I will show later (Faye 2014; Ece 2015). So, there are management plans imposed by the forest service, elected rural councils with rights to sign off on these plans (rights they are denied by dint of being forced to sign), and there are fines and confiscations for those who do not obey the array of regulations in place.

So, you have a system that is embedded in policy, but of course it is also embedded in social relations. Identity and other social relations mediated by all kinds of institutions, historical ones like nobility and captivity that relate merchants to woodcutters, who migrated to Senegal when the French attacked the Fulbe aristocracy of Upland Guinea in 1906 and when Sékou Touré attacked them in the 1950s. Or the relations between the religious brotherhoods and the merchants that end up eking out quotas and permits and influence who gets to sit on the regional and rural councils through their relations with political parties as well.

So, a market is not outside of its broader context. International agencies and large NGOs put the next layers on – via their projects. The projects, such as those of USAID and the World Bank, often say that government is 'bad bad bad bad'.

They go into the area and they create the 'Business Model' and rename the quota the contract – despite that the contract is exactly the same thing as the quota: it says who can produce, the amount they get to produce and when and where. So, development, conservation and adaptation become the priority of projects, despite that goals were supposed to be set by the rural council. Management plans and 'participation' are instituted with new institutions like committees, NGOs and chiefs and 'participatory projects and processes' because those are the ones who the projects believe should be making decisions. The rural council was elected and has the legal right to make these decisions, but the councils are what Papa Faye calls 'de-recognised' by these outside institutions (Faye 2014). Their powers are allocated to committees and thus they are disempowered and lose recognition.

There are many laws and regulations and programmes and projects to, ostensibly, help increase rural incomes. Participatory forestry Code of 1993 that became what I call 'participatory corvée' (Ribot 1996), forced labour – and that is what 'participatory' often is (see also Cooke and Kothari 2001; Galvin and Haller 2008). Democratic decentralization that empowered the rural council but gets taken back; contracts that I have already described were created to replace the quota; all to 'empower' local people, but they do not end up doing this. Forestry projects, USAID and the World Bank are very well meaning, but they did not end up doing what they claim.

Basically, with these projects, a lot of things unfolded in between design and implementation that had to do with the actual in-place negotiations amongst various hierarchies and following different laws, customs and conventions concerning who gets what, when and how. For whom these programmes would be lucrative is already pre-determined by the previous generation of patterns. They

Figure 2.4 Painting by the Senegalese reverse-glass painter Mor Gueye.
Source: Ribot 2017.

are ethnically defined, party defined, they are defined by religious brotherhoods in various ways, but they are also defined by gender, by place of origin and so on. So, each policy has a micro-politics that I cannot fully develop here, yet of which I have mentioned a few elements. These are important, and we need to understand them. They are often shrouded in interesting discursive sorts of mechanisms. Each local income improvement was in this case supported and fought for by the elected local government. By the way USAID fought to support the local government in many of these instances, and in certain instances helped local government to help rural people; but again, ultimately failed for a variety of reasons.

As an example of the kind of discourse, here is a tongue-in-cheek caricature of a forester speaking to, or lecturing, villagers from a book I wrote (Ribot 2017; also see Ribot 1997):

> Look, here in the rulebook – which you must obey,
> you have rights to the things that we don't take away.
>
> But we can't take the wood without taking the trees,
> so you'll have to make do, with the stumps, and some seeds.
>
> You can grow woodlots, eucalyptus or pine,
> we'll help you to manage them, through incentives or fines.
>
> If you listen, look, learn, and do as we say,
> even democratisation will be on its way.
>
> We must protect forests from people like you,
> so that people with business will have business to do.

That is the kind of language that forest villagers have heard over and over again while being 'empowered'; and they are sick of it. They know it and they hear it, and they even know how to play projects – by just saying 'yes' back (Scott 1985). But all the good intentions of projects do not take away this kind of discursive deceit – a deceit that project staff and forests may not even be aware of since they believe their own contorted narratives.

The precarity of sector by sector exploitation does not fall from the sky

This is a quick sketch of part of what is going on – in domains of law, social relations and project interventions – that contribute to the production of the inequalities we see in the charcoal sector (see also Ribot 1996; Faye, Haller and Ribot 2018; Faye and Ribot 2017). We see these skewed distributions that I have described, however, not only in charcoal but in all of the sectors. In the cotton sector, the international price goes up but the producer price stays flat. For good reasons, producer prices are fixed by intermediaries and corporations in cahoots with government and so on. Forest villagers remain at subsistence with no hope and no profit – regardless of international prices. They are also strapped with debt from equipment and input advances. Out-migration is one option; but as you can see, these migrants are not climate migrants – they are refugees of a system that excludes them from a livable income (Ribot, Faye and Turner forthcoming [2020]).

Virtually no Senegalese farmer will tell you that they suffer due to the climate, nature is not their problem. Farmers knowingly place responsibility in taxes, permits, licences, fixed prices, failed representation, market access, access to government services and so on. These are what need fixing and adapting. Elected local governments have tried to improve benefit retention but again, they have been thwarted by line ministries, which are the single most important obstacle to any decent democratic decentralization effort worldwide.[6] They hang onto the funds they should have transferred to local authorities. It is time to 'adapt' the line ministries: take their money away and transfer it to local governments.

I have described some of the major causes of vulnerability and resulting 'climate-related' crises. Why, however, are they so difficult to identify and treat? Why don't we engage in causal analysis and treat the causes?

On causes of vulnerabilities: Weber, Douglas, Rose and risk under a changing sky

The second part of this chapter touches on the causes of vulnerability as they have been treated by social sciences historically. The case I presented shows that suffering and crises that climate modelers usually attribute to rainfall and drought cannot be explained outside of social and political-economic relations. Most serious social scientists know this – it is not a mystery. But why can't this kind of explanation even enter into our understanding of causes of the climate-related crises that the climate change community is ostensibly so worried about. Why, for example, in the IPCC 5th assessment report, has vulnerability been reduced to a focus on indicators of *who* is vulnerable rather than an analysis of why (based on my search of the word 'vulnerability' and its use in the report). We need to know why people have become and remain vulnerable if we are going to change things and fix the problem.

Analysts seem to want to know who will be damaged and how much, but attention to cause receives a lot of resistance – or ends at 'adaptive capacity' (see Ribot 2011). Why is the question of 'why' so complicated? Why are social sciences that attend to political and economic structure and causality written off in climate science and adaptation studies, as well as being underutilized in common property work. I suggest that we avoid explaining the causes of risk because causality is linked to two major issues: one is the issue of responsibility, blame and liability, and the other is the maintenance of authority. I think that vulnerability is difficult because risk is actually a very special category of social interaction and meaning. It is not just any phenomenon, vulnerability is distinct. It has very special characteristics that make it difficult to talk about and respond to.

The British sociologist Iain Wilkinson (2010) argues, as do many sociologists from Max Weber on, that risk plays key roles in organising society, and in doing so produces very political divide that make risk so important and so contentious in social debate. They all postulate – and this is interesting – that risk is different from other disputed and political ideas. Rooted in Wilkinson's reading of risk sociology, I will briefly go through three key theories and what they have to say about this. I am going to start with Max Weber on risk and enlightenment and then say a bit on Ulrich Beck, Mary Douglas and Nicholas Rose.

According to Weber, pain and suffering are the foundations of atheism (Wilkinson 2010). Religious leaders spent a great deal of their energies on the practice of theodicy, that is, the practice of explaining why there is pain and suffering if God is so good. When rationalization of suffering in religious systems failed, people moved toward secular understanding. Theodicy was an important branch of theology, since it helped to retain believers. It also secured the authority of the church. As Wilkinson (2010: 28) notes,

> … all human cultures are faced with the task of dealing with what appears to be an excessive amount of suffering in the world. In all times and all places people struggled to find solutions to this pain and suffering and are compelled to work at making sense of this experience in terms of the overall meaning of their lives.

And, of course, I would add, they must also explain in order to cope with or reduce suffering, or even just to sense that they have some control over their world and lives.

From authority, people expect protection or explanation, they have a kind of moral-economy relation with those who govern (Scott 1985). The rationalization, and thus containing and explaining, of suffering is a foundational element of social organization and of any governing system. Suffering is the enemy, God (or, post-enlightenment, government) should protect us. The authority of recent governments rests on their ability to rationalize and protect (look at the loss of legitimacy of the Bush regime after its failure to even appear to protect people during and after Katrina and the showy efforts of warning that US state and federal authorities now make before any storm). For Weber, the link between rationalisation, cause and blame are key elements in the study of risk, and they are key elements in how risk is understood. In short, when in need of release from suffering, as Wilkinson argues in his reading of Weber, people devote themselves to rationalization. They explain cause and thus attribute blame.

A little side observation here, which Weber does not make, in the course of response, responses are guided by different theoretical and philosophical stances, that is different rationalizations that inform notions of causality. Theory and historiography matter then in understandings of causality. Hence, they too become contested by those who stand to gain or lose based on explanations. So, explanation itself becomes contested terrain. This is for me a central pillar of a politics of knowledge (knowledge of cause) – and I would argue that it is central to any politics of risk; including the politics of including the causes of risk within scientific models of climate impact.

Wilkinson (2010: 29) explains, getting back to Weber, that there is an 'imperative of consistency', a kind of compulsion, that forces people to make sense of the world in terms of their cultural outlooks. They try to maintain the status quo. But the changing world also forces them to change cultural outlooks in order to maintain consistency with the world. This is basically Weber's theory of social change. In a sense, explaining of disaster, where disasters represent changes in people's worlds, is, for Weber, a driver of cultural change; thus, explanation threatens the status quo. We, that is climate risk vulnerability analysts, note this when we analyze causality of vulnerability and find ourselves up against resistance in multiple camps. In short, Weber argues that theodicy (pre-enlightenment) and rationalizing of risk (post) are foundation of social change.

Then comes Ulrich Beck (again see Wilkinson 2010 for his fabulous summary). Beck, with his 'risk society' approach, makes another set of arguments placing risk at the center of social organization. He asserts that risk perception, as risks have become global and potentially destructive across any human borders, has a political-mobilization force. He believes that pending global catastrophe pushes people to come together and organize across national boundaries. The result of his reflections is a utopian cosmopolitanism that enables NGOs and other groups to reshape global institutions. People becoming more 'world-risk conscious' will organize to save themselves. This wishful thinking has many shortcomings, but it does place risk at the center of social and political organization and change and launches numerous debates on risk and society.

Next are Mary Douglas's ideas of risk, danger and society. Again drawing on Wilkinson (2010), for Mary Douglas, as a cultural anthropologist, the reality of risk that people fear is not the issue. Reality does not matter. She only focuses on the social function of risk. She argues that when social bonds that hold people together are weak, people tend to become obsessed with disaster. Perhaps the weak bonds of late capitalism are a condition of our own sense of risk in the fragmented societies of late capitalism. She argues that preoccupation with disaster plays a positive role for society. By finding a common threat, communities come together and organize around common social objectives. They protect their group from harm. This may seem self-evident, but simultaneously, as part of groups' survivals, comes the search for explanation and blame, which helps further consolidate the group by defining insiders and outsiders. It defines and identifies those who belong and creates boundaries that define the community. It also defines the other. Certainly, a point common property theorists might want to take up, when it comes to boundary-making.

In agreement with Weber, Mary Douglas sees discourses of risk as having taken the place of discourse of theodicy. The language of 'sin' had been supplanted by language of 'risk', both being languages of blame. Where people used to talk about being 'sinned against', they now speak of being 'at risk'. '...The language of sin appeals to the authority of priests…, risk appeals to the authority of science and modern rationality' (Wilkinson 2010: 49–50). Populations defined as 'at risk' carry the moral stigma of sinners since they are burdens on society. Douglas would argue that in secular consciousness, nature has replaced God. In so doing, people can rationalize crises through probabilities and attribute causality (and thus blame) to the vagaries of nature. They can sidestep blame – since 'nature' cannot be guilty (unless you want to attribute agency to it as does Bruno Latour – a very problematic proposition [see Braunstein 2018 on the moral philosophy of humanism]).

This brings us to an interesting question in the present. What happens in the current era of anthropogenic nature, when nature is caused by human action, and no longer a mysterious random blame-free force? What happens when storms, floods, droughts are blamed on industry and SUV drivers in the North? Then a new structuring of insiders and outsiders forms, and responsibility and blame are re-arranged on a global scale. A new causality behind nature itself suddenly erases nature as a sink for blame. When God is dead, and nature can no longer replace God because it has become a social product, there is no longer room for non-human explanation of the experience of suffering. Society is turned back on itself. God and nature are dead, God dies with enlightenment, and nature, as sink for blame and perhaps as an entity, dies with the Anthropocene – at least for those who can cope with social explanation. The Anthropocene infuses social explanation into the forces we usually call nature. Perhaps we should call it the Sociocene. Welcome to the Sociocene (that's the scene I like to be in)!

With this death of nature, the above social theories would imply that we are entering a transformative moment of social re-organization. The first scary question is who will the priests be? Political scientists? Social scientists? I do not think so. But in any case, Mary Douglas (1996) also talks of the public assessment of risk, as being embedded more in the trust of sources of information than in what is or is not truth, making it inherently political. Here, trust, authority and credibility become another foundation of the contentious politics of risk. Clearly, we are seeing the reconfiguration of credibility in the Trump era – and it is playing on denial of all kinds of blame, liability and responsibility. It is creating dueling causal arguments along with dueling social spheres – each with its different boundary construction.

I cannot develop this line of thought here, it implies, however, that with a proliferation of authorities (related to God, nature and society), since none seem to fully die, problems of trust will deepen and so will associated social fragmentations. Douglas casts the politics of knowledge as one of trust, I would cast it as one of interest as well (being somewhat of a materialist). Of course, the two are intertwined: we trust those, whose positions resonate with our interests. This explains partly why God keeps rearing up and why nature seems to be such a comforting sink for blame. When we displace blame from state and society onto God and nature, we protect the trustworthiness of existing state and scientific authorities, since nobody can trust or mistrust a transcendental God or an authorless nature. This is precisely why they are preferred over social explanations, which usually indicate blame and undermine these very same authorities. This is why we need a sociodicy of the Anthropocene – a sociodicy being an attempt to explain pain and suffering in human terms and through human/social cause. Those causes can be found when looked for.

We need a little note here about causality, responsibility and anthropogenic climate change – for climate change makes climate un-natural. As human agency now arcs across the sky, why is it that many government institutions are accepting explanations that acknowledge anthropogenic climate change – making it possible to place blame on the industrial world? The acknowledgement of anthropogenic climate change is still, of course, incomplete – the Trumps of the world deny it – but many within various governments around the world accept the link between human action and climate stress. It seems that part of what makes this possible is that climate-change causality remains so abstract and diffuse that it blames everyone and thus no one. Industries spread worldwide are blamed, individual car drivers are indicated, tropical forests are brought in, fuelwood burning is brought in, India and China and Brazil's development are brought in to the causes – thus including and blaming both the industrial and developing world. Cause is being constructed as common, a commons. It is being constructed as an effect of being human (*anthropos*) rather than the outcome of specific social structures or of specific actors in specific places whose profligate waste of resources and energy are undermining the opportunities and potentials of others whose access is restricted. Responsibility in this representation remains diffuse – despite even the acknowledgement of responsibility (limited responsibility) constructed under the notion of 'additionality' in UN conventions on climate change (see Ribot 2014).

Lastly, a few words about Foucault. This is important because it is a very different frame. A Foucauldian frame implies another track posing key roles for risk as a discourse in the service of domination. Foucault is interested in how individuals and groups are corralled into taking up behaviours that they 'choose' as their preferred ways of being. Governmentality – basically a 'conduct-of-conduct' type of argument. For Nicholas Rose, a Foucauldian, certainly also inspired by Mary Douglas, risk is a different kind of turning point in the history of government (Rose 1996). Risk analysis and communication produce categories of 'at-risk' people. These categories make them aware of themselves as being exposed in the face of stressors like storms or disease. It causes a shift from a collective sense of hazard as something that might be responded to collectively, to a category of specific people and sub-groups at risk. In short, discourses make people believe that they are responsible for their own risk. This enables the blaming of the victim by the victim herself or himself and by society. It internalizes causality to the individual and group, it becomes another way in which blame is diverted from government authorities. It is the individualization or neo-liberalization of risk.

The term 'adaptation' fits right in. It is natural (like natural selection), and it does not blame society (see Ribot 2011). Instead, it focuses its attention on the individual's qualities needed to innovate and sustain in a difficult and changing world. It takes attention away from circumstances, placing responsibility on the affected party rather than on a larger political economy or government. Indeed, adaptation terminology carries an implicit survival-of-the-fittest social-Darwinist ethic that blames the victim by locating failure within the adjusting or affected unit – in their adaptive capacity. By locating cause within the victim, risk reduction becomes divorced from debates over political economy and is addressed as a mere matter of personal education and moral reform of the individual – or it can take the form of social assistance that can gain governments and do-gooders more credit.

Under these individualizing scenarios, political economy is occluded, broad social explanations vanish. Rose (1996) calls this 'the death of the social,' since social explanation is erased. I do not like this terminology since I believe that the social is very much alive and well, and it is only buried by those who wish to ignore it or benefit by hiding it. Hence, part of my project is the continuous bringing of social and political explanation back into analysis and into public debate. My project is to develop a sociodicy of risk – one that shows how and where responsibilities lie within human society.

Conclusion

Having imagined the death of God, nature and society, where do we stand? We stand in the Anthropocene, a time of mistrust and fragmentation of authority, a time of diffusion of blame and a time of the occlusion of causality in the service of non-responsibility. I think that what is at mortal risk today is knowledge itself. Knowledge is contingent on the politics of responsibility, cause and blame. If anything is being killed by the current moment it is knowledge, erased by those it does not serve. We live in an era where risk is accepted as individual, adaptation proceeds, God, nature and the social are gravely wounded. How do we bring knowledge back in to assert responsibility so humans can take their place caring for each other – and thus for the world we all need?

I do not believe that God, nature or society are really dead, it depends on who you ask. They live in different ways in each person. I do, however, believe that risk plays an important role in how we all view our options and responsibilities, as well as our identities and our belongings. If we are to understand why natural and social science notions of risk are disputed, believed and denied, we need to better understand the role of risk in shaping society. If we are able to understand the causes of vulnerability, we also must understand how risk itself recursively shapes the ways we generate and avoid risk – as a causal element itself.

By understanding why certain causes of vulnerability are inadmissible in science, policy and public debates, we can inform a process of bringing knowledge of causality into public discourses and debate, thus changing vulnerability itself. In an emancipatory polity, the discursive conduct of conduct will be challenged, and the social revived from its injured retreat. God and nature will be held at bay. In a democratic polity, which few of us seem to occupy today, science would be continuously pressured to reflect experience and need. It may not seem plausible, but I do think we have to try. For me, the challenge to the social sciences is to bring into our analysis of the causes of vulnerability the missing reflexive relations of emancipation. If it is in our analytics it has more chance of emerging in practice.

In the end, the relations between outcome, cause, judgement and responsibility need to be integrated into the causal chain and our explanations of vulnerability. These cannot be separated from the relations of representation of knowledge and representation of citizens. To end, two emancipatory recursive reflexive loops need to be attended to: the democratization of science, that is the bringing in of knowledge and cause into vulnerability theory and climate modelling, and the democratization of practice, that is as Sen (1981, 1984, 1997) and Watts (1991; also see Watts and Bohle 1993) call for, the enabling of people to shape the political economy that shapes their entitlements. As Haller et al. (2016) put it – although I do not see it as a 'game' – for security, commoners must be able to define their own 'rules of the game'. These are neglected elements of climate-related and broader security, so we still need to develop an emancipatory sociology of risk, a humanist sociodicy for the Anthropocene.

Notes

1 This chapter is based on a talk by the author and represents the edited version of the presentation by the editors in consensus with the speaker. The talk was a keynote address at the International Association for the Study of Commons in Bern, Switzerland in May 2016. This was delivered as a talk and this transcription may omit some references or contain errors in transcription.

2 Galtung appears to me to be an anti-Semite (although some disagree: www.transcend.org/tms/2012/05/johan-galtung-an-anti-semite-i-dont-think-so/). I am uncomfortable citing him. But, he has had his influence already and not citing him would be an omission. For the accusations see: www.haaretz.com/1.5218261.

3 Papa Faye is Executive Secretary of the Centre d'Action pour le Développement et la Recherche (CADRE, www.cadre.sn) and 2017–2018 Open Society Foundation Non-Resident Fellow.

4 The boat and bodies have since been recovered.

5 Out of Africa video: www.youtube.com/watch?v=SfTswlvYXz4.

6 The re-emergence of customary authorities and the onslaught of NGOs have also been major factors in undermining elected local governments. They have given alternatives to intervening agents, facilitating their circumvention of democracy (Ribot, Chhatre and Lankina 2008).

References

Bourdieu, P. (1998) 'La précarité est aujourd'hui partout' ['Precariousness is Everywhere Nowadays'], in Contre-feux, *Propos pour servir à la résistance contre l'invasion néolibérale* (pp 96–102). Paris: Raisons d'agir.

Blaikie, P., Cannon, T., Davis, I., Wisner, B. (1994) *At Risk: Natural Hazards, People's Vulnerability and Disasters*. Routledge: London.

Braunstein, J.-F. (2018) *La Philosophie devenue Folle: Le genre, l'animal, la mort*. Paris: Bernard Grasset.

Butler, J. (2009) *Frames of War: When Is Life Grievable?* Brooklyn: Verso.

Cooke, B. and Kothari, U. (2001) *Participation: The New Tyranny?* New York: Zed Books.

Danish Refugee Council (2016) 'Mixed Migration in the Horn of Africa and Yemen Region', in Nairobi, Kenya. *Mixed Migration Monitoring Mechanism Initiative*, https://reliefweb.int/sites/reliefweb.int/files/resources/4mi_Mixed_Migration_Monthly_Summary_June_2016.pdf (accessed 8 October 2017).

Douglas, M. (1996) *Purity and Danger*. London: Routledge.

Ece, M. (2015) 'Representation Through Privatization: Regionalization of Forest Governance in Tambacounda, Senegal', Responsive Forest Governance Working Paper No. 23. CODESRIA, IUCN and University of Illinois.

Faye, P. (2014) 'Democratizing Forestry: Decentralization and Constitutionality in two Interventions in Senegal' PhD thesis, University of Bern.

Faye, P. and Ribot, J. (2017) 'Causes for adaptation: access to forests, markets and representation in Eastern Senegal', *Sustainability*, vol 9, no 311, p 20.

Faye, P., Haller, T., and Ribot, J. (2018) 'Shaping rules and practice for more justice. Local conventions and local resistance in Eastern Senegal', *Human Ecology*, vol 46, no 1, pp 15–25.

Foote, W. (2016) 'Climate Refugees are Leaving the Farm Behind', *Forbes*, 12 July 2018, 20 July 2016, www.forbes.com/sites/willyfoote/2016/07/20/climate-refugees-are-leaving-the-farm-behind/#73541f211a8e (accessed 1 November 2018).

Franke, R.W. and Chasin, B.H. (1980) *Seeds of Famine: Ecological Destruction and the Development Dilemma in the West African Sahel*. Montclair, NJ: Allanheld and Osmun.

Friedman, T.L. (2016) 'Out of Africa', *The New York Times*, 13 April 2016, www.nytimes.com/2016/04/13/opinion/out-of-africa.html?mcubz=0 (accessed 13 January 2018).

Galtung, J. (1969) 'Violence, peace, and peace research', *Journal of Peace Research*, vol 6, no 3, pp 167–191.

Galvin, M. and Haller, T. (eds.) (2008) *People, Protected Areas and Global Change: Participatory Conservation in Latin America, Africa, Asia and Europe*. Perspectives of the Swiss National Centre of Competence in Research (NCCR) North-South, University of Bern.

Haller, T. (2015) 'Climate Change Impacts, Institutional Resilience and Livelihoods Vulnerability: A Comparative Analysis of Climate Change and Adaptation Strategies in Asia, Africa and Latin America', in Rosendhal, J., Zanella, M. and Weigelt, J. (eds.) *Pro-poor Resource Governance under Changing Climates*. Potsdam: IASS and IFAD.

Haller, T., Acciaioli, G. and Rist, S. (2016) 'Constitutionality: conditions for crafting local ownership of institution-building processes', *Society and Natural Resources*, vol 29, no 1, pp 68–87.

Larson, A. and Ribot, J. (2007) 'The poverty of forestry policy: double standards on and uneven playing field'. *Journal of Sustainability Science*, vol 2, no 2. http://pdf.wri.org/sustainability_science_poverty_of_forestry_policy.pdf (accessed 1 November 2018).

Latour, B. (1988) *The Pasteurisation of France*. Cambridge: Harvard University Press.

Nixon, R. (2011) *Slow Violence and the Environmentalism of the Poor*. Cambridge: Harvard University Press.

Ostrom, E. (2010) 'Polycentric systems for coping with collective action and global environmental change', *Global Environmental Change*, vol 20, no 4, pp 550–557.

Peluso, N. (1992) 'Rich forests, poor people', *Political Economy*, vol 51, pp 9–26.

Pope Francis (2015) 'Laudito Si', http://w2.vatican.va/content/francesco/en/encyclicals/documents/papa-francesco_20150524_enciclica-laudato-si.html (acccessed 1 November 2018).

Ribot, J. (1995) 'The causal structure of vulnerability: its application to climate impact analysis', *GeoJournal*, vol 35, no 2, pp 119–122.
Ribot, J. (1996) 'From exclusion to participation: turning Senegal's forestry policy around?' *World Development*, vol 23, no 9.
Ribot, J. (1997) 'An ode to the Lorax: the business of sustainable development, an African forest tale', *Africa Today*, vol 44, no 2.
Ribot, J. (1998) 'Theorizing access: forest profits along Senegal's charcoal commodity chain', *Development and Change*, vol 29, no 2.
Ribot, J. (2011) 'Vulnerability before adaptation: toward transformative climate action' from affirmative to transformative climate action', *Global Environmental Change*, vol 21, no 4.
Ribot, J. (2014) 'Cause and response: vulnerability and climate in the Anthropocene', *Journal of Peasant Studies*, vol 41, no 5, pp 667–705.
Ribot, J. (2017) *The Business of Sustainable Development, An African Forest Tale*. Self published.
Ribot, J. and Peluso, N.L. (2003) 'A theory of access', *Rural Sociology*, vol 68, no 2, pp 153–181.
Ribot, J., Faye, P., and Turner, M. (Forthcoming [2020]) 'Climate of Anxiety in the Sahel: Emigration in a Xenophobic Anthropocene', *Public Culture*, vol 32, no 1.
Rose, N. (1996) 'The death of the social? Re-figuring the territory of government', *Economy and Society*, vol 25, no 3, pp 327–356.
Scott, J.C. (1985) *Weapons of the Weak: Everyday Forms of Peasant Resistance*. New Haven, CT: Yale University Press.
Sen, A. (1981) *Poverty and Famines: An Essay on Entitlement and Deprivation*. Oxford: Oxford University Press.
Sen, A. (1984) 'Rights and Capabilities', in Sen, A. (ed.) *Resources, Values and Development* (pp 307–324). Oxford: Basil Blackwell.
Sen, A. (1997) 'Editorial: Human Capital and Human Capability', *World Development*, vol 25, no 12, pp 1959–1961.
Sen, A. (1999) *Development as Freedom*. New York: Knopf.
Standing, G. (2014) *A Precariat Charter: From Denizens to Citizens*. London: Bloomsbury Press.
Watts, M.J. (1983) *Silent Violence*. Berkeley, CA: University of California Press.
Watts, M.J. (1991) 'Entitlements or empowerment? Famine and starvation in Africa', *Review of African*, no 51, pp 9–26.
Watts, M.J. and Bohle, H. (1993) 'The space of vulnerability: the causal structure of hunger and famine', *Progress in Human Geography*, vol 17, no 1, pp 43–67.
Wilkinson, I. (2010) *Risk, Vulnerability and Everyday Life*. London: Routledge.

3 Disruption, community, and resilient governance

Environmental justice in the Anthropocene

David Schlosberg

Introduction

By now, it is practically a requirement in any area of environmental studies that you have something to say about the Anthropocene. This is an edited volume on commons in a "glocal world," and the new thinking in this context also entails the notion of Anthropocene. As commons literature is concerned with environmental issues it cannot escape the debate on how humans now have a major impact on all systems, local and global commons included. However, and specifically in relation to the way commons management is discussed, this chapter argues that one needs to link this debate much more to issues of social justice, which the reality of the Anthropocene threatens in many ways.

Unfortunately, the Anthropocene also undermines many concepts underpinning environmental and commons management, because, frankly, it undercuts much of the normative underpinning for conservation, preservation, and even restoration of such places and commons. Those concepts have been the anchor of much environmental management, commons management, and certainly much classic environmental advocacy, for over a century, and the Anthropocene signals the inadequacy of such justifications (Schlosberg 2016). That is simply one example of how the very idea of the Anthropocene can be disruptive.

The point of this chapter is to examine the relationship between the idea of the Anthropocene, this reality of its disruption, and the impact on thinking about environmental justice (EJ). I have spent much of my career working on the issue of environmental justice (Schlosberg 1999, 2007, 2013), and so have been reading and thinking about the relationship between justice and the Anthropocene as a response to the growing focus and literature on the topic. Rather than make a single argument, this chapter addresses four distinct, but interrelated, points about the relationship between environmental justice and/or in the Anthropocene.

The first point is, frankly, a rant about the state of the literature on the Anthropocene. There is an empty space, or even a negative space, in much Anthropocene writing in relation to justice and environmental justice. It is not completely invisible, but it is clearly problematic how little considerations of justice have come up in the literature on this widely-addressed concept.

The second point is the context of justice impacts. My own thinking and focus on this intersection between justice and the Anthropocene starts with two key manifestations of disruption – both the sudden and immediate disruptions of single events, impacts, and emergencies that much adaptation planning is addressing, as well as the kind of gradual, slow violence of environmental impact that Rob Nixon (2011) has written about. We face an age of intense shocks as well as ongoing and increasing stresses.

Third, I introduce what I see to be the key concept in thinking about the relationship between the Anthropocene and justice. My central concern here is the potential disruption of community and attachments, including commons. Anthropocene disruptions, the argument goes, undermines our attachment to place, our relationality and entanglements with the human and nonhuman communities in which we are immersed. And that is a type of environmental injustice.

Finally, I will touch on some implications and applications for environmental governance – including governance of commons – given not only injustice and the Anthropocene, but also in light of these specific types of disruptions to environment, community, and capabilities.

The Anthropo-scenery

Before getting to those, however, we need a definition of exactly what it is we are discussing in the Anthropocene. I do not want to fully explain the concept of the Anthropocene, as it has been quite the popular term lately, moving from the sciences to the social sciences and humanities.

There has, of course, been much written on the idea. Bill McKibben (1989), in his first foray into writing on climate change, foretold *The End of Nature* – meaning that the natural world is now so fully impacted by human activity that we just cannot call it natural any more, if it were ever the case that such a pure nature existed (Ellen 1982; Haller and Galvin 2011; Haller et al 2013, Haller this volume). Crutzen and Stoermer (2000) put forth the original scientific argument for a new human-directed geological age, and some of the best work on the whole range of visual geological indicators has been done by Jan Zalasiewicz (see, for example, Zalasiewicz et al. 2015) and Will Steffan (for example, Steffan et al. 2015). In their presentations on the topics, Zelasiewicz often shows pictures of plastic, of concrete, of massive mines, and of the decimation of the sea floor by trawlers to illustrate human impact. Steffan, on the other hand, uses illustrations of "The Great Acceleration" to show how just about any indicator of human impact has vastly increased since the mid-20th century.

These visualisations of disruption tell a very clear story of a changed planet. Ultimately, the idea of the Anthropocene is that human beings now influence, if not control, the ecological makeup and direction of the planet – and that future geologists will look at the physical evidence of such a shift baked into the earth. The point is that we now, and geologists in the future, will be able to clearly see a demarcation between a point when human beings impacted individual places, to where we, as a species, altered global environmental systems.

As Dipesh Chakrabarty (2009) argues, human beings now act with the power of a geophysical force, like tectonic plates or volcanoes. "We" are the *Earthmasters*, as Clive Hamilton insists (2013) – and we have already geoengineered our way into a climate-changed planet (though, as noted above, the universalizing "we" is clearly inaccurate and unfair). While there is obviously some hyperbole there (there is quite a difference between clear global impacts and "mastery" of our actions on the planet), the impact of the behaviours of some key humans and human practices on the trajectory of earth systems is undeniable.

Donna Haraway (2015, p. 159), even though she does not like the term Anthropocene, has a pretty solid definition:

> It's more than climate change; it's also extraordinary burdens of toxic chemistry, mining, depletion of lakes and rivers under and above ground, ecosystem simplification, vast genocides of people and other critters, etc, etc, in systemically linked patterns that threaten major system collapse after major system collapse after major system collapse.

The key shift is that human beings have moved from being serial depleters of environments, despoiling and moving on, to something a bit more serious, global, and constant in scale and impact.

There are a lot of important and underexplored questions about the Anthropocene that I am not going to address in depth here. Most pertinent to this discussion is the fact that it is not really fair to call it the age of human in universal terms, as there are some very particular humans and practices to blame, and I will come back to that question of the politics at the heart of such a title. But the focus of this chapter is not about the definition of the Anthropocene, where it starts, who is responsible, better names for it, whether it is good or not, etc. The point here is to examine what the reality of the age means – that it is now a whole suite of global environmental systems that are being impacted, interrupted, undermined – for considerations of environmental and social justice.

Justice in considerations of the Anthropocene

In the vast literature on the idea of the Anthropocene, there is very little on the consequences for environmental or social justice – which is odd for something that is basically a result of an economic system driven by access to cheaper resources and the labour to process it. There is, instead, an empty space around the idea of justice in the Anthropocene literature, or more specifically, the responses to the growing concept in the humanities and social sciences. The idea of the Anthropocene is primarily a physical and visible one. The point of the stratigraphers and geologists is that there is nothing on the planet that is untouched by human impact and, crucially, we can see that clearly, as will future geologists looking at carbon, or concrete, or plastic, or radionuclides. It is not really the job of the stratigraphers to link something like a radiation signature or bioturbation with unjust human impacts; that is our job in the social sciences and humanities. But that response has been anemic, or worse.

For example, the initial critical responses to the Anthropocene, from the environmental humanities in particular, are devoid of environmental justice concerns or implications – and often just exacerbate embarrassing absences of a concern for justice in numerous fields. In addition, as Giovanna Di Chiro (2016) argues in her great overview of various keynotes on the Anthropocene, even the proponents of the idea have some key blind spots when it comes to social justice implications of the age.

Disputing the term Anthropocene

So, to start, we have a host of critics from environmental humanities that see the very term Anthropocene as an offense. They want to deny that we are doing what we are doing because of an idealised view of an untouched nature they see as endangered by the implications of the word "Anthropocene." Eileen Crist (2013), for example, complains that the very idea of the Anthropocene simply submits and affirms human centrality and domination. Worse, it shrinks the discursive space for challenging this domination of the biosphere, and blocks discussion of alternative forms of human life on earth. Similarly, the philosopher Ned Hettinger (2014, p. 4) insists that the Anthropocene is "an arrogant overvaluation of humans' role and authority", that it undermines "the importance of nature preservation, restoration, and rewilding," and it will have us simply become managers of the earth we have created, promoting ecosystem invention and geoengineering.

It may be tragic, and certainly destabilising, but between the acknowledgment of the reality of human beings impacting global ecological systems, and the reality of the changing nature of those systems, the historical ideas of conservation, preservation, or restoration really do not work anymore. Preserve what, if ecosystems can no longer reproduce? Restore to what, if local conditions can no longer support species that have lived there for generations? The preservationist ideal is dead. But I digress.

On the question of justice, this denialist approach to the Anthropocene – and their very old-school idea of an external, independent, wild nature to be preserved – lacks any way to address the actual intersection of humans and our environments, or, crucially, the dysfunctions of this relationship that are brought to the forefront by considerations of environmental justice. As many have argued, this insistence makes invisible those environmental justice issues that come with human/nonhuman entanglements and immersions. But many of these nature advocates do not really see justice issues as "environmental" concerns at all.

Worse, the point of this kind of opposition to the term Anthropocene by some is simply to put humans back in a minor place as opposed to the greater nature out there – which is really a head-in-the-sand response in the face of the reality of human impact on global ecological systems. The Anthropocene is all about attachments, entanglements, immersions, and yet some insist on preserving this fiction that we can have nature over here and humans over there. The point is that much of the argument around the reality of the Anthropocene really does not help us understand what this new era of human impact means for social and environmental justice.

The embracers of Anthropocene

Then we shift from those that deny the Anthropocene to those that embrace it just a bit too heartily. There are vast differences here, but there is the same indifference to the reality of injustice. Here I include the self-proclaimed "Ecomodernists" – the faithful techno-futurists who want to take advantage of the Anthropocene idea that we influence all natural processes and radically upscale it by insisting that we can control those processes. So they have argued that we can have not just a good Anthropocene, but a great one (Asafu-Adjaye et al 2015).

First of all, this is not a new or original position – Stewart Brand, in the very first Whole Earth Catalog in 1968, insisted that "We are as gods and might as well get good at it." Of course, we are not very good at it. More importantly, with the Ecomodernists we have exactly the same problem as with the Anthropocene denialists: the desire for a false sense of human separation from nature. Ecomodernists want more and more economic growth. But they conflate the goal of many environmental economists – the idea of decoupling economic growth from environmental damage – with decoupling the human and nonhuman. They argue that economic development can happen over "here" while we leave more supposed unmolested nature over "there." The Ecomodernists want to "make more room for nature" but they do not want to "harmonise" with it (ibid, p. 6).

But how can we recognise the reality of the Anthropocene without recognising the way that human and nonhuman are tied together, co-embedded, entangled in these changing global processes? This is the essence of the Anthropocene – that our actions and practices are linked to broad scale impacts on the processes of the nonhuman world in which we live and on which we depend. But the Ecomodernists just do not see that; instead, the manifesto calls for the ecologically impossible – more tech, development, and growth here, so that we can, basically, leave nature alone over there. The Ecomodernist Manifesto really is the basic statement of a feral promethean ideology.

The larger problem for our purposes here is that justice simply is not mentioned in the Ecomodernist Manifesto, nor is equity. Neither, not once. Search the text, and the term "fair" is only there as part of "laissez-faire" – which they do not critique. There is no mention of vulnerability, nor of inequality. I may be biased, but I do not understand how anyone can write an environmentalist manifesto without even acknowledging such growing concerns, or engaging at all with the broad discourse of environmental justice.

Unfortunately, there is a long history of the Breakthrough Institute (those behind the Ecomodernist Manifesto) and its inability to understand or engage with environmental justice. Maybe it is an improvement that they have gone from criticising the movement to just ignoring it. It is difficult to understand why this Ecomodernist school is so anti-environmental justice. In reality, their purposeful detachment of an active embrace of technology with any concern for justice is obviously unnecessary.

There is plenty of potential common ground with green tech development, efficiency, jobs in new greener industries, healthier tech processes, etc. Someone like Van Jones (2008) – who has been writing about green collar jobs as a response to environmental injustice for a decade – could be worthy partner. The Ecomodernists are clearly more interested in being belligerent to generate PR and funding than they are with what a just Anthropocene would look like. In the lead-up to the Encyclical on climate change, one of Pope Francis' environmental advisors noted that "Nowadays man finds himself to be a technical giant and an ethical child" (Revkin blog 2014). That is a good description of the Ecomodernists, and their refusal to engage with the idea of justice.

That all said, I do think it is possible to think of ways of encompassing a concern for justice with an agenda for a better Anthropocene. Unfortunately, even well-meaning attempts to broaden a discussion of the social impacts of the Anthropocene are often woefully lacking. So, for example, while there are academic approaches that focus on the important interplay of social and ecological systems in the Anthropocene, they still tend to ignore ethics, including any aspects of justice. An example of this is a piece from Xuemai Bai and others on a research agenda for what they call a plausible and desirable Anthropocene. On the one hand, the piece crucially shifts the thinking from a geological past and present to the design of a future research agenda in the social sciences (related to the Future Earth project). But even this full-scale discussion of a "desirable" future Anthropocene only mentions a very anaemic notion of justice in passing (Bai et al 2016, p. 5). A bit about equity, a bit about who decides, but very little attention to this crucial impact. In addition, the set of essays edited by Clive Hamilton and colleagues on rethinking modernity in the new epoch (Hamilton et al 2015) has lots on post-naturalism, catastrophism, and human destiny in the Anthropocene – it is a very interesting collection – but there is absolutely nothing on equity or justice.

Multiple-cenes?

That is not to say that this space is completely empty. Giovanna Di Chiro (2016) has been writing specifically on environmental justice and the Anthropocene, and Kyle Whyte, along with a number of other feminist, indigenous, Marxist, and anti-neoliberal critics have noted that what we are really talking about is the MANthropocene, the CAPITALcene, the NEOLIBERALcene, or even the EUROcene (Haraway 2015; Moore 2015; Whyte 2017). The key here is that there is a crucial distributional issue of justice to note – that it is not the full ANTHROPOs that is the cause of this shift in earth systems, but rather very specific humans and practices. The Anthropocene, here, is not properly about the anthropos or the universal "we," but rather some very specific "thems" imposing a wholly new planetary condition on a whole lot of other peoples and defining the rules of the game. Except, of course, it is not a game – and the rules are self-destructive.

In this kind of questioning of the very term Anthropocene, there is often an environmental justice focus – but, again, sometimes not. Jason Moore, one of the originators of the term CAPITALcene over two long articles and a book (Moore 2015), and one of the best critics of the long ecologically destructive practices of capital, does not use the term justice, or equity, or vulnerability even once in laying out this crucial concept.

Oddly enough, one of the most powerful arguments for attention to environmental justice in the Anthropocene is not by an academic, but by the science fiction author China Miéville (2015) in a probing piece in the online journal *Salvage*. He addresses classic environmental justice findings, and critiques neo-liberalism, profit and accumulation, colonialism, fossil fuel corporations, and more. It is a thorough, articulate, angry, and agenda-setting piece that gets at the intersection of environment, justice, and this broad notion of the Anthropocene.

On the one hand, then, it is clear that the shift to a more engaged social science and humanities focus on the epoch does raise some hope for broadly interdisciplinary and critical approaches to the epoch – and for the broader engagement with a change in the intellectual climate, as Noel Castree et al (2014) have argued. But, on the other, that agenda is not quite there yet, and there is work to do on recognising the environmental justice implications of the impacts of the Anthropocene. But for the most part, as for academics accumulating around the idea of the Anthropocene, we have the denialists, the feral prometheans, and early indications of some well-intended social scientists missing this intersection of justice and the new age of human/nonhuman relations in the Anthropocene. Given the increasing circulation of environmental justice discourse across social movements and into a broad range of environmental and climate politics, it is disappointing to see such a lack of attention so far.

The Anthropocene, disruption, and environmental justice

If the vast majority of literature on the new age of humans is inadequate to the task, how might we think more thoroughly about the environmental justice implications of the Anthropocene? And how do we find the connection, or the mortar, to bind the geological and social ideas, to understand the justice impacts of the era? In a word, I think the context of this intersection is all about the reality and implications of disruption. The Anthropocene is not about human control, but about human impact on systems, mostly to disrupt or undermine their previous functioning.

But, of course, there are at least two kinds of impact – two temporally different kinds of disruptions of the Anthropocene – both with justice implications (Nixon 2011). First, the violence of the Anthropocene will come in spectacular events – the fast punches, the devastating impacts of the rapid undermining of life. Punctuated disturbances, or disequilibria – the shocks of storms, floods, fires, and more.

Disruption and disequilibria

Events like Hurricanes Katrina or Sandy, for example, were turning points for environmental justice in the US. Those events brought a recognition not just to new environmental vulnerabilities, but more broadly to the income and race-based vulnerabilities of the sudden impacts of climate change – and, I will argue, to the threats to place attachment (see also Haller 2015). These intense and punctuated disturbances have already included sudden onslaughts of flooding in Pakistan, fires in Indonesia, typhoons in the Philippines, the collapse of harvests in Syria, and heatwaves in various countries, including my new home of Australia. These are the obvious, media-covered fast punches that radically disturb environments and undermine communities. And they immediately impact in a range of unjust ways (see, for example, Bullard and Wright 2009 on Katrina).

The EJ community has taken on and responded to these devastating, disrupting events. While official emergency and disaster planning rarely examines justice and inequity, the community certainly sees the impacts of climate change disasters as justice issues. After Sandy, for example, there was much input from the EJ community on the development of New York City's plan to adapt to climate change. There has been a push to increase the recognition of, and response to, community concerns around vulnerability, disturbance, and functioning in planning for such sudden disruptive events. Rosemary Lyster's (2015) recent work focuses specifically on this relationship between disasters and climate justice. The simple fact, she notes, is that "climate disasters fundamentally destroy and undermine capabilities" (p. xviii). And it is also key that protecting those capabilities – reducing exposure and vulnerability on the one hand, and building resilience and response on the other – is crucially based in procedural justice, the inclusion of the public in reflecting on the real impacts of such events on their everyday lives. Adaptation and disaster response will also be part of the post-Paris management of the green climate fund, but it will be interesting to see what normative frameworks are developed for funding and evaluation of projects – whether capabilities, participation, and justice will be part of the discourse of governing the fund.

Nixon's slow violence

Second, and beyond the shock events, is the "slow violence" that Rob Nixon (2011) describes in his book of that title. This is

> a violence that occurs gradually and out of sight, a violence of delayed destruction that is dispersed across time and space, an attritional violence that is typically not viewed as violence at all … a violence that is neither

> spectacular nor instantaneous, but rather incremental and accretive, its calamitous repercussions playing out across a range of temporal scales.
> (2011: 2)

Nixon's list of slow environmental violence includes climate change, of course, but also toxic drift, deforestation, the radioactive aftermaths of wars, acidifying oceans – what he calls "long dyings – the staggered and staggeringly discounted casualties, both human and ecological..." (ibid.). The focus here is on long-term degradations to environment, culture, resource, food, land, health.

Admirably, I think, this is what many resilience efforts are getting at – not just bouncing back from shock events, but identifying and addressing these creeping stressors. And clearly, this slow violence is based in inequality as well – Nixon (2014) elsewhere argues that great acceleration and great divergence occur simultaneously. Environmental and economic injustice are tied together, even slowly, in the Anthropocene.

We often focus on climate change as the major marker of the Anthropocene, but there is much more to the idea than just climate. One place where environmental justice has addressed other human disruptions have to do with the implications of impacts on the human body – this is one of the things we have remade, disturbed, and undermined. Rachel Carson's warnings about DDT in the classic *Silent Spring* (1962) were, in retrospect, warnings about the slow violence of the Anthropocene. And the EJ movement has long focused on reproductive health, the impacts of chemicals, asthma rates, lead poisoning, uranium mining and poisoning, the lack access to healthy foods. These leave markers, literally visual ones, in and on human bodies, "normalized toxic bodies" (Gabrielson 2011). As Di Chiro (2008) has written, these impacts undermine the very functioning and social reproduction of a wide range of communities.

There are many impacts which simultaneously undermine the functioning of our bodies and our communities – slowly, over the long term. This slow violence and injustice is experienced as disruption whether it comes from rising oceans drowning lands, uranium contamination threatening indigenous cultures, or lead poisoning in African American communities from Flint to Philadelphia. The anticipation of a range of slow violence impacts have been central to environmental justice and to climate justice organisations and movements. We have heard them in discussions of indigenous movement arguments about the impacts of climate, and in fears about ongoing lead poisoning in African American communities.

So, we have these two kinds of social and environmental disturbances or disruptions to address in the Anthropocene, the immediately disrupting and the long-term slowly disrupting. I should note here that much adaptation planning – and that means policy planning rather than the broader field of adaptation studies well-represented here – focuses on the first type of disturbance. Many adaptation plans are risk management, emergency response, and disaster planning in the face of the increase of sudden events. One of my points here is that it is not only those, but also the slow impacts of climate change (in particular cultural and ecological, and their ties) and other slow violent impacts of the Anthropocene (such as toxins, air pollution, biodiversity loss and mass extinctions, migrations, conflicts, the ideological stranglehold of neo-liberalism – and, again, all of their ties to and impacts on identity and place) that are crucial for environmental justice studies in the Anthropocene age.

Disruption and detachment

Now let me clarify a bit more how, exactly, this wide variety of potential disruptions and detachments are an environmental injustice. Vulnerability of *who* is a key question; vulnerability exacerbated by *what* specific social and economic conditions is another. But I am interested in the *how* of vulnerability – how injustice is actually experienced. We have much work being done on rising waters and inundation, on storms and heat impacts, on psychological effects. But I think the key vulnerability in the Anthropocene will be disruption and detachment from place and community – the disruption of commons and of attachments and interactions with those commons. Attention to potential disruptions to community and commons gets at the intersection of the physical impacts of the Anthropocene, the disruptions to the current entanglements between and across human and nonhuman realms it will bring, and, hopefully, a potentially reflexive, productive, and just response.

There is a growing interest in climate displacement – the migrations that will inevitably occur, both within existing national borders (which is where the majority of such migration will occur), and also across borders (which seems to be where most popular thinking is focused). But my point is that we need to think about disruption in place as well – we can have significant disruption *to* place without being *dis*-placed. Detachment, in place, to and from commons and common environments, will be one of the central and recurring experiences of the Anthropocene.

As I have long argued, there are a range of aspects of environmental justice (Schlosberg 2007, 2013). Equity will always remain crucial, as Flint and so many other examples remind us. But environmental justice also encompasses recognition/misrecognition, inclusion/participation, and the capabilities and functioning of both individuals and community. I have also made the argument that EJ is as much about community functioning as individual (Schlosberg and Carruthers 2010), at both the human and nonhuman level (Schlosberg 2012) and I think climate change and the Anthropocene bring that reality and relationship into clear focus.

I want to push the application of both the idea of recognition and of capabilities a bit further here, in light of the specifics and scale of the disruptions and dislocations expected due to climate change and other impacts of the Anthropocene. This means I want to focus on the co-constructive quality of justice and place – and, related, injustice and the experience of displacement. In other words, I want to make an argument for why these disruptions and place detachments are injustices now, and will be for the foreseeable future.

Form and experience of environmental injustice

Environmental justice is not only or primarily about justice as equity, but also about where and how we live in an environment; it understands people's experience of and relationship to place as an important element of broader questions about environmental justice (or injustice). Here, justice hinges on a sense of a positive place attachment – and avoidance of negative impacts on commons and shared place, such as pollution, threats to environmentally-based, culturally-valued practices – and displacement. Similarly, Groves (2015) has argued for the inclusion of a conception of "attachment" in the consideration of environmental justice – and the recognition that threats to our attachment to place will impact both that place and our identities. He cites longstanding research that people who are affected by land-use decisions often frame their activism in terms of the links between their attachment to place and the everyday experiences of socio-environmental degradation (for example, Burningham and Thrush 2003; see also Anguelovski 2013; Devine-Wright 2013). Attachment to place is a process, a relationship, and a constitutive element of our identity and functioning (ibid.), and EJ has long lamented – and acted upon – disruptions to that attachment. The saving of attachments, and the rebuilding of commons and communal spaces – and shared commons management with both traditional knowledge of place and new knowledge of change to come – is key to environmental justice in the Anthropocene.

Neil Adger (Adger et al. 2011; Adger 2016) also highlights the way that places give meaning to lives, and how place attachment is a core element of individual well-being. Where Groves', Adger's, and my own interpretations intersect is in the idea that attachment – to both people and place – is a basic capability. Attachment is a basic need, essential for a functioning life. This is how the loss of attachment to both others and our common and communal places can be seen as an injustice (see also Schlosberg, Rickards, and Byrne 2017).

In a capabilities approach (Sen 2001; Nussbaum 2011), justice occurs with the availability of all of the basic capabilities of human life. To deny, disrupt, or interrupt those capabilities is, then, the definition of an injustice. So we create environmental injustice by imposing environmental bads, removing valued environmental goods, forcing one to move elsewhere, disrupting commons and our attachments to it – any number of possible interruptions of basic capabilities.

Attachment, as Grove notes, occurs only in relationality. The harm is not just a lack of recognition, but the "loss of the implicit interdependence of individual, community and place – a perhaps largely unspoken condition of 'solidarity' between humans and the socio-natural places they inhabit" (Grove 2015, p. 858). The point here is that these are the relations being wrenched by climate change and the Anthropocene. Human beings construct attachments to both other people and places – and the intersection of the two is crucial. In the Anthropocene, it is not just various human and nonhuman communities that are threatened, but also our constructed relationship with each other and the spaces (built and otherwise) in which we are both embedded. Thus, all our relations are threatened.

There is an important connection to be drawn between this idea of the undermining of the capability of attachment to nonhuman places and Glenn Albrecht's conception of Solastalgia – a feeling of distress that comes "when your endemic sense of place is being violated" (Albrecht 2005). If attachment to place is a basic capability, and various types of disruptions undermine both place and that relationship, one's own individual – and, crucially, collective – identity is undermined. What Albrecht insists is that we feel that disruption as a loss. *The Guardian* used Albrecht's notion to frame a whole long form story on "Generation Anthropocene" (Macfarlane 2016). "A worldwide increase in ecosystem distress syndromes," they quoted Albrecht, is "matched by a corresponding increase in human distress syndromes."

These detachments, and in particular detachment from place – the undermining of our attachments that will come both quickly and slowly in the Anthropocene – are also an injustice, the interruption of the basic capability of attachment. I think it is clear that one of the reasons we feel this loss so acutely is because it also feels like an injustice. This combined feeling and experience of loss, of disconnect, of detachment from the familiar will be an ongoing marker of the Anthropocene. The institutionalisation of an Anthropocene that detaches human beings from commons – that encloses, captures, and/or destroys the places that give us meaning – will continue to be a violence and the basis of environmental injustice.

In addition, if that sense of place, and that crucial relationship and attachment often found in common property institutions of resource governance, are not recognised in governance processes that are built to respond to the Anthropocene, then we also risk a second type of injustice, one of mis- or nonrecognition. We risk a combination of disruption of place, detachment from place and the identity it provides, the undermined functioning of individual and community – human and nonhuman, and a lack of recognition of all of the above. One of the tragedies here is that the necessity and impact of these attachments and our embeddedness with each other and our environment becomes obvious only as they are lost. Their visibility comes with their disappearance. For me, that is a key element of the Anthropocene, and one of the real challenges of avoiding environmental injustices of detachment.

Implications for the intersection of environmental justice, the Anthropocene, and environmental governance of commons

As a political theorist looking at this intersection of the Anthropocene, disruptions, and environmental justice, one key question for me is how we can govern for detachment in ways that are aligned with the demands of justice. What do justice implications of disruption have to do with environmental governance, and with the commons and commons management?

The focus of most conceptions of both something like climate justice and environmental governance is too often on top-level global governance, or multi-level governance, or Ostrom's polycentric governance (2010). I would suggest that in order to respond to the kind of potential displacements and disruptions I have discussed, governance studies will need to mimic the kind of downscaling, forecasting, and policy development we are seeing in the best examples of adaptation planning. But, as I have said, adaptation planning itself needs to address both the fast/immediate impacts like disasters, as well as the slow violence of being gradually separated from place, relationships, and community. We need to engage with what environmental change in the Anthropocene actually entails on the ground in terms of both disruptive experience and the potential of just governance. This is the point of the "glocal" in commons management at the heart of this volume.

Key to a just environmental governance is an engagement with the relationship between people and place, people and environment, culture and identity and place. That is, importantly, attention and response to the experience of disruption. This attention, of course, is not just about the human community, but the intersection, entanglements, and assemblages between human and nonhuman. As I noted, Neil Adger and colleagues have been writing on the relationship between the importance of and attachment to place, well-being of individuals and communities, and adaptation planning, so there is some important prodding in that direction happening in this literature. Adger argues that detachment undermines adaptive capacity, so that adaptation governance needs to recognise and include attachment. I would simply link that more explicitly to understandings of environmental justice – and to our understanding of the commons and the threats it faces.

One example of a response comes from the Australian Aboriginal scholar Steve Kinnane, who has proposed a conception of the "sustainacine" (Kinnane 2014). The idea, at its core, contains a recognition of the link between human and nonhuman community functioning. This is based in a realisation that the functioning of human communities, ecosystems, and environmental systems that provide for human communities are all tied together and all threatened with disruption. Such a notion of commons, of course, is long understood in indigenous communities, and a decolonising approach brought to our treatment of the nonhuman in which we are immersed is key to a more just approach to human, nonhuman, and commons (Whyte 2017).

Not surprisingly, when you actually engage community members in something like adaptation planning, as we did in the city of Sydney (Schlosberg, Collins, and Niemeyer 2017), you hear many examples or iterations of this sustainocene idea, or what I have elsewhere called ecological reflexivity – the range of knowledges and experiences of environmental change that impact on everyday life for individuals and communities. These kinds of knowledges and linkages are key to adaptive capacity and just governance in the Anthropocene.

Jeremiah Purdy's (2015) *After Nature: A Politics for the Anthropocene* has received some deserved attention. His main argument is that we should make the politics of the Anthropocene more universally democratic. "[I]f Anthropocene ecologies are a political question," he writes, "then no one should be left out of the decisions that shape them" (p. 49). Purdy says a more fortified democracy is necessary to survive, adapt, and thrive in the Anthropocene. I do not disagree, and have clearly argued the importance of procedural justice as an integral part of environmental justice. But I am also much more inclined to include a sense of just what that democratic participation would be focused on. For me, the key to incorporating environmental justice in the Anthropocene comes with a focus on the protection of community functioning and community relationship to, and attachment to, place – and thus to the commons as well.

This is not just an academic argument. The concerns raised by climate justice movements, new environmental movements around everyday life, and the public involved in Sydney adaptation work illustrate that there is a public focus on the functioning and vitality of community and connections, entanglements, and assemblages with each other and environment – even as we face multiple forces of disruption. Research that I have been doing on both public interest in adaptation (Schlosberg, Collins, and Niemeyer 2017), and a range of new materialist environmental movements around new food and energy systems (Schlosberg and Coles 2016, Schlosberg and Craven 2019), illustrates the importance of the attachment to both community and the broader environment, and the desire to build relationships and practices of governance that are more resilient against disruption and detachments. Interestingly, these movements focus on the creation or re-establishment of attachment to place and to commons, in response to the alienation of neo-liberal practices – for example around food and food systems. But environmental and climate justice movements in the Anthropocene are also about preserving existing attachments to place, in the face of potential disruptions.

In the process of engaging the public in adaptation planning for the city of Sydney, I found a set of environmental justice concerns in the public. It was not to be solely an examination of risks, as the professional risk assessors had done. Instead, the residents of the city wanted more engagement with questions of justice – in particular differential vulnerabilities, the protection of cultural practices, the inclusion of the broad community – all focused on the protection of the existing functioning of the community as community, neighbourhood as neighbourhood, and all as connected to, attached to, and entangled with place and the commons in the city.

This citizen engagement with the Anthropocene was heartening and constructive. I think in the coming years we will, inevitably, have to see more of that. More democratic – and there I mean more engaged and deliberative – governance processes can address both the threat of detachment and disruption and the potential conflicts between individual and community functionings. So the key to both environmental justice and a just environmental governance in the face of the range of disruptions and disconnections the Anthropocene includes engaging those disruptions in public discourse. That also means attention to disruption, to the potential of displacement, to the reality of slow changes and quick hits. And this is where the idea of the Anthropocene, adaptation (or transformation) planning, attention to the role and reality of the commons, and environmental justice all intersect.

For the commons, the meaning of a new form of resilience and justice in the Anthropocene is to have more bottom-up institutions being built in a more democratic way, and around an attachment to a material and emotional world which is recognised and "co-owned." Local communities clearly rely heavily both materially and emotionally on communally-owned territories and their landscape ecosystems linked to them – even as those spaces change. Many of these ecosystems are maintained by communal, coordinated, collective actions, from African contexts (Haller et al 2016, 2018), to Swiss Alpine pastures, to the songlines and "caring for country" of Aboriginal communities in Australia (see also cases in this volume on Spain and Norway).

I mentioned earlier the novelist China Miéville's (2015) discussion of this new age of environmental politics. He focuses on the limits of utopian thinking, but he ends by insisting that "we must learn to hope with teeth… We should utopia as hard as we can." I would just add that we should defend attachment to our places as hard as we can as well. That is the way toward embracing and constructing environmental justice in and for the Anthropocene.

References

Adger, N. (2016) 'Place, well-being, and fairness shape priorities for adaptation to climate change', *Global Environmental Change* vol 38, pp A1–A3.

Adger, N.W., Barnett, J., Chappin III, F.S. and Ellemor, H. (2011) 'This must be the place: underrepresentation of identity and meaning in climate change decision-making', *Global Environmental Politics* vol 11, no 2, pp 1–25.

Albrecht, G. (2005) '"Solastalgia". A new concept in health and identity', *PAN: Philosophy Activism Nature* vol 3, pp 41–55.

Anguelovski, I. (2013) 'From environmental trauma to safe haven: place attachment and place remaking in three marginalized neighborhoods of Barcelona, Boston, and Havana', *City & Community* vol 12, pp 211–237.

Asafu-Adjaye, J., Blomqvist, L., Brand, S., Brook, B., Defries, R., Ellis, E., Foreman, C., Keith, D., Lewis, M., Lynas, M., Nordhaus, T., Pielke Jr., R., Pritzker, R., Roy, J., Sagoff, M., Shellenberger, M., Stone, R., Teague, P. (2015) 'An Ecomodernist Manifesto', www.ecomodernism.org (accessed 30 October 2018).

Bai, X., van der Leeuw, S., O'Brien, K., Berkhout, F., Biermann, F., Brondizio, E.S., Cudennec, C., Dearing, J., Duraiappah, A., Glaser, M., Revkin, A., Steffen, W., Syvitski J. (2016) 'Plausible and desirable futures in the Anthropocene: a new research agenda', *Global Environmental Change* vol 39, pp 351–362.

Bullard, R. and Wright, B. (eds.) (2009) *Race, Place, and Environmental Justice after Hurricane Katrina.* New York: Routledge.

Burningham, K. & Thrush, D. (2003) 'Experiencing environmental inequality: the everyday concerns of disadvantaged groups', *Housing Studies* vol 18, no 4, pp 517–536.

Carson, R. and Darling, L. (1962) *Silent Spring.* Boston: Houghton Mifflin.

Castree, N., Adams, W.M., Barry, J., Brockington, D., Büscher, B., Corbera, E., Demeritt, D., Duffy, R., Felt, U., Neves, K., Newell, P., Pellizzoni, L., Rigby, K., Robbins, P.,

Robin, L., Rose, D.B., Ross, A., Schlosberg, D., Sörlin, S., West, P., Whitehead, M., and Wynne, B. (2014) 'Changing the intellectual climate', *Nature Climate Change* vol 4, pp 763–768.
Chakrabarty, D. (2009) 'The climate of history', *Critical Inquiry* vol 35, no 2, pp 197–222.
Crist, E. (2013) 'On the poverty of our nomenclature', *Environmental Humanities* vol 3, no 129–147.
Crutzen, P. and Stoermer, E.F. (2000) 'The Anthropocene', *Global Change Newsletter* vol 41, pp 17–18.
Di Chiro, G. (2008) 'Living environmentalisms: coalition politics, social reproduction, and environmental justice', *Environmental Politics* vol 17, no 2, pp 276–298.
Di Chiro, G. (2016) 'Environmental justice and the Anthropocene meme', in Gabrielson, T., Hall, C., Meyer, J.M. and Schlosberg, D. (eds.) *Oxford Handbook of Environmental Political Theory* (pp 362–381). Oxford: Oxford University Press.
Devine-Wright, P. (2013) 'Think global, act local? The relevance of place attachments and place identities in a climate changed world', *Global Environmental Change* vol 23, pp 61–69.
Ellen, R. (1982) *Environment, Subsistence and System: The Ecology of Small-Scale Social Formations*. Cambridge: Cambridge University Press.
Gabrielson, T. (2011) 'The Normalized Toxic Body: A Site for Theorizing an Environmental Politics. Western Political Science Association 2011 Annual Meeting Paper', available at SSRN: https://ssrn.com/abstract=1766653 (accessed 30 October 2018).
Groves, C. (2015) 'The bomb in my backyard, the serpent in my house: environmental justice, risk, and the colonisation of attachment', *Environmental Politics* vol 24, no 6, pp 853–873.
Haller, T. (2015) 'Climate change impacts, institutional resilience and livelihoods vulnerability: a comparative analysis of climate change and adaptation strategies in Asia, Africa and Latin America', in Rosendhal, J., Zanella, M., and Weigelt, J. (eds) *Pro-Poor Resource Governance under Changing Climates* (pp160–181). Potsdam: IASS.
Haller, T., Acciaioli, G., and Rist, S. (2016) 'Constitutionality: conditions for crafting local ownership of institution-building processes', *Society and Natural Resources* vol 29, no 1, pp 68–87.
Haller, T., Belsky, J.M., and Rist, S. (2018) 'The constitutionality approach: conditions, opportunities, and challenges for bottom-up institution building'. *Human Ecology* vol 46, no 1, pp 1–2.
Haller, T., Fokou, G., Mbeyale, G., and Meroka, P. (2013) 'How fit turns into misfit and back: institutional transformations of pastoral commons in African floodplains', *Ecology and Society* vol 18, no 1, p 34. http://dx.doi.org/10.5751/ES-05510-180134.
Haller, T. and Galvin, M. (2011) 'Challenges for participatory conservation in times of global change: lessons from a comparative analysis and new developments', in Wiesmann, U. and Hurni, H. (eds), *Research for Sustainable Development: Foundations, Experiences, and Perspectives* (pp 467–503). Perspectives of the Swiss National Centre of Competence in Research (NCCR) North–South, University of Bern, Vol. 6. Bern: Geographica Bernensia.
Hamilton, C. (2013) *Earthmasters*. Sydney: Allen and Unwin.
Hamilton, C., Bonneuil, C., and Gemenne, F. (eds.) (2015) *The Anthropocene and the Global Environment Crisis – Rethinking Modernity in a New Epoch*. London: Routledge.
Haraway, D. (2015) 'Anthropocene, Capitalocene, Plantationocene, Chthulucene: making kin', *Environmental Humanities* vol 6, pp 159–165.

Hettinger, N. (2014) 'Valuing naturalness in the "Anthropocene": now more than ever', in Wuerthner, G., Crist, E., and Butler, T. (eds.) *Keeping the Wild: Against the Domestication of the Earth* (pp 174–179). Washington, DC: Island Press.

Jones, V. (2008) *Green Collar Economy*. New York: Harper Collins.

Kinnane, S. (2014) 'Sustainability, country and social reconstruction in the Fitzroy Valley', Paper given at the conference on *Encountering the Anthropocene*, Sydney Environment Institute, Sydney Australia, 26–28 February.

Lyster, R. (2015) *Climate Justice and Disaster Law*. Cambridge: Cambridge University Press.

Macfarlane, R. (2016) 'Generation Anthropocene', *The Guardian*, 1 April. www.theguardian.com/books/2016/apr/01/generation-anthropocene-altered-planet-for-ever (accessed 30 October 2018).

McKibben, B. (1989) *The End of Nature*. New York: Random House.

Miéville, C. (2015) 'Limits of Utopia'. *Salvage*, http://salvage.zone/in-print/the-limits-of-utopia/ (accessed 30 October 2018).

Moore, J. (2015) *Capitalism and the Web of Life: Ecology and the Accumulation of Capital*. London: Verso.

Nixon, R. (2011) *Slow Violence and the Environmentalism of the Poor*. Oxford: Harvard University Press.

Nixon, R. (2014) 'The great acceleration and the great divergence: vulnerability in the Anthropocene', *Profession*, https://profession.mla.hcommons.org/2014/03/19/the-great-acceleration-and-the-great-divergence-vulnerability-in-the-anthropocene/ (accessed 30 October 2018).

Ostrom, E. (2010) 'Polycentric systems for coping with collective action and global environmental change', *Global Environmental Change* vol 20, no 4, pp 550–557.

Purdy, J. (2015) *After Nature: A Politics for the Anthropocene*. Oxford: Harvard University Press.

Revkin, A. (2014) 'Exploring Academia's Role in Charting Paths to a "Good" Anthropocene', http://dotearth.blogs.nytimes.com/2014/06/16/exploring-academias-role-in-charting-paths-to-a-good-anthropocene/ (accessed 30 October 2018).

Schlosberg, D. (1999) *Environmental Justice and the New Pluralism: The Challenge of Difference for Environmentalism*. Oxford: Oxford University Press.

Schlosberg, D. (2007) *Defining Environmental Justice: Theories, Movements, and Nature*. Oxford: Oxford University Press.

Schlosberg, D. (2012) 'Justice, Ecological Integrity, and Climate Change', in Thompson, A. and Bendik-Keymer, J. (eds.) *Ethical Adaptation to Climate Change: Human Virtues of the Future* (pp 165–183). Cambridge, MA: MIT Press.

Schlosberg, D. (2013) 'Theorising environmental justice: the expanding sphere of a discourse', *Environmental Politics* vol 22, no 1, pp 37–55.

Schlosberg, D. (2016) 'Environmental Management for the Anthropocene', in Gabrielson, T., Hall, C., Meyer, J.M., and Schlosberg, D. (eds.) *Oxford Handbook of Environmental Political Theory* (pp 193–208). Oxford: Oxford University Press.

Schlosberg, D. and Carruthers, D. (2010) 'Indigenous struggles, environmental justice, and community capabilities', *Global Environmental Politics* vol 10, no 4, pp 12–35.

Schlosberg, D. and Coles, R. (2016) 'The new environmentalism of everyday life: sustainability, material flows, and movements', *Contemporary Political Theory* vol 15, no 2, pp 160–181.

Schlosberg, D. and Craven, L. (2019) *Sustainable Materialism: Environmental Movements and the Politics of Everyday Life*. Oxford: Oxford University Press.

Schlosberg, D., Collins, L.B., and Niemeyer, S. (2017) 'adaptation policy and community discourse: risk, vulnerability, and just transformation', *Environmental Politics* vol 26, no 3, pp 413–437.

Schlosberg, D., Rickards, L., and Byrne, J. (2017) 'Environmental justice and attachment to place: Australian cases', in Holifield, R., Chakraborty, J. and Walker, G. (eds.) *The Routledge Handbook of Environmental Justice* (pp 591–602). London: Routledge.

Steffen, W., Richardson, K., Röckstrom, J., Cornell, S., Fetzer, I., Bennett, E.M., Biggs, R., Carpenter, S.R., de Vries, W., de Wit, C.A., Folke, C., Gerten, D., Heinke, J., Mace, G.M., Perrson, L.M., Ramanathan, V., Reyers, B. and Sörlin S. (2015) 'Planetary boundaries: guiding human development on a changing planet', *Science* vol 347, no 6223.

Whyte, K.P. (2017) 'Our ancestors' dystopia now: indigenous conservation and the Anthropocene', in Heise, U., Christensen, J. and Niemann M. (eds) *Routledge Companion to Environmental Humanities* (pp 206–215). London: Routledge.

Zalasiewicz, J., Waters, C.N., Williams, M., Barnosky, A.D., Cearreta, A., Crutzen, P., Ellis, E., Ellis, M.A., Fairchild, I.J., Grinevald, J., Haff, P.K., Hajdas, I., Leinfelder, R., McNeill, J., Odada, E.O., Poirier, C., Richter, D., Steffen, W., Summerhayes, C., Syvitski, J.P.M., Vidas, D., and Wagreich, M. (2015) 'When did the Anthropocene begin? A mid-twentieth century boundary level is stratigraphically optimal', *Quaternary International* vol 383, pp 204–207.

4 A definition of the commons, between human rights, resistance, and social change

Elisabetta Cangelosi

Introduction

The debate about the commons is evolving and growing at global and local level. This growth is uneven and involves a variety of disciplines (Goldman 1998; Dietz, Dolsak, Ostrom and Stern 2001; Dolšak and Ostrom 2003; Linebaugh 2008; Hardt and Negri 2009; De Angelis 2010; Federici 2010; Mattei 2011; Bollier and Helfrich 2012; Marella 2012; Dardot and Laval 2014; De Angelis 2014; De Angelis 2017). Content, focus and participants of the most recent conferences of the International Association for the Study of the Commons offer a very clear overview of such an evolution. Interesting elements emerge: an increasing participation of scholars from different disciplines (beyond economics and political economy, consistently with rated by the list of authors mentioned above); a remarkable opening to practitioners, and the recognition of the relevance of the commons, at local and global level, from a transformative point of view. The themes of these conference are somehow self-explanatory: 'Practicing the commons: Self-governance, cooperation, and institutional change' (XVI Biennial IASC-Conference-2017), 'The commons amidst complexity and change' (XV Biennial IASC-Conference-2015), 'Commons in a "glocal" world: Global connections and local responses' (4th Regional European IASC-Conference – 2016), 'The city as a commons' (1st Thematic IASC-Conference on Urban Commons – 2015), 'From generation to generation – the use of commons in a changing society' (3rd Regional European IASC-Conference – 2014).

On the other hand, the commons are slowly but constantly gaining attention among policy makers (representatives, local authorities and international organisations), and social activists. For example, the Land Rights Now campaign (www.landrights.org), launched in 2016 by a platform of organisations, aims at the recognition of collective land rights as essential condition for Indigenous Peoples and local communities. Most of the community resources touched by the campaign (land, forestry, mountains, lakes and rangelands) correspond to the most pertinent definition of common property (Ostrom, Gardner and Walker 1994) as owned collectively by a given group that is entitled to it and has the possibility and the capacity to exclude non-members from access. This same definition is relevant, even if not crucial, for the broader debate about the commons.

Meanwhile, more and more scholars, from a variety of disciplines, started debating about the commons in their own field, exploring definitions, functions and impacts of the commons at local and global level, often envisaging interactions among disciplines. The result is a stimulating complexity where ideas and opinions constantly converge and diverge.

In fact, the issue of defining the commons is not solved, and this could be considered a direct consequence of the diversity of disciplinary approaches: while there is some agreement on basic aspects of what the commons are, from an historical, ecological, juridical, economic, and social perspective, experiences, practices and analysis may vary significantly.

In addition, the debate is further complicated by the fact that, beyond the theoretical level, the commons are mostly a practice, which is affected by cultural, historical, social and economic elements. Traditional commons (mostly natural resources) and new commons (mostly digital, but also urban) take a different shape; legal systems might be more or less open to recognise the commons; the commons could represent a socio-economic reaction to the economic crisis or an indigenous value to be defended from grabbing and expropriation; rural and urban approaches might also differ. Eventually, defending the commons might appear concurrent with defending other rights, in particular when we look at women's equal rights.

From a purely traditional and customary form of managing natural resources, the commons now represent a key topic for the debate on economic, social and environmental justice. Therefore, the debate about the commons could play a relevant role at the global level, connecting experiences of social change and resilience around the world.

While this chapter does not aim at finding a conclusive and global definition of the commons, it claims that focusing on this debate through a combination of a human rights approach, resistance (and resilience), and social change could contribute to a better understanding of the essence and the function of the commons, in a rapidly changing and interconnected world.

The goal of this overview and analysis, that starts from publicly available data collected by the network Remix the Commons and reaches a theoretical level, is to identify a few core elements of the commons, while highlighting some key differences that deserve attention at social and political levels. The legal framework, despite challenging, is a fundamental element of such a debate; in particular if we make reference to the provisions coming from international human rights law, rather than focusing only on civil law. Approaching the commons from a human rights perspective, rather than focusing exclusively on the debate about private and public property, could open interesting paths of analysis and produce some positive impact on social and economic change.

The analysis builds on different components, matching with the diversity of the phenomenon, in order to draw coherent conclusions on the relevance of human rights, resistance and social change (both for traditional and new commons).

The evolution of the debate about the commons

While E. Ostrom's work (Ostrom 1990) still represents the core foundation, or at least the lowest common denominator of any discussion and analysis about the commons, the debate has clearly expanded beyond classical political economy. Ostrom's focus on commons management as 'neither Market nor State' (Ostrom 1990; Ostrom 1994) served as a basis for an increasing and articulated debate on social components. Such a debate takes place both at theoretical and practical levels. Often, but not always, an interaction between the two worlds exists. In particular, this is the case of a broad range of analysis focusing on the connection between the commons and social and economic justice, and the role of the commons as tools for social transformation (Linebaugh 2008; De Angelis 2010; Federici 2010; Mattei 2011; Marella 2012; Coriat 2013; Dardot and Laval 2014; De Angelis 2014, 2017).

A particularly interesting case of convergence between scholars and practitioners is represented by the discussion about the action of *commoning*, whose elaboration (even if it differs in terms of root-causes, description and impact) is present both in De Angelis's theoretical political analysis (2007–2017) and in the practitioners' approach of Bollier and Helfrich (2012). Both focus on the action of creating something as a commons, and doing so in common, based either on potentiality or need. Concepts such as sharing, participation, inclusion, responsibility and reciprocity are crucial to the process through which a place or space is made 'common' (De Angelis 2010, 2014), which in fact consists in the process of creating a community. In 2007, De Angelis, started elaborating on alternative economic and social models from a commons perspective, presenting the idea of pcommoning as horizontal process involving a plurality of people. Despite the process of 'commoning' being clearly mostly relevant for the debate about the new commons (and in particular urban ones), and while definitions of the commons are not necessarily consistent nor homogeneous, all those analysed in this framework include the idea of the commons as a trigger for a model of society inspired by social and economic justice

Definitions are often influenced by context and political tradition. Nevertheless, a quest for a common understanding of the commons is clear and the relevance of the topic is slowly but constantly increasing, both at local and global levels. In fact, while presenting significant differences in terms of regional coverage, approach and content, several organisations and networks include the commons in their political and social action (among others: Transform!, Remix the Commons, Commonsnetwork). Some of them cooperate and federate at the national level (for example the network 'Echanges' in France) or even at a regional one (for example the European Assembly of the Commons).

The political inspiration is definitely not a constant: some activists, for example from Transform!, analyse the commons from an explicit leftist or Marxist perspective and promote the commons as central for left policies (in particular in Europe) (Delmas 2016; Baier 2017); while the others, such as Remix the Commons, have a less politically influenced approach to the commons. All of them, however, highlight the alternative component as a core characteristic of their work and analysis.

At the political level at least one case deserves attention, as it confirms an important change in the approach to the commons as tools for social transformation and alternative. In 2014 an intergroup within the European Parliament was created, with the aim of debating policies and actions on commons and public goods and convening exchanges on different issues related to the commons from a social change perspective (from urban agendas, to natural resources and digital rights).

The variety of approaches and the increasing interest for the topic are definitely a richness of the debate. The different elements contribute to a possible definition, which, if not 'universal' could at least serve to identify the fundamentals.

Legal definition of the commons: human rights beyond property

So far no State law provides for any legal definition aimed at protecting the commons as such, with the exception of legislations that recognise community, indigenous and/or customary rights (for example Australia, Canada, Brazil, Mexico and other countries in Latin America). Nevertheless, an increasing number of legal analysis is being developed in Europe (Marella 2012; Micciarelli 2014; Napoli 2014; Rodotà 2016), which matches with a recent interest of political actors. In international law there are a few legal reference to the commons: on the one hand the idea of Common Heritage (Baslar 1998), on the other hand the recognition of indigenous rights, by ILO Convention 169, which represents a good legal ground, in particular with regard to ancestral land (Candelaria 2012).

Despite similar to the image of "common good of the mankind", used in particular with regard to political struggles about water as a commons (Alternatives Sud 2002), the common heritage approach does not really match with the actual experiences of *commoning* as they present a different scale in terms of creation of a community. The concept could at the very best apply to digital commons, but even in this case the community that participates in the process of *commoning* could never be as broad as the whole humankind (also considering the digital divide). The recognition of indigenous rights, on the other hand, might play a relevant role at least with the regards to those commons that are managed by indigenous communities.

An exceptional case exists with regard to legal analysis as an attempt of including the commons in national law not with regard to indigenous people rights. In 2007, a committee of legal scholars is charged of reviewing the Italian civil code with a specific regard to the norms on public goods (Mone 2017). The draft proposed by the Commitee (Commissione Rodotà, Disegno di Legge Delega 2007) included a clear legal definition of the commons:

> goods expressing functional utilities to the exercise of fundamental rights and to free human development. Common goods are to be protected and safeguarded by the legal system, also for the benefit of future generations. Right holders of the common goods can be public legal entities or private actors. In any case their collective use must be granted, according to limits and modalities established by law. When right-holders are public legal entities, common goods are managed by a public subject and are put out of trade; concessions are regulated by law, for limited duration and without extension.[1]

The text builds the definition of the commons around the ideas of fundamental rights, human development and future generations; it also provides details about property regimes, specifying that commons could belong to either public or private entities but that their collective use be guaranteed. It also adds that, when public entities are involved, the commons market regime is out of trade. In fact, the discussion about property regimes is crucial for the contemporary debate about the commons.

Because of its intrinsic nature of shared resources the commons contribute to rethink the node between property and sovereignty, and promote a shift from the centrality of the idea of property (which, as shared common property, is, as previously said, a core component of the definition of traditional commons) to the relevance of the idea of access (Rodotà 2016). As a consequence, according to legal scholars involved in the debate, discussing property regimes seems crucial to analyse the commons from a legal perspective (Marella 2016).

The proposal did not enter in actual legislation, as it failed to be discussed due to a government change by the time it was to be debated. However, this legal definition represented the basis for further elaboration, conducted at the local level in Italy.

While the lack of coherent legal frameworks is an important characteristic of the 'new' commons, it is interesting to note that the connection with social and economic rights often emerges in the debate. Among the most significant elaborations, which contribute to the legal analysis, it is worth to mention the case of Naples, with regard to the management of a common-pool resource, water, as well as of urban spaces. Naples might even be considered as a benchmark. Since the Referendum on public water in 2011, the debate on the idea of commons ('beni comuni' in Italian) progressively increased among Italian activists and political actors. In fact, the very slogan of the Referendum was '2 Sì per l'Acqua Bene Comune'. More broadly, the referendum focused on the idea of democracy, human rights, public water and commons. In the aftermath of the referendum the municipality of Naples took the decision to remunicipalise the water supply services. The text of this decision makes the explicit reference to the text produced by the Commissione Rodotà. Following the same approach, the same Municipality takes an analogous position a few years later, in 2016, recognising some occupied spaces as commons.

Definition of the commons 'around the world': the data from 'Remix the Commons'

The search for a definition guided or influenced the work of several actors working on the commons. One of the above-mentioned organisations, the French Remix the Commons, launched a specific project 'Definir les biens communs', in order to find one. They describe this attempt as 'a multilingual project sharing definitions of commons [...] over several years and in different places around the world'.[2]

The project consisted in a mapping of short and spontaneous definitions, provided by a variety of actors: activists, practitioners and scholars (including a few key authors as U. Mattei and S. Helfrich and D. Bollier). The definitions are recorded on short video-clips, mostly in English and French (but also in some other languages, as the participants were asked to define the commons in their own language). The first interviews were recorded during the International Commons Conference, organized in Berlin in November 2010 by the Heinrich Böll Foundation and the Commons Strategies Group.[3] The collection of videos continued during other meetings and events in the following year. Around 100 videos have been collected from slightly less than 100 different participants (in a few cases interviews have been replicated at two different moments) from 35 countries (with a prevalence of European Countries). Even if the website of the project does not mention the criteria used by the organisers to select participants, it seems likely that they did not use theoretical criteria but based their choice on accessibility and opportunity: all the interviews have taken place during events and are on a voluntary basis.

The mapping exercise was intended, as the title says, to define the commons, but the organisation never conducted any actual data analysis to elaborate such a definition. However, the data collected remained available and hence represent a rich, though not exhaustive, database of what activists and scholars interviewed think about the commons, and which is representative of informants' perspective and backgrounds.

An analysis of the responses offers some significant elements, which contribute to the broader exercise of identifying key aspects of the commons. Among the 35 countries represented I selected 18 countries, inspired by a principle of geographical balance: Belgium, France, United Kingdom, Italy, Greece, Spain, Poland and Germany for Europe; Algeria, Tunisia and Egypt for North Africa; Peru, Brazil and Colombia for Latin America; plus India, Senegal, United States and Australia. I then proceeded with an analysis of the definitions. Of course, the diverse cultural, social and political background of the respondents influenced their responses. In addition, as the videos are very short, the definitions provided are not exhaustive. Nevertheless, it is possible to group the definitions by four main categories: resilience/resistance, reciprocity, human rights and democracy. The respondents are mentioned by their country of origin in the following overview.

The first category, resilience/resistance, included mentions of: political validity (France); protection and claim of sovereignty over community resources

(Algeria); livelihoods (India); and more broadly on resources (United States); creation of an innovative answer to competition-oriented economic model (Belgium); alternative to market relationship elaborating an antidote to commodification (Brazil). From a perspective of social change and resilience, it is particularly interesting to highlight the connection identified between the commons and the possibility to reboot the imaginary (Brazil) and the desire for a fair society (United States).

A second crucial element, which is constantly highlighted, is the idea of reciprocity. This includes reference to: sharing (Tunisia, Egypt, India, Brazil) and community relations and management (Germany, Brazil, Peru, Colombia, Mali). Furthermore, some respondents mentioned: co-creation (Tunisia and Poland), network perspective (Spain) and collaborative process (United States).

Geographical coverage is broad in all the categories, therefore geographical background seems to have no or very little influence on the definition of the commons. Rather, it is interesting to remark that respondents from almost all the areas mentioned elements such as the role played by the community and the importance of sharing. Protecting and claiming resources, however, seem slightly stronger in areas where traditional commons (land, forestry and natural resources) prevail over new commons.

A further element, even if less explicitly mentioned, is the connection between the commons and human rights and democratic participation: respondents from Brazil, Greece, Italy and the UK included at least one of the two elements in their definition of the commons. Consistent with what was described above concerning the attempt to produce a grounded legal framework, several respondents from Italy mentioned the idea of caring for future generations (making a clear and open reference to the work of the Rodotà committee).

The key elements identified by the different respondents seem interconnected and consistent. Almost all identified the commons as tools for resistance, resilience and social change, linking them to struggles against commodification of nature and resources, and, in some cases, explicitely mentioning current market and economic models. In this framework, which for many respondents implies the construction of alternative models of society, cooperation, and hence reciprocity play a crucial role in defining the commons. This also implies, according to the respondents, community participation and better democratic models.

A few respondents also make a reference to the rights of future generations, while this aspect is not fully developed. In particular this idea is mentioned in mirroring the idea of protecting resources in order to preserve them against commodification and grabbing and keep them available for future members of the community (mostly within indigenous communities). However, the idea of the importance of future generations with regard to recognising, claiming, protecting or creating the commons, does not apply, according to the respondents, solely to natural resources, but also to urban areas, cultural heritage and digital goods. Thus, this could be identified as a bridge between traditional and new commons.

Defending, claiming and creating the commons

Several studies on the diversity of the language, the experiences of the commons, and their implications (Barbato et al 2013; Cavallo 2013; Coriat 2013; Marella 2013; Cangelosi 2014; Cornu et al 2017; De Angelis 2017) also testify to an existing variation in terms of the 'action' that it is linked to the commons. They can refer to the idea of defence and protection, to the idea of claiming and to the process of creating the commons, defined by De Angelis (2014) as 'commoning'. While these three actions might clearly coexist, they respond to specific needs and are influenced by the context. 'Defending the commons' is in the first place a reaction to a threat of different forms of grabbing (privatization, commodification, expropriation) of resources, which traditionally belong to a community. 'Claiming the commons', on the other hand, implies a sort of re-appropriation of something the community is or should be entitled to. Compared with the previous expression the focus switches from the protection from an external threat to the idea of resistance against a social or economic situation which affects the enjoyment of a given resource, space or right, which is therefore to be claimed. This is particularly relevant for urban commons. Eventually, the process of 'commoning' plays a further different role. In this case, the focus is on the action of building a commons through a dedicated set of actions, which includes as a first step the creation of a community, based on reciprocity and solidarity bonds.

In some cases, however, the process of commoning is part of the defence or claiming strategy. This applies when this affects a resource (or a place) which was not a common in itself (as it was not managed by a community), but is recognised as such as a direct consequence of the threat (in case of the defence approach) or of the ongoing situation (in the case of the claim approach). This applies to several struggles against major infrastructural projects affecting communities' resources and livelihoods, in particular through evictions.

Urban and other new commons

The urban commons represent an excellent example of the convergence of claiming the commons and commoning, and significantly contribute to the analysis from a legal perspective. Several studies have been conducted on the specificities of urban commons (Iaione 2012; Colding et al. 2013; Bresnihan and Byrne 2014; Borch and Kornberger 2015; Cangelosi 2015; Foster and Iaione 2015; Festa 2016). Some focusing on the potential for change and the role of social struggles (Festa 2016), others on the idea of polycentric governance strategy (Foster and Iaione 2015). Indeed, the urban commons take different shapes, and respond to a diverse range of claims and needs. Hence, they are described and addressed in different ways. Anyway, they always respond at least to the ideal of the 'right to the city' (Lefebvre 1968), if not to the idea of urban revolution (Harvey 2012).

While in most of the cases commoning remains a self-organised process at community level, there are a few cases where this action encounters some sort of legal recognition and manages to influence law and policy makers.

In July 2016, the Municipality of Naples deliberated in favour of an official recognition of seven occupied spaces as commons (Delibera 446/2016) under promotion of left parties in the municipal council and with the support of the mayor. The text of the Delibera 446/2016 clarifies that the decision is based on the use of those spaces for social, cultural and political purposes and motivated by the relevance of their ultimate purpose: contributing to the enjoyment of existing, and claiming for further, social and economic rights. This choice was part of a broader process of recognition of the commons and their relevance for the city, that included the above-mentioned efforts to manage water supply as a commons, through the remunicipalised water company (Lucarelli 2017). The interest for the commons of the Municipality of Naples is testified by a number of political decisions after the remunicipalisation of water supply put in place by the Deliberation 740/2011. In 2012, the Municipality approved the '*Regolamento delle Consulte per la Disciplina dei beni comuni*', which established crucial elements concerning governing and management of the commons included afterwards in the Deliberation 17/2013. In 2013, the Municipality adopted, via the deliberation 521/2013, the *Carta dello Spazio Pubblico* focusing on the use of public spaces; in 2014 a new deliberation (258/2014) established the procedures to identify and govern public goods that could be linked to civic uses and collective welfare. This last deliberation is considered fundamental for the legal theory about the commons in Italy (Micciarelli 2014).

All these policy documents testify to this increasing interest, strongly embedded in the debate about the commons in the city of Naples, and clearly consistent with the debate described in the previous paragraphs. The experiment of Naples is particularly innovative, as it contributed to build different institutional imagery and opened the path for a legal recognition of the commons (Micciarelli 2014). As already highlighted, in all of these policy documents, there is an explicit reference to the definition of the commons proposed by the Rodotà Committee and a focus on the role played by the citizens. Hence, these documents offer some legal basis for a broader recognition of the commons at the local level.

A few elements emerge and contribute to rethink the relation between institutions and citizens, inspired by grounded bonds of reciprocity: access, collective use and participatory actions contribute to define these urban commons (Festa 2016). The experience of Naples combines two different actions. On the one hand, it consists in a claim for the commons: places which are subtracted from a situation of social, economic and cultural crisis are re-used for better purposes. On the other hand, those involved initiate a truly innovative process of commoning. The core element of this process is that it is conflictual in its nature and relies on self-organisation. The decision of the Municipality fills the gap in law in recognizing the intrinsic value of the nature of these urban commons, as a response to the needs of the community and part of a process of self-organisation and claiming.

Something similar happens in other cities (Bresnihan and Byrne 2014; Iaione 2012), but the specific features of the case of Naples makes it the most advanced one from a legal point of view.

Another typical example of urban commons is represented by the urban gardens that flourish worldwide. With few exceptions that analyse urban gardens from an enclosure perspective and highlight the potential risk for exclusion (Kurtz 2013), urban agriculture is mostly presented as an alternative to enclosure and commodification (McClintock 2009). However, as these are experiences that have a very long history and significantly differ on the basis of the context, not all of them identify themselves as commons. Nevertheless, even when the actors involved are not completely aware of the existing debate, some concepts identified as core features on urban gardening experiences appear remarkably consistent with all the theoretical debate about the commons and their definition (Cangelosi 2015): the key elements are reciprocity, joint management and shared responsibility, which are considered essential for governing the commons.

While the positioning of the Municipality of Naples represents a significant step-forward towards a proper legal definition of the commons, other experiences of urban commoning, such as the gardens, provide an important contribution to the debate. Indeed, they reinforce the idea of reciprocal and mutual exchange, people/citizens involvement, and struggle for a (more or less radical) change in existing social and economic frameworks of commodification, as a fundamental component of the commons. It must be noted that these experiences are often recognised and supported by the local authorities (Cangelosi 2015), hence implicitly contributing to framing the political debate on the commons.

An entire field of analysis has been developed about the most explicitly 'new' commons: digital ones, which represent a clear and prominent example of how commoning applies to immaterial resources. Because of their intrinsic nature digital commons have the very specific feature of creating a community that goes beyond space limits and relies on peer-to-peer knowledge exchange. Several concrete examples of digital commons exist, and include most of the collaborative, open-access, non-proprietary digital resources or platforms, among others Wikipedia, LibreOffice, Linux and Loomio (Bollier and Helfrich 2015).

However, even more openly than in other cases, the debate about digital commons can take very diverse shapes from a political theory perspective (Papadimitropoulos 2017). One of the most significant elements is that it does not always resonate with the situation of commoners who depend on material resources. Furthermore, they might even compete with them: being dependent on IT components, they require natural resources (commons), which are eventually appropriated to ensure the functioning of those infrastructures (Harvie and De Angelis 2014). Nevertheless, from a theoretical perspective, it is fundamental to point out that they play a relevant role in the broader debate (Clément-Fontaine 2017). Considering that digital commons are much more integrated in regulatory frameworks dealing with digital markets, knowledge sharing and use of digital platforms, they can definitely contribute to enrich the legal debate.

In analysing urban commons, but also more broadly in analysing new commons, it would be interesting to identify gendered differential impact; however, to date, there are not yet enough available data to conduct such an analysis in depth. More information and more room for analysis is available when focusing on traditional, natural resources-related, commons.

Gender and commons[4]

While gender issues are now mainstreamed in almost any field (which does not automatically imply a relevant increase in gender awareness, responsiveness and sensitivity), very few analysis exists on the issue of commons from a gender perspective. Two significant exceptions are Silvia Federici's work, who approached the debate about the commons from a feminist perspective (Federici 2010), and the research on Indian community-based initiatives conducted by Soma Kishore Parthasarathy (2015).

First, it must be noted that some of the difficulties of this dialogue between commons claims and women's right claims are partially due their intrinsic nature. This emerges clearly when we look at the commons from a traditional point of view, linked with natural resources: 'in the rush to provide secure land tenure for communities, there is a risk that women's rights will not be documented or secured, thus weakening their rights to the collective land' (Giovarelli et al 2016: 1).

In fact, while the protection of community land rights is perfectly coherent with the idea of resistance (in particular against land grabbing) and protection of rights, a point of fracture might emerge from a gender perspective. The crucial knot where the claim for women's land rights and the claim for community land rights might collide is the role played by, and the value given to, customary practices in governing the commons. This analysis does not aim at demonstrating that a commons approach would automatically represent the best solution for women's rights, within their communities and families. It rather highlights the gendered differentiated impact of the claims for the protection of commons, as well as the crucial role played by women in claiming their, and their communities', rights.

The issue of access and rights to land and resources is a particularly controversial subject, and its impact on women is even more complex and differentiated. Diverse perspectives interact and co-exist in approaching land grabbing, tenure security, land titling, land reforms, commercial pressure, evictions, large scale vs small scale (and/or family) farming, food security and food sovereignty. Despite such diversity, and while it is not possible to generalise, a few common elements emerge.

Women represent a significant portion of the labour force in the rural world, however,

> the pattern that women own less land than men, regardless of how ownership is conceptualized, is remarkably consistent [and] women are disadvantaged relative to men in nearly all measures of landownership and bundles of rights.
>
> (Doss et al 2013)

This is partially due to discriminatory laws, and partially to customary practices and traditions. Some scholars, such as the Indian economist Bina Agarwal, identify property rights as the best fitting solution (Agarwal 1994). However, literature demonstrates (Bruce et al 2008; De Schutter 2011; Latorre 2015) that land titling is not automatically the best nor the sole option to secure land rights, in general and for women in particular. In fact, from a gender perspective land titling measures might produce further insecurity in a variety of situations: gendered impact of corruption linked to land registration (Randria Arson et al 2018), land registered exclusively under men's names or, in case of joint titles, a gap between de jure and de facto enjoyment of land rights. On the other hand, it cannot be denied that women's rights are often undermined by customary practices and traditions.

This dynamic produces a core tension between the claims for women's rights and the claims for community rights: as customs and traditional ways of governing resources are reinforcing these customary practices in the effort to protect the commons, this might have an impact on women's equal rights (Fuys et al 2008; Oxfam, ILC, RRI 2016).

In fact, community management systems, when implemented in a traditional structure, could reinforce situations of discrimination (since most of the community chiefs are men) and hence raise an issue in terms of accumulation of power and discrimination against women (Peters 2013: 550). The result would be that because of the 'social and cultural mechanisms that operates the channel of access to the commons, women [might] find themselves at a disadvantage and discriminated' (Aier 2011: 177). As a consequence, when community land rights are at stake, 'women under customary tenure regimes face a fight on two fronts: [...] they are battling to be treated as equals while also defending their customary land rights to protect their communities and identity' (Oxfam, ILC, RRI 2016).

However, this same quote partially provides the response to the issue: the core element is the level of participation and inclusion of women in decision-making processes. From this point of view the concept of 'participatory exclusion', elaborated by B. Agarwal (2011), which consists in the exclusion of women from decision making roles within institutions that appear participatory, provides a meaningful insight: looking participatory is not a guarantee of actual inclusivity, hence ensuring inclusiveness is crucial.

However, the case of the commons seems to offer concrete opportunities to challenge this power dynamic while ensuring protection of the resources: this is due to the fact that commons are not static and 'flexibility provides a measure of security [...] by creating reciprocal expectations of resource sharing between groups' (Meinzen-Dick et al 2004). In protecting the commons as community struggle power relations might be challenged and new dynamics put in place: 'Indigenous women and those from local communities share a history of struggles and activism – speaking out in defense of collective land rights, cultural identity and social change as part of the global movement for women's liberation' (Cunningham 2006: 55).

While limited to the very specific, but crucial, category of traditional commons, women's active role would lead to a broader enjoyment of women's rights, both individually and collectively. Women's engagement and proactive behaviour could favour the synthesis of the two claims for women's land rights and commons (in particular community land). Even more, women's agency could contribute to reinforce commons claims, and vice versa. This specific case proves that it is 'crucial that the discourse on the commons engage with issues of gender justice' (Aier 2011: 177). And this applies far beyond the case of natural resources and traditional commons.

The feminist perspective, used by both Federici and Soma Kishore Parthasarathy, highlights that women are the first and most affected by commodification and privatisation of resources (described as a direct consequence of capitalist and neo-liberal economic models), because of their lower bargaining power and their role as care givers, but also the key actors of resistance, re-appropriation and claim. The two scholars reach the same conclusion despite the differences in context. Federici develops her analysis focusing on struggles in the western world (in particular in the US) and on the 'commoning' of the material means of reproduction as key element for reinforcing mutual bonds. She argues for a leading role of women in the collectivization of reproductive work in order to build an alternative society (Federici 2010). Kishore Parthasarathy, on the other hand, contextualises the role of women in traditional community based society in India, where women play the role of custodians of the resources and of the relevant knowledge. The scholar demonstrates that women can become key actors in fighting against commodification and privatisation of the commons, and, therefore, gain a crucial role within their communities (Kishore Parthasarathy 2013).

The role played by women in challenging a system of discrimination and lack of rights emerges as part of the voices of those involved in struggles to protect human rights, both as individual rights-holder and in terms of the rights of their own communities. In the struggles for community land rights, while reshaping internal power dynamics, these women act as human rights defenders.

Human rights, resistance and social change

The connection between human rights and the commons builds upon a reflection about alternative economic models, based on inclusion rather than exclusion, and matches with the idea of social function and participative models (Rodotà 2016), from the perspective of a bundle of rights that applies to both new and traditional commons.

Whether this is about forestry, water, urban spaces or the internet, the commons exist as an alternative to the current commodification process and socio-economic model: in some cases they respond to traditional practices, in other they represent a new strategy to fight social injustice.

Naples municipal deliberation – focusing on the creation of social and relational capital – the work of the Rodotà committee – grounding the commons on fundamental rights – the local experience of urban gardens and the global experience of digital commons, as well as the feminist analysis – focusing on differential impact and potential for change – concur to highlight different facets of the commons.

Rethinking citizenship (Rodotà 2016) and building solid reciprocal bonds (Festa 2016) are identified as core elements of this commoning process, which aims at producing a meaningful response to existing crises. The elements of resistance against commodification and neo-liberal policies, and, in some cases, of actual social struggle, are embedded in the idea that commons and human rights are strongly interconnected.

A human right based approach to the commons, inspired by the prevalence of access over property, could find clear correspondences with UN documents and International Law (in particular, with the International Covenant on Economic Social and Cultural Rights, the UN Declaration on the Human Right to Water and Sanitation, and the UN Declaration on the Rights of Indigenous People). This is notably clear for traditional commons, such as community land or forestry; but beyond natural resources, this also applies to urban and even to digital commons. For instance, the access to internet is to be considered a human right as suggested the UN Special Rapporteur on the promotion and protection of the right to freedom of opinion and expression in 2011 (A/HRC/17/27). The same approach was implied in the 2016 resolution of the Human Right Council, on the promotion, protection and enjoyment of human rights on the Internet (A/HRC/32/L.20), focused on freedom of expression and the digital divide.

While the approach to the commons might vary significantly in terms of political theory (Papadimitropoulos 2017), human rights claims, resistance and social change emerge as consistent core issues.

The human rights discourse is clearly mentioned by the work of the Rodotà Committee and finds its correspondence in International law. The reference to the rights of future generations integrates the debate and matches with the context of more traditional commons, where the claims aim at ensuring that community resources are preserved and their management is adapted to new models, in order to transmit them to its future members.

Resistance against privatisation, commodification and exploitation of resources, whether material or immaterial, emerges as a key feature when commons are at stake. Finally, social change is particularly relevant in the feminist approach to the commons, as women's agency should be a key component in claiming the commons and women's rights could be reinforced by the commoning process. Furthermore, social change and alternative economic and social models are what the commons put in practice, as demonstrated by the example of Naples.

This attempt to define the commons, builds upon key concepts that are identified as crucial by activists (from political activists dealing with urban commons and social justice to indigenous activists defending community resources) and scholars, in particular political-economists, lawyers and sociologists. These concepts include: reciprocity, responsibility, inclusion and resistence against commodification.

On a further level, the commons and human rights also appear intertwined. This applies with regard to future generations (as per the legal definition promoted by the Rodotà committee and the indigenous approach to preservation of resources), as well as to women's rights and indigenous people rights.

As clarified in applying a gender lens to the debate about community land and commons, it is not possible to affirm that the commons represent the sole response to women's rights claims to land, nor that they would hamper such rights. Similarly, it cannot be affirmed that collective rights should prevail over individual rights, nor that they would automatically ensure a better or stronger enjoyment of human rights.

However, what makes the commons particularly interesting from a human rights and socio-economic justice perspective, is their two-fold impact. On the one hand they seem to offer an opportunity for empowerment within the groups and communities involved (this is particularly true for women, but can apply to marginalised groups in other contexts). On the other hand, independently from the context, from the city of Naples to rural communities in India, they contribute to build an imagery of resistance against different phenomena of commodification of resources.

Notes

1 Translation is mine. Original in Italian: 'cose che esprimono utilità funzionali all' esercizio dei diritti fondamentali nonché al libero sviluppo della persona. I beni comuni devono essere tutelati e salvaguardati dall' ordinamento giuridico, anche a beneficio delle generazioni future. Titolari di beni comuni possono essere persone giuridiche pubbliche o privati. In ogni caso deve essere garantita la loro fruizione collettiva, nei limiti e secondo le modalità fissati dalla legge. Quando i titolari sono persone giuridiche pubbliche i beni comuni sono gestiti da soggetti pubblici e sono collocati fuori commercio; ne è consentita la concessione nei soli casi previsti dalla legge e per una durata limitata, senza possibilità di proroghe'.

2 www.remixthecommons.org/en/projet/definir-les-biens-communs/

3 https://wiki.p2pfoundation.net/International_Commons_Conference_-_2010

4 Part of this analysis has been presented as joint contribution to a dedicated panel on Gender and Commons promoted as International Land Coalition in the framework of IASC 2018 Global Conference.

References

Agarwal, B. (1994) A *Field of One's Own: Gender and Land Rights in South Asia*. Cambridge: Cambridge University Press.

Agarwal, B. (2011) 'Participatory Exclusions, Community Forestry, and Gender: An Analysis for South Asia and a Conceptual Framework', *World Development*, vol 29, no 10, pp 1623–1648.

Aier, A. (2011) 'Women and Commons: Engaging with Gender Justice', in Cheria, A. and Edwin, D. (eds), *Vocabulary of Commons*. Foundation for Ecological Security, pp 167–178.

Alternatives Sud – various authors (2002) *Acqua Bene Comune dell'Umanità*, Milano: Punto Rosso.

Baier, W. (2017) *The Case for a European Plan B – Thinking Europe as a Commons*. Brussels: Transform!Europe.
Barbato, M., Berdini, P., Colosimo, C., Corradini, P., Di Pietro, G., Gallo, S., Gesualdi, F., Gualtieri, V., Marella, M.R., Mattei, U., Pascali, F., Settis, B., Soscia, D., Stampacchia, M., Veneziano, D., and Viale, G. (2013) *Rebelpainting, Beni Comuni e Spazi Sociali: una creazione collettiva*! Pisa: Rebeldia edizioni.
Baslar, K. (1998) *The Concept of the Common Heritage of Mankind in International Law*. Dordrecht: Martinus Nijhoff.
Bollier, D. and Helfrich, S. (ed.) (2012) *The Wealth of the Commons: A World Beyond Market and State*. Amherst, MA: Levellers Press.
Bollier, D. and Helfrich, S. (ed.) (2015) *Patterns of Commoning*. Amherst, MA: Levellers Press.
Borch, C. and Kornberger, M. (ed.) (2015) *Urban Commons: Rethinking the City*. London: Routledge.
Bresnihan, P. and Byrne, M. (2014) 'Escape into the City: Everyday Practices of Commoning and the Production of Urban Space in Dublin', *Antipode*, vol 2014.
Bruce, J.W., Giovarelli, R., Rolfes, L,. Bledsoe, D., and Mitchell, R. (2008) *Land Law Reform: Achieving Development Policy Objectives. World Bank Law, Justice, and Development Series*. Washington, DC: The World Bank.
Cangelosi, E. (2014) 'Commons: Practices of Spaces and Social Change', *Comparative Law Review*, vol 5, no 2.
Cangelosi, E. (2015) 'Reshaping Spaces and Relations: Urban Gardening in a Time of Crisis', *Partecipazione e Conflitto*, PACO, vol 8, no 2, pp 392–416.
Candelaria, S.M. (2012) *Comparative analysis on the ILO Indigenous and Tribal Peoples Convention No. 169, UN Declaration on the Rights of Indigenous Peoples (UNDRIP), and the Indigenous Peoples' Rights Act (IPRA) of the Philippines*. Geneva: ILO.
Cavallo, R. (2013) *I beni comuni tra vecchi cliché e nuove sfide*. Micromega. Il Rasoio di Occam 26 Aprile 2013, accessible online: http://ilrasoiodioccam-micromega.blogautore.espresso.repubblica.it/2013/04/26/i-beni-comuni-tra-vecchi-cliche-e-nuove-sfide/ (retrieved November 2018).
Clément-Fontaine, M. (2017) 'Communs numériques', in Cornu, M. Orsi, F. and Rochfeld, J. (eds), *Dictionnaire des biens communs*. Paris: PUF.
Colding, J. et al. (2013) 'Urban Green Commons: Insights on Urban Common Property Systems', *Global Environmental Change*, vol 23, no 5, pp 1039–1051.
Coriat, B. (2013) *Le retour des communs*, Revue de la régulation, 14 | 2e semester.
Cornu, M., Orsi, F. and Rochfeld, J. (2017) *Dictionnaire des biens communs*. Paris: PUF.
Cunningham, M. (2006) 'Indigenous Women's Visions of an Inclusive Feminism', *World Development*, vol 49, no 1, pp 55–59.
Dardot, P. and Laval, C. (2014) *Commun. Essai sur la Révolution au XXIe siècle*. Paris: La Découverte.
De Angelis, M. (2010) 'The Production of Commons and the "Explosion" of the Middle Class', *Antipode*, vol 42, no 4.
De Angelis, M. (2014) 'Crisis and Commons Today', in Brincat S. (ed.), *Communism in the 21st Century*. Santa Barbara, CA: Praeger.
De Angelis, M. (2017) *Omnia sunt Communia*. London: Zed Books.
De Angelis, M. and Harvie, D. (2014) 'The Commons', in Parker, M., Cheney, G., Fournier, V., and Land, C. (eds), *The Routledge Companion to Alternative Organizations* (pp 280–294). Abingdon: Routledge.
Delmas, C. et al. (2016) *Socialisation and Commons in Europe*. Brussels: Transform!Europe.

De Schutter, O. (2011) 'How not to think of land-grabbing: three critiques of large-scale investments in farmland', *Journal of Peasant Studies*, vol 38, pp 249–279.

Dietz T., Dolsak N., Ostrom E., and Stern P.C. (2001) *The Drama of the Commons, Comittee on the Human Dimensions of Global Change*. Washington, DC: National Academy Press.

Dolšak, N. and Ostrom, E. (2003) *The Commons in the New Millennium: Challenges and Adaptation*. Cambridge, MA: MIT Press.

Doss C., Kovarik, C., Peterman, A., Quisumbing, A., and van den Bold, M. (2013) *Gender Inequalities in Ownership and Control of Land in Africa. Myths versus Reality*, IFPRI Discussion Paper 01308, Washington, DC: IFPRI.

Federici, S. (2010) 'Feminism and the Politic of the Commons', in Hughes, C., Peach, S. and van Meter, K. (eds.), *Uses of a World Wind, Movement, Movements, and Contemporary Radical Currents in the United States*. Oakland, CA: AK Press.

Festa, D. (2016) *Les communs urbains. L'invention du commun*, Tracés. Revue de Sciences humaines, #16 | 2016.

Foster, S. and Iaione, C. (2015) 'The City as a Commons' (August 29, 2015), *Yale Law & Policy Review*, 34, pp 281–349.

Fuys, A., Mwangi, E. and Dohrn, S. (2008) *Securing Common Property Regimes in a Globalizing World*. Rome and Washington, DC: ILC and CAPRi.

Giovarelli, R., Richardson, A. and Scalise, E. (2016) *Gender and Collectively Held Land, Good Practices and Lessons Learned from Six Global Case Studies*. Washington, DC: Resource Equity and Landesa.

Goldman, M. (ed.) (1998) *Privatizing Nature: Political Struggles for the Global Commons*. New Brunswick: Rutgers University Press.

Hardt, M. and Negri, A. (2009) *Commonwealth*. Cambridge, MA: Belknap Press of Harvard University.

Harvey, D. (2012) *Rebel Cities: From the Right to the City to the Urban Revolution*. Ann Arbor, MI: MPublishing, University of Michigan Library.

Iaione, C. (2012) 'Governing the Urban Commons', *Italian Journal of Public Law*, vol. 7, no 1.

Kishore Parthasarathy, S. (2013) *A Feminist Perspective on the Commons as Process*. Heinrich Boell Stiftung.

Kishore Parthasarathy, S. (2015) 'Our Ways of Knowing: Women Protect Common Forest Rights in Rajasthan' in Bollier, D. and Helfrich, S. (eds), *Patterns of Commoning*. Amherst, MA: The ommons Strategies Group.

Knight, R., Adoko, J., Auma, T., Kaba, A., Salomao, A., Siakor, S. and Tankar, I. (2012) *Protecting Community Lands and Resources: Evidence from Liberia, Mozambique and Uganda*. Rome and Washington, DC: Namati and IDLO.

Kurtz, H. (2013) 'Differentiating Multiple Meanings of Garden and Community', *Local Environment*, vol 16, no 6.

Latorre, S. (2015) 'The Making of Land Ownership: Land Titling in Rural Colombia – a Reply to Hernando de Soto', *Third World Quarterly*, vol 36, no 8, pp 1546–1569.

Lefebvre, H. (1968) *Le droit à la ville*. Paris: Anthropos.

Linebaugh, P. (2008) *The Magna Carta Manifesto: Liberties and Commons for all*. Berkeley, CA: University of California Press.

Lucarelli, A. (2017) 'Naples (experience de bien commun)', in Cornu, M., Orsi, F., and Rochfeld, J. (eds), *Dictionnaire des biens communs*. Paris: PUF.

Marella, M.R. (ed.) (2012) *Oltre il pubblico e il privato*. Verona: Ombre Corte.

Marella, M.R. (2013) 'Pratiche del comune. Per una nuova idea di cittadinanza', *Lettera Internazionale*, vol 116, no 2, pp 40–44.

Marella M.R (2016) *La propriété reconstruite: conflits sociaux et catégories juridiques*, Tracés. Revue de Sciences humaines, #16 | 2016.

Mattei, U. (2011) *Beni Comuni. Un Manifesto.* Roma-Bari: Laterza.

McClintock, N. (2009) 'Why Farm the City? Theorizing Urban Agriculture through a Lens of Metabolic Rift', *Cambridge Journal of Regions, Economy and Society*, vol 3, pp 191–207.

Meinzen-Dick, R., Pradhan, R., and Di Gregorio, M. (2004) 'Understanding Property Rights', in Meinzen-Dick, R. and Di Gregorio, M. (eds), *Collective Action and Property Rights for Sustainable Development.* Brazil: IFPR.

Micciarelli, G. (2014) 'I beni comuni al banco di prova del diritto. La soglia di un nuovo immaginario istituzionale', *Politica & Società*, vol 3, no 1, pp 123–142.

Mone, D. (2017) 'Commissione Rodotà', in Cornu, M., Orsi, F. and Rochfeld, J. (eds), *Dictionnaire des biens communs.* Paris: PUF.

Napoli, P. (2014) 'Indisponibilité, service public, usage. Trois concepts fondamentaux pour le "commun" et les "biens communs"', *Tracés*, no 27, pp 211–233.

Ostrom, E. (1990) *Governing the Commons: The Evolution of Institutions for Collective Action.* Cambridge: Cambridge University Press.

Ostrom, E. (1994) *Neither Market nor State: Governance of Common-pool Resources in the Twenty-first Century.* Washington, DC: IFPRI Lecture Series, International Food Policy research Institute.

Ostrom, E., Gardner, R. and Walker, J. (1994) *Rules, Games and Common-Pool Resources.* Ann Arbor, MI: Michigan Press University.

Oxfam, International Land Coalition, Rights and Resources Initiative (2016) *Common Ground. Securing Land Rights and Safeguarding the Earth.* Oxford: Oxfam.

Papadimitropoulos, V. (2017) 'The Politics of the Commons: Reform or Revolt?', *Triple C*, vol 15, no 2.

Peters, P. (2013) 'Conflicts Over Land and Threats to Customary Tenure in Africa', *African Affairs*, vol 112, no 449, pp 543–562.

Randria Arson, H.P. et al. (2018) *Women, Land and Corruption.* Berlin: Transparency International.

Rodotà, S. (2016) *Vers les biens communs. Souveraineté et propriété au XXie Siècle*, Tracés. Revue de Sciences Humaines, #16 | 2016.

5 Towards a new institutional political ecology

How to marry external effects, institutional change and the role of power and ideology in commons studies

Tobias Haller

Introduction

This conceptual chapter proposes to marry two very different approaches: On the one hand it looks at the political ecological (PE) analysis of divergent actors' powers and interests in the governance and use of common-pool resources (CPRs). On the other hand, it tries to relate to the new institutionalism (NI) approach in social anthropology dealing with CPRs, showing how external and internal, as well as market driven, dynamics of a resource context lead to justification strategies and institution shopping (selection of rules and regulations) by the different actors involved, and finally to behaviour and distributional effects related to the commons. The fusion of these two approaches is then called the New Institutional Political Ecology' (NIPE) (Haller 2017).

The reason to bring these two approaches together is that NI provides a well-defined concept of interrelations of factors leading to institutional change and explaining use and overuse of CPRs, while PE brings in a solid analysis of power relations. But why use NI for this and not the so-called old institutionalism? As Olivier de Sardan (2013) argues in an overview paper on institutional approaches in economics, social sciences and humanities, there is no link between old political economy and inspired institutional and neo-institutional approaches. The former supposedly address issues of power while the latter focus on enhancing efficiency and reducing transaction costs in economic activities. However, if we look at the work of Douglass North (1990), it becomes evident that the idea that new institutionalism does not contain issues of power is misleading. North as well as Ensminger (1992) looked explicitly not just at institutions as a means to reduce transaction costs, but at actors' power as bargaining power and their options to select rules and transform as well as ideologically legitimise multiple institutional settings – for example, forms of property rights of CPRs (see also Haller 2013 for a revised version in NI in social anthropology).

I propose a further extension of this aspect in order to unpack the ideologies that contain the discourses (in a Marxist framework) and narratives (in a post-structuralist Foucauldian frame), which provide both the legitimacy and the possibilities to increase bargaining power of particular actors over others. This process, I argue, facilitates institution shopping within a context of institutional pluralism governing access and use of CPRs and containing not just local but also national and international institutions in a 'glocal' world (Haller ed 2010).

This version of the model, however, still lacks the capacity to deal with the question of how actors gain access to resources through a historical political process of contestation and alienation (see e.g. Blaikie and Brookfield 1987; Bryant and Bailey 1997; Zimmerer and Bassett 2003; Robbins 2004). This 'basic' focus gave way to two PE orientations: One is a more Marxian political economy-oriented version. The other is a more post-structural version, focusing on construction and deconstruction of certain categories such as nature and culture. However, now there is a new wave of 'post-constructivist' political ecology, which criticises the anthropocentrism with its dualism as well as the lack of a concrete ontological reflection of the discipline itself (Escobar 1999, 2011). Oscillating between Descola (2005, 2013) and Haraway (2008, 2016), and by always continuing to ask the basic question of, "Who has the power to define the rules of the game?", the challenge will then be: What shall be used from these debates and resulting concepts? I will argue that integrating elements of the political and power issues as well as issues of construction and ontologies from PE improves the conceptualisation of NIPE.

The paper is organised as follows: After an introduction to PE core thinking, the issue of how to locate the power issue in the debate as well as what can be used from the post-structural and post-humanism approach shall be discussed. I will then turn to the NI model of CPR management and of institutional change, showing what NI has to offer in this marriage. In a related next step, the paper will give a short sketch of ontologies and epistemologies regarding land in three historical African time frames related to the commons. This part provides an understanding of political, economic and ecological conditions, constructions of meaning and ideology as (re)source of legitimacy. It also shows how identity and power driven institutional and material aspirations of actors are contested. A history oriented and institutionally informed analysis on litigations over a pasture to be transformed into an irrigation scheme in Zambia shall illustrate the way NIPE might be used.

The conclusion will then tease out how the model of new institutionalism can be married to the dynamic of power constellations in the human-environment interaction stemming from PE and its new critical turns: I will argue that NI provides a concise way in which different variables are interacting, while PE often lacks the analysis of external factors leading to basic economic changes and the way this impacts local constellations, shapes power relations, selection of institutions and processes of organisation and legitimacy production. PE is often not very well structured on how these global as well as local fine-tuned mechanisms are working and thus can profit from the NI model in the social anthropology version. PE then helps NI to analyse power elements of path dependent developments, which NI is often lacking. It sheds light on the way relative price developments are shaped by power constellations, the way this impacts bargaining power of actors and groups, who then interactively and strategically select institutions (who has access to which institution) and seeks ideological legitimacy (who has the power to define hegemonic and counter discourses and narratives) leading to specific kinds of distributional power, strategic behaviour and the outcome of the use of CPRs.

This new approach is also suitable to discuss many of the papers in this volume, on the one hand the European commons' cases of historical and contemporary institutional changes and outcomes in the management of the commons in Europe, including bottom-up institutional innovations. Or on the other hand, it helps to explain the impacts of European investments in contexts of CPRs worldwide (such as processes of land, water and green grabbing leading to resilience grabbing). And, last but not least, this new approach allows for a discussion analysing interactions between international institutions and local commons.

'Core thinking' in political ecology

PE deals with critical views on environmental degradation discourses and materialities: Grounded in Marxist or Neo-Marxist perspectives in social anthropology (see Wolf 1982) and human geography, the focus is not on the fact that degradation occurs, but on the reasons why so-called land managers degrade environments. This is not just visible in demographic increase, greed or ignorance, but in political processes and power asymmetries within communities and between actor groups. At the same time, the focus was also laid on wealthy land-owners overexploiting resources for capital accumulation, who, however, never took the blame regarding the overuse of natural resources as the poor people did in the degradation discourse (see Blakie and Brookfield 1987; Robbins 2004). Robbins shows that early political ecologists argued that land managers engaged in unsustainable uses of natural resources such as CPRs because they basically lacked the power to make their own decisions. Interestingly, Alexander von Humboldt was one of the first authors between the 18th and 19th centuries using power specific explanations for overuse of resources such as water levels, deforestation and wildlife in Latin America. He accused the Spanish colonial elites of dominating indigenous peoples and subordinating African slaves while also overexploiting and destroying cultural landscapes created by indigenous peoples. Such elements can be found in many publications of Humboldt such as his diaries as well as in his seminal book project "Cosmos" (see Robbins 2004). Later thinkers such as Marx and Engels also belong to the early fathers of PE based on their analysis of capitalism as an economy of exploitation but with less emphasis on environmental issues (ibid).

Human geography and social anthropology scholars after them provided key elements of the basic PE thinking which can be summarised in the following way: The nature of power relations in political ecology is defined by the question of who is able to control access, actions and activities of others via exclusive property rights on and access to natural resources or via dictating conditions of inclusion and exclusion of the use of these resources (Blaikie and Brookfield 1987; Robbins 2004; see also Ribot and Peluso 2003).

These conditions have historical reasons from which the power asymmetries in a local context stem. The approach also starts from a consideration of the so-called 'natural conditions' of a resource or resource context and advocates that local actors are knowledgeable and do not act in an 'irrational way'. While this does not encompass the view of a narrow concept of rational choice, the basic idea is that land managers, for example degrading a pasture, would do otherwise if they could decide on how to organise the use of resources via secured property and mobility. But restricting mobility from one resource area to another or hindering the use of alternative resources leads to increased pressure on a resource context (see Blaikie and Brookfield 1987; Zimmerer and Basset 2003). Inequality or unclear institutional contexts then lead to overuse of renewable resources and to land degradation.

This constellation provokes so-called positive feedback loops as described in systems theory. Poverty leads to degradation, which further increases poverty. Many parts in African and Australian drylands illustrate these processes. However, these processes are hidden in the dominant discourse of state and national as well as international powerful actors using the narrative of local groups degrading their environment (Blaikie and Brookfield 1987) and needing development. James Ferguson (1994) and Arturo Escobar (2011) both argue that the concept of development as well as degradation acts as an 'anti-politics machine' hiding the historical, political and ideology driven processes of dependency, exploitation and subordination by colonial and neo-colonial powers. Discourses for mending the problem and delivering development range from demographic reduction of the world population, to technological innovation, cultural change and a more market-based involvement of actors. Bailey and Bryant (2005) thus argue that since the mid-1990s, the range of explanations for such processes increasingly included not just the local, but referred to several levels of scales including the global *and* the local scale (thus the 'glocal' view), a point taken up as well by Robbins, who illustrates this in several cases.

So, work in the 2000s also included an analysis of local reactions triggering social movements and resistance against dominant groups of actors owning land and social capital for defining the conditions of top-down management of resource use. It further incorporated analyses of the way dominant narratives and discourses are used to construct views in degradation and sustainable use of natural or CPRs (see Zimmerer and Bassett 2003; Robbins 2004) related to protected areas and conservation (Brockington et al 2008; Galvin and Haller 2008; Fletcher 2010), Neo-Malthusianism (Zimmerer and Basset 2003) and institutional change (Haller ed. 2010). Different theoretical orientations viewed the reasons for the same process driving degradation and marginalisation, environmental conflicts, conservation of biodiversity as well as social environmental movements as a reaction to capitalist appropriation, for example, in mining and land grabbing processes.

They also looked at the so-called extension of the colonial frontier in the world system: Marxism focuses on processes of exploitation and primitive accumulation (see Harvey 2005), feminist studies on male dominance of control of resources and gendered division of labour (Narotzky 1997), environmental history by unpacking discourses of degradation reflected on natural environmental conditions (see Neumann 1998), postcolonial and subaltern studies on the construction of environmental knowledge and a view from 'below', while hiding biases in the presentation of the environmental conditions (see Biersack and Greenberg 2006) and deconstructing views of pure nature by natural science with the concept of cultural landscape ecosystems and engagement in local ecological knowledge (see Fairhead and Leach 1996; Berkes 1999; Haller et al 2013). However, the issue still is where the political in political ecology views are located.

Trying to locate the issue of power

While the issue of power is addressed in Marxist and also Weberian thinking, such as the power to be able to pursue primitive accumulation (see Harvey 2005) and actor-oriented bundles of power (see Ribot and Peluso 2003) excluding all other actors from the use of CPRs (see Paulson et al 2003; Paulson and Gezon 2005; Robbins 2004; Fletcher 2010), there is a new reflection on contestations over meaning in ecological resource use and management contexts. Thus, it is about locating the political in the way actors are able to make their economic rights and symbolic views legitimately dominant compared to other actors with the same claims (see also Robbins 2004). Taking this further by using the views of Foucault and Butler, it is the ability to be able to form subjects and linking this to the perceptions of environments, their claims, their aspirations and their role (external and self-inscriptive). Power is thus this notion of being able to define the construction of 'the' reality and who 'the' actors are in a dominant way, acting in a legitimate way (see also Paulson et al 2003). Politics then would be: "the practices and processes through which power, in its multiple forms, is wielded and negotiated" (Paulson et al 2003: 209). Power and its enactment via politics is thus about the way environments are perceived and how knowledge on the environment is constructed, including a whole range of different actors with different power relations and positions (from the local to government to national and international organisations' non-governmental representatives) (see Escobar 1999; Paulson et al 2003).

Political ecology thus also highlights the agency of marginal actors (see Scotts' *Weapons of the Weak* 2008) and underlines that not ecological reasoning alone should be the starting point of analysis, but political disputes and institutional changes, which might have nothing directly to do with an ecological problem, but later turn out to have environmental consequences.

These are reactions to a debate, which Vayda and Walters triggered in 1999 by arguing that political ecology is not at all concerned with ecological issues – an argument, which can also be found in later debates, where authors having a political orientation are said to have no notion of ecological contexts and should thus not speak up on ecological issues (see Robbins 2004; chapter "Against Political Ecology"). This means that the debate on what is specifically political about political ecology in the sense of locating power issues is not over yet. Perhaps the debate has not even really started, although the *Journal of Political Ecology* and many authors have underlined that the combined interest of political *and* environmental issues are equally of interest in the approach.

Furthermore, I would argue that political ecology raises the question of power to be able to hegemonically (see Simon 2015) define the labelling of environments, the construction of nature for example as pure nature, often hiding local common property rights (see Haller ed. 2010, 2013; Haller and Galvin 2011). This is interrelated to other features of power, which contain the ability to define what is at stake in a concrete setting and who shapes these settings and the power relations (see also Tsing 2001; Robbins 2004). That also echoes Fletcher's argument that state and neo-liberal actors do define what is at stake and enact this as different environmentalities (Fletcher 2010), which are then moved up and down different scales (ibid, see Paulson et al 2003) in multiple causalities (see Gardner 2003). Furthermore, other authors such as Swyngedouw (2009), analysing hydro-power projects, highlight how different scales such as the so-called global and the local interact in the relations of production and are metabolised in often unexpected ways in biological processes. These then lead to conflicts, which again are subject to discourses in order to legitimate different strategies shaping control of access to resources. Political ecology thus also looks at neo-liberal discourses that focus on *the* market as an institutional solution and hiding the question to whom, for example, water as a CPR belongs to, meaning the contestation over notions of property rights and the power to define and defend these.

The focus on social power in its different forms and how it is manifested is what political ecology literature regards as being the most important issue (Swyngedouw 2009). This is lacking in other approaches such as social ecological systems and institutionalism approaches (see Berkes 1999; Berkes and Folke 2000; Ostrom 2005). Moreover, the sole focus on institutions seems to create blind spots regarding how agency in politically heterogeneous communities is strategically reproduced, developing unplanned outcomes and this further indicates that the messiness of local constellations does not enable so-called rational choice actions (see Cleaver and Frank 2005). It is true that Ostrom did not address or even downplay politics and different power relations (see Crane 2010; Haller ed. 2010; Fabinyi et al 2014). But this work, especially Cleaver's (Cleaver 2003) approach focusing on institutional bricolage and critical institutionalism, does pay too little attention to the means, reflections and strategic action of actors themselves, which is hidden behind the notion of messiness. Actors are portrayed as if they did not calculate – not meaning that they are always able to anticipate the outcomes, which indeed are too often too messy (see also Haller 2017).

This brings us to the issue of strategic planning in environmental contexts and thus to the question of how to address resilience under power asymmetric conditions: The challenge is on how to bring in political and power specific issues in resilience studies which are so vital for political ecology as Fabinyi et al (2014) outline. The question is resilience of what exactly and who is affected? I would argue that using an historical perspective is vital to discuss the issues of speed of external changes and the power specific capability to adapt to these changes dealing with risk and uncertainty. It is at this level that I argue that problems in resilience studies should be looked at: For example, commons systems were far more widespread in pre-colonial times than they are today. Where they are still in use, they provide access to vital resources for marginal and gender specific groups, while dismantling the commons undermines local access and maintenance of cultural landscape ecosystems and the resilience of social *and* ecological systems (Haller et al 2013; Haller 2016).

This clearly moves away from ecological determinism and reductionism (see Orlove 1980), which was the critique of ecological anthropology since the work of Rappaport (1968). Again, the power asymmetries in local communities, transformed by local and external actors who in turn shape internal and external contestations over resources, are key to newer political ecology analysis (see Netting 1993; Robbins 2004; Brosius et al 2005; Haller ed. 2010). This also indicates that the management of resources and related contestations occur through time and are a product of power relations defining processes of inclusion and exclusion to the CPRs.

The power issue is discussed differently in the two main orientations in political ecology: the one focussing on political economy, the other on poststructuralist social theory (Brosius 1999; Fabinyi et al 2014). The first discussion draws on neo-Marxist approaches referring to market and capitalist relations of power leading to poverty and overuse of resources. The other one draws from Foucauldian poststructuralist insights regarding power and the way narratives, discourses and representations hide operating power relations. These also shape how people perceive their environments (Fabinyi et al 2014, referring to Brosius, Agrawal, Tsing, Li and others, I would also count Robbins 2004 to that group). Actors thus become subjects of more powerful state or other dominant actors on the global scale (NGOs, GOs and Global Organisations such as UN and World Bank, etc.). In PE, both streams should be reunited in order to understand that, for example, issues of resilience cannot leave out discussions of power relations in order to define what resilience means in a concrete system and for whom.

In a recently published overview on power theories in political ecology, a similar approach is taken by Svastad et al (2018). They propose a combination of, first, actor-oriented approaches making actors and their power visible (see Ribot and Peluso 2003; Brockington et al 2008), adding, second, neo-Marxist approaches (see also discussion above; showing how political structures are changed by powerful actors, while disempowering others in the context of capitalism, see Harvey 2005), and third, the Foucauldian notion of discursive power, including hegemonic power, governmentality and biopower issues (Svanstad et al 2018). This attempt to fuse several notions of power is similar to what is outlined above, but brings in more differentiation in structuralist and poststructuralist notions. Still, it ignores the relation between ideology (Marx) and discourses and narratives as well as the insight that power configurations change depending on external and internal contexts. I would also maintain that while we are dealing with three versions of power, the actors and neo-Marxist views are very close/similar and what we are dealing with only two basic views that are interrelated.

PE and the ontological, epistemological and posthumanist turn

However, these two current streams of critical PE and their notions of power are themselves strongly criticized by several new approaches labelled 'post-constructivist political ecology' or post-humanism approaches. These state that the discipline needs to be decolonised from humanist and male centred views. In several publications in the *Journal of Political Ecology*, *Geoforum* and elsewhere, the argument is stressed that the discipline is still anthropocentric and oriented in colonial ontologies of sustainability and development (Srinivasan and Kasturirangan 2016; Sullivan 2017). The main argument of this critique is that PE is based on problematic ontologies. A reflection on the way power structures shape these ontologies is needed; how are their uses shaped by neo-liberal orders also inherent in the discipline itself. The 'nature beyond the human' approach rooted in a combination of Harawayian and Latourian visions of the world (cyborg spaces and quasi-objects as well as views of the non-human), deals with other ideas of what we call 'nature' (see Sullivan 2017), providing room for a concept that actually comes close to Descola's views on ontologies among so-called indigenous groups (Descola 2005, 2013), such as animism and totemism.

In these ontologies, the divide between 'humans and other living beings' does not exist in the same way as in the anthropocentric views in the capitalist and neo-liberal order, a fact that is also highlighted in much older ecological anthropology literature (see Haller 2007b for an overview). Ontology in this critical view of PE as the study of being and what being means, has to be embedded in political realities, which are again contested and multiple (see Graeber 2015). Sullivan argues that we need to see which ontologies are used in a multiple and interacting way and in what kinds of contexts/under what kind of circumstances, also in order to bring so called 'non-humans' (again a dichotomy) into focus. Therefore, on the one hand, it is not just about the indigenous peoples in the 'Amazon' or the 'African rainforest', but also about the animals and plants and, last but not least, all other people labelled as so-called non-indigenous (thus labelled in political terms).

The neo-liberal capitalist order, on the other hand, has a naturalist ontology (see Descola 2013) constructing 'pure nature' as a controllable and commodifiable entity (see Haller 2007b, also Sullivan 2017). One could argue that this is in line with the PE idea of construction and with NI's concept of ideology. But Sullivan indicates that the struggles over ontologies and epistemologies are at the same pace as struggles over land (ibid), in which we all have our place of being but from which not all face the same consequences (see Schulz 2017). While agreeing on this point, it begs the question of where the issue of multiple ways of being and knowing – requested by the critical and decolonising political ecology – leads the different actors, quasi-actors/objects and subjects regarding the power to control and use CPRs?

In addition, there is a much older view of Escobar (1999) than the current 'turn' that encompasses the analysis that the same context – for example deforestation in the Amazon – entails three different ontological and epistemological views on the environment: a) *organic nature* (including local knowledge and views and animistic as well as totemistic ontologies, see Descola 2005, but also including transformations of 'nature' by subsistence production and management of CPRs; see Ellen 1982; Fairhead and Leach 1996; Haller ed. 2010); b) *capitalist nature* (including views and knowledge needed for market and commodification specific management of the environment); and c) *techno nature* (which views the environment to be manipulated and transformed radically from large scale infrastructure projects to genetic engendering). These different views also entail different values and perceptions as well as strategies regarding the way these three different forms of nature are perceived and claimed by whom. It also contains the debate on which views are more important and thus more powerful than others in a specific context.

The same is true for Ferguson's notion of the 'anti-politics machine', used for showing that development discourses hide power issues on a national as well as an international scale (Ferguson 1994, 2006). I would argue that this approach is more valuable theoretically than the new turn because the issue is still the old basic PE question: Who has the power to define hegemonic views that are used by concrete groups of people to control, manipulate and legitimate the flow of material, technological and other world realities across the planet? And who defines and enacts the rules of the game from economics to climate issues? These are in fact ever emerging issues in PE and related fields in social anthropology (see the work of Ferguson). These critiques of mainstream PE can be accommodated in a combination of the older approaches outlined above. Finally, however, these have to be rooted in an analysis of power constellations so as to be useful for NIPE.

I propose that these elements can be incorporated fruitfully particularly by showing how existing ontologies and epistemologies are replaced or pushed back by powerful actors over time with their respective ontologies and epistemologies. The interesting aspect is – related to NI, as we will see – that older ontologies and epistemologies might also be re-instrumentalised by powerful actors as they see fit (see for example the conservation discourse; Brockington et al 2008; Galvin and Haller 2008).

This will be illustrated in one of the next sections by a look at the meaning of land after I have outlined NI in social anthropology as a new partner of PE. The basic issue has not yet been addressed in PE, the question being: How does power systematically unfold in interactions on several levels and scales in relation to political and economic pressure? This is, as I would argue, much better addressed by NI.

New institutionalism in social anthropology: commons between transaction costs and power relations

A good starting point to bring in an NI analysis is the detected intensification of land use and land related resources worldwide based on the narrative of repeated hunger crises in African contexts since the 1970s. This created the image of land being degraded due to local people's overuse of natural resources held in common. This in turn triggered a debate on the relation between land and property not just from an economic angle, but from an ecological point of view as well. Reference was made to Hardin's *Tragedy of the Commons* (1968). In this essay, which was not based on any empirical evidence but was a Neo-Malthusian polemic against the freedom of human population growth, Hardin did not want to deal with land issues per se. He painted a picture of pastoralism in which actors are only interested in increasing their herd size, thus illustrating the process of freedom in reproduction (here of cattle or more generally of people, which was his main interest). This freedom then degrades the land because it is governed as common property (not as state or private property) and thus underlies open access conditions (see Acheson 1989; Ostrom 1990; Haller 2007a). This had a strong effect on state policies, especially in supporting justifications of states to control natural resources within their boundaries (see Feeney et al 1990). And it was a welcome ideological legitimacy for stricter state governance and later for neo-liberal privatisation policies.

The arguments picked up by mainstream science and by governments regarding the economic and ecological flaws of traditional land tenure then triggered a process of revision and rethinking. This process was led by Elinor Ostrom (1990), who devoted her Nobel Prize winning work to craft a new picture of CPR management in common property regimes. The main argument was that renewable CPRs, which are subtractable (what is taken away cannot be used by others for the moment) and difficult to defend (but possible by a group that can organise collectively), could be managed in a sustainable way by so-called robust common property institutions. Foremost, she and other scholars highlighted that Hardin erred in his view on the common property as open access, but that resources held in common are the property of a group and not no one's property. Eight design principles for well working institutions in her book *Governing the Commons* (1990) were deducted from mostly social anthropological case studies. They indicated that institutions address the problem of freedom and free riding by reducing transaction costs (information, monitoring and sanctioning). Ostrom's work related to environmental issues and followed the line of argument of the new institutionalism idea on property rights proposed by Douglass North (1990) himself following Roland Coase's theory on the firm and labour contracts to illustrate that institutions do reduce transaction costs.

The amount of work coming out of Ostrom's approach to new institutionalism as related to the commons is immense. There is now a digital library of the commons, as well as journals linked directly to the issue or picking it up. This is proof of the broad scholarly interest, also manifest in the growing participation

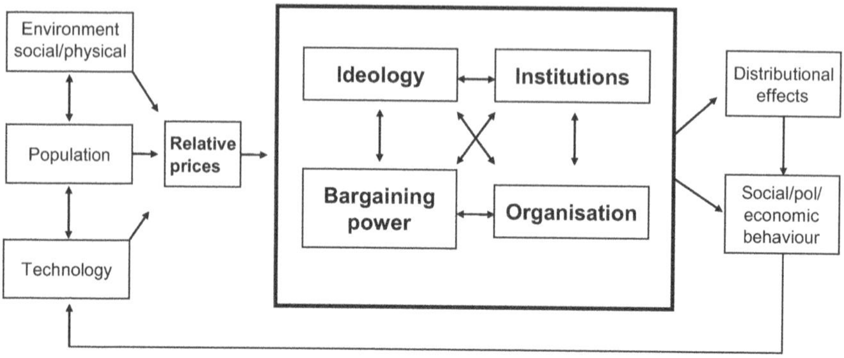

Figure 5.1 Modelling institutional change.
Source: Jean Ensminger (Ensminger 1992:10), see also Haller ed. 2010, Haller 2013.

at the *International Association for the Study of the Commons* (IASC), previously named *International Association for the Study of Common Property*. However, some variables that the economic historian Douglass North had integrated in his model were missing in Ostrom's work. She was more concerned with the possibility of self-organisation and sustainable use of CPRs, neglecting a historical embedment or issues of politics and of power. In spite of recognising that institutions are also embedded in larger systems her primary focus was thus not multi-layered as viewed by political economy and ecology. In these perspectives, local systems are often not articulated, but still related parts of state constitutions, legal systems and state elites, as well as of international global governance regimes and global markets with all the power constellations these scales entail (Haller ed. 2010, 2013).

While many studies focusing on the way self-organisation is made possible as a puzzle stemming from game theory (see Ostrom et al 1994), a new approach from social anthropology emerged that was not often paid attention to but which I consider as the most valuable approach in order to combine NI and PE. Starting from models in economic anthropology, Jean Ensminger, an US social anthropologist, proposed an interrelated NI model, in which external factors (environment, demography and technology) lead to changes in relative prices of goods and services and have a local impact (Ensminger 1992).

In her model, local context variables (in the black box) are influenced by the externally driven changing prices and relative value that a specific resource or a region is gaining. In line with the work of North, Ensminger (1992) argues that the bargaining power of actors, the way they organise and the way they select and craft institutions and legitimate institutions selected by ideologies, impacts the distribution, use and reproduction of CPRs (see feedback loops). Therefore, she does – in line with North – not follow other economists who predict that the market will choose the optimal institutions (see Williamson 1989). Rather, institutional settings and later distribution are shaped by bargaining power and ideology, as a resource of legitimacy. This means that, for example, a rise in the relative price of land will not automatically trigger immediate privatisation, but can have different outcomes depending on the bargaining power of actors and the way they are able to produce legitimacy (ideology) for their strategic choice of a certain institutional option (institution shopping). This model has been enlarged, looking closely at issues of bargaining power and ideology (including constructivist approaches regarding discourse and narrative to produce legitimacy, often in either so called traditional or modernization ideologies) (Haller 2013).

This point will be taken up in the next sections, in which the marrying of the two approaches will take place. A central issue where the two approaches merge is the question on how land and land related CPRs are governed and managed. While PE looks at power constellations, NI looks at the structured way institutions such as property rights are transformed and selected over time. In a much more structured way than PE, NI focuses on the interactions of political and economic forces at several scales. NI looks at how these then change the commodification of a context and how this in turn structures the power constellation, the institutions selected and crafted as well as the ideologically legitimated production in a fine tuned and interrelated way as proposed by the NI model of institutional change. It thus contributes to the understanding of structural dynamic processes, which PE lacks. It brings in especially the concept of institutional pluralism and the way the selection of institutions and strategies of legitimacy production works out. Therefore, I argue that we have a valuable link between the newer strands of PE and NI as PE can profit from NI by using these structured interrelations and detect where power, defined in the above sections, unfolds.

Before turning to the case study to illustrate this dynamic process, I will pick up the third of the three strands from PE and draw attention to the way pre-colonial, colonial and post-colonial ontologies and epistemologies of what the environment is and who interrelates with it in an ideologically legitimacy producing way. I argue that these ontologies and epistemologies provide an ideological basis for discourses to legitimate ownership of resources in contemporary contexts and for NIPE to work out. These ontologies and epistemologies contribute massively to the unfolding of the concept of ideology in the NI model. As I will argue, this is well related to critical positions and non-human to human interactions in political ecology, however with a different focus than the ontological change proposes. But I will use a historical and plural ontological approach illustrating the change of institutional meaning of land in African contexts focusing on different timelines of actors with their corresponding paradigms.

Historical changes of ontologies and epistemologies of the meaning of land

What land means as a central aspect in PE and NI depends – very much in line with newer versions of PE – on ontologies and epistemologies. In this section of the paper I will show how these ontologies and epistemologies have changed but still contribute to sources of ideology. I distinguish three processes in cases studied in African floodplains: the ontological meaning of land including the non-human in pre-colonial times, the ontology of the colonial and post-colonial state with epistemologies of disconnections and the ontology of neo-liberalism including the paradox of the state being present and absent at the same time.

Ontology of the meaning of land in pre-colonial contexts

Several authors emphasise that the notion of land in Africa is shaped by numerous factors, which involve tenure systems, type of land use and what ontologies and identities land generates. They argue that land is not – and never has been – based on the notion of state and market related private property. Instead, it was always based on an interrelated mix of private and communal property under the governance of leading offices (elders, specialists, etc.) in more politically symmetric or leaders in more asymmetric powerful groups. And it was related to the so-called first settlement of such groups, or, as a consequence of conquest, on feudal or more powerful systems (see Benjaminsen and Lund 2003; Toulmin 2009; Chanock 1985, 2005; Haller ed. 2010, 2013; Peters 2013). These authors then argue that the 'traditional tenure systems' we see today do no longer represent pre-colonial tenure but are mostly the result of imposed colonial transformations of property rights labelled 'customary law' (see Chanock 2005; Haller ed. 2010). Regarding the emic perception of land in pre-colonial times, comparative research in African floodplains (African Floodplain Wetlands Project, AFWeP, see Haller ed. 2010) revealed that the category 'land' is embedded in views of territory and landscape with all its resources, which are perceived as being interconnected and used for securing economic and political existence. Four issues are of importance to understand what 'land' as we understand the term means as seen through the different socio-cultural lenses:

1 First-comer – late-comer relations: In all cases there are groups who claim to have arrived first in the area and that land is just part of a territory, often controlled by spiritual beings. The territory is transformed into cultural landscapes through the use of these first comers. They take the lead when other groups join.
2 Coordination: they do this in order to coordinate and organize reciprocal use of the territory containing interlinked CPRs (pasture, fisheries, forestry water, wildlife, etc.), which are not perceived as separate but as part of a whole entity.
3 Spiritual relations: related to this, we often find organised spiritual landscape relations for successful use. The relation with these beings, usually via rituals that legitimise the first comers in their management role as they argue that they have a special relationship with the spiritual beings for land and related resources.
4 Institution building: thus they craft institutions for CPR management, which the different groups usually create based on their bargaining power

constellations. Like this conflicts and contestations can be resolved in new institutional forms often combined with older rules. The emerging institutions – for example access rules to a pasture – can thus be seen as a compromise (Haller ed. 2010).

In addition, the adaptation to seasonality was one driving force for developing these rules. Another one was conflict and conflict resolution. The institutions, established for the use of pasture, wildlife, fisheries and land use, as for example in the Kafue Flats in Zambia or in other floodplain cases studied comparatively (see Haller ed. 2010), were often established *after* conflicts occurred. They are based on myths and oral history, well remembered and contributed to the establishment of a complex, but well working governance system. This, in turn, reproduced and created the cultural landscapes found by the colonisers. To conclude, I argue that in these pre-colonial contexts land is part of a larger, symbolically viewed complex of a cultural landscape ecosystem. It is inhabited by spiritual beings that influence people's production and consumption. The occupants, who arrived at various stages in time, occupy different (hierarchical) identities. Moreover, the different resources that we view separately are in fact closely interconnected. The institutions show a certain flexibility across the annual cycle needed in complex cultural landscape ecosystems. This 'legal pluralism' is of a different order than the one externally established during colonial and post-colonial times (see below).

The colonial and post-colonial disconnect

This analysis is important in order to understand the impact colonization had on the way land and land related issues are perceived by powerful actors in state and administration until today, thus creating a state and naturalist ontology. Although there have been changes during precolonial times in the interaction between Europe and Africa, highlighting global trade and slavery also in other parts of the world (see Wolf's seminal work "Europe and the People Without History" 1982, and Bodley's "Victims of Progress" 1975 [2014]), the changes brought about by the colonial administration cannot be underestimated: It served as – despite all the differences in national colonial policies and formal procedures – a blueprint leading to legal pluralism of another order (see Mamdani 1996). Scott's work (1998) on the way that states see resources, reveals the basic perspective adopted by colonial administrations to control and measure resources for colonial profit. Previously created cultural landscape ecosystems were labelled as 'pure nature' to be controlled by colonial and post-colonial policies leading to four disconnections. These are based on the central institutional change from common to state property of land and land related resources:

First was the separation of areas into a) those of special interest to the state under state law; and b) areas of lesser colonial interest (or only for reserve labour army) under indirect rule and managed by selected leaders (for example, African Chiefs) to satisfy the colonial power's demands (see also Moore 1986;

Berry 1993; Benjaminsen and Lund 2003; Chanock 2005; Peters 2013). Legally, there was the dualism between 'modern' European Roman law for land (applied by the state and by white private owners linked to the colonial power) and the 'traditional' but state controlled, formalized 'customary' laws (Benjaminsen and Lund 2003; Lenz 2006; Haller ed. 2010; Peters 2013). The *second* source of pluralism as outlined above, refers to the way the state looks at resources in a fragmented way. This fits nicely into market, global and neo-liberal logics later on, changing the more integrated spiritual view (including the living and the dead and mystical powers influencing all elements) to a perception of different isolated resources which can be commodified. This leads to a *third* form of disconnections: The colonial states separated the resources of the cultural landscape ecosystems, made them into state property and managed/administered them in different departments such as agriculture, fisheries, wildlife, veterinary, water and energy, tourism, etc. All these departments continue to base their actions on their own legislation and legal settings, leading to a disconnect in management extended into the so-called independent states (Haller ed. 2010; Mhlanga et al 2014). The problem is therefore not a regard or disregard for formal or informal institutions, but an undermining of existing local institutions – which (inter)connected CPRs used in complex ecosystems – through a set of external pluralistic rules. These external rules then lead to institutional disconnections of interrelated CPRs with a territory, creating confusion and contradictions. This pluralism, I argue, then leads to institution shopping by those powerful actors who know how to use this plurality in their interests. From the post-structural and post-naturalist debate in critical PE, one could argue that this echoes a naturalist ontological position (see Descola 2005).

The present-absence of the state and the paradox of neo-liberalism

Most institutional pluralism was taken over by the new independent states, but was also additionally enlarged into international and global economic trade networks. Several international arrangements were installed, influencing the management of natural resources. The new, post-independence African states tried to continue the raw material export oriented economic structures they inherited from the colonisers. In all cases studied in AFWeP, states depended on one or two basic resources for cash and foreign exchange revenues. This was the economic basis with which the elites in power wanted to finance what they saw as the economic basis for imports in order to modernise the countries. Large infrastructure buildings and services (road networks, dams, large plantations, green revolution and agrarian subsidies, etc.) were set up. Often, they were paid for with foreign loans, anticipating high revenues from the raw material exports (mines, plantations, etc.) in the future. Between 1975 and 2000, changes in relative prices, such as the oil price increase (imports) and the decrease of other raw materials (metals, food and fibre cash crops for export, etc.) then led to a financial crisis and to weak state structures as staff and infrastructure could no longer be financed. Suddenly, governments were in the position that 'making state' got too expensive, while, at the same time, state activities could not contribute to the project of modernisation any more (see also Ferguson's book *Expectations of Modernity* 1999).

So, due to the lack of financial means, governments faced the financial inability to 'make state'. This means that state institutions responsible for the management of all the disconnected resources could not be enforced since neither staff nor infrastructure could be financed (see Gibson 1999 for a detailed study on Kenya and Zambia, but generally Haller ed. 2010). In most of the cases studied, this situation led to an institutional change from state property to a de facto open access situation of CPRs because the state was present (via laws and notions of citizenship, giving access to resources according to specific rules) and absent at the same time since it is too poor to finance the enforcement. This contributed to the overuse of most CPRs studied in the AFWeP cases (see Haller ed. 2010; Haller 2016).

This historical change giving way to the paradox of the state being present and absent at the same time, is rooted in a wider paradox of neo-liberalism of wanting and not wanting the state: James Ferguson's work on neo-liberalism (2006), referred to above in the section on PE, illustrates the issue nicely. Ferguson describes an international policy that tries to cut state costs by delegating management of activities to the market and to lower levels of action arenas under the state and into the private sectors. This strategy seemed to be a central remedy to cope with state debts all over the continent – and elsewhere as well (we observed this also in Europe). And herein lies the paradox of neo-liberalism: the idea of a state that, ever since colonial times, has been the driving vessel for protecting capitalism, now is much too expensive and needs a cure based on the medicine of dismantling the state.

The discourse of efficiency, also in relation to Africa (see Ferguson 2006), develops in the direction of dismantling state services, as these seem to be too costly. Hegemonic internalised discourses on the waste of resources, on corruption of elites, on mismanagement and inefficiency in formal legal processes, merge with the notion of the state as a colonial construct. As Harvey (2005) points out from a Neo-Marxist perspective, neo-liberalism can be seen as a political process of 'accumulation by dispossession', by which the state reduces its legal capacities and controls, favouring the economically and politically strong elites as it creates cheap room for their manoeuvrings. But these powerful actors still need the state because the state is the body that restructures the legal order to create what I would call open opportunities. At the same time, the state provides a security-net for powerful market actors and provides legitimacy for their actions ("all actions are made in the name of a democratic state"). This again opens up a space for external as well as internal investment and for privatisation and/or further open access constellations of the commons to be illustrated by the case studies from Zambia.

Case study from Zambia: irrigation on the former pasture

The case study was carried out in Southern Province in Zambia among the Ila agro-pastoralists of the chiefdom Nalubamba (see Haller 2013 for full details; see also Haller and Merten 2008, 2010; Haller et al 2013). It shows how external and internal processes are historically interlinked and, in combination with specific power relations, contribute to the selection of institutions. Also, ideological justifications play a pivotal role in the selection process, leading to the undermining of local as well as ecological resilience in the management of pastures. This case illustrates a land dispute in which a former commons in the Kafue Flats pasture in the Nalubamba chiefdom should be transformed into an irrigation scheme (Haller 2013; Haller et al 2013).

The basis for the conflict was a New Land Act from 1995, which gives the president and the local chiefs the power to hand out leasehold titles for 99 years within the chief's territories. It was based on the neo-liberal ideology demanded by the World Bank to create private property as incentives in order to attract investments. Such lands were about to be allocated in the best pastures in the Kafue Flats, a fact that led to opposition and created great conflicts. The chief had leasehold titles himself and also wanted to attract agricultural projects from the government and foreign investors in a time of a hunger crisis in 2002–2004. The context of this crisis was that in the 1990s many households suffered a great loss of cattle through a *theileria parva* epidemic (also called East Coast Fever; leading to 50 to 80 per cent loss of cattle among the Ila) due to changed flooding by a dam (the area got dryer creating better conditions for ticks that transmit the disease) and a lack of veterinary services (financial crisis of the state). Due to governmental privatisation campaigns based on a law declaring that common property such as cattle should be the private property of orphans and widows, many individuals sold cattle. Under local, older rules it was forbidden to sell animals if a group had less than about 40 heads, numbers being based on notions of securing subsistence and being resilient.

Thus, the change of this older rule led poorer households to divide their small herds of cattle even more and to sell them due to their constant need for quick cash (issues of so-called catastrophic health and other expenditures). Finally, they sold too many animals, remained poor *and* ended up with no cattle at all. The richer households ignored the new law, and tried to keep the cattle herds as an extended family commons. Nevertheless, relative prices for cattle were nationally and locally on the rise due to increasing demands in urban centres. At the same time, pastures were reduced as a consequence of the environmental changes due to the dams discussed above (less inundation in dry years). This, in turn, led to an increase of the value of pastures and coincided with increased demands of individual cattle camps, where cattle did not move, overusing the areas around these camps.

Adding to the degradation were absentee herd owners, who increased their herd size following the rise in relative prices for cattle and pasture. These herd owners tried to make arrangements with individual herders to let their foreign cattle graze for several months. Thus, local users lost access to pastures due to the de facto open access situation of the resource and their diminished power to react to this institutional change. Last but not least, decreasing rains and the loss of subsidies for seeds and inputs led to decreasing yields of maize, while access to fisheries became difficult because commercial users and traders gained more power and thus access to this CPR. This means that the four basic subsistence strategies (three of them based on common property) were severely undermined.

In this context, Chief Nalubamba of the Mbeza territory proposed an irrigation scheme (Mbeza Irrigation Project). An area of 3,600 hectares of pasture land should be converted for irrigation (water from the Kafue River should be used for the production of rice, wheat etc., following the national Poverty Reduction Scheme of the government at the time). But the chief's proposition triggered opposition from a local leader, an ex-major of the Zambian army, who became the spokesperson of the richer cattle owners fearing to lose the pasture. This group questioned the authority of the chief, threatened to kill him and organised counter-rallies to the chief's rallies in support of the project. In this context the chief as well as the opposition leader were acting like pre-colonial first-comer big men, attracting supporters by promising to distribute land, resources and money as well as relief food.

The two sides used different strategies regarding institutions and ideologies for legitimacy. The opposition leader followed an Ila tradition ideology, based on the commons institutions as being threatened. In his discourses he used ideas evoking the traditional way of life and the narrative of a harmful state development. He argued that the Ila had been rich in the past and had a secure system of cattle wealth before the outbreak of *theileria* because the Itezhitezhi dam dried up the floodplain, creating good conditions for the ticks which function as a vector of the disease. He further stated that the project would threaten the use of the commons and the traditional way of life of the cattle breeders by undermining the possibility to secure a livelihood and increase social security via 'tradition'. Here, the basic narrative was ethnicity: The Ila had been successful cattle herders in the past, the problems came from the government and the irrigation project would violate commons and identity of the Ila as cattle herders and undermine their future as an ethnic group. This ethnicity discourse was coupled with the narrative that the chief was not a son of the soil – his ancestors were not former big men ritually and spiritually linked to the territory, but had been installed by British powers during colonial times based on the strategy of indirect rule. Via these discourses and narratives, the opposition leader tried to attract more followers against the irrigation scheme.

The chief himself followed a modernity ideology, also containing different discourses and narratives: His major discourse was based on the notion of modern development. The chief argued that he wanted to improve the livelihoods and social security of 'his' people. Irrigation would reduce their vulnerability to governmental relief food and was portrayed as the alternative to cattle because of the *theileria* outbreak and bad yields of rain fed agriculture. He also argued that this would be done in a participatory way as the pasture caretaker group was on his side (he selected a village headman who was not the acknowledged representative regarding pasture issues).

His narrative was that the conflict is based on a 'class-conflict' pushed by actors stuck in 'tradition'. The rich (those who have cattle), he argued, are against irrigation and do not want the poor to produce food. The chief's narrative could be summarised as follows: "We, the Ila, are poor because of traditional ways of production and because agro-pastoralism is out-dated as a model in the changing ecological situation".

The hidden agendas of both main actors were neither to keep the commons alive nor to provide development to the area, but their institutional choice and the ideologies interlinked with discourses and narratives were influenced by their own interests. The opposition leader had title deeds in the pasture area and was longing for more. He manipulated the caretaker group with the aim to privatise the commons and get payment for the access to cattle camps (one cow each year from every family). In addition, he wanted to get rid of the chief who would never grant leasehold titles to him. The chief, on the other hand, had the power to grant leasehold titles (together with the President) via the new land act and so decided land issues would solve his financial problems. He had no cattle and needed project funds to have financial means to redistribute among his followers to regaining prestige in the community.

Thus, the New Land Act only increased the conflict. Both actors had high bargaining power and central interests, but the chief had more power, forcing the opposition leader to act. The conflict was fought at local, regional and national level (media and government). In the end, the government withdrew its support to the chief, arguing that the project was too conflictive. The irrigation project finally failed. The opposition leader, however, did not receive his leasehold title either (Haller 2013).[1]

Discussion: towards a new institutional political ecology

The case study serves to illustrate how the fusion of NI and PE could be made by using the structural, scalar, historical and feedback-mechanism related analysis of NI and the analysis of how power is constructed and recreated from the three different branches of PE. The case study summarised in the adapted Ensminger model (see Figure 5.2) shows how external factors such as environments (legal, political, economic and natural environment changes), demography (increase in people interested in the area) and technology (road networks, new communication means – both bringing the area closer to urban centres) rise relative cattle prices, which again change the bargaining power of local actors. Outsiders get more power to use pastures; the chief gets more bargaining power via the increase of land value due to privatisation laws; opposition leaders of the groups rich in cattle argue that they want to maintain the commons, which are under price pressure due to the scarcity and increase of relative prices for cattle. In a first step the opposition leader wanted to get rent money for the use of the commons, actually representing a kind of privatisation by itself. But then, in opposition to the chief and against his own high bargaining power based on the new land act, he selected common property as an institution to be defended. This was a strategic move to organise his followers.

These actors thus drive either for common property, open access or privatisation as part of institution shopping processes. This can be legitimated as follows: On the one hand, the chief argues that the planned irrigation project is part of national and international legal schemes to fight hunger and that this can be done on the legal basis of private property like leasehold titles. To further boost his bargaining power, he selected ideologies of modernity, discourses of citizenship and modern development and a narrative of tradition as being backwards. The same is done by the absentee herd owners, who see themselves as citizens who should have access to the state resources. The opposition leader, on the other side, legitimates his claims with the pre-colonial institution of pastures as common property, rooted in the ideology of 'tradition' and the discourse of this being part of 'the Ila people's identity' in which the agro-pastoral traditional way of life – as a positive narrative – is glorified.

Now, what are the triggering factors of this interactive process and how can NIPE provide a better form for analysis? I will focus on how external variables of state related political, legal, economic and ecological environmental factors as well as demography and technology impact changes in relative prices of an area and its resources. Based on these changes, NI can show how the main property rights in land and land related CPRs have historically changed: In pre-colonial times, big men as leaders managed the land and land related resources like spiritual, cultural landscape ecosystems seen as organic nature, based on an ontology and epistemology of an interconnected world with spirits, with whom humans have to arrange themselves. The big men coordinated access and gained prestige in distributing CPRs; they also attracted followers in a political environment, in which one was safer in a larger settlement. This ontology was transformed in colonial times. Chiefs were installed as part of the indirect rule system and had much more power than the former leaders. They still managed access, but no longer to the commons, but to what has become the property of the state – land and land related CPRs, fragmented into several administrative departments with their corresponding institutions.

Pasture was still in the chiefs' hands, but water and all other related resources such as fisheries and wildlife are fragmented and disconnected from the land and pastures. In the first phase after independence we see an increase in state power in Zambia, a socialist African nation. But then, the reduction of copper prices – the main state revenue – finally contributes to high debts for the country, which is then forced to adopt a neo-liberal order of privatisation and open access. However, already during the decline of state control, state property already deteriorated to de facto open access for several CPRs, including pasture in some areas. Since the neo-liberal order, increasing privatisation tendencies developed (private cattle camps, private ownership of cattle), both leading to degradation of pasture areas as a result of uneven distribution processes and strategic behaviour of herders (see part of distributional effects in Figure 5.2). The final step in this privatisation process is the New Land Act that allows for a 99-year leasehold titling in order to attract investments, while for other resources and areas open access prevails. In combination with environmental changes (locally human induced via dams or globally via climate change), the quality of pastures is diminished. Sometimes pastures are also overused due to absentee herd owners and in spite of cattle diseases having reduced the number of animals in the area. It also means that people are experiencing commons grabbing tendencies because the powerful are trying to openly or clandestinely transform pastures into parts of private property via irrigation or by demanding rent.

In order to understand these processes a combination of NI and PE helps to analyse *external factors* and the way these shape *changes in relative prices* as one of the most important triggers for change. The way NI sees external factors operating has been outlined above (see Figure 5.2 as well), while PE helps to understand what dynamics of power operate in the environmental, political, legal and economic aspects in which African states such as Zambia find themselves in. In an inherited and mining resource dependent economy, which gets 'too expensive', as revenues from copper as a main cash source are reduced, a neo-liberal order that calls for privatisation is forced on the country as a remedy for development. From a PE analysis, making use of Ferguson's 'anti-politics machine', these external factors create incentives for dismantling the commons by hiding the economic and political dependencies the country has been in for many decades. Thus, the neo-liberal order influences the change in relative prices from a PE analysis and leads to a commodification of CPRs, which are facing the paradox of the state being present and absent at the same time. These are state resources that are not state controlled and which give elites and powerful actors the options for legal and informal private claims. This is the first element leading to a rise in prices for cattle and pastures. PE helps to analyse the paradox of a neo-liberal order which at the same time reduces state power and increases it by providing tools on all levels for the elites ("repertories of domination", see Poteete and Ribot 2011) to appropriate resources once these are of economic interest.

This is the context that clearly shows the combination of PE and NI. NI outlines how different resources suddenly gain more value via such a process, while PE shows how this is legitimated via powerful discourses and mechanisms of 'anti-politics machines'. The other two external variables are also important as a contribution of NI to the new model, while PE shows the power related issues of in- and out-migration in an area following economic and political state policies and decisions where and what type of technology is implemented. The decision for a dam or a road as well as for technologies and how to use resources – big plantation irrigation schemes or an extended vet service are processes, which do not just increase relative prices, but include power dynamics in the decision making in favour of (or against) such infrastructure. The dam was not discussed with local interest groups and subsidies/grants for veterinary services are given or stopped in a top-down manner without involving local communities and

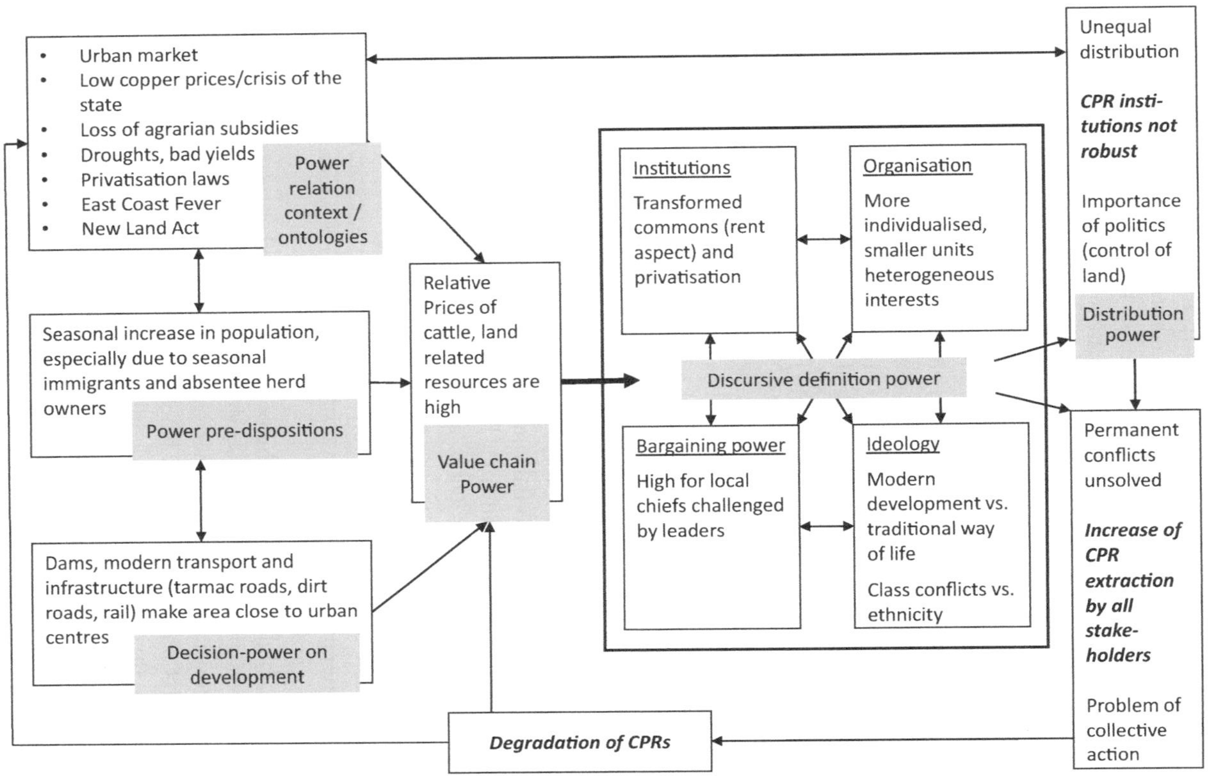

Figure 5.2 NIPE analysis of irrigation conflict.

representatives. These processes lead to an increase in value of pasture and cattle and also need to be looked at in terms of power relations stemming from PE.

In a further step this impacts *bargaining power of local actors* from a NI viewpoint. PE can contribute considerably to the analysis on what this power was based in the past and on what powerful actors rely on now and in the future. From PE, the pre-configurations of power are rooted in colonial contexts of indirect rule. Chiefs as well as other wealthy and powerful actors outside the area can rely on the notion of state citizenship or on the idea of pre-colonial autochthony. These issues combine with how changes in relative prices trigger changes in the power constellations of all these actors. The important point is that a specific actor's power is to be understood in relation to other actors' power in the field. While PE helps to conceptualise power, NI sheds light on this interrelated notion of *bargaining* power, important for the analysis.

In a further step, NIPE represents a better analysis on how the institutional change happens on the basis of *institutional pluralism* and the *process of institution shopping*. It shows which rules of ownership and access to land and land related resources are selected by whom and based on what interests. It also shows how institutions compete, highlighted by several researchers through the issue of how land and belonging are interlinked (see Benjaminsen and Lund eds. 2003; Kuba and Lenz eds. 2006; Derman et al eds. 2007). Interestingly, it seems that different institutions, starting with the basic option of property rights ranging from common, state, private as well as open access constellations, are linked to a large variety of sub-institutions. These have been crafted from pre-colonial times until today: starting from operational rules on the use of the commons in a floodplain from pre-colonial times (traditional legal flexible pluralism in a seasonally changing environment). But since common property is undermined because of colonial state property, a variety of formal and informal rules are created, ranging from different state rules to international rules for development and environment (from veterinary service to the protection of wildlife or fisheries). This produces an institutional pluralism as options and basis for institution shopping. Thus, the case study nicely shows which institutions are selected and transformed.

Again, PE highlights who has the power to select what type of institution on different levels and scales. The case shows that in the local context the chief and the opposition leader use other institutions than those they make use of on the district and national level, where they also have to be present in order to secure the investment or to undermine it. While it is clear that there might also be a process of not really strategic mixing rules called institutional bricolage (Cleaver 2003), I would argue that this case and others our research team studied in comparative research projects, indicate a clear strategic orientation based on the available information of actors. Using the term 'bricolage' might underestimate political manoeuvring and power specific strategic interrelated action. For example, while the chief selects and activates the new land act, the opposition leader strategically focuses on the old common property institutions to attract the richer followers.

This strategic selection process needs *legitimacy based on ideologies including discourses and narratives*. Literature on African land tenure debates and CPR management (Benjaminsen and Lund 2003; Kuba and Lenz 2006; Haller ed. 2010) show that from the past to the present, identity is built and constructed in relation to state institutions, meaning that a scarcity of resources changes relations between 'first-comers' and 'late-comers' in a local context. Here identity production is an aspect brought in by PE, while NI further shows how this institution shopping necessitates ideologies, including discourses and narratives, to legitimise the selection of institutions in a concrete setting, based on concrete economic and political options and preferences. These options and preferences are relational and also dependent on price changes. Therefore, the new institutionalism model of Ensminger/Haller can be further adapted to political issues and informs the analysis of ideologies used.

An obvious ideological dichotomy that is often used is the one between 'modernity' and 'tradition' (Haller 2013), which is similar to Ferguson's 'cosmopolitan' and 'local' (Ferguson 1999). Newer versions of PE also refer to the need to analyse multiple ontologies and epistemologies produced since pre-colonial times in order to strategically employ discourses and narratives for the production of legitimacy (see Escobar 1999; Haller et al 2013; Sheridan 2016). Following are some examples: 'Sons of the ancestors in a living landscape' versus citizenship in a state with fragmented ownership of resource by chiefs; pre-colonial big men versus customary, but colonially transformed leadership; modernity (neo-liberal 'development' based on private property) versus tradition ('traditional way of life') might be ideological and discursive elements to be picked by actors. These are linked to issues of belonging and indicate that autochthony (as discussed in Kuba and Lenz ed. 2006) are also discourses that actors might choose. This is especially relevant in contexts of neo-liberal ontologies and epistemologies, in which local proof of autochthony and belonging signifies belonging to the ones who can legitimately select the institutions used for ownership and access rights to land and related CPRs. While NI provides guidelines on which aspects to look at, PE helps to analyse the power specific context: Neo-liberal and decentralising policies give more power to the construction of the local as they provide legitimacy to the construction of private ownership of local actors, mostly colonially constructed local elites.

Following the *overlapping* areas of the different *PE branches* (the political economy/Marxian and the post-structuralism and the posthuman/feminist branches) power is inherent in pre-existing political economy differences, which are manifest in and backed up via *legitimacy and symbolic production and reproduction of power* of actors. Bringing in the three branches from PE, we can argue that in the Mbeza irrigation project and related conflicts we have a political economy analysis of neo-liberal ideologies, focusing on private ownership as a modern way of development. It matches the construction of first and late comers and makes use of ontologies and epistemologies of powerful actors, either by playing an important role in a spiritual landscape, or for modern capitalist development. This power stemming from using different ontologies and epistemologies is then further used to shape and make use of the institutional pluralism via selection and transformation. This is one of the aspects where NI contributes to a better understanding of systemic interrelations.

In a further step, NI and PE can be married, also in regard to the *distributional and behavioural outcome leading to commons and resilience grabbing*. The chosen institutions can be looked at via Ostrom's design principles as to whether they are robust and lead to sustainable resource use or not (Ostrom 1990). These give an indication of the ecological outcome from the use of CPRs. But again, the use is depending on an institutional and power specific distribution of these resources – not analysed by Ostrom but by Ensminger and better conceptualised by PE – between actors thus impacting their further behaviour (see Figure 5.2). However, as criticised by the newer PE approach, this analysis is based on an anthropocentric type of analysis, leading to a critique of the perception of general commodification of resources. These are analysed as being subjugated to a detachment process from the wider landscape ecosystem context and free for sale, backed by the above-mentioned notion of modernity by which not only foreign, but also local investments are justified. These investments, I argue, undermine local livelihoods because CPRs in cultural landscapes become legally and also physically fragmented and access to them is undermined for the less powerful.

At the same time, 'expectations of modernity' do not come true as there are fewer jobs and less income, while subsistence crop production and access to the commons is restricted or made impossible. Already existing asymmetries are strengthened and new ones are created because many households loose access to buffer resources – which common property represented – important sources of everyday survival during times of crisis, while having to take higher risks in troubled times to be able to make a livelihood. Thus, we do not just see commons, but resilience grabbing as well – not just the removal of access to the commons, but undermining the ability to recover after a shock. In PE, this process is related to the old debate on degradation and marginalisation with a link to loss of resilience rooted in political constellations rather than in a mere *Tragedy of the Commons*.

What is missing so far – and this answers the question of where to take the debate – is to clearly analyse how local actors view this process, how they view the deals and conceptualize aspirations. Again, this is where NI can be married to PE studies: It is about the way that environmental contestations and conflicts can lead to the building of *new environmental identities and social or identity movements and constitutionality*. Ideas of being local and defending local resources as well as ideas of a new deal and a new way of sharing are rooted in this process (see Robbins 2004; see Haller and Merten 2008, 2018). It is interesting to see, then, what type of strategies and resistance as well as institution shopping local actors adopt in order to buffer the problem of the commons grabbing. This is then vested in strategies to change and devise own institutions that might lead to a bricolage (Cleaver 2003), but also to a strategic selection of options based on bargaining power and possible within institutional pluralism. As is shown in a case not covered here on the problem of overfishing in the Kafue Flats, a crisis can also lead to what is called constitutionality, the conscious bottom-up institution building process that enables all actors to participate and thus create a sense of ownership of the institution building process (see Haller et al 2016, 2018).

Conclusion

To conclude, NIPE as a new approach indicates that NI in social anthropology profits from the debate on power relations and the production of power on several scales stemming from PE, while PE profits from NI as it outlines more concisely and in a systemic model, how changing relative price constellations occur. NI then continues to look at how this leads to changing interactive bargaining power positions of actors, their perceptions and selection of institutional options backed by ideological justifications, which are interrelated in a systemic way, leading to distribution and behavioural strategic constellations. This does not just apply to the powerful, but to all actors, even the less powerful ones. It will be important to study such processes and to discuss what the dominant strategies selected by the less powerful local actors are in a regaining of resilience in the use of cultural landscapes. Communal titles and mapping of resources, new discussions on the local level about institutions for the management of resources (by-laws, local conventions) as well as refocusing on pre-colonial ontologies might be important strategies in a 'glocal' world. I do not speak in favour of formalisation, but of true participation. Despite local political asymmetries, most actors will thus gain a sense of ownership of the institution building process, that has been framed as constitutionality (Haller et al 2016, 2018). Studying such processes can also lead to policy driven ideas to support the protection of local resource users from the commons grabbing and to strengthen resilience of their livelihoods and their cultural landscape ecosystems.

Furthermore, NI brings in a better analysis of: a) the history of power relations and outcomes in the use of CPRs in context; b) external changes, relative price developments and power relations; c) how actors and actor groups shape and select institutions as well as legitimise processes (how institution shopping interrelates with ideology via discourses and narratives of different ontologies and epistemologies as resources of legitimacy); and d) how this triggers distribution, consumption and impact on the resource base of cultural landscapes and their environmental status. NI thus can help explain path dependency and strategic behaviour, while PE adds the power specific element of such behaviour leading to 'sustainable use' or 'overuse' of CPRs. In the NI model this feeds back on the external variables, mostly on the environment. But as I have outlined elsewhere (Haller 2013), it will feed back on relative prices as well as on the internal constellation as we have also seen in the case presented (i.e. increased price for land). For these systemic elements and interactions, PE contributes power analysis and the way how discourses and narratives are produced and employed.

However, the outcome cannot be predicted conclusively: We are still facing not the tragedy, but the drama of the commons: The irrigation project itself did not materialise, but the project might still be waiting on the shelves of the next chief to be reactivated if land prices increase ... NIPE also means that the 'show' will go on, but with certain optional predictions and not really in the form of institutional bricolage (Cleaver 2003), twilight institutions (Lund 2007) or environmentalities (Agrawal 2005) as state embodied subject-building.

Note

1 For reason of space not all details of the case can be presented here. For the full discussion, see Haller 2013, Chapter 8.

References

Acheson, J. (1989) 'Management of Common Property Resources', in Plattner, S. (ed), *Economic Anthropology* (pp 351–378). Stanford: Stanford University Press.

Agrawal, A. (2005) *Environmentality, Technologies of Government and the Making of Subjects*. Durham and London: Duke University Press.

Bailey, S. and Bryant, R. (2005) *Third World Political Ecology: An Introduction*. London: Routledge.

Benjaminsen, T.A. and Lund, C. (2003) 'Formalisation and Infromalisation of Land and Water Rights in Africa. An Introduction', in Benjaminsen, T.A. and Lund, C. (eds), *Securing land Rights in Africa* (pp 1–10). London and Portland, OR: Frank Cass.

Berkes, F. (1999) *Sacred Ecology. Traditional Ecological Knowledge, and Resource Management*. Philadelphia: Taylor & Francis.

Berkes, F. and Folke, C. (eds) (2000) *Linking Social and Ecological Systems*. Cambridge: Cambridge University Press.

Berry, S. (1993) *No Condition is Permanent. The Social Dynamics of Agrarian Change in Sub-Saharan Africa*. Madison, WI: The University of Wisconsin Press.

Biersack, A. and Greenberg, J.B. (eds) (2006) *Reimagining Political Ecology*. Durham: Duke University Press.

Blaikie, P.M. and Brookfield, H. (eds) (1987) *Land Degradation and Society*. London: Methuen.

Bodley, J.H. (2014 [1975]) *Victims of progress*. Lanham, MD: Rowman & Littlefield.

Brockington, D., Duffy, R., and Igoe, J. (2008) *Nature Unbound: The Past, Present and Future of Protected Areas*. London: Earthscan.

Brosius, J.P. (1999) 'Green dots, pink hearts: displacing politics from the Malaysian rain forest', *American Anthropologist*, vol 101, no 1, pp 36–57.

Brosius, P.J., Tsing, A., and Zerner, C. (eds) (2005) *Communities And Conservation: Histories and Politics of Community-Based Natural Resource Management*. Walnut Creek, CA: Altamira.

Bryant, R.L. and Bailey, S. (1997) *Third World Political Ecology*. New York: Routledge.

Chanock, M. (1985) *Law, Custom, and Social Order: The Colonial Experience in Malawi and Zambia*. Cambridge: Cambridge University Press.

Chanock, M. (2005) 'Customary Law, Sustainable Development and the Failing State', in Oerebech, P., Bosselman, F., Bjarup, J., Callies, D., and Chanock, M. (eds), *The*

Role of Custimary Law in Sustainable Development (pp 338–383). Cambridge: Cambridge University Press.

Cleaver, F. (2003) 'Reinventing Institutions: Bricolage and the Social Embeddedness of Natural Resource Management', in Benjaminsen, T.A. and Lund, C. (eds), *Securing land Rights in Africa* (pp 11–30). London and Portland, OR: Frank Cass.

Cleaver, F. and Frank, T. (2005) *How Institutions Elude Design: River Basin Management and Sustainable Livelihoods*. Bradford: University of Bradford.

Crane, T.A. (2010) 'Of models and meanings: cultural resilience in social-ecological systems', *Ecology and Society*, vol 15, no 4, art 19.

De Sardan, O. (2013) 'Embeddedness and informal norms: Institutionalisms and anthropology', *Critique of Anthropology*, vol 33, no 3, pp 280–299.

De Schutter, O. (2011) 'How not to think of Land Grabbing: three critiques of large scale investments in farmland', *The Journal of Peasant Studies*, vol 38, no 2, pp 249–280.

Descola, P. (2005) 'On anthropological knowledge', *Social Anthropology*, vol 13, no 1, pp 65–73.

Descola, P. (2013) *Beyond Nature and Culture*. Chicago, IL: University of Chicago Press.

Derman, B., Odgaard, R., and Sjaastad, E. (eds) (2007) *Conflicts over Land and Water in Africa*. Oxford/Pietermaritzburg/East Lansing, MI: Oxford/KwaZulu-Natal Press/ Michigan State University Press.

Ellen, R. (1982) *Environment, Subsistence and System. The Ecology of Small Scale Social Formations*. Cambridge: Cambridge University Press.

Ensminger, J. (1992) *Making a Market. The Institutional Transformation of an African Society*. Cambridge: Cambridge University Press.

Escobar, A. (1999) 'After nature: steps to an antiessentialist political ecology', *Current Anthropology*, vol 40, no 1, pp 1–30.

Escobar, A. (2011) *Encountering Development: The Making and Unmaking of the Third World*. Princeton, NJ: Princeton University Press.

Fabinyi, M., Evans, L., and Foale, S.J. (2014) 'Social-ecological systems, social diversity, and power: insights from anthropology and political ecology', *Ecology and Society*, vol 19, no 4, art 28.

Fairhead, J. and Leach, M. (1996) *Misreading the African Landscape. Society and Ecology in a Forest-Savannah Mosaic*. Cambridge: Cambridge University Press.

Feeney, D., Berkes, F., McCay, B.J., and Acheson, J.M. (1990) 'The Tragedy of the Commons: twenty-two years later', *Human Ecology*, vol 18, no 1, pp 1–19.

Ferguson, J. (1994 [1990]) *The Anti-Politics Machine. "Development" and Bureacuratic Power in Lesotho*. Minneapolis, MN: University of Minnesota Press.

Ferguson, J. (1999) *Expectations of Modernity. Myths and Meanings of Urban Life on the Zambian Copperbelt*. Berkeley, CA: University of California Press.

Ferguson, J. (2006) *Global Shadows: African in the Neoliberal World Order*. Durham and London: Duke University Press.

Fletcher, R. (2010) 'Neoliberal environmentality: towards a poststructuralist political ecology of the conservation debate', *Conservation and society*, vol 8, no 3, pp 171–181.

Galvin, M. and Haller, T. (eds) (2008) *Perspectives of the Swiss National Centre of Competence in Research (NCCR) North-South*. Bern: University of Bern.

Gardner, A. (2003) 'The new calculus of Bedouin pastoralism in the Kingdom of Saudi Arabia', *Human Organization*, vol 62, no 3, pp 267–276.

Gibson, C.C. (1999) *Politicians and Poachers: The political economy of wildlife policy in Africa*. Cambridge: University of Cambridge Press.

Graeber, D. (2015) 'Radical alterity is just another way of saying "reality". A reply to Eduardo Viveiros de Castro', *HAU: Journal of Ethnographic Theory*, vol 5, no 2, pp 1–41.

Haller, T. (2007a) Understanding Institutions and their Links to Resource Management from the Perspective of New Institutionalism. *NCCR North-South Dialogue* 2. Bern: NCCR North-South.

Haller, T. (2007b) 'Is there a culture of sustainability? What social and cultural anthropology has to offer 15 years after Rio', in Burger, P. and Kaufmann-Hayoz, R. (eds), *15 Jahre nach Rio – Der Nachhaltigkeitsdiskurs in den Geistes- und Sozialwissenschaften: Perspektiven – Leistungen – Defizite* (pp 329–356). Bern: Schweizerische Akademie der Geistes- und Sozialwissenschaften Bern.

Haller, T. (ed.) (2010) *Disputing the Floodplains: Institutional Change and the Politics of Resource Management in African Floodplains*. Leiden and Boston, MA: Brill.

Haller, T. (2013) *The Contested Floodplain. Institutional Change of the Commons in the Kafue Flats, Zambia*. Lanham, MD: Lexington Books.

Haller, T. (2016) 'Managing the commons with floods: the role of institutions and power relations for water governance and food resilience in African Floodplains', in Ostegard, T. (ed.), *Water and Food – Africa in a Global Context* (pp 369–397). Uppsala: The Nordic African Institute. I.B. Tauris.

Haller, T. (2017) 'Perceptions and control of assemblage in a "Glocal" World'. *Dialogues in Human Geography*, vol 7, no 2, pp 207–211.

Haller, T. and Galvin, M. (2008) 'Introduction: the problem of participatory conservation', in Galvin, M. and Haller, T. (eds), *People, Protected Areas and Global Change: Participatory Conservation in Latin America, Africa, Asia and Europe* (pp 13–34). Bern: Perspectives of the Swiss National Centre of Competence in Research (NCCR) North-South, University of Bern, *Geographica Bernensia*, 3.

Haller, T. and Galvin, M. (2011) 'Challenges for participatory conservation in times of global change: Lessons from a comparative analysis and new developments', in Wiesmann, U. and Hurni, H. (eds), *Research for Sustainable Development: Foundations, Experiences, and Perspectives. Perspectives of the Swiss National Centre of Competence in Research (NCCR) North-South* (pp 467–503). Vol. 6. *Geographica Bernensia*. Bern: University of Bern.

Haller, T. and Merten, S. (2008) '"We are Zambians – don't tell us how to fish!" Institutional Change, Power Relations and Conflicts in the Kafue Flats Fisheries in Zambia?' *Human Ecology*, vol 36, no 5, pp 699–715.

Haller, T. and Merten, S. (2010) '"We had cattle and did not fish and hunt anyhow!?" Institutional Change and Contested Commons in the Kafue Flats Floodplain (Zambia)', in Haller, T. (ed.), *Disputing the Floodplains: Institutional Change and the Politics of Resource Management in African Floodplains* (pp 301–360). Leiden and Boston, MA: Brill.

Haller, T. and Merten, S. (2018) 'Crafting our own rules: constitutionality as a bottom-up approach for the development of by-laws in Zambia', *Human Ecology*, vol 46, no 1, pp 3–13.

Haller, T., Belsky, J.M., and Rist, S. (2018) 'The constitutionality approach: conditions, opportunities, and challenges for bottom-up institution building', *Human Ecology*, vol 46, no 1, pp 1–2.

Haller, T., Fokou, G., Mbeyale, G., and Meroka, P. (2013) 'How fit turns into misfit and back: institutional transformations of pastoral commons in African floodplains', *Ecology and Society*, vol 18, no 1, p 34. http://dx.doi.org/10.5751/ES-05510-180134.

Haller, T., Galvin, M., Meroka, P., Alca, J., and Alvarez, J. (2008) 'Who gains from community conservation? Intended and unintended costs and benefits of participative approaches in Peru and Tanzania', *Journal of Environment and Development*, vol 17, no 2, pp 118–144.

Haraway, D.J. (2008) *When Species Meet* (vol 224). Minneapolis, MN: University of Minnesota Press.

Haraway, D.J. (2016) *Staying with the Trouble: Making Kin in the Chthulucene*. Durham, Duke University Press.

Hardin, G. (1968) 'The tragedy of the commons', *Science*, vol 162, no 3859, pp 1243–1248.

Harvey, D. (2005) *A Brief History of Neoliberalism*. Oxford: Oxford University Press.

Kuba, R. and Lenz, K. (eds) (2006) *Land and the Politics of* Belonging. Leiden: Brill.

Lenz, K. (2006) 'First-comers and late comers: indigenous theories of land ownership in the West-African Savanna', in Kuba, R. and Lenz, K. (eds), *Land and the Politics of Belonging* (pp 34–57). Leiden: Brill.

Li, T.M. (2010) 'Indigeneity, capitalism, and the management of dispossession?', *Current Anthropology*, vol 51, no 3, pp 385–414.

Lund, C. (ed.) (2007) *Twilight Institutions: Public Authority and Local Politics in Africa* (Development and Change Books). London: Blackwell Publishing.

Mamdani, M. (1996) *Citizen and Subject; Contemporary Africa and the Legacy of Late Colonialism*. Princeton Series in Culture/Power/History. Cape Town: David Philip.

Mhlanga, L., Nyikahadozi, K., and Haller, T. (eds) (2014) *Fragmentation of Natural Resources Management: Experiences from Lake Kariba*. Berlin: LIT.

Moore, S.F. (1986 [2000]) *Law as Process: An Anthropological Approach*. Münster: LIT.

Narotzky, S. (1997) *New Directions in Economic Anthropology*. London: Pluto Press.

Netting, R. McC (1993) *Smallholders, Householders: Farm Families and the Ecology of Intensive, Sustainable Agriculture*. Cambridge: Stanford University Press.

Neumann, R.P. (1998) *Imposing Wilderness. Struggles over Livelihood and Nature Preservation in Africa*. Berkeley, CA: University of California Press.

North, D. (1990) *Institutions, Institutional Change and Economic Performance*. Cambridge: Cambridge University Press.

Orlove, B.S. (1980) 'Ecological anthropology', *Annual Review of Anthropology*, vol 9, no 1, pp 235–273.

Ostrom, E. (1990) *Governing the Commons. The Evolution of Institutions for Collective Action*. Cambridge: Cambridge University Press.

Ostrom, E. (2005) *Understanding Institutional Diversity*. Princeton, NJ: Princeton University Press.

Ostrom, E., Gardner, R., and Walker, J. (1994) *Rules, Games, and Common-Pool Resources*. Ann Arbor, MI: University of Michigan Press.

Paulson, S. (2003) *Political Ecology Across Spaces, Scales, and Social Groups*. New Brunswick, NJ: Rutgers University Press.

Paulson, S., Gezon, L.L., and Watts, M. (2003) 'Locating the political in political ecology: An introduction', *Human Organization*, vol 62, no 3, pp 205–217.

Paulson, S. and Gezon, L.L. (2005) *Political Ecology across Spaces, Scales, and Social Groups*. New Brunswick, NJ: Rutgers University Press.

Peters, P. (2013) 'Conflicts over land and threats to customary tenure in Africa', *African Affairs*, vol 1, pp 1–20.

Poteete, A.R. and Ribot, J.C. (2011) 'Repertoires of domination. Decentralization as process in Botswana and Senegal', *World Development*, vol 39, no 3, pp 439–449.

Rappaport, R. (1968) *Pigs for the Ancestors: Ritual Ecology of a New Guinea People*. New Haven, CT: Yale University Press.

Ribot, J.C. and Peluso, N.L. (2003) 'A theory of access?', *Rural Sociology*, vol 68, no 2, pp 153–181.

Robbins, P. (2004) *Political Ecology: A Critical Introduction*. Malden, MA: Blackwell.

Schulz, K.A. (2017). 'Decolonizing political ecology: ontology, technology and "critical" enchantment', *Journal of Political Ecology*, vol 24, no 1, pp 125–143.

Scott, J. (1998) *Seeing Like a State: How Certain Schemes to Improve Human Condition Have Failed*. New Haven, CT: Yale University Press.

Scott, J. (2008) *Weapons of the Weak: Everyday Forms of Peasant Resistance*. New Haven, CT: Yale University Press.

Sheridan, M. (2016) 'The politics of rain: Tanzanian farmers' discourse on climate and political disorder', in Crate, S.A. and Nuttall, M. (ed), *Anthropology and Climate Change: From Actions to Transformations* (pp 228–240). New York and London: Routledge.

Simon, R. (2015) *Gramsci's Political Thought: An Introduction*. London: Lawrence and Wishart.

Srinivasan, K. and Kasturirangan, R. (2016) 'Political ecology, development, and human exceptionalism', *Geoforum*, vol 75, pp 125–128.

Sullivan, S. (2017) 'What's ontology got to do with it? On nature and knowledge in a political ecology of the "green economy"', *Journal of Political Ecology*, vol 24, no 1, pp 217–242.

Svarstad, H., Benjaminsen, T.A., and Overå, R. (2018) 'Power theories in political ecology', *Journal of Political Ecology*, vol 25, no 1, pp 350–363.

Swyngedouw, E. (2009) 'The political economy and political ecology of the hydro-social cycle?', *Journal of Contemporary Water Research & Education*, vol 142, no 1, pp 56–60.

Toulmin, C. (2009) 'Securing land and property rights in Sub-Saharan Africa: the role of local institutions', *Land Use Policy* vol 26, pp 10–19.

Tsing, A.L. (2001) 'Nature in the making', in Crumley, C. (ed.), *New Directions in Anthropology and Environment* (pp 3–23). Walnut Creek, CA: Altamira.

Vayda, A. and Walters, B. (1999) 'Against political ecology', *Human Ecology*, vol 27, no 1, pp 167–179.

Williamson, O.E. (1989) 'Transaction cost economics', in Schmalensee, R. (ed.), *Handbook of Industrial Organization*, vol 1 (pp 135–182). Amsterdam: North-Holland.

Wolf, E. (1982) *Europe and the People Without History*. Cambridge: Cambridge University Press.

Zimmerer, K.S. and Basset, T.J. (2003) *Political Ecology*. New York and London: Guilford Press.

Zoomers, A. (2010) 'Globalisation and the foreignisation of space: seven processes driving the current global land grab', *The Journal of Peasant Studies*, vol 37, no 2, pp 429–447.

Part II

European examples from past and present

Top: Pasture and farmland on the "Kirchenfeld" of the Bernese Civic Corporation. Aquarelled pen and ink drawing by Wilhelm Stettler. Replica provided by the Historic Museum of Bern. For a detailed analysis of the common fields of the Bernese Civic Corporation in the 17th century, see Chapter 9.

Bottom left: Fisherman of Sami origin during cod fisheries. Picture by Angelika Lätsch. For more information on bottom-up institution building and identity among Coastal Sami in Northern Norway, see Chapter 11.

Bottom right: Common alpine pasture in Sumvitg, Switzerland. Picture by Gabriela Landolt. For a detailed analysis of the use of alpine pasture management in Sumvitg, Switzerland, see Chapter 12.

6 Common challenges, different fates. The causal factors of failure or success in the commons

The pre-modern Brecklands (England) and the Campine (Southern Low Countries) compared

Maïka De Keyzer

Introduction

Why do commons fail?

> Each man is locked into a system that compels him to increase his herd without limit in a world that is limited. Ruin is the destination towards which all men rush, each pursuing his own best interest in a society that believes in the freedom of the commons. Freedom in a commons brings ruin to all.
>
> (Hardin 1968: 1244)

This paradigm that was vocalised the best by Garrett Hardin, has long held sway. Communal property and unsustainability were considered inextricably linked. This point of view has been fundamentally criticised and overruled by now. Common-pool resources (CPRs) were managed intensively by common-pool institutions (CPIs). Open access situations, which Hardin actually described, were extremely rare. CPIs were a common feature in preindustrial Europe and were able to manage natural resources sustainably for decades or even centuries. From Finland to Spain, case studies can be found where local communities were successful in exploiting natural resources in a sustainable manner (Sundberg 2002; Casari 2007; Lana Berasain 2008; Olivier 2013). The key to success were strict management and delimitation of users. Nobel Prize laureate Elinor Ostrom showed that freedom in the commons does not exist. She discovered that most societies created CPIs to regulate and monitor the use of CPRs, therefore preventing 'freedom in the commons' (Ostrom 1997). Joachim Radkau took this reasoning even a step further and claimed that:

> the opposite of Hardin's argument can be true, namely that the care of future generations is in better hands with communities than with individuals who

> think only of their own lifespan. They show a frightening willingness to sacrifice themselves for larger communities and distant goals.
>
> (Radkau 2008: 73)

Following Ostrom, the current academic debate is strongly focused on discovering the key to success of communal property and action. Ostrom proposed seven design principles which must be met in order to reach a sustainable management of common resources and successful CPIs (Ostrom 1997). Her model is constantly tweaked, corrected, ameliorated and complemented by countless examples and case studies (Berkes 1989; De Moor, Shaw-Taylor, and Warde 2002; Sundberg 2002; Anderies, Janssen, and Ostrom 2004; Casari 2007; Haller and Chabwela 2009; Rodgers et al. 2011; Laborda Peman and De Moor 2013; Bollier and Helfrich 2015).

Nevertheless, in the new paradigm that has replaced Hardin's thesis, the pendulum has swung too far to the other extreme. Commons are consistently portrayed as robust and successful, apart from some exceptional failures (De Moor 2015). This is predominantly the result of a biased selection of case studies and a lack of a clear definition of successful CPIs. Most studies do not use clear conditions of success. Others consider commons as successful as long as the institution lasts longer than two centuries or if the landscape is not fundamentally degraded. This rather ad hoc appreciation of commons is too vague or inconsistent and includes a large share of CPIs that have a dubious claim to success. Commons should be tested using uniform criteria, including ecological resilience, an equitable distribution of communal benefits and the robustness of the common property regime (De Keyzer 2018b).

In historical reality several CPIs all over the world and through time experienced serious degradation of resources, faced ecological problems and disasters because of mismanagement and were confronted with stifling free rider and non-compliance issues. Most of these CPIs did not fail because of freedom in the commons. Other factors could and did lead to failing commons. Looking into the reasons of failure is equally as important as analysing success in order to understand collective action.

Therefore, this chapter explores two regions with very similar challenges and different fates, in order to analyse what makes communal action successful and what makes them vulnerable and likely to fail. I will look into the ability of both regions to manage resources sustainably and fend off environmental hazards, since these two factors are the core business of CPIs. Other factors such as equitable access and enclosure struggles, which are important as well, will not be discussed in this particular chapter (for this debate see De Keyzer 2018b). The case studies selected are the Brecklands in East Anglia (Norfolk and Suffolk, England) and the Campine in the Southern Low Countries. Both regions were located in the European coversand belt. As a result they possessed the same ecosystem and climatic conditions, which confronted them with similar challenges. Both were labelled marginal economies, since the sandy infertile conditions did not provide the circumstances for intensive arable production. The landscape was dominated by vast heathlands, suitable for extensive grazing, but vulnerable and prone to degradation, which led to disastrous sand drifts if managed unsustainably. Both opted for collective management from the later Middle Ages onwards and these common property regimes survived until the nineteenth century.

In the Campine, the CPIs have proven to be successful in managing the different ecological resources sustainably between 1250 and 1750, while at the same time coping with the constant hazard of sand drifts. The diverse heathlands were maintained and no fundamental sand drifts occurred between the later Middle Ages and the eighteenth century. The Brecklands are located on the opposite side of the spectre and can be labelled a 'tragedy of the commons'. After a period of more or less sustainable management of the landscape between the later Middle Ages and the end of the sixteenth century, the CPIs failed in their task. During the seventeenth century, the landscape was degraded fundamentally, diverse resources were significantly reduced, and consecutive waves of sand drifts affected the Brecklands. This chapter will explore, which institutional and societal factors played a role in determining the outcome of collective action, by comparing both regions' CPIs and the socio-economic and political constellations of both societies.

The Campine versus the Brecklands

The most important task of CPIs is the management of common resources in a sustainable manner and preventing a 'tragedy of the commons'. 'Sustainable' refers to a development that meets the needs of the present without compromising the ability of future generations to meet their own needs (WCED 1987: 43). In the European coversand belt, the resources in need of protection were the diverse heathlands. Although formerly wooded, the wood cover had already diminished significantly before the development of mature CPIs during the thirteenth century because of active clearance for construction and exploitation of new areas and grazing land (Bastiaens and Deforce 2005). The heathlands during the Middle Ages were not the typical purple heath fields nor barren bracken fields that are most often depicted in nineteenth century paintings and engravings. Pre-modern heathlands can be labelled as green heathlands, because of the diverse flora that could provide fodder for different types of animals, sods and clippings for manure, thatches and building material and peat or wood for fuel (Spek 2004). This fragile ecosystem was quite easily disturbed by human exploitation and mismanagement. The heathlands could rapidly be reduced to ecologically degraded bracken or heather fields, or even to open, barren, sandy landscapes, that started to drift. Sand drifts are a good proxy for a failed management of heathland landscapes in the past, since they leave clear trails and are clear indicators of a deteriorated ecosystem (Koster 2007; Fanta and Siepel 2010).

These landscapes are called landscapes of risk (Mauelshagen 2007; Bankoff 2009). They pose a continuous and recurrent risk. Even though these ecological circumstances are threatening, they do allow societies to anticipate and respond to these challenges. In the coversand belt, communities learned the hard way

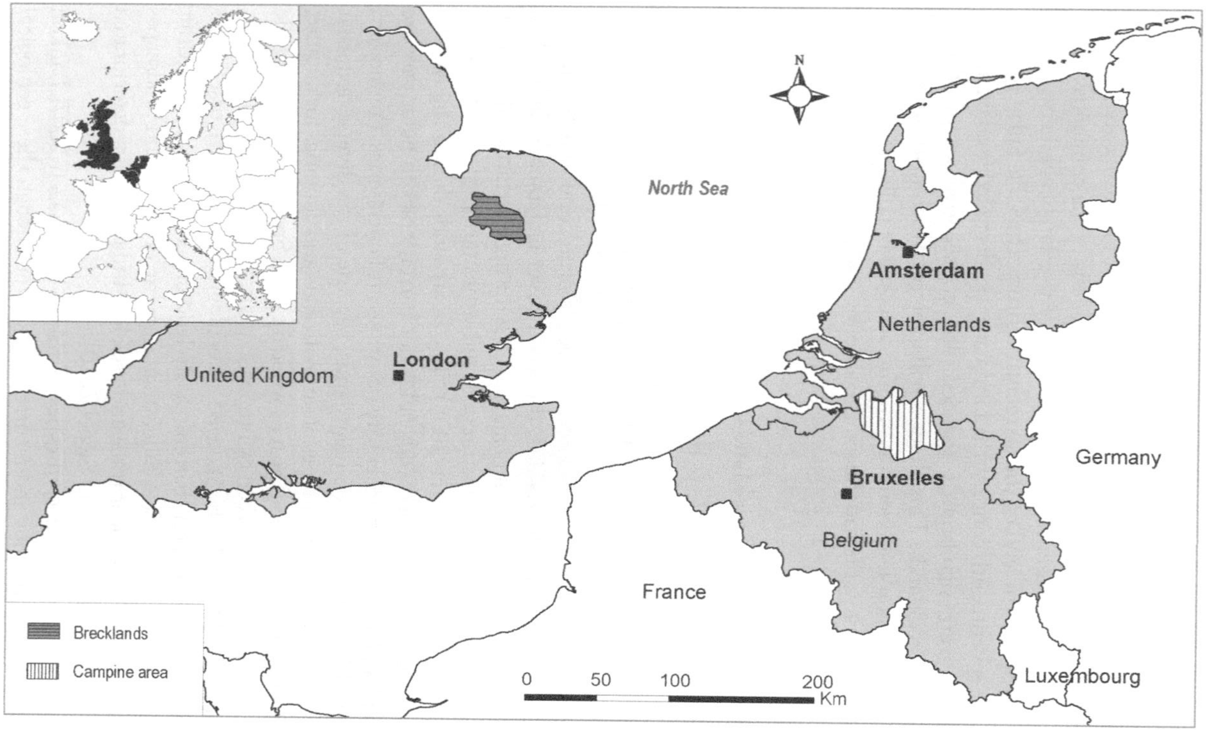

Figure 6.1 Map showing the selected case studies: The Brecklands and the Campine.
Source: Map compilation and design by Ulla Gämperli Krauer (2018), CDE, University of Bern, Switzerland. Map base: Iason Jongepier (2014), GIStorical Antwerp, Centre for Urban History, University of Antwerp, Belgium. Further geodata source: OSM data by OSMaxx (2018). Geometalab, HSR University of Applied Sciences Rapperswil, Switzerland.

that uncovering the soil without windbreakers or constructions to halt the sand led to disastrous sand drifts. The first rural communities that arrived between the Roman period and the eleventh century were still exploring and adapting their exploitation strategies; they were caught in the sand, leading to the destruction and abandonment of several sites (Bateman and Godby 2004; Derese et al. 2010; Heidinga 2010; De Keyzer and Bateman 2018). Some communities in the coversand belt, nevertheless, developed coping mechanisms. These comprised building wind breakers, such as woodlands, hedges and fences preventing the reduction of plant cover, limiting grazing on fragile spots such as the dunes and collective planting on drift sand locations (De Keyzer 2016).

In the Campine, the management of the heathlands was done sustainably by the CPIs. Before the arrival of CPIs and formal village communities, which started to develop from the late twelfth and thirteenth century, the communities learned about the challenging landscape. After the disastrous failures during the tenth century with sand drifts such as the one in Pulle, destroying a part of the village and the productive land of the site, the communities and CPIs invested time, effort and money in maintaining the vital heathlands (Eggermont et al. 2008; Derese et al. 2010). The landscape was not stabilised entirely, as the layers of sand in the late medieval and early modern sand dunes show. However, the communities living in the eighteenth century were able to use the commons in the same way as their predecessors living in the fourteenth century, and no disastrous sand drifts destroying villages and productive land have been registered, neither in the written nor the geological source material (De Keyzer 2016).

In the Brecklands, however, the management of the landscape was less successful. The heathlands became degraded to such an extent that increasingly less agricultural activities were viable in the long run. By the seventeenth century, plots of arable land were abandoned, heathlands were overstocked and vegetation cover reduced. The smallholders and tenants complained about the reduced possibilities and the reduction of fodder for their cattle (Postgate 1960; Williamson 2007; Whyte 2011). Seventeenth- and eighteenth-century observers, in addition, recorded almost apocalyptic descriptions of the Breckland landscapes. John Evelyn wrote in 1677 that 'travelling sands have so damaged the country, rolling from place to place, and like the sands in the deserts of Libya, quite overwhelmed some gentlemen's whole estates' (Evelyn 1952). Even though these agronomists had their motives to paint a dire picture of the Brecklands in order to introduce their agricultural strategies, their point of view is to some extent supported by the geological evidence. At least four disastrous periods of acute sand drifts occurred between Anglo-Saxon times and the twentieth century. The first occurred 1400 years ago when the Anglo-Saxon communities started to exploit the land more intensively, the second during the High Middle Ages around 900 and 800 years ago, the third during the seventeenth century and finally during the nineteenth century (De Keyzer and Bateman, 2018). The lessons from the past did not lead to a society that learned to cope with the constant threat in the Brecklands.

It remains to be questioned, why these two societies and CPIs diverged so fundamentally with regard to managing the landscape and coping with the environmental hazards. In this chapter, I will explore several factors that could have contributed to the diverging paths. First, the design principles of the formal CPIs are analysed. Second, the social and political barriers to engage in collective action are investigated. Finally, the interests of the different stakeholders in both societies are discussed.

Common pool institutions: formal versus informal

The management of the landscape in common-pool regimes was the responsibility of CPIs. While some CPIs operated independently from village governments and political structures, both regions had CPIs that were firmly embedded within the existing village structures (Hoppenbrouwers 2002; Shaw-Taylor 2002; Allison 1957). In the Campine area, the villages had received privileges from the Dukes of Brabant to form village governments. These institutions were organised partly from the bottom-up with village representatives such as aldermen, and partly from the top-down with the local lords and their representatives such as the bailiff and steward. These village communities in addition received the privilege to use, control and manage the commons, which constituted between 75 and 90 per cent of the village land (Enklaar 1941; Steurs 1993; Verboven, Verheyen, and Hermy 2004). In the Brecklands, similar structures existed that had the same responsibilities. During the pre-modern period, the key institution in regulating the exercise of use rights on common land in England were local seigniorial courts or manor courts. These were local meetings called by the lord of the manor and presided over by his steward, which those holding land in the manor were required to attend (Rodgers et al. 2011).

Seven design principles defined by Elinor Ostrom (1997)

1. clearly defined boundaries (physical boundaries as well as delimitation of community of users);
2. congruence between appropriation and provision rules and local conditions;
3. collective choice arrangements;
4. monitoring;
5. graduated sanctions;
6. conflict resolution mechanisms;
7. minimal recognition of rights to organize.

These institutions for collective action implemented the design principles of Elinor Ostrom (see Figure 6.2). The community of users was delimited, even though in fundamentally different ways. In the Campine, outsiders such as immigrants and members of neighbouring communities were excluded, but all the members of the village community, regardless of the socio-economic status, were included (De Keyzer 2018b). In the Brecklands, access was more complicated and exclusive. The most important communal benefits were grazing rights and this was organised in the fold course system. Access had to be granted by the lord, to put sheep in the communal flocks that could graze on the open fields, brecks and waste lands. Even though in theory nobody was formally excluded, a barrier on access was introduced. By the seventeenth century, this barrier had become significant, since lords reserved the valuable spots for their own leaseholders, therefore excluding most of the tenants in the village (Allison 1957; Postgate 1960; De Keyzer 2013).

The CPIs merged together with the village structures and therefore were recognised by the local lords as well as the state. The manorial courts or village aldermen benches could be used to solve conflicts, and peasants had access to complementary courts, such as the royal courts, to plead their case (Whyte 2009). The penalty or sanctioning of offences was mostly in the hands of the lord, via their courts, while the monitoring was transferred to representatives. In the Campine, these were officials that were appointed by the village government (Enklaar 1941). In the Brecklands, this remains more unclear.

Rules were defined by collective choice arrangements. According to Tobias Haller et al. decisions have to be made not only by consent of the community, but by active involvement of the members of the community of users (Haller, Acciaioli, and Rist 2012). In both regions, however, environmentality was applied, i.e. a couple of representatives can make up the rules that are then accepted and internalised by the community as a whole (Agrawal 2005). The Campine community members could introduce rules, when they were supported by a group, but had to be confirmed by the local aldermen bench and the representative of the lord. Most rules were introduced by that council itself (De Keyzer 2018b). In the Brecklands, decisions were taken by a jury composed of tenants of the manor; their deliberations grounded in the concept of the custom of the manor (Rodgers et al. 2011). These are, however, the theoretical structures concerning rule making. Who in the end defined the rules is up for discussion and will be dealt with later in this chapter.

Finally, we arrive at the congruence between the carrying capacity of the ecosystem and the appropriation rules. This principle is connected to sustainability in the most direct manner. Both regions, as most CPIs, were very much focussed on these issues. In the Brecklands, the focus was firmly on grazing pressure. The number of sheep per fold course, the location of the hurdles, and the management of the sheep movements were the prime concern (Bailey 1989, 1990). They applied stints, or a numerical limitation on sheep, in the communal flocks. Sheep were not allowed to roam freely, but were gathered in certain locations. Rules concerning sod or peat digging, wood collecting etc. are more difficult to trace. Bye-laws were not published as a separate document, but have to be discovered between the countless manorial court records. Therefore it is difficult to reconstruct all rules (Shaw-Taylor 2002; Rodgers et al. 2011). In the Campine, a large part of the rules explicitly deals with the congruence between the natural conditions and what could be harvested. No stints were introduced on cattle, but the amount of resources that could be collected or gathered from the commons was monitored strictly. Loam, peat, clippings and sods were only to be collected from specific appointed places and only for subsistence use. Every household could collect communal resources during a restricted period and with a limited amount of people (Verhoeven 1907; Meeusen 1932; Prims 1944; Helsen 1949; Ernalsteen 1935; Koyen 1958). These detailed rules immediately show that the communities were well aware of the carrying capacity of the landscape and portray the willingness to adjust the rules so as to maintain these valuable resources sustainably.

Social barriers and political monopolies

Design principles and formal rules, however, do not show the full picture. Formal rules often hide more than they reveal. These are theoretical frameworks that have to be applied in real life. This application is dependent on societal factors and restrictions. It is on this level that the Campine and the Brecklands diverged fundamentally, which can explain the different fates of both regions.

Access, inclusion and collective choice arrangement

Collective action and management should be supported and performed by all the members of the community of users. Inclusion of all interest groups is vital for the management of communal resources (Ostrom 1997; Agrawal and Gibson 1999; Haller, Acciaioli, and Rist 2015). In the Brecklands, however, the largest part of society was excluded from the control and management of the commons, which allowed for a skewed management that only took into account certain interests that had negative effects on the overall sustainability of the landscape. In pre-modern Europe, communities of users within CPIs and the total population of a village did not correspond entirely. Segments of society were progressively excluded from gaining access. The possession of land, cattle, a tenancy or privileged farmsteads were required (Neeson 1993; Shaw-Taylor 2001; De Moor, Shaw-Taylor, and Warde 2002; Casari 2007). As mentioned by Rodgers et al. (2011), the tenants of the lord were supposed to come to meetings and decide on rules and issues regarding the commons. This group of people was, however, shrinking compared to the total population in England during the pre-modern period. Landless labourers, cottagers without animals and subtenants constituted a growing segment of society (Whittle 2000; Shaw-Taylor 2012). Tim Soens has shown that the shift from local landownership towards absentee landlords and temporary tenancies fundamentally impeded the influence and power of smallholders in the water boards in Coastal Flanders. As a result the flood awareness, dyke management and resilience towards floods deteriorated significantly, with a surge in floods as a consequence (Soens 2011).

The same happened to the Breckland smallholders. The number of tenants of the manorial lords decreased significantly after the Black Death (Spufford 1965; Bailey 1989). The population dwindled and the manorial lords shifted towards labour extensive sheep breeding and rabbit rearing (Bailey 1988, 1990). As a result, the number of tenants decreased. Leigh Shaw-Taylor made a strong point that the declining number of manorial tenants veils the growing number of subtenants and labourers that constituted a greater part of the village community (Harrison 1979; Hipkin 2000; Shaw-Taylor 2012). Their rights to the commons, however, are much more obscure. While they might have received access, or could obtain a licence for the fold course, their voice will probably not have been heard in the local meetings of the manor court. The common-pool regime was therefore increasingly tailored to the needs of the elites rather than to the tenants of the Breckland villages. The commons and open fields functioned predominantly as grazing land for the manorial flocks, to the detriment of the tenants that required productive arable land and fodder for their cattle (Allison 1957; Postgate 1960; Whyte 2009). A diverse heathland, providing multiple benefits for the community members, was increasingly reduced to barren heath lands that could sustain only rabbits and sheep. This degradation was, however, no problem for the flockmasters and manorial lords, because this turned sand into gold (Bailey 1990). The introduction of enclosed fields, to intensify arable production and limit the sheep from grazing too much that was demanded by the tenants, was prohibited by the lords to protect their fold course system (Whyte 2009). As a result, fences and hedges, which are the most important windbreakers and preventative measures against drifting sand, were banned.

In the Campine, the opposite happened. All members of the community were allowed access to commons and could influence the decision making, either indirectly or directly (De Keyzer 2018b). This was due to a different socio-economic development in this region. Throughout the Middle Ages and the pre-modern era, smallholders remained the most dominant group and possessed a remarkable power base (van Onacker 2017). Village elites were scarce and not overpowering. The manorial lords had lost most of their power after the arrival of the Dukes (Steurs 1993; Dekkers 1995). Strong liberties were issued to the peasant communities, allowing them to form village governments, control and manage the commons, and defend their communal rights. These privileges were bestowed on all the community members, even the poorest and weakest members of the community. As a result, 98 per cent of the Campine village communities actively used the commons during the sixteenth century.[1] More importantly, however, all the smallholders were able to defend their interests and steer the management of the commons. Even though it is highly unlikely that all the members of the community were actively involved with the rule making, these Campine smallholders were good at defending their interests. One of their interests was retaining an ecologically diverse and sustainably managed landscape. In contrast to the rural elites, they were unable to purchase grazing land, fodder, fuel or building materials on the market. They needed all these resources from the commons, which could only be obtained if the heathlands and meadows were managed sustainably.

When looking at the organisation of the CPIs (see Figure 6.2), it becomes clear that certain important rules and strategies were entirely based on the interests and demands of the smallholders of the community that opposed the elite's interests. These smallholders were therefore powerful enough to defend their interests and

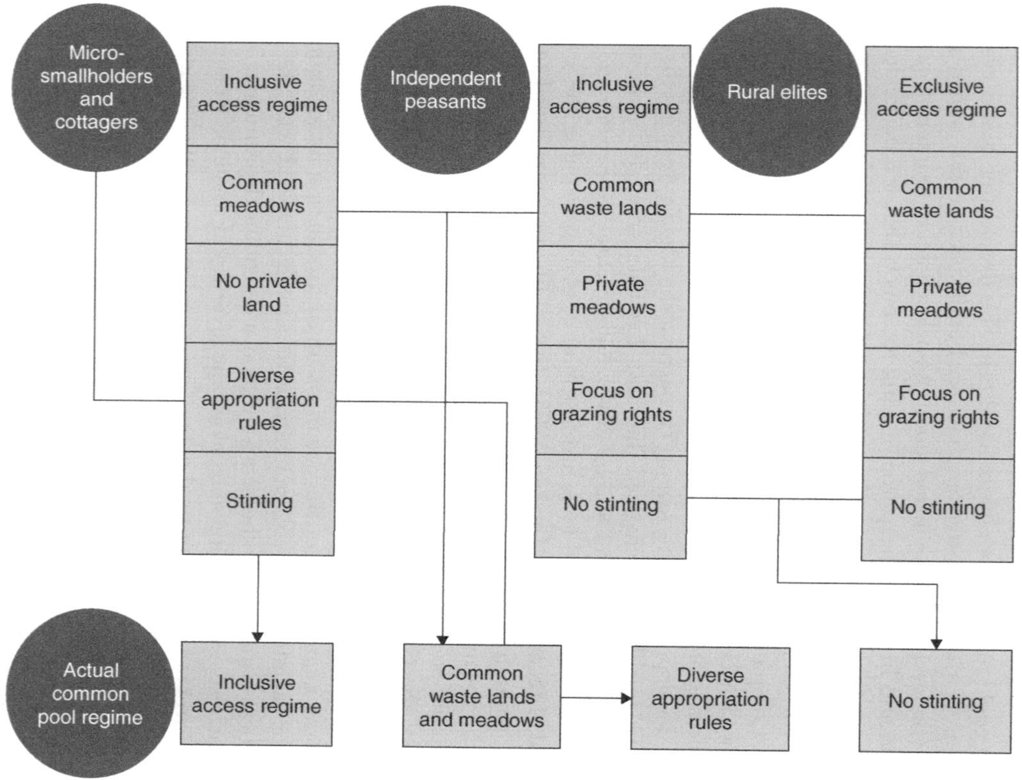

Figure 6.2 Overview of the interests of the different stakeholders and the translation of these interests in the actual structure of the common-pool regime.

were able to make sure that a common denominator of the different interests was obtained, protecting the ecosystem at the same time (De Keyzer 2018b).

Collective management and abidance by the rules

Common resources require collective action, social control and abidance by the rules. Free rider issues are one of the biggest problems of commons. Even though the institutional framework was set up to prevent such behaviour to limit the options for overstocking and to punish any trespassers, this system was not functioning in the Brecklands. Almost every bye-law in Europe has very similar rules and penalties for trespassers. Rodgers et al. (2011) call this a common custom that developed in north-western Europe. Enforcing these rules and making sure that the formal rules are actually applied is, however, a different matter. Beatrijs Augustyn, for example, discovered clear rules against degradation and overexploitation of the coastal dunes. But they were not enforced, because the trespassers and enforcers were one and the same group (Augustyn 1992). Checks and balances and control from different interest groups are important for a community to be able to sustainably manage an ecosystem. According to Renée van Weeren and Tine De Moor, this type of social control and force from within the village community is more important than high fines and harsh penalties (van Weeren and De Moor 2014). A fundamental power balance is required for these checks and balances to work.

This was, however, missing in the Brecklands. As stated before, the position of the smallholders deteriorated fundamentally after the Black Death. The manorial lords with their flock masters and rabbit warreners became the wealthiest but especially the most powerful members of society. They were the ones that started to monopolise the use of the commons, while at the same time they called the shots in the manor courts. Complaints about overstocking, grazing pressure, trespassing and offences were constantly overruled. Nicola Whyte discovered countless complaints from the smallholders and tenants against lordly transgressions through their flock masters and warreners, but they were not heard during the seventeenth century. Cases in the manor courts, but even in higher courts where peasants could plead their case, were almost every time won by the lords (Whyte 2009, 2011). While manorial power was rather weak during the later Middle Ages because of small and multiple manors in the Breckland villages (Bailey 1989, 2002), their pre-modern counterparts were much more dominant. Peasant resistance or calls for adjustments to the use of the commons were mostly ignored (Postgate 1960; Whyte 2009).

Peasant resistance and power to withstand elites in general was becoming less potent. While peasants were often threatening as a social group during the Middle Ages, as small scale acts of rebellion, symbolic violence and even uprisings showed, successful collective action against lords was increasingly scarce from the sixteenth century onwards (Hilton and Aston 1984; Blickle 1986; Goddard, Langdon, and Müller 2010; van Bavel 2010, 2016; Müller 2012). This was especially the case in the Brecklands, where proletarianised land labourers replaced smallholders and the middle groups. Their bargaining power was seriously reduced, which was felt in the CPIs as well.

Bargaining power was vital though. Even in egalitarian commons, such as those in the Campine, a serious and frank discussion was often needed. In the Campine, even the smallest cottagers were effective in protesting against infractions and defending their rights of usage and taking part in the commons. When their access to the commons was threatened, when the commons were used inappropriately or general conflicts arose, they reacted both within the formal courts, as with informal justice. In contrast to the Brecklands, the Campine peasants often, even though not always, won (De Keyzer 2018a). Not one group in the Campine society was strong enough to dominate the other interest groups. As a result, social control was very effective and checks against unsustainable or unfavourable behaviour were in place.

Collective management and action

Commons do not only require prohibiting certain actions. Equally as important is the willingness to act together. To maintain fragile heathlands, constant management is required, which cannot be done by a couple of individuals. When the land is exploited as intensively as it was during the Ancien Régime, maintenance and preventive measures had to be taken. In order to prevent degradation of the ecosystem, grazing had to be monitored, digging for sods and harvesting clippings had to be controlled and limited, zones of activity had to be appointed, fragile zones had to be protected and planting woodlands or scrubland could restore vulnerable spots. CPIs could be ideal for this type of responsibilities, since the entire community divides the work. Time-consuming efforts that do not generate a lot of extra profit or economic gains are not likely to be performed by individuals, but the costs can be spread among the community members. Nevertheless, not all CPIs managed to take those responsibilities.

In the Brecklands, several issues impeded this type of collective management. A lot of factors have been discussed that could hamper collective action, such as religion, ethnic conflicts, trust issues, inequality etc. One of the most important issues here was the dominance of the manorial lords and their unwillingness to cooperate or even allow collective action to happen. A lot of the vital policies to manage a fragile sandy landscape, such as plantations on fragile zones and windbreakers, were prohibited because of the fold course system (Allison 1958; Whyte 2009). In addition, the incentive to invest in costly maintenance was missing, because the degradation of the landscape did not negatively affect the most important economic activities of the manorial lords, namely sheep and rabbit breeding. Land that was fundamentally degraded was simply added to the already extremely extensive warrens (Postgate 1960; Williamson 2007). Next to this, peasants and smallholders did not have the power or means to compensate for such detrimental behaviour, which seriously impedes collective action to arise. As a result, the CPIs in the Brecklands were more or less powerless in this respect.

This collective behaviour was necessary, however, as the Campine proves. During the entire later Middle Ages and early modern period, the communities engaged in collective action to secure a sustainable management. The common heathlands were used intensively, and the community members had chosen not to exclude any groups from gaining access to the commons. As a result, the pressure on the landscape was high. The CPIs that had to monitor and control the commons were not content with preventing illegal and unsustainable behaviour, but actively supported actions to maintain and restore the commons on a permanent basis. The common wastelands were planted with trees and shrubberies. Practically every village government obligated its inhabitants to help with the planting and maintenance of these plantations (De Keyzer 2016). The villages of Retie, Kalmthout, Ravels, Geel and Arendonk all refer to small plantations or wooded areas, called "*heibossen*" (heather forests), that were constructed on the wastelands in order to prevent or limit drifts (Verhoeven 1907; Meeusen 1932; Ernalsteen 1935; Prims 1944; Helsen 1949; Koyen 1958). From at least 1554 onwards, but probably also before that date, the community of Retie could appoint two men who were responsible for the protection of the community against the sand and for planting trees.[2] In addition, once a year, community members were required to perform communal tasks such as controlling boundaries, clearing ditches and brooks, but also maintaining woodlands planted on the common heathlands. The bye-law of Ravels and Eel, for example, stated that 'the wood, needed to stop the sand, will have to be repaired by everyone on the punishment of 6 stuiver'.[3] Besides, several villages had received the right to plant trees on the commons, ten feet behind their private land.[4] This 'pootrecht' secured their basic needs for timber, but equally functioned as a barrier against drifting sand.

Conclusion

Historical CPIs did not always succeed in sustainably managing communal resources. Contrary to the paradigm of Hardin, this was mostly not due to freedom in the commons. A free for all situation after all was extremely rare. Pre-modern CPIs were all remarkably similar with regard to the institutional design. The design principles that Ostrom defined as vital were adopted almost everywhere, despite some practical differences to cope with local circumstances.

Commons, however, still managed to fail, when socio-political restrictions and barriers prevented the efficient functioning of CPIs. The existence or development of institutions themselves is not sufficient. Institutions are the instruments of societies, and therefore do not have agency by themselves to secure sustainable and efficient management of resources. Some socio-political constellations make it extremely difficult to have successful commons. In the Brecklands, a 'tragedy of the commons' occurred because of a set of restrictions. The first was the appearance of a plutocratic system. A monopoly of power in the hands of the manorial lords prevented the largest part of society to engage in collective action and to steer the management of the commons. As a result, the fold course system that organised collective grazing became orientated towards the elites' interest, to the detriment of the smallholders and the environment. Second, this power imbalance prevented protests and social control, which is vital to stop or prevent unsustainable actions in the commons. Finally, in this troubled society, collective action to manage and maintain the landscape was missing.

In the Campine, on the contrary, the socio-political organisation of society was highly conductive to foster efficient and sustainable collective action. Here power was not monopolised, and not one group could dominate or impose a one-sided management of the landscape. As a result of the common denominator, the landscape had to be managed sustainably to provide all possible benefits for the different stakeholders. Social control was strong because unsustainable practices could be stopped thanks to judicial and symbolic actions. Finally, this egalitarian climate, with a convergence of interests, created the ideal conditions for the community members to engage in collective action to maintain the landscape effectively.

Notes

Acknowledgements: This work is funded by the ERC advanced grant N° 339647 (COORDINATINGforLIFE) and the FWO.

1 Rijksarchief Antwerpen-Beveren, OGA Zandhoven, 148.
2 The oldest version of Retie's byelaws dates back to 1554, but oral regulations predated this first written document (Helsen 1949).
3 'Het hout ook voor indien het nodig mocht wezen om het zand te stoppen, zal iedereen komen en repareren op alle mogelijke manieren, op de straf van 6 stuivers' (Koyen 1958).
4 See for example: Rijksarchief Antwerpen-Beveren, OGA Rijkevorsel, 12.

References

Agrawal, A. 2005. *Environmentality: Technologies of Government and the Making of Subjects.* Durham: Duke University Press.

Agrawal, A. and C.C. Gibson. 1999. Enchantment and Disenchantment: The Role of Community in Natural Resource Conservation. *World Development* 27 (4): 629–649.

Allison, K.J. 1957. The Sheep-Corn Husbandry of Norfolk in the Sixteenth and Seventeenth Centuries. *The Agricultural History Review* 5 (1): 12–30.

Allison, K.J. 1958. Flock Management in the Sixteenth and Seventeenth Centuries. *The Economic History Review* 11 (1): 98–112.

Anderies, J.M., Janssen, M.A., and Ostrom, E. 2004. A Framework to Analyze the Robustness of Social-Ecological Systems from an Institutional Perspective. *Ecology and Society* 9 (1).

Augustyn, B. 1992. *Zeespiegelrijzing, transgressiefasen en stormvloeden in maritiem Vlaanderen tot het einde van de XVIde eeuw.* Brussels: Algemeen Rijksarchief.

Bailey, M. 1988. The Rabbit and the Medieval East Anglian Economy. *The Agricultural History Review* 36 (1): 1–20.

Bailey, M. 1989. *A Marginal Economy? East Anglian Breckland in the later Middle Ages.* Cambridge: Cambridge University Press.

Bailey, M. 1990. Sand into Gold. The Evolution of the Fold-Course System in West Suffolk, 1200–1600. *Agricultural History Review* 38: 40–57.

Bailey, M. 2002. *The English Manor c. 1200 to c. 1500*. Manchester: Manchester University Press.

Bankoff, G. 2009. Cultures of Disaster, Cultures of Coping. Hazard as a Frequent Life Experience in the Philippines. In *Natural Disasters, Cultural Responses: Case Studies Toward a Global Environmental History*, edited by C. Mauch and C. Pfister, 265–284. Lanham: Lexington Books.

Bastiaens, J. and Deforce, K. 2005. Geschiedenis van de heide. Eerst natuur en dan cultuur of andersom? *Natuur.focus* 4 (2): 40–44.

Bateman, M.D. and Godby, S.P. 2004. Late-Holocene Inland Dune Activity in the UK: A Case Study from Breckland, East Anglia. *The Holocene* 14 (4): 579–588.

van Bavel, B. 2010. Rural Revolts and Structural Change in the Low Countries, Thirteenth-Early Fourteenth Centuries. In *Survival and Discord in Medieval Society. Essays in Honour of Christopher Dyer*, edited by R. Goddard, J. Langdon, and M. Müller, 249–268. Turnhout: Brepols.

van Bavel, B. 2016. *The Invisible Hand? How Market Economies have Emerged and Declined Since AD 500*. Oxford: Oxford University Press.

Berkes, F., ed. 1989. *Common Property Resources: Ecology and Community-Based Sustainable Development*. London: Belhaven Press.

Blickle, P. 1986. The Criminalization of Peasant Resistance in the Holy Roman Empire: Toward a History of the Emergence of High Treason in Germany. *The Journal of Modern History* 58: 88–97.

Bollier, D. and Helfrich, S., eds. 2015. *Patterns of Commoning*. Amherst: Levellers Press.

Casari, M. 2007. Emergence of Endogenous Legal Institutions: Property Rights and Community Governance in the Italian Alps. *The Journal of Economic History* 67 (1): 191–226.

De Keyzer, M. 2013. The Impact of Different Distributions of Power on Access Rights to the Common Waste Lands: the Campine, Brecklands and Geest compared. *Journal of Institutional Economics* 9 (4): 517–542.

De Keyzer, M. 2016. All We Are Is Dust in the Wind. The Social Causes of a 'Subculture of Coping' in the Late Medieval Coversand Belt. *Journal for the History of Environment and Society* 1 (1): 1–35.

De Keyzer, M. 2018a. Access versus Influence. Peasants in Court in the Late Medieval Low Countries. In *Handbook of Rural Life*, edited by M. Müller. London: Routledge.

De Keyzer, M. 2018b. *Inclusive Commons and the Sustainability of Peasant Communities in the Medieval Low Countries*. London: Routledge.

De Keyzer, M. and Bateman, M.D. 2018. Late Holocene Landscape Instability in the Breckland (England) Drift Sands. *Geomorphology* 323 (December 15, 2018): 123–134.

De Moor, T. 2015. *The Dilemma of the Commoners: Understanding the Use of Common-Pool Resources in Long-Term Perspective, Political Economy of Institutions and Decisions*. Cambridge: Cambridge University Press.

De Moor, M., Shaw-Taylor, L., and Warde, P. eds. 2002. *The Management of Common Land in North West Europe, c. 1500–1850*. Turnhout: Brepols.

Dekkers, P.J.V. 1995. Brandend zand. Hoe de hertog van Brabant zijn heerschappij op de kempense zandgronden verwierf ten koste van de lokale en regionale adel. *Noordbrabants historisch jaarboek* 12: 10–40.

Derese, C., Vandenberghe, D., Eggermont, N., Bastiaens, J., Annaert, R., and Van den Haute, P. 2010. A Medieval Settlement Caught in the Sand: Optical Dating of Sand-Drifting at Pulle (N Belgium). *Quaternary Geochronology* 5: 336–341.

Eggermont, N., Annaert, R., Bastiaens, J., Derese, C., Vandenberghe, D., Van den Haute, P., Haneca, and Van Strydonck, K. 2008. Nederzettingssporen uit de ijzertijd en de vroege middeleeuwen onder een stuifduin langs de Keulsebaan te Pulle, edited by VIOE. Brussels.

Enklaar, D.T. 1941. *Gemeene gronden in Noord-Brabant in de middeleeuwen*. Utrecht: Kemink.

Ernalsteen, J. 1935. Keuren van Gheel. *Oudheid en Kunst* 26 (2): 19–66.

Evelyn, J. 1952. Diary Entry for 10 September 1667. In *The Diary of John Evelyn*, edited by W. Bray. London: J.M. Dent and Sons.

Fanta, J., and Siepel, H. eds. 2010. *Inland Drift Sand Landscapes*. Zeist: KNNV Publishing.

Goddard, R., Langdon, J., and Müller, M. eds. 2010. *Survival and Discord in Medieval Society: Essays in Honour of Christopher Dyer*. Turnhout: Brepols.

Haller, T., Acciaioli, G., and Rist, S. 2012. Constitutionality: Emic Perceptions of Bottom-up Institution Building Processes (Conference Paper, "Design and Dynamics of Institutions for Collective Action", Utrecht, 29 November to 1 December, 2012).

Haller, T., Acciaioli, G., and Rist, S. 2015. Constitutionality: Conditions for Crafting Local Ownership of Institution-Building Processes. *Society & Natural Resources* 28 (9): 1–20.

Haller, T. and Chabwela, H.N. 2009. Managing Common Pool Resources in the Kafue Flats, Zambia: From Common Property to Open Access and Privatisation. *Development Southern Africa* 26 (4): 555–567.

Hardin, G. 1968. The Tragedy of the Commons. *Science* 162 (3859): 1243–1248.

Harrison, C.J. 1979. Elizabethan Village Surveys: A Comment. *The Agricultural History Review* 27 (2): 82–89.

Heidinga, H.A. 2010. The Birth of a Desert. The Kootwijkerzand. In *Inland Drift Sand Landscapes*, edited by J. Fanta and H. Siepel, 65–79. Zeist: KNNV Publishing.

Helsen, I. 1949. Het dorpskeurboek van Retie. *Bijdragen tot de geschiedenis* 1 (1): 85–107.

Hilton, R.H. and Aston, T.H. ed. 1984. *The English Rising of 1381*. Cambridge: Cambridge University Press.

Hipkin, S. 2000. Tenant Farming and Short-Term Leasing on Romney Marsh, 1587–1705. *The Economic History Review* 53 (4): 646–676.

Hoppenbrouwers, P. 2002. The Use and Management of Commons in the Netherlands. An Overview. In *The Management of Common Land in North West Europe, c. 1500–1850*, edited by M. De Moor, L. Shaw-Taylor, and P. Warde, 87–109. Turnhout: Brepols.

Koster, E. 2007. Aeolian Environments. In *The Physical Geography of Western Europe*, edited by E. Koster, 139–160. Oxford: Oxford University Press.

Koyen, M. 1958. Keuren van Ravels. *Oudheid en Kunst* 41 (2): 3–19.

Laborda Peman, M. and De Moor, T. 2013. A Tale of Two Commons. Some Preliminary Hypotheses on the Long-Term Development of the Commons in Western and Eastern Europe, 11th–19th Centuries. *International Journal of the Commons* 7 (1): 7–33.

Lana Berasain, J.M. 2008. From Equilibrium to Equity. The Survival of the Commons in the Ebro Basin: Navarra from the 15th to the 20th Centuries. *International Journal of the Commons* 2 (2): 162–191.

Mauelshagen, F. 2007. Flood Disasters and Political Culture at the German North Sea Coast: A Long-Term Historical Perspective. *Historical Social Research* 32 (3): 133–144.

Meeusen, G. 1932. Keuren van Esschen, Calmpthout en Huybergen. *Oudheid en Kunst* 23: 112–124.

Müller, M. 2012. Arson, Communities and Social Conflict in Later Medieval England. *Viator* 43 (2): 193–208.

Neeson, J.M. 1993. *Commoners: Common Right, Enclosure and Social Change in England, 1700–1820*. Cambridge: Cambridge University Press.

Olivier, S. 2013. Peasant Property, Common Land and Environment in the Garrigues of the Languedoc from the Seventeenth to the Twenty-First Centuries. In *Rural Societies and Environments at Risk. Ecology, Property Rights and Social Organisation in Fragile Areas (Middle Ages – Twentieth Century)*, edited by B. van Bavel and E. Thoen, 85–110. Turnhout: Brepols.

Ostrom, E. 1997. *Governing the Commons. The Evolution of Institutions for Collective Action*. Cambridge: Cambridge University Press.

van Onacker, E. 2017. *Village Elites and Social Structures in the Fifteenth and Sixteenth-Century Campine Area*. Turnhout: Brepols.

Postgate, M.R. 1960. *Historical Geography of Breckland, 1600–1850*. London: University of London (M.A. thesis).

Prims, F. 1944. Keuren der vreyheyt van Arendonk. In *Feestbundel H. J. Van De Wijer, den jubilaris aangeboden ter gelegenheid van zijn vijfentwintigjarig hoogleeraarschap aan de R. K. universiteit te Leuven 1919–1943*, edited by H. Draye, 273–283. Leuven: Instituut voor Vlaamse topynomie.

Radkau, J. 2008. *Nature and Power. A Global History of the Environment*. Cambridge: Cambridge University Press.

Rodgers, C.P., Straughton, E.A., Winchester, A.J.L., and Pieraccini, M. 2011. *Contested Common Land: Environmental Governance Past and Present*. London: Earthscan.

Shaw-Taylor, L. 2001. Labourers, Cows, Common Rights and Parliamentary Enclosure: The Evidence of Contemporary Comment c. 1760–1810. *Past and present* 171: 95–126.

Shaw-Taylor, L. 2002. The Management of Common Land in the Lowlands of Southern England circa 1500 to circa 1850. In *The Management of Common Land in North West Europe, c. 1500–1850*, edited by M. De Moor, L. Shaw-Taylor, and P. Warde, 59–86. Turnhout: Brepols.

Shaw-Taylor, L. 2012. The Rise of Agrarian Capitalism and the Decline of Family Farming in England. *The Economic History Review* 65 (1): 26–60.

Soens, T. 2011. Threatened by the Sea, Condemned by Man? Flood Risk, Environmental Justice and Environmental Inequalities Along the North Sea Coast 1200–1800. In *Environmental and Social Justice in the City. Historical Perspectives*, edited by G. Massard-Guilbaud and R. Rodger, 91–111. Cambridge: The White Horse Press.

Spek, T. 2004. *Het Drentse esdorpenlandschap. Een historisch-geografische studie*. Utrecht: Stichting Matrijs.

Spufford, M. 1965. *A Cambridgeshire Community. Chippenham from Settlement to Enclosure*. Leicester: Leicester University Press.

Steurs, W. 1993. *Naissance d'une région. Aux origines de la Mairie de Bois-le-Duc, Recherches sur le Brabant septentrional aux 12e et 13e siècles*. Vol. III, *Memoire de la Classe des Lettres*. Brussels: Académie royale de Belgique.

Sundberg, K. 2002. Nordic Common Lands and Common Rights. Some Interpretations of Swedish Cases and Debates. In *The Management of Common Land in North West Europe, c. 1500 1850*, edited by M. De Moor, L. Shaw-Taylor, and P. Warde, 173–193. Turnhout: Brepols.

Verboven, H., Verheyen, K., and Hermy, M. 2004. *Bos en hei in het Land van Turnhout (15de-19de eeuw). Een bijdrage tot de historische ecologie*, edited by Monumenten & landschappen. Leuven: Ministerie van de Vlaamse gemeenschap, Monumenten & Landschappen en het Vlaams Instuut voor het Onroerend Erfgoed.

Verhoeven, P.J. 1907. Keuren van Calmpthout. *Oudheid en Kunst* 3: 45–46.

van Weeren, R. and De Moor, T. 2014. Controlling the Commoners: Methods to Prevent, Detect, and Punish Free-Riding on Dutch Commons in the Early Modern Period. *Agricultural History Review* 62 (2): 256–277.

WCED (United Nations, World Commission on Environment and Development). 1987. *Our Common Future*. Oxford: Oxford University Press.

Whittle, J. 2000. *The Development of Agrarian Capitalism: Land and Labour in Norfolk 1440–1580*. Oxford: Clarendon Press.

Whyte, N. 2009. *Inhabiting the Landscape. Place Custom and Memory, 1500–1800*. Oxford: Oxbow Books.

Whyte, N. 2011. Contested Pasts: Custom, Conflict and Landscape Change in West Norfolk, c. 1550–1650. In *Custom, Improvement and the Landscape in Early Modern Britain*, edited by R.W. Hoyle, 101–126. Farnham: Ashgate.

Williamson, T. 2007. *Rabbits, Warrens and Archaeology*. Stroud: Tempus.

7 For the common good

Regulating the Lake Constance fisheries from 1350 to 1800

Michael Zeheter

Introduction

Garret Hardin's "Tragedy of the Commons" (1968) does not make the news too often, but when it does, it is quite regularly in the context of overfishing. Be it local fisheries off the coast of Savannah, Georgia (Goggins 2017), the impact of China's growing prosperity on global fish stocks (Jacobs 2017), the possibility of a global framework to limit exploitation of the resource (Governing the Oceans 2014), or the failings of the European Union's common fisheries policy (Worstall 2017), such reports use Hardin's thought experiment as a point of reference in their arguments. That the political positions and intentions of these arguments diverge in substantial ways, and that they employ the "Tragedy of the Commons" to very different ends, must not surprise. The fact that Hardin's argument is present in the news at all, however, is testament to both its value as an analytical tool with which to comprehend and explain real world problems, and the urgency of the problem of the current overexploitation of marine fish stocks.

The prominence of fisheries as a common property resource in the news is also mirrored in the commons studies research. Among the earliest examples in the field's seminal texts are several that explored on and off shore fisheries in Canada, Turkey and Sri Lanka (Ostrom 1990: 144–178). Inland fisheries, however, have received far less attention, with few exceptions (Haller 2012). This disproportion is not restricted to sociological, anthropological or economic studies of present and very recent uses of fish as a common resource, but is true of the environmental history of fisheries as well. Most of the writings on the subject are focused on marine fisheries, and only very few on lakes (Bogue 2000). While this might be justified by the economic importance of marine fisheries, and their impact on larger populations, the neglect of inland fisheries is a missed opportunity to study the use and management of commons in a circumscribed space by a local population over a long period of time. Exploring the history of the Lake Constance fisheries from the late Middle Ages until the turn of the nineteenth century allows me to do just that. After introducing the political as well as environmental landscape of the Lake Constance region during the period under consideration, I will present the legal and institutional conditions of the local fisheries, and the development and workings of the resource management regime, and end with an assessment of its effectiveness.

Lake Constance and its surroundings

It is by no means obvious that Lake Constance is a single entity. Located in the Alps' northern foothills, the lake is traditionally considered to consist of three distinctive parts: the Upper Lake, the Lower Lake and the Lake Rhine connecting the two. The Upper Lake is the largest and deepest, with a surface of 473 square kilometres and a maximum depth of 254 meters. Fed by the clear and cold water of numerous rivers and creeks, the most important of which is the Rhine, the Upper Lake is oligotrophic and the home of fish species such as lake trout and whitefish among others. The Lake Rhine is the Upper Lake's only outlet and connects it to the Lower Lake. Its length comprises less than four kilometres with a difference in altitude of only 0.3 metres; it attracts species preferring a weak current such as barbel and chub. With a surface of 63 square kilometres and a maximum depth of 46 metres, the Lower Lake is much smaller and shallower than the Upper Lake. Its waters are warmer and richer in nutrients, making it eutrophic and providing a habitat for prized species such as pike and perch, but also a great number of less popular cyprinids including carp, roach, rudd and bream (Berg 1993; Internationale Gewässerschutzkommission für den Bodensee 2004: 8–11). This coarse description of the lake's three main parts, however, neglects the fact that both Upper and Lower Lake feature many islands, bays and shallows, providing local habitats for species that are not considered typical of their part of Lake Constance. Many fish species are also not static over the course of the year, but migrate according to the season or for spawning. Thus, cyprinids can also be found in the Upper Lake, and a certain kind of whitefish, *Gangfisch*, are spawning in the Lake Rhine in great numbers. For the fishermen of the past as well as for those of today, the complexities of geography, ecology and the weather formed a formidable challenge that made their craft dangerous and experience an asset.

Even after the end of the Holy Roman Empire (1806), the political landscape of the Lake Constance region has been almost as heterogeneous as its ecology and this complexity still exists today, with Austria, Germany and Switzerland sharing the lake's shore. But this current status is only a mild reflection of the fractured political map that characterised the region from the Middle Ages to the first decade of the nineteenth century. Among the oldest and most prestigious principalities were the Benedictine abbeys of Reichenau and St. Gall, as well as the bishopric of Constance. The monasteries were founded by monks who were later canonised, and they quickly developed into prestigious centres of power and learning, which translated substantial properties. Reichenau came to own most of the land surrounding the Lower Lake, and St. Gall dominated the southern shore of the Upper Lake (Feger 1975).

The bishopric was not quite as successful. It managed to gain possessions along the Lake Rhine and on the northern shore of the Upper Lake, where

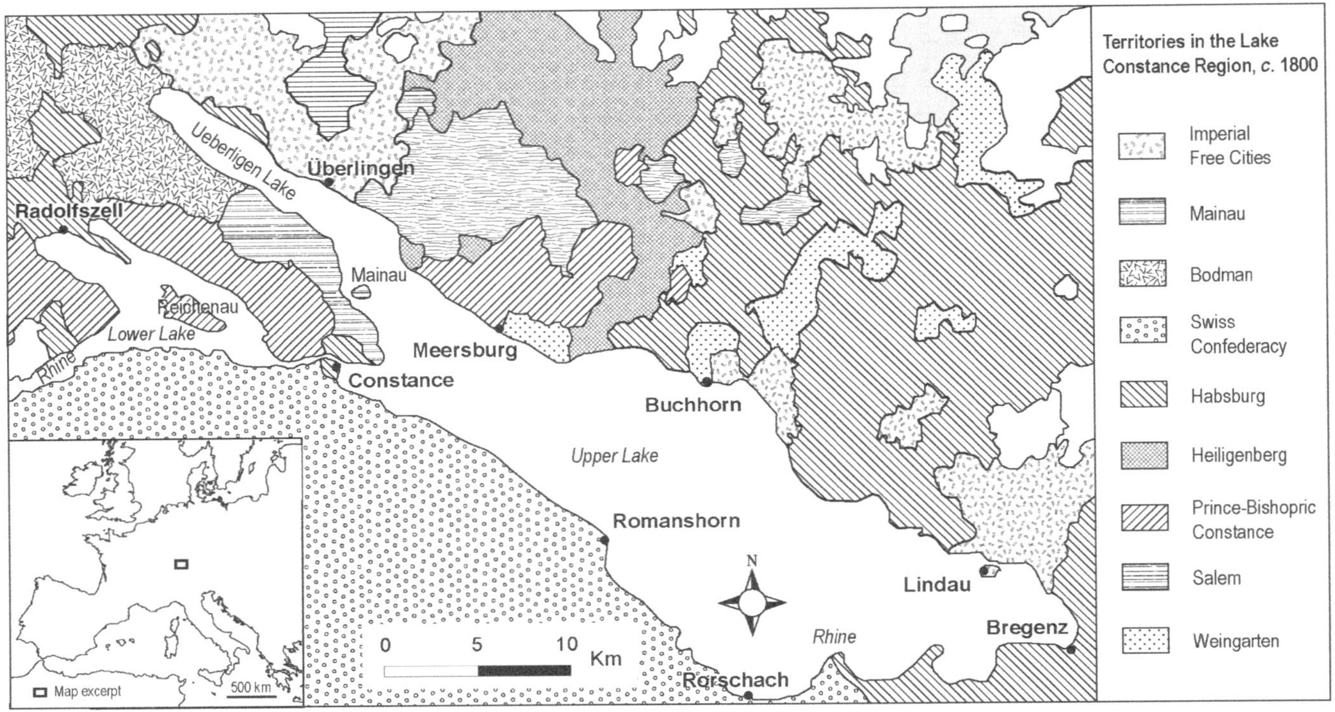

Figure 7.1 The Lake Constance region, *c.*1800.
Source: Map compilation and design by Ulla Gämperli Krauer (2018), CDE, University of Bern, Switzerland. Map base: Hölzle (1938).

Constance was the most populous town and developed into the most important market place in the region. As Lake Constance lay at an intersection of long distance trade routes from Germany to Italy across the Alps, and along the alpine foothills from Burgundy to Bavaria and Austria, the market towns around the lake began to prosper in the twelfth and thirteenth centuries, putting some of them in the position of being able to shake off their feudal overlords. Buchhorn (today part of Friedrichshafen), Constance, Lindau and Überlingen thus became free imperial cities and gained political independence (Eitel 1982; Feger 1983; Kirchgässer 1988).

This new position was, however, almost immediately under threat as two larger and more powerful political entities extended their reach towards the Lake Constance region, the Habsburgs and the Swiss Confederacy. From the thirteenth century onwards they were entwined in a struggle that lasted until the early sixteenth century, and led to the establishment of two political spheres of dominance around Lake Constance. The Habsburgs had their power base in Vorarlberg at the eastern end of the Upper Lake and over time acquired several other territories on the northern shore of the lake, while the Swiss integrated all of the southern shore into their confederacy. The remaining independent principalities on the northern shore, including the free imperial cities, the Commandry of the Teutonic Order on the island of Mainau, the Cistercian abbey of Salem and the principality of Heiligenberg, among others, came to accept Habsburg dominance with the exception of the city of Constance, which tried to maintain an independent position in between and was subsequently conquered by Habsburg troops in 1548 (Feger 1981; Press 1989; Dobras 1991). This territorial situation remained mostly stable until the era of the Napoleonic wars when the Grand Duchy of Baden as well as the kingdoms of Bavaria and Württemberg absorbed the smaller principalities due to secularisation and mediatisation, and extended their territory to the northern shore of Lake Constance (Wieland 2002).

Legal and institutional foundations of the fisheries

While the borders of the territories surrounding Lake Constance have not changed for almost 150 years, the main body of the Upper Lake remains disputed territory. Although Germany and Switzerland have negotiated a border for the Lower Lake and the Lake Rhine, the Upper Lake's three neighbours have never managed to come to an agreement. They treat it as a tridominium with shared sovereignty (Schuster 1951). In many respects, this oddity continues an older legal status. Since medieval times, most of Lake Constance was considered a commons. The deep lake was beyond its neighbours' legislative and judicial control and its fish stocks unregulated. The various principalities that shared Lake Constance's shores extended their authority only over the littoral. Only there were the neighbouring principalities allowed to regulate access to the fish resources, and were private fishery rights possible. This regulatory authority, however, did not exclude the general population from the fish stocks. Every person who obeyed the rules, regardless of their status as a subject of the local authority or not, had the right to fish, except in those places where others owned private fishing rights (Stoffel 1906: 3–5). In practice, these legal principles applied differently to the three parts of the lake. The deep parts of the Upper Lake were unregulated, the Lake Rhine was under the authority of its respective neighbours, and the shallow Lower Lake was the exclusive domain of the abbey of Reichenau (Kunz 1994: 22–23). Nonetheless, and despite the best fishing grounds such as those of Lake Rhine being almost entirely private property, much of Lake Constance's fish stocks were legally a common use resource accessible to all.

For most of the population, this right was moot. Although every person in the region was allowed their own catch it was mainly the fishermen who exploited the fish stocks to make a living. They were the only ones with the necessary equipment – seines, gill nets, fish traps and lines to which several hooks were attached (Klunzinger 1892) – and knowledge to navigate the vicissitudes of Lake Constance and the behaviour of its inhabitants. The Lake Constance fisheries were not a subsistence economy, they relied on the local market. Fish was, and still is, a highly perishable food. Before the invention of modern refrigeration technology, the fishermen had to sell their catch quickly to the local population. Demand was high, as the Catholic Church prescribed around 140 days of fasting[1] per year and most people substituted forbidden meat with fish. The many monasteries around Lake Constance used the fish markets to satisfy their even higher demand due to stricter dietary rules. The cultural significance of fish caused the local authorities to closely regulate their fish markets in order to ensure the availability of affordable fish for their population and the quality of the product by issuing price lists that set maximum prices for most species[2] and by controlling the sale of fish on the market through officials (Strigel 1910).[3] Keeping the fish market well stocked with a high quality product was a crucial social responsibility for the authorities.

All fishermen in the Lake Constance region had access to the fish markets regardless of their place of origin and their legal status there. Fishermen from the rural areas around Lake Constance were subjects of their lords, and tasked with exploiting their fishing rights. For this privilege, they had to pay a certain portion of their catch either in coin or in kind. The rest they could sell to earn their living, which was usually supplemented by the cultivation of a plot of land.[4] In the cities of Buchhorn, Constance, Lindau and Überlingen the various trades, including the fishermen, as in the free imperial cities in general, had shed their subjection to a feudal lord and formed guilds. Within the cities, the guilds had a monopoly on their trade or craft after they had taken over from the patrician families, and gained political control over the city council around 1350. Membership required a period of apprenticeship, the payment of a membership fee and the acceptance of the city's ordinances that regulated the trade (Leipold-Schneider 2002).

The ordinances were designed to curb competition and ensure each member's economic survival. In the case of the fishermen's guild, the ordinances regulated six different aspects of the craft. The first concerned equipment, its design, mesh size and shape, and was supposed to guarantee that all members of the guild used identical equipment. A second set of rules restricted or prohibited the use of certain equipment at different times of the year, for example during the spawning season. Third, the ordinances protected certain species of fish by setting closed seasons, by prohibiting the catching of young fish or by prescribing a minimal length for fish caught. A fourth area governed the conduct of the fishermen on the water, a fifth the selling of fish on the market, and a sixth determined the penalties for violating the rules, which ranged from fines to imprisonment (for a transcript of the fisheries treaty negotiated between Constance and Überlingen in 1536 see Zeheter 2014: 117–138).

The ordinances were issued by the city councils, but given their detailed content they must have been based on an intimate knowledge of the fishermen's craft, the lake's properties and the behaviour of the fish. This indicates that the guilds were heavily involved in negotiating the regulations that regulated their members' livelihood. The ordinances, however, did not apply solely to the members of the guild. In order to prevent competition on the city's fish market from fishermen who were subjects of other authorities, all fish on the market had to be caught according to the rules, not only of the fish market but of the local guild. The cities with the most important fish markets, primarily Constance and Lindau, were thus able to extend the influence of their fishermen's ordinances far beyond the lake's littoral, not only onto the deep lake but into the areas under the authority of other principalities. This improved the authorities' control over the fishermen whose compliance to the rules they could survey only on the market, and not on the lake. They therefore relied on the denunciations of other fishermen if they wanted to punish behaviour on the lake that was impossible to detect from land. There was undoubtedly a certain solidarity among members of the same guild, or fishermen from the same village or town, but fishermen from different places were highly competitive. If they saw others gaining an illegal advantage, they were ready to denounce them to the authorities (Schuster 1997: 133–134).[5]

The fishermen's ordinances proved to be a highly successful tool that regulated the fisheries not only in Lindau and Constance, where it was first developed around the year 1350, but which spread in the following decades to the other free imperial cities – Buchhorn and Überlingen – with their fishermen's guilds, and was also adapted around the year 1450 by the abbeys of St. Gall and Reichenau, although their fishermen were serfs and their fish markets were only of minor importance.[6] Apparently, both abbeys recognised the importance of a regulatory regime for their fisheries that took the opinions, interests and the experience of their fishermen into account.

By the time of the adaptation of the fishermen's ordinances by St. Gall and Reichenau, four regulatory regimes had emerged that covered most of Lake Constance. The abbots of Reichenau abbey issued their ordinances for the Lower Lake. In Lindau, the fishermen's guild had integrated all fishermen located in the easternmost part of the Upper Lake, thus expanding its membership far beyond the city's territory and population (Stoffel 1906: 152–154). Constance had chosen another way to extend its influence, by negotiating fisheries treaties with neighbouring principalities, thus creating a common regulatory regime for the western part of the Upper Lake (Stoffel 1906: 106–107, 129). St. Gall again followed the example of Constance and negotiated a fisheries treaty that covered the southern Upper Lake (Stoffel 1906: 8).[7] The fishermen's guild of Buchhorn and its ordinances existed outside this framework, but was apparently of only local importance (Stoffel 1906: 131–136).

These larger regulatory regimes had clear advantages for the fishermen. A single set of rules simplified their lives considerably. They could design and fabricate uniform equipment, and they could be certain that all the fishermen in their part of the lake conformed to the same rules. It was easier to identify those who neglected or infringed on the rules. A common set of rules also gave them the certainty that they would be able to bring their catch to any market covered by the treaty. They could thus decide later rather than sooner where to offer their catch, or try two or more markets on the same day. The advantages for the principalities were even more striking. The regulatory regimes reinforced the position of their fish markets or their status as lords over their subjects, improving their control over the fishermen. Although the fishermen had a say in the regulatory process, the authorities were still in charge of proceedings and the results bound both the fishermen and themselves to the common rules. This minimised the potential of conflict between the fishermen as a group and the authorities, and instead designed a potential fault line between those fishermen conforming to the rules and those who did not.

Resource management

The regulatory tool of fisheries treaties and fishermen's ordinances gave both the fishermen and their authorities the opportunity to actively manage Lake Constance's fish stocks. The ordinances and treaties were read by the town clerk and discussed by those present at the fishermen's annual meetings (Amacher 1999). Apparently, the effectiveness of the different rules contained in the legal text was debatable. New treaties and ordinances were issued by the authorities in irregular sequences containing some changes to the previous rules. Some paragraphs disappeared as new ones were added and existing ones were altered.

A serial analysis of the changes to the ordinances and treaties reveals that they were neither arbitrary nor a product of neglect or ignorance. Rather we can identify many gradual and some abrupt changes that can be categorised in hindsight under four distinct strategies: continuity, tightening, loosening and differentiation. In order to demonstrate their occurrence in practice I have chosen four particularly clear-cut examples from the ordinances issued by the abbey of Reichenau for the Lower Lake.

An example for the strategy of continuity concerns the rule that governed the mesh size of drift nets, which was constantly set at 37 "mesh per Constance twisted yarn cubit" (*Konstanzer Zwillichelle*)[8] from its first appearance in 1455 to its last in 1774.[9] Such consistency over the course of centuries was far from being a singular occurrence, but frequent changes to a rule were far more common. The mesh size for seines gives a good impression. It was first introduced in 1522 at 24 "mesh per Constance twisted yarn cubit" and remained unchanged until 1594 when it was reduced to 21. This increased the mesh size and made the seines less efficient. The following ordinance did not change this regulation but added a new one for the seine specifically designed to catch roach, which was set at 24 "mesh per Constance twisted yarn cubit". The mesh size for the roach seine remained at 24 until to the last ordinance in 1774 while the mesh size for the other seines was increased again to 20 in 1717 and then reverted to 21 in 1774.[10] This example clearly demonstrates two different strategies. First, the rule was tightened by increasing the mesh size from 24 to 21 in 1594. This was apparently too sweeping a change for the roach seine, which made differentiation necessary. By introducing a specific rule for this particular kind of equipment, the authorities acknowledged that their previous tightening had been too general and a partial loosening had been the consequence. This subtle change can only be explained by both the authorities and the fishermen coming to the conclusion that the previous change had gone too far. The changes must have been the result of a lengthy process of negotiation between both parties that were only codified when both sides agreed. Yet there were also more straightforward cases of loosening restrictions. In the Lower Lake ordinance of 1542, a new rule was introduced that limited the number of lines set over the course of a year to 400. This specific regulation was repeated unchanged in the ordinance of 1550 until it was halved to 200 lines in 1583. This tightening was confirmed in 1594 until it was partially reversed in 1613 and set at 300 lines. This compromise was regularly renewed until 1774.[11]

These examples of continuity, tightening, loosening and differentiation are by no means extraordinary. There are numerous other changes to the rules both for the Upper and Lower Lake that demonstrate that these were not random occurrences or the consequence of neglect but the result of deliberate considerations that involved both the authorities and the fishermen. The former brought legal expertise and their knowledge of past regulations to the negotiating table, the latter their awareness of recent changes in the lake, their economic situation and their practical experience. Unfortunately, we lack sources that could clarify the negotiation process or the intentions behind every specific change to the rules; however, the vast number of alterations over the course of centuries within several regulatory regimes demonstrates that all parties to those negotiations regarded them as important. They were conducted in good faith, and everyone believed that the changes to the rules manifested themselves in the lake. The fishermen and their authorities tried to play an active role in managing the resources of fish.

This belief is also evident in the preambles of several fisheries treaties, as well as in the correspondence between fishermen's guilds. The preamble of the treaty between Constance and Überlingen negotiated in 1536 states that its purpose was "the common benefit in the lake" and that it intended to "restore the lake" by protecting the spawn and young fish, while attributing the necessity of such measures to the misconduct of some fishermen.[12] The social aspects of the regulatory regime were highlighted by the preamble of a treaty between

Constance and Mainau from 1566. It also complained of the pitiful state of the lake and of the negative consequences not only for the fishermen but also for "the common poor man" who had to suffer a shortage of fish.[13] Although we have to take such statements with more than a grain of salt, they nonetheless show an awareness of the vulnerability of the fish stocks, especially in the catch of young fish and the destruction of roe, and the detrimental consequences such practices could have for those relying on fish as food or for their income.

This language not only indicates a belief in the effectiveness of the fishermen's regulatory measures, but also a concept of what we would today call sustainability. Although that term only came into existence in the eighteenth century, deliberate action that was intended to ensure the long-term viability of a resource was much older (Grober 2010). This is not to say that the regulation of the Lake Constance fisheries was the result of planning based on scientific assessment, but rather a framework that allowed for the search for viable compromise between diverging interests through trial and error.

Measuring effectiveness?

The question remains, however, what impact the regulatory regime had on the Lake Constance fisheries. The equipment used by the fishermen was not particularly effective, especially when compared with today's, and they had to rely on their own muscles in order to haul in their catch. There were also numerous other factors that could affect the fish stocks both positively and negatively, such as climate and weather, water levels and nutrient content, epizootics and many more. Since there are no catch statistics for Lake Constance before the late nineteenth century, it is impossible to exactly quantify the effect of the pre-modern fisheries on the lake, let alone what impact the individual rules contained in the ordinances and treaties had on certain species. Nonetheless, the sources give us some indications that are worth contemplating.

The fishermen themselves were certainly aware that the lake was subject to many effects beyond their control. In some years certain species were rare while others abounded, but we have no indication that the fishermen were able to attribute these changes to a pattern. These at times drastic fluctuations were part of the vicissitudes of their job. Other occurrences were easy to identify, such as epizootics. In 1536, a plague among roach in the eastern part of the Upper Lake was reported from Überlingen to Constance, and disease decimated pike in the Lower Lake in 1790. In both cases, the authorities reacted, but their measures were short-term and had apparently no lasting influence on the ordinances and treaties.[14]

In addition to such natural factors, human actions other than the fisheries also could have a detrimental effect on the fish stocks. In 1580, the city of Constance built a new mill on the Rhine bridge connecting the city with the suburb of Petershausen to the north. In order to concentrate the strength of the weak current of the Lake Rhine on the millwheel, a funnel made of wattling was constructed. A few years later, there began to be complaints that the funnel was inhibiting the migration of *Gangfisch*, a variety of whitefish, to its spawning grounds in the Lake Rhine. Since this was the most profitable season for those fishermen who had purchased the much sought-after right to exploit the private fishing rights in the Lake Rhine with huge seines, the economic effects were immediately evident. From 1580 onwards, those fishermen accrued immense debts in kind, which they owned to the proprietors of the fishing rights, and they repeatedly asked for either a reduction of their dues or their conversion into coin. Only in 1606 did the city of Constance agree to convert a debt of 40,000 *Gangfish* into 40 guilders, payable in two instalments, while the annual duty was reduced from 3,000 to 2,000 per year, but only for the following six years. This arrangement soon proved to be inadequate, as the fishermen accrued further debts of 44,000 *Gangfisch* between 1629 and 1650.[15]

Although other factors certainly played a role in the accumulation of the fishermen's debt, and the transformation of payments in kind into payments in coin was part of a larger trend that was in the interest of both the fishermen and the owners of the fishing rights,[16] the sources leave little doubt that the new mill on the Rhine bridge had a negative effect on the fisheries. The city of Constance would not otherwise have compensated the abbey of Petershausen with two fishing rights for their lost income[17] and the value of a share in the large seine used for catching *Gangfisch* during spawning season would not have collapsed from 46 to 13 guilders between 1594 and 1602, at a time of inflation across most of Europe.[18]

The mill on the Rhine bridge demonstrates how human actions could affect fish stocks locally, but only the fisheries had the potential to harm them as a whole. The case of the Zeller See in Austria provides an instructive example of the devastating effects pre-modern overfishing could have. In the mid-fourteenth century, the local fishermen agreed to pay the lake's owner, the Archbishop of Salzburg, 27,000 whitefish and eighteen lake trout per year for the right to exploit the fish stocks. It took them only one human generation to collapse the whitefish stocks. The introduction of pike in order to compensate for their losses only harmed the trout stocks. In the end, the authorities had to ban all fishing for three years, and afterwards introduced strict regulations (Hoffmann 1996: 648).

The overfishing of the Zeller See shows the destructive potential of the pre-modern fisheries. Constant pressure on the fish stocks without any restrictions could have devastating consequences, even if the equipment used was relatively ineffective. Although it is certainly inappropriate to draw too many conclusions from this case – Zeller See is far smaller than Lake Constance – it does lend some credibility to the observations of overfishing contained in the fishermen's ordinances and fisheries treaties. The authors of these texts knew the lake well enough to identify at least the most obvious effects of the fisheries. It was definitely no coincidence that they paid particular attention to the one practice that had the most destructive potential for the fish stocks: The catching of young fish that had not yet reproduced. Young perch and whitefish in particular were considered a delicacy by local consumers who were willing to pay a good price on the market. Since the fishermen depended on the market for their livelihood, it must have been more than tempting to emphasise short-term profits and neglect the long-term viability of the fisheries.

It is therefore only natural that the fishermen's ordinances and fisheries treaties backed up their preambles with rather strict regulations concerning the catch of young perch.[19] Completely banning the practice, however, was apparently impossible, whether due to customer demand or the fishermen's dependency on the income. That the only attempt to negotiate a fisheries treaty for the whole Upper Lake in 1790 was instigated by the perceived overfishing of young perch fits into this picture. Although the negotiations among the parties from the northern shore of the lake came to a successful conclusion, which included a ban on the catching of young perch for an unprecedented ten years, the treaty failed because the Swiss refused to join. The agreement then collapsed because the signatories could not compel their fishermen to obey such a harsh rule that did not apply to everyone.[20]

The failure of the comprehensive fisheries treaty of 1790 should not distract us from the successes of the regulatory regimes for the Lake Constance fisheries. Even if they did not completely outlaw the most harmful practices, such as the catching of young whitefish and perch, they managed to curb them in a way that evidently succeeded in maintaining the fish stocks of both species and all the others. That this was not an accident is underlined by an analysis of the temporal distributions of single rule changes contained in ordinances and treaties. Most occurred in the years between 1450 and 1550 and then again in the eighteenth century. These periods of accelerated population growth pushed the agricultural capacities of Central Europe, including the Lake Constance region, to their limits (Kissler 1992: 13–21; Pfister 1994: 8–24). As the number of fishermen increased alongside the general population, the temptation must have been strong to satisfy both their economic needs and the increasing demand for foodstuffs at the same time by relaxing the rules. Apparently, the opposite happened. The authorities and their fishermen put the long-term viability of the fish stocks first, even if it meant that demand in the short-term could not be fully met.

Considering the factors that enabled these regimes to function for such a long time, we find a striking congruence with the "design principles" for the sustainable management of commonly used resources, which Elinor Ostrom had developed almost 30 years ago. Although the boundaries of the resource were not explicitly defined, the difficulty of extracting the resources reduced access in practical terms, and excluded everyone but the professional fishermen. Since fish was a highly perishable food, they had to rely on the local markets for selling their catch, thus making it impossible for foreigners to extract the resource outside the regulatory regime. The rules were developed by the fishermen in cooperation with the local authorities and therefore designed for the local environmental, social and political conditions, and were adapted according to changing circumstances. Consequently, the voices of all fishermen would be heard in the regular debates regarding the rules, and the outcome would be the result of a collective decision that included the perspectives of the authorities as well as the resources' users. The collective responsibility for the development of the rules also extended to the monitoring of their observance, with the authorities having to rely on the fishermen for information on infractions. If punishment was necessary, it was along a gradient according to the severity. Conflicts between fishermen were solved through the local authorities even if that required the cooperation of different principalities. Finally, the fishermen's right to organise in guilds was never inhibited by the authorities of the free imperial cities or, as in the case of post-conquest Constance, by the Austrian government. The only design principle that is not easily discernible is nesting, although the failed comprehensive treaty of 1790 could be described as an attempt to achieve a more integrated regulatory system for the Upper Lake (Ostrom 1990: 88–102).

Reflecting on the local development of the regulatory regimes for the pre-modern Lake Constance fisheries that so neatly fits Ostrom's analysis, it is impossible to overestimate the role of the guilds in developing and maintaining a system of rules based on resource management that lasted for centuries (De Moor 2008). They provided the institutional framework in which the fishermen's ordinances were first developed. Since they both represented the local fishermen who were their members and were part of the free imperial cities' councils, the highest local political authority, they incorporated both the perspective of the regulator and the regulated. They established the space for negotiations that would do justice to the interests of the fishermen, the city and its inhabitants. The abbeys of Reichenau and St. Gall adopted the fishermen's ordinance as a regulatory tool from the cities, although that their fishermen were subjects who had no say in legal affairs is telling. The abbeys thus gave their fishermen responsibility and a say, and both profited as a result.

Notes

1 The literature disagrees on the exact numbers of fasting days prescribed by the Catholic Church during the late medieval period, so I chose the average (cf. Lampen 1997: 150; Hoffmann 2007: 301).

2 Numerous price lists can be found in the archives: for Constance, cf. Stadtarchiv Konstanz (hereafter StA KN), DI Fasc. 28, 36, 44 and DI Band 43; for the Lower Lake, cf. StA KN, DI Fasc. 38, 39, 40, 42 and Generallandesarchiv Karlsruhe (hereafter GLA KA), 96/405; for Lindau, cf. Stadtarchiv Lindau (hereafter StA LI), A III 55,2; and for Überlingen, cf. Stadtarchiv Überlingen (hereafter StA ÜL), C 988.

3 Fish market ordinance Constance, 1433; Fishermen's ordinance Constance, 1491, both StA KN, DI Fasc. 44.

4 The fishermen's fields were mentioned in a letter by the reeve of Ittendorf to the city of Überlingen of 1513, StA ÜL, C 976/1.

5 Report by Constance sent to Überlingen in 1531 on the arrest of a fisherman from Überlingen, complaint by Überlingen sent to Salem in 1560 about a fisherman from Maurach, both StA ÜL, C 976/1; complaint by Constance sent to Reichenau in 1606 about the fishermen from Ermatingen, StA, KN DI Fasc. 37; complaint by the fishermen from Constance sent to the bailiff of Thurgau about the fishermen of Landschlacht, StA KN, DI Fasc. 32; complaint by the fishermen of Lindau sent to Constance in 1716 about the fishermen from Staad, StA LI, A III 110,16.

6 Fishermen's ordinance from Constance, undated (14th century?), StA LI, A III 101.8; fishermen's ordinance for the Lower Lake, 1450, StA KN, DI Fasc. 44; fishermen's ordinance from St. Gall, 1497 (Stoffel 1906: 18). On ordinances from Lindau see Leipold-Schneider 2002: 60.
7 Fisheries treaty from St. Gall, 1544, StA KN, DI Fasc. 44.
8 There were two different kinds of cubits in use in Lake Constance in the pre-modern period, and it has been impossible to ascertain which one was referred to (Göttmann 1989: 38–40).
9 Fishermen's ordinances for the Lower Lake, 1455 and 1613, both StA KN, DI Fasc. 44; 1522, StA KN, DI Fasc. 37; 1542 and 1550, both StA KN, DI Fasc. 37; 1594, StA KN, DI Fasc. 26; 1707, StA KN, DI Fasc. 39; 1717, StA KN, DI Fasc. 40; 1774, StA KN, DI Fasc. 42.
10 Fishermen's ordinances for the Lower Lake, 1522; 1532, StA KN, DI Fasc. 37; 1542; 1550; 1594; 1613, StA KN, DI Fasc. 44; 1707; 1717; 1774.
11 Fishermen's ordinances for the Lower Lake, 1470 and 1583, both StA KN, DI Fasc. 44; 1542; 1550; 1594; 1613; 1707; 1717; 1774.
12 Fisheries treaty between Constance and Überlingen, 1536, StA KN, DI Fasc. 44 and StA ÜL, C 976/1.
13 Fisheries treaty between Constance and Mainau, 1566, StA KN, DI Fasc. 35.
14 Letter from Constance to Überlingen, 11 September 1536, StA ÜL, C 976/1; letter from Reichenau to Constance, 8 May 1790, StA KN, DI Fasc. 29.
15 Memorandum 1604, GLA KA, 209/306; compilation of debts, GLA KA, 95/47.
16 Several treaties, GLA KA, 5/398.
17 Report from 1581, GLA KA, 95/44.
18 Contracts, 1594, 1595, 1603, StA KN, DI Fasc. 29 (cf. Gerhard 1993).
19 Specific rules restricting the catch of perch are present in almost all fishermen's ordinances and fisheries treaties.
20 Correspondence between the principalities of the Lake Constance region, 1789 and 1790, StA KN, DI Fasc. 36, StA LI, A III 55,2, and GLA KA, 209/328; minutes of the fisheries conference, 1790, StA KN, DI Fasc. 45; letter from Lindau to Constance, 24 July 1790, StA LI, A III 55,2, letter from Constance to St. Gall, 30 July 1790, and letter from Münsterlingen to Constance, 3 January 1791, both StA KN, DI Fasc. 36.

References

Amacher, U. 1999. Die Fischermaien. Die Gerichtstage der Fischer zwischen eigener Reglementierung und herrschaftlicher Machtausübung. In *Wirtschaft und Herrschaft: Beiträge zur ländlichen Gesellschaft in der östlichen Schweiz (1200–1800)*, edited by T. Meier and R. Sablonier, 279–294. Zurich: Chronos.

Berg, R. 1993. Über die Fische des Bodensees. In *Bodenseefischerei: Geschichte – Biologie und Ökologie – Bewirtschaftung*, edited by B. Wagner, H. Löffler, T. Kindle, M. Klein, and E. Staub, 58–72. Sigmaringen: Thorbecke.

Bogue, M.B. 2000. *Fishing the Great Lakes: An Environmental History, 1783–1933*. Madison, WI: University of Wisconsin Press.

De Moor, M. 2008. The Silent Revolution: A New Perspective on the Emergence of Commons, Guilds, and Other Forms of Corporate Collective Action in Western Europe. *International Review of Social History* 53: 179–212.

Dobras, W. 1991. Konstanz zur Zeit der Reformation. In *Konstanz in der Frühen Neuzeit. Reformation, Verlust der Reichsfreiheit, Österreichische Zeit*, edited by M. Burkhardt, W. Dobras, and W. Zimmermann, 11–146. Constance: Stadler.

Eitel, P. 1982. Die Städte des Bodenseeraums: Historische Gemeinsamkeiten und Wechselbeziehungen. In *Der Bodensee*, edited by H. Maurer, 577–596. Sigmaringen: Thorbecke.

Feger, O. 1975. *Geschichte des Bodenseeraumes*, vol. 1: *Anfänge und Größe*. 4th ed. Sigmaringen: Thorbecke.

Feger. O. 1981. *Geschichte des Bodenseeraumes*, vol. 3: *Zwischen alten und neuen Ordnungen*. 2nd ed. Sigmaringen: Thorbecke.

Feger, O. 1983. *Geschichte des Bodenseeraumes*, vol. 2: *Weltweites Mittelalter*. 3rd ed. Sigmaringen: Thorbecke.

Gerhard, H.-J. 1993. Ursachen und Folgen der Wandlungen im Währungssystem des Deutschen Reiches 1500–1625. Eine Studie zu den Hintergründen der sogenannten Preisrevolution. In *Geld und Währung vom 16. Jahrhundert bis in die Gegenwart. Referate der Gesellschaft für Sozial- und Wirtschaftsgeschichte vom 9. bis 13. April 1991 in* Dortmund, edited by E. Schremmer, 69–85. Stuttgart: Steiner.

Goggins, B. 2017. Looking for Pearls: "Shifting Baselines" to Save a Way of Life on the Water. *Savannah Now*, 14 August 2017, www.savannahnow.com/accent/column/news/2017-08-14/looking-pearls-shifting-baselines-save-way-life-water (last accessed 9 November 2018).

Göttmann, F. 1989. Altes Maß und Gewicht im Bodenseeraum. Systeme und Kontinuitäten. *Zeitschrift für württembergische Landesgeschichte* 48: 25–68.

Governing the Oceans. The Tragedy of the High Seas. *The Economist*, 22 February 2014, www.economist.com/leaders/2014/02/22/the-tragedy-of-the-high-seas (last accessed 9 November 2018).

Grober, U. 2010. *Die Entdeckung der Nachhaltigkeit. Kulturgeschichte eines Begriffs*. Munich: Kunstmann.

Haller, T. 2012. *The Contested Floodplain: Institutional Change of the Commons in the Kafue Flats, Zambia*. Lanham, MD: Lexington Books.

Hardin, G. 1968. The Tragedy of the Commons. *Science* 162: 1243–1248.

Hoffmann, R.C. 1996. Economic Development and Aquatic Ecosystems in Medieval Europe. *American Historical Review* 101: 630–669.

Hoffmann, R.C. 2007. Footprint Metaphore and Metabolic Realities: Environmental Impacts of Medieval European Cities. In *Natures Past: The Environment and Human History*, edited by P. Squatriti, 288–325. Ann Arbor, MI: University of Michigan Press.

Hölzle, E. 1938. *Der Deutsche Südwesten am Ende des Alten Reiches*, Stuttgart: Württembergisches Statistisches Landesamt.

Internationale Gewässerschutzkomission für den Bodensee (ed.) 2004. *Der Bodensee: Zustand, Fakten, Perspektiven*. 2nd ed. Bregenz: Internationale Gewässerschutzkommission für den Bodensee.

Jacobs, A. 2017. China's Appetite Pushes Fisheries to the Brink. *New York Times*, 30 April 2017, www.nytimes.com/2017/04/30/world/asia/chinas-appetite-pushes-fisheries-to-the-brink.html (last accessed 9 November 2018).

Kirchgässer, B. 1988. Strukturfragen von Handel und Verkehr des Bodenseeraums im Mittelalter. In *Bernhard Kirchgässer. Wirtschaft, Finanzen, Gesellschaft: Ausgewählte Aufsätze zu seinem 65. Geburtstag*, edited by J. Wysocki, 235–259. Sigmaringen: Thorbecke.

Kissler, A. 1992. *Die Bevölkerung der Stadt Radolfzell am Bodensee des 17. und 18. Jahrhunderts. Demographische Strukturen einer "Ackerbürgerstadt" vor Beginn der Industrialisierung.* Constance: Hartung-Gorre.

Klunzinger, C.B. 1892. *Bodenseefische, deren Pflege und Fang.* Stuttgart: Enke.

Kunz, R. 1994. *Fischereirechte im Untersee und Seerhein. Eine rechtshistorische Untersuchung über die Entstehung, Ausbildung und Weiterentwicklung von Fischereirechten.* PhD thesis. Fribourg: Université de Fribourg.

Lampen, U. 1997. *Fischerei und Fischhandel im Mittelalter. Wirtschafts- und sozialgeschichtliche Untersuchungen nach urkundlichen und archäologischen Quellen des 6. bis 14. Jahrhunderts im Gebiet des Deutschen Reichs.* Husum: Matthiesen.

Leipold-Schneider, G. 2002. Die Lindauer Schiffer- und Fischerzunft von den Anfängen bis ins 18. Jahrhundert. In *Das Recht im kulturgeschichtlichen Wandel. Festschrift für Karl Heinz Burmeister zur Emeritierung*, edited by B. Marquardt and A. Niederstätter, 59–85. Constance: UVK.

Ostrom, E. 1990. *Governing the Commons: The Evolution of Institutions for Collective Action.* Cambridge: Cambridge University Press.

Pfister, C. 1994. *Bevölkerungsgeschichte und historische Demographie.* Munich: Oldenbourg.

Press, V. 1989. Vorderösterreich in der habsburgischen Reichspolitik des späten Mittelalters und der frühen Neuzeit. In *Vorderösterreich in der frühen Neuzeit*, edited by H. Meier and V. Press, 1–41. Sigmarinen: Thorbecke.

Schuster, B. 1951. *Die Entwicklung der Hoheitsverhältnisse auf dem Bodensee seit dem Dreißigjährigen Krieg unter besonderer Berücksichtigung der Fischerei.* Constance: Merk.

Schuster, P. 1997. *Eine Stadt vor Gericht: Recht und Alltag im spätmittelalterlichen Konstanz.* Paderborn: Schöningh.

Stoffel, F. 1906. *Die Fischereiverhältnisse des Bodensees unter besonderer Berücksichtigung der an ihm bestehenden Hoheitsrechte: Historisch-dogmatische Studie.* Bern: Stämpfli.

Strigel, A. 1910. *Die Fischereipolitik der Bodenseeorte in älterer Zeit mit besonderer Rücksicht auf Überlingen.* PhD thesis. Freiburg im Breisgau: Universität Freiburg.

Wieland, G. 2002. Die Integration der Städte in die neuen Staaten. In *Zum Übergang schwäbischer Reichsstädte vom Kaiser zum Landesherrn. Begleitband zur Ausstellung "'Kronenwechsel'. Das Ende reichsstädtischer Freiheit"*, edited by D. Nohrath, G. Weig, and M. Wettengel, 56–110. Ulm: Kohlhammer.

Worstall, T. 2017. Brexit Allows Us to Solve the Haddock Problem by Leaving the CFP. *Forbes*, 19 March 2017, www.forbes.com/sites/timworstall/2017/03/19/brexit-allows-us-to-solve-this-haddock-conservation-problem-by-leaving-the-cfp/#2d86855a6384 (last accessed 9 November 2018).

Zeheter, M. 2014. *Die Ordnung der Fischer. Fischerei und Nachhaltigkeit am Bodensee (1350–1900).* Cologne: Böhlau.

8 The commons in highland and lowland Switzerland over time

Transformations in their organisation and survival strategies (seventeenth to twentieth century)

Anne-Lise Head-König

Introduction

For several centuries, despite changing political regimes and the varying territorial configuration of some Swiss cantons, access to the commons and even their very existence have frequently been the subject of virulent disputes both at a local and cantonal level. The bone of contention varied accordingly. Should access to the common resources be based on rights linked to the possession of landed property or rather to individual rights? During the Ancien Régime, and in the public interest, should the peasant corporations share the political power up to a certain degree with the emerging peasant communities? Similarly, after the Helvetic Revolution (1798–1803), to what extent should corporations of burgesses[1] waive part of their economic and political power in order to share it with the newly created municipalities (political communities)? For decades, the question of the transfer of the landed assets of the peasant corporations and of the communities of burgesses to the municipalities (political communities) remained a major controversial issue. A conclusive solution acceptable for all parts of Switzerland was in fact never arrived at due to the extreme variation of situations from one canton to another, especially when the situation in the mountain regions is compared to that of the lowlands.

In this chapter, then, I will discuss the following points: the reasons for the continuing existence of common land in lowland and upland Switzerland; the characteristics of the diverse organisations owning common resources in the Swiss lowlands and highlands in the seventeenth and eighteenth centuries; the introduction of a host of stricter rules being implemented due to the increase of population and the measures taken to hinder an increase in the circle of beneficiaries; the involvement of several cantonal governments in the sometimes very lengthy process of appropriating common property by either facilitating its transfer to the municipalities or its allocations to individual citizens during the nineteenth century; and finally, the central issue involving the forces of communal traditionalism on the one hand and those of cantonal centralism on the other.

The continuing diversity in the very existence of common land in lowland and upland Switzerland

Switzerland is a country of great geographical variety and this explains the emergence of very different types of organisations to manage and regulate common land resources, which will be dealt with below. The differences were partly the result of different altitudes – lowlands compared to mountain regions – but also the result of climate and the concomitant water regimes, the size of the forests and the extent of meadows providing the possibility to winter livestock. With regard to the governance of common land, topography and remoteness were decisive factors at the root of the significant differences between lowland and highland areas. From the fifteenth and sixteenth centuries, peasant communities in the lowland regions were never entirely free to administrate their common land as they wished; they were always subject to some extent to the state, after state authority had come to supersede seigniorial and ecclesiastical domination. In the pre-alpine and alpine regions, on the other hand, there was a progressive emancipation from part or all external domination in the case of a number of rural communities, which became possible by their greater market involvement serving to finance their emancipation. These self-governing bodies often had a large degree of autonomy up to the nineteenth century and even beyond.

The continuing diversity in the organisation of the commons is also the result of yet other factors, such as the historical process involving the creation of new territories. Some cantons, such as St. Gall, Aargau or Ticino, created at the beginning of the nineteenth century, are a patchwork of former territorial entities: Each entity has kept its own rules up to the present day with regard to common land and highland pastures. However, even an old cantonal territory such as Bern, created at the end of the Middle Ages, had to acknowledge that the provisions regarding the access to ownership of mountain pastures in the districts of the Bernese Oberland and going back to their late medieval statutes were still valid in the nineteenth century. Thus, in the valley of Grindelwald and in the Haslital attempts by external entrepreneurs at the end of the nineteenth century to buy upland pastures with a view to developing a more commercial pastoralism were successfully countered. The cantonal government was obliged to recognise the strict rules concerning the appropriation of the upland pastures.

Even after the Helvetic Revolution with the newly created municipalities, the common land and its resources owned by burgesses' communities or rural corporations in their various guises remained a significant reality despite the fact that they were the cause of escalating disputes, especially where there was a determined government policy to abolish such self-governing bodies. The controversies in many regions often lasted for decades and frequently necessitated the intervention of cantonal Courts of Justice. One is struck, however, by the resilience of the common property regime of the old corporations of burgesses,[2] later the civic corporations, not only in the mountainous areas, but also both in the rural and even the urban lowlands, despite the ferocious opposition they met with, particularly in the last decades of the nineteenth century. Nowadays, of the

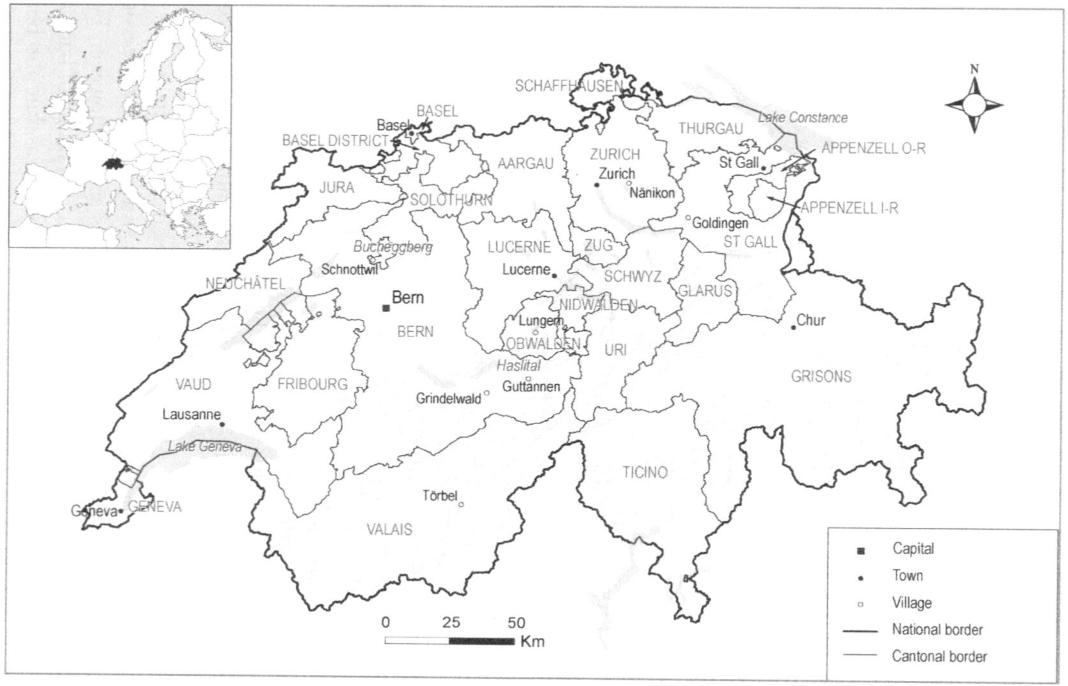

Figure 8.1 Map of Switzerland showing the present-day cantons, towns and villages mentioned in this chapter.
Source: Map compilation and design by Ulla Gämperli Krauer (2018), CDE, University of Bern, Bern, Switzerland. Geodata source: OSM data by OSMaxx (2018). Geometalab, HSR University of Applied Sciences Rapperswil, Switzerland. Hillshade: Swiss Hillshade by Juerg Krauer (2017), CDE, University of Bern, Bern, Switzerland.

nineteen cantonal capital cities situated in the lowlands, only five of them no longer possess a burgesses' corporation, and the corporations, which have survived up to the present day as civic corporations, continue to possess substantial areas of the countryside. In fact, eight of them own between 500 hectares and 5000 hectares, which always include forests and often farmland, urban properties as well as building land and sometimes even alpine pastures.

Whatever tensions emerged in relation to the precise definition of those who should benefit from the common resources, the acceptance of the varying forms of governance is also due to the acknowledgement by the Federal Government of its own inability to impose a common policy at national level, which would be acceptable to all. The 1874 Swiss Federal Constitution merely settled the nagging issue of the creation of municipalities, which became compulsory for all the cantons, which so far had carefully avoided implementing this new type of political organisation. The significance of this development cannot be underestimated since this decision concerned many Alemannic cantons with large rural corporations. Nevertheless, the Federal State used a pragmatic and circumspect approach with this imposed dualism. The solution adopted in relation to collective organisations whatever their denomination, as art. 59 of the 1907 Swiss Federal Constitution states, was to leave all decisions concerning their status within the cantonal jurisdiction, since as was argued at the time of its adoption, 'commons associations' (*sociétés d'allmends*) and their inherent characteristics were closely linked to the economy of the relevant Swiss regions. The influence of cantonal legislation on the continuing existence of common resources belonging to a rural corporation is not to be underestimated either. The examples of Vaud, Zurich and Neuchâtel – all cantons with considerable migratory movements in the second half of the nineteenth century and with citizens no longer living in their community of origin – bear witness to the fact that the intervention of a determined government with the aim of transforming resources previously belonging to a burgesses' community into collective resources belonging to the relevant political municipality could effectively be successful.

Up to the beginning of the nineteenth century, whatever the size of a community was, nearly all Swiss communities possessed common land, be it urban communities, small and large towns, or rural communities. There were just a few rare exceptions, as in the canton of Thurgau, where some tiny villages or isolated farms did not form a part of any community at all in this period. Communal land might belong to a peasant community or a hamlet (*fraction de commune*), which was often the case in the French speaking part of Switzerland as in the canton of Vaud, but also in the Grisons. However, it might also be co-owned by specific families or might be the joint property of neighbouring villages or belong to a whole valley.

Nevertheless, communal land owned by a corporation, especially alpine pastures, might also exist at a considerable distance from the village territory, when, for example, a peasant community had wanted to increase its usable land and had been unable to expand at the expense of its neighbouring communities. Such policies can be observed already in the sixteenth century as well as in later years. In 1514, the peasant corporation of Törbel purchased the Alp Oberaar from the community of Guttannen (under Bernese jurisdiction) with its 350 hectares, which was a three days walk away across the Grimsel Pass (2164 metres). It provided pasture for several hundred sheep, plus cattle and horses. Right-of-way and overnight grazing privileges had to be bought from other villages along the route (Netting 1972: 135). Expansion outside their communities' boundaries was also a necessity for some peasant communities in the canton of Glarus in the eighteenth century. There, the communities bought land from other communities in order to be able to allocate a plot of land to each of the households belonging to their corporation.

Perhaps the more speaking examples of expansion outside one's territory are to be found in the strategies of urban communities, such as Bern, St. Gall and Lucerne, which systematically acquired land, forests and sometimes even alpine pastures from the sixteenth century onwards. Over time, even the small town of Chur, with wealth accumulated from transalpine transit, managed to purchase alpine pastures from three different peasant corporations, which, by the end of the nineteenth century, could provide pasture for 450 cows (Dönz 1943; Jörimann 1970; Buchmann 1997). Thus, the land owned by the burgesses' corporations did not and still does not necessarily coincide with the geographical location of the present political municipalities that were created in the nineteenth century.

The legal basis for the rights related to commons up to the beginning of the nineteenth century

The disparate origins of the different types of corporation and collective organisation possessing common land and resources explain to a large extent the complexity of property right arrangements in respect of the access, management and the types of common resources. However, the element that all these different forms of rural corporations/associations have in common is that their assets, that is to say the common property resources, were possessed in common by a group of co-owners. The decisions concerning the management of the common land were taken by an assembly of their members or by a council, which was elected by the members of the association. In practically all cantons, though, the users of the common land, in return for the benefits they obtained from the common resources, had to perform statute labour (*Frondienst, corvée*). In upland regions, this was carried out according to a schedule based on the number of LUs/*UGB* (Livestock Units/*Unités de gros bétail*), which were being grazed. In such cases the main activity was removing stones, bush and shrubs from the pastures, and later the installation of fences, drinking troughs and the distribution of manure.

The variety in the form of the rural corporation brings with it numerous problems relating to the terminology applicable, albeit rather less in the lowland regions but then all the more so in the highland regions, where the governance of the common resources could vary considerably from one valley to the next.

For simplicity's sake, it is possible to distinguish between two main groups of entitlement rights for access to common land resources. First, there was a system based on the possession of immovable property, such as a house or a farm holding. The owners of such rights (*Rechtsame, Gerechtigkeit*) form a peasant corporation called *Rechtsamegemeinde* (*genossenschaftliche Gemeinde*). This system can be traced back to the Middle Ages and is a similar prerequisite to that of towns which required a right *in rem* (immovable property) to be able to enjoy access to common resources. In this case, by virtue of possessing a specific house or a farm holding, its owner was automatically regarded to be a true member of the community in accordance with the statute. This type of organisation reflects an agricultural system more focussed on cereal production than one on livestock farming typical of higher regions. It gradually spread throughout the rural regions from the end of the sixteenth century in order to protect the rights of such owners at a time, when it was thought that the wasteland available for increasing the common had all been used up (Stadler 1982; Pfister 1993; Ineichen 1996). Statutory regulations were implemented specifying which owners of immovable property were entitled to use the resources of the common land. Further, this type of organisation was typical of regions where rural corporations had only limited autonomy in the management of their resources. They existed mainly in Alemannic regions often specialised in arable farming, as was the case for the lowland regions of the cantons of Bern, Zurich, Solothurn, Zug and Aargau, but was scarcely in evidence in the French and Italian speaking parts of the country.

The second system is based on the possession of personal rights, usually transmitted by inheritance from father to son. The owners of such rights form a corporation, an association or a co-operative, which is characterised by a high degree of autonomy, at least until the nineteenth century. In fact, the enormous variety in the designations for these long-standing local organisations also reflects their specific relationship to a territory. First, there are corporations of owners, which relate to the entire communal territory. *Tagwen, Bürgerschaft, Bürgergemeinde, Burgergemeinde, 'eine ganze Gmein', Dorfgemeinde, Nachbarschaft, Teilsame, Korporationsgemeinde, Geschlechterkorporation, bourgeoisie, commune bourgeoise, vischnancam, vicinanza* belong all to this first category. Second, there are corporations of owners, private associations of families and co-operatives of landowners, which relate to only a part of the territory. *Nutzungsgemeinde, Geteilschaft, Allme(i)nkorporation, Korporation, Genossame, Genossenschaft, Bergschaft, Taleinung, Bäuert, Bauernzunft, Rhode, Fleckengenossenschaft, Halbrhode, patriziato, consortage, fraction de commune* belong to the second category. Third, there are the corporations of owners of rights associated with a specific resource: *Alpgenossenschaft, Alp- und Allmendgenossenschaft, Flurgenossenschaft, Nutzungsgenossenschaft, Alp-, Holz-, Wald-, Weidekorporation, Geteilschaft, consortage d'alpages, de forêts, de bisses.* In the event the distinction between the three categories is not infrequently blurred, and especially between the second and the third type.

With the institutional changes following the Helvetic Revolution and the creation of the municipalities (political communities), new designations were sometimes introduced for some of the former institutions, in particular those for the old *Bürgergemeinde/bourgeoisie*,[3] which could become an *Ortsgemeinde*, *Ortsbürgergemeinde* or *vischnanca burgaisa*, *Korporationsbürgergemeinde*.

Whatever the system of access to common resources was, there were always strict regulations to limit the number of the holders of rights, and this was especially so during the Ancien Régime when the population was increasing.

Even within the first system mentioned above, when the use of rights was appended to immovable property, the relevant economic and demographic implications could vary to a considerable extent. This would depend on whether – in addition to the appending rights related to a house in a village or to a farm with its land – there was a further requirement for the owner to be a full member of the peasant community. In many cases, the number of *Gerechtigkeiten*, that is to say the number of houses, seems not to have always been fixed once and for all in times of population pressure, as the village of Nänikon (canton of Zurich) demonstrates. The rigidity of the system was circumvented at the end of the seventeenth century when newcomers, who had to pay an entrance fee (*Einzugsgeld, droit d'entrage*) to reside in the *Dorfgemeinde*, and after paying an additional fee were allowed to build a house there. As the necessary amount for the fee could be borrowed and no requirements as to the size and the quality of the house to be built were specified, there was a considerable increase in the number of poorly built houses ('bad shacks'), which nevertheless qualified their owners for land-use rights. Hence, after a few decades, at the beginning of the eighteenth century, the building of new houses had to be prohibited to prevent the overuse of the common resources; in other localities, the building of houses was not prohibited, but their owners were not entitled to use the common resources (Kläui 1964: 138; Pfister 1993: 434). This system could even co-exist together with other types of corporation within a village community (*Dorfgemeinde*) and mostly as from the seventeenth century for reasons we develop later. In the second type of arrangement, where the use of the commons was viewed as appendant to a specific farmholding, the share of the relevant resources was no longer egalitarian as from the middle decades of the seventeenth century. It often depended on the extent of the land to be farmed; the modalities were the result of local or regional practices often introduced at this time. However, it is evident, though, that those who did not possess a dwelling with some land officially had no right to make use of the commons. An interesting example of the variation in the size of the allocations, depending on the extent of land and the needs of the farm, can be observed in the very strict rulings approved by the Council of State of Solothurn for the common property in the community of Schnottwil in the Bucheggberg district in 1675 (Renaud 1845: 43–44; Jäggi 1953). The rights for using the commons were divided into five classes according to the size of land: the first four going from a full right to a quarter right, the fifth class being for those possessing only a small house with a little plot, which resulted in a one eighth of a right and permitted only a very restricted use of common resources. The possession of approximately 11 hectares of land was the prerequisite to the possession of one full right of the entitlement to send 40 pigs in the forest, to make use of an entire portion of wood from the forest and pasture one's cattle on the common land. Additionally, the requirement for the land was that three quarters should be devoted to tillage and one quarter should be meadow.

Important changes occurred in the system of the *Rechtsamegemeinde* in the last decades of the seventeenth century and in the course of the eighteenth century. It then became possible to sell the farmstead with its common rights, in some regions even to buyers from outside the community. It was also possible to split only the right (and not the farm), sometimes even in such a way that the previous owner of the farm lost all entitlement to the commons since membership of the village community did not necessarily go hand-in-hand with access to common resources.

When comparing this type of access based on the size of farm land to that of some regions in the Swiss highlands, one can detect a number of similarities, in particular in the system which prevailed to limit the overuse of scarce resources, such as in Obwalden or in the Grisons (Dosch 2002: 332–334). It was the rule that only the cattle, which the owner had fed during the winter with the straw and hay he had harvested on his own holding within the boundaries of the *Gemeinde* or *Teilsame*, could be pastured on the common alp. This rule later became even more restrictive. Additionally to the 'straw and hay rule' for pasturing cattle on the common land, the landowner also needed to be a member of the corporation (community) and to reside within its boundaries.[4] In the *Bergschaften* of Grindelwald, it was the number of the land plots owned privately in the valley, which were the determinants for the number of LUs being allowed to graze on the summer pastures (Tiefenbach 2013).

Let us consider the implications of the second system for accessing common land resources. It was a distinctive characteristic of most mountain regions that the rights to use common resources were personal rights – with sometimes some specific rules regarding the use of alpine pastures – with the prerequisite of being a member of the village corporation or of the association of owners of specific resources. At a time of low population pressure, newcomers were welcome and received the entitlement to use the resources of the common land when helping with the colonisation of the land, the regulation of the course of mountain torrents and the building of dams, or helping with the building of *suonen/bisses* (water channels) for irrigation purposes. In the course of the fifteenth and sixteenth centuries, however, in most parts of highland Switzerland, a combination of institutional and economic factors were to result in a new and more restrictive management of property rights relating to the common resources. This led in turn to the emergence of closed corporate communities with their own precise requirements: descent from an old, established lineage, residence within the *Gemeinde*, to be male[5] and the head of a household. Over time, rules dealing with the acquisition of local citizenship – and its loss – became ever more restrictive, and in the course of the seventeenth and eighteenth centuries, the acquisition of local citizenship at one's place of residence was ultimately almost impossible due to the high cost involved. Consequently, those whose ancestors had failed to acquire local citizenship (*droit de bourgeoisie/Bürgerrecht*) at the time when it was created remained excluded from all the vital resources generated by the common-rights economy, a point to which we will return below.

Regulations for sustainable use

Beginning in the seventeenth century, population growth threatened the sustainability of the common resources in nearly all regions of Switzerland. Numerous restrictions had to be adopted progressively by all types of corporations, associations, co-operatives, whether they were in lowland or upland Switzerland, and whatever their economic environment, rural or proto-industrial, cereal-growing or pastoral, or a mixture of these, and even by those with extensive common resources. The number of corporations able to increase the extent of their common land was limited and to do so was not without risk, especially when clearance of forests could lead to damage to the environment or to conflicts with neighbouring corporations due to the complex mix-up of ownership's rights involved. This was despite the fact, as mentioned before, that some corporations managed to acquire land outside their boundaries to increase the extent of their common property. It is obvious that many others lacked the means to do so. The demand for common land resources was on the increase: more people wanted to graze cattle after the harvest and also in the forest when hay was in short supply; more people needed wood (firewood and timber) and in times of crisis also plots to grow hemp, flax and vegetables, and even cereals as we already see in the case of lowland Lucerne in the sixteenth century. At the same time, the risks of overuse and deforestation both in lowland and upland regions were striking, becoming visible in the eighteenth century with a considerable increase of marshy land at the bottom of valleys as caused, for example, by the river Linth in the canton of Glarus, or by the river Glatt in lowland Zurich.

The pressure on the resources was manifold not only within the corporations but also from outside. Within their own boundaries, a double, endogenous problem had to be reckoned with: the question of those living in the corporation with no or very limited rights and that of a natural increase in the population. In addition, there was an exogenous pressure on resources due to immigration. The difficulties were particularly acute in the expanding industrial communities situated in the lowland and hilly districts and yet more so, where rural corporations were subject to the growing influence of state power with regard to the management of their affairs as is evident in Zurich and Bern. In the Alpine zone proper, in contrast, the corporations' greater autonomy facilitated the implementation of institutional barriers and thus prevented an influx of an extraneous population.

Clearly, the methods available for regulating the use of common resources to avoid their overuse varied depending on the relevant system of entitlement. We shall now examine the steps that were taken to restrict resources, first, in the system relating to rights *in rem*, next in the system with personal rights, and finally those taken within each of these systems in order to restrict the group of these users.

Restricting the allocation of resources

Restricting the allocation of resources was a possible method. It is quite evident that in some cases the restrictions on resources to avoid their overuse were inherent to the management of the commons, as mentioned before in the case of Grindelwald. However, where resources depended on a right *in rem*, the policy of limiting the building of new houses was an attempt to restrict the overuse, but was often not very successful, as is demonstrated by the example of the *Genossame* of Goldingen (canton of St. Gall) with its growing pauperisation (Stadler 1982: 68–69).

Restrictions of wood allocations were a further possibility, as were the various restrictive uses of the forest, the banning of the goats from grazing on the pastureland, and the limitation of the number of cattle allowed on the common land (*Allmend*).

In regions with rights *in rem*, the frequent decisions to changes in wood allocation not only for the farmers with full rights, but also for those with limited rights of access, is a clear indication of the increased concerns over the degradation of the forest. One instructive example among others is to be found in Nänikon, where the wood allocation was reduced by half from 1665 to 1674 (Kläui 1964: 138). On the other hand, in the lowland one can observe the creation of a number of specific corporations, the *Holzkorporationen*, each with a fixed number of users, for helping their members to avoid any restriction in the allocation of wood (Bluntschli 1839: 78ff).

A second restriction concerned the customary rights for the use of forest resources. Not only were wood allocations reduced, but so were also rights to pasture cattle in the underwood and to gather grass and dry leaves. This was even more often the case in lowland regions, where the central state could make its authority felt more easily, such as in Bern (Schüpbach 2012: 233ff). The numerous and lengthy negotiations at local level with the aim of reaching an agreement reveal the many difficulties involved in this process. In fact, forest degradation was a problem everywhere in Switzerland, and the restrictions on wood allocations and the attempts to restrict other uses of the forests can also be observed in regions where rights of access were personal rights.

A third restriction concerned the common pastureland (*Allmendweide*). There, as from the seventeenth century onwards, we can similarly observe several changes due to population pressure: for example, the frequent decisions to bar goats from having access to the pastureland and to reserve the use of pastureland to one type cattle, mostly to cows. Even more restrictive was the setting of a maximum number for the LUs allowed to graze on the pastureland, a measure strictly observed in some places such as in the rural areas of Bern (Schüpbach 2012: 231) or in the canton of Glarus, although much less so in Obwalden (Head-König 2016: 238). In addition, there were gradual changes in the use of the *Allmend* proper, which, from the point of view of the cattle owners, drastically restricted the land available for grazing their cattle. These changes occurred sometimes over several decades as the landless corporation members, those possessing no cattle but perhaps just one or two goats or sheep, demanded ever more insistently the introduction of a permanent policy for the distribution of land plots for individual use. The *Tagwen* of Glarus is a characteristic example of the lengthy process of change, which lasted for over a hundred years from the end of the seventeenth century right up to the end of the eighteenth century.

Limiting the group of users

The methods for restricting the groups of users of the common resources varied significantly from one system to another at a time, when corporations were confronted with both the growing problem of in-comers into existing rural corporations and the one of endogenous demographic growth of the corporations themselves.

We have seen that in regions, where access was dependent on rights *in rem*, the corporations resorted to the prohibition to construct houses with appending rights in order to avoid the overexploitation of the available resources, but that was often only the first step to limiting their use. Such corporations were often situated in regions under the control of territorial governments (city-states), which, for economic reasons, were opposed to preventive measures for prohibiting the mobility of their subjects within their own boundaries; the corporations had only limited power to restrict the influx of in-migrants. At the same time, decisions were also taken at Swiss and cantonal levels according to which local rural communities were responsible for providing relief to their members in case of need. The combination of these factors explains why there was often no way entirely to exclude those not possessing *in rem* property and who were not members of a corporation from some of the benefits, but who had nevertheless originally paid an entrance fee, which was a normal prerequisite for all new in-comers into a *Dorfgemeinde*. This was also true for their descendants who had subsequently become local citizens. The consequence of this was the emergence, in many regions, of a new practice, that of the division of the houses into several parts. This procedure had the effect of simultaneous splitting the entitlement rights to be ascribed to each new owner and also drastically limited his access to benefits (Pfister 1993: 432–434). Thus, apart from considerable overcrowding in the houses, a large proportion of the households in the canton of Zurich possessed only a half, a third or a quarter of a right, whilst in the assemblies of the corporations the rule of only one vote per house continued to be applied. In other cantons, however, the splitting of entitlement rights could be more extreme, as in the cantons of Solothurn (Trevisan 1983: 67) and Aargau (Dubler and Siegrist 1975: 266–268, 288–289).

At the other end of the spectrum are the regions mostly in the pre-alpine and alpine districts, where there was a high level of autonomy and the rights of access to common resources were personal rights. Each local administration was bent on protecting its economic resources from the neighbours and had its own procedures concerning the stringency of immigration restrictions. In-comers were often anything but welcome. There remained the problem of those living in the corporation with no or very limited rights, the *Beisass* and *Hintersass*, often the descendants of families that had not been able to afford the necessary fee to become a member of a corporation. Their number varied very significantly from one region to another depending on the measures taken by the relevant authority to prevent an increase (for instance, by not allowing such people to marry).

The second potential source of conflicts within such corporations based on personal rights was the problem of intergenerational issues, when there was insufficient land to cater for the needs of all adult sons. It must be remembered that, as from the mid-sixteenth century, the condition of local citizenship was passed on from father to son and daughter (*ius sanguinis*) and was a *conditio sine qua non* for becoming a member of a rural corporation. In times of land scarcity, these requirements were not enough and the corporations created new conditions for becoming what could be described as an active member of the corporation with access to common resources. Each corporation solved these problems in its own distinctive way, some of them very harshly.[6] The discussions of the assemblies of the self-governing corporations were often long and arduous and sometimes lasted for decades, before a decision could be reached to agree the conditions for the admission of adult sons as full active members of the corporation with access to the common resources.[7] Until the nineteenth century, a corporation member had to be the head of a household, to have a specified minimum age and to live within the corporation. Although opinions differed in respect of the necessity to be married, most corporations ruled that it was an essential. Only a few corporations made compromises for accepting older single men as full members (Head-König 2016: 238).

Communal dualism: the lengthy process of resolving the issue of rights

With the institution of the municipalities – a political body that was to include all the citizens settled within a specific territory – the intentions of the politicians of the Helvetic Republic (1798–1803) were clear. The institution of the municipality was intended to resolve two old and recurrent problems still existing in many regions: the limited or total absence of political and economic rights for a number of the people living in rural corporations on the one hand, and the problem of the assistance to be provided for the poor who did not belong to the corporation or to the community of the burgesses on the other hand.

With the emergence of what has been called communal dualism, both systems were subject to important changes over time. In the lowlands, the further existence of the corporations was often called into question. The interference of the central territorial power and the demographic imbalance between the holders of rights and those with hardly any entitlement to common resources were factors, which were to determine the fate of the rural corporations. In some cases, cantonal governments can hardly be said to have been favourable to the maintenance of these old institutions, intent as some of them were on claiming for themselves the ownership of parts of the common resources, especially of the forests. The upland regions with their more decentralised organisation and their high degree of autonomy, on the contrary, were able to limit the impact of the new legislation of the Helvetic Republic.

The 'not entirely new' conception of communal dualism of the Helvetic Republic

It must be remembered that, in some cantons, the idea of a greater participation on the part of those living in a peasant community (*Dorfgemeinde*) in the decisions affecting the community as a whole was not a new conception at all. The relationship of the *Dorfgemeinde* with the old peasant corporations (*genossenschaftliche Gemeinden*) in the canton of Zurich is a case in point, as Johann Caspar Bluntschli (1839: 54–90) has demonstrated. There, during the seventeenth and eighteenth centuries, mobility and population growth changed the imbalance between those owning nearly all of the rights to common land and those who hardly possessed any rights or none at all, but, nevertheless, possessed local citizenship through inheritance or payment for its acquisition. At the same time, as the access to the peasant corporation (*genossenschaftliche Gemeinden*) became more and more restricted, the size of the peasant communities (*Dorfgemeinden*), which included all heads of household with citizenship status, increased. Whilst the common resources of the corporations consisted of immovable property (forests, pastures, meadows, waste land, etc.), those of the peasant communities consisted mostly of movable property accumulated from entrance fees for residence, acquisition of citizenship and accrued with fines for poaching and wood theft. Such peasant communities also granted citizenship and their assemblies decided on the allocation of resources for the infrastructure and the growing public needs.

The exacerbation of conflicts

The conflicts where the system of rights *in rem* (*Rechtsamen*) dominated were numerous up the mid-nineteenth century. In 1830, the memoranda addressed by more than 100 municipalities in the canton of Bern to their government, requesting the suppression of the privileges of the *Rechtsamen*, bear witness to the virulent antagonism, which still existed. Several factors appear to have contributed to the partial demise of this system of common land benefits. First, the new municipalities claimed for ownership of a portion of the communal resources, since they had to provide a number of services for which no funding had been provided. Second, the perversion of the system provided arguments for those opposed to this system. In many regions, it had become possible to buy the right *in rem* 'on its own', even without the appending property. Thus, the accumulation of considerable resources by rich landowners contributed to the polarisation of the social groups. Third, the political power of decision remained in the hands of the possessors of rights *in rem*, since only they could sit on the communal boards to decide how the communal assets were to be used. Fourth, this power structure permitted the increased commercialisation of common resources: the leasing out of pastureland instead of providing plots for individual use and the auctioning of the surplus wood from the forests, which the owners of rights did not need instead of providing those with insufficient land with a minimum quantity of wood and plots to cultivate. This situation was also the result of the transformation of the agriculture production. Artificial meadows and stable-feeding made it unnecessary for the wealthier farmers to keep part of the common land for the grazing of their own cattle.

The complex rights situation and the solutions arrived at to disentangle them

During the nineteenth century, as mentioned before, the timing of the process for disentangling the mix-up of rights inherent in the different systems of access differed widely, as did the solutions adopted to solve the problem of dualism not only from one canton to another, but also within a canton where different systems had coexisted since earliest times.[8] Frequently, solutions for the diverging points of views could only be reached after the cases had been referred to the cantonal Courts of Justice, or after 1848 even to the Federal Court of Justice.

The total dissolution of the rural corporations was one of the alternatives. The canton of Lucerne, in the first decade of the nineteenth century, was one of the first cantons to start privatising the *Allmend* and the forests. In its cereal-growing regions, the land allocated was based on the size of an individual's privately owned land, in the pre-alpine region on an individual's personal rights. Up to the 1850s, in this canton the distribution of *Allmend* for individual use totalled 2,800–3,300 hectares and 11,500 hectares for the forest (Lemmenmeier 1983: 137–138).

Yet another solution was the partial maintenance of the corporations with their rights *in rem* organisation, but with a transfer of a part of their resources to the civic communities which then later became municipalities. This procedure was adopted, for example, in the canton of Zug, after lengthy negotiations, which persisted until a satisfactory apportionment for the civic corporations was reached in 1848.

Finally, in many mountain regions as a reaction to the Federal regulation of 1874, there was the privatisation of the corporation *in toto* with the retention of the common ownership and management. The aim was to keep up the long practice of restricting access rights for outsiders. The result was that in some cantons there were even more corporations than municipalities, precisely because the old types of corporation continued to be recognised as such.

Conclusion

The importance of common rights cannot be stressed enough for the period up to the nineteenth century. In the first phases of industrialisation, the resources gathered from the commons were extremely valuable, especially when the population was growing. The systems of access to the common land, however, varied considerably: the rights *in rem* system to be found in the cereal-growing regions being much more discriminatory than the system based on citizens' rights characteristic of mountain regions and the towns. The implementation of stricter rules both for preventing an increase in the circle of beneficiaries and to avoid an overuse of the commons considerably restricted the advantages for the lower strata of the population, especially where the rule of rights *in rem* existed. Since there were also huge differences in the wealth and the resources of the corporations, the benefits for each household varied accordingly. In the course of the nineteenth century, the growing interference of some cantonal governments to allocate some or all of the resources coming from common land to the municipalities was the consequence of the still existent and unsatisfactory poor laws created in the sixteenth century. These laws had given the *Dorfgemeinde* the responsibility to assist its poor, the role of central state institutions in this matter being only subsidiary. In the mountain areas, many rural corporations wanted to maintain the traditional organisation. Their underlying motivation presumably was to limit the overexploitation of their resources. For to do so it was necessary to keep the old rule of restricting the access of outsiders to the common resources. Not only did their income enable them to provide the necessary public services for the corporation members, but they had also always been in charge of the assistance for their poor. Hence, many corporations decided to register as private corporations in accordance with the relevant cantonal law in the second half of the nineteenth century.

As the economy developed, especially after World War II, the interest for the property owned in common on the part of those working outside the primary sector began to fade. Today, corporations and rural corporations are still very numerous, but the economic significance of the different types of common resources differs widely. Plots of land are less in demand, summer pastures remain important for pastoral farming, but the rapid decrease in the number of farms affects their management and use (under-use of grazing rights). Water corporations, though, remain important in the driest parts of the alpine districts. With regard to forests, the Federal forest policy, since the end of the nineteenth century, has brought about significant changes in their management and the rural corporations have to conform to the relevant requirements. For the time being, the forest corporations are not very profitable. On the other hand, there are numerous corporations with a large spectrum of resources, which can reinvest their profits in socially compatible activities for the benefit of all the inhabitants of the political commune.

Notes

1 At least up to the end of the eighteenth century, the term of 'burgess', of medieval urban origin, is far more appropriate here than that of 'citizen' when it comes to defining the members of the *Bürgergemeinde/commune bourgeoisiale*. It refers to those inhabitants of a town or borough and later to those belonging to a peasant community who possessed full rights inherited in most cases from their forbears who were the 'original' inhabitants.

2 Hereafter when referring to the *Bürgergemeinde/bourgeoisie* in the nineteenth and twentieth centuries I shall use the term civic communities. See note 1.
3 See note 1.
4 The *Teilsame* Lungern in Obwalden was an exception. There, even the *Beisassen* with their limited rights could use the alpine pastures.
5 A rare exception was the corporation of Lachen in the canton of Schwyz. There, in 1780, the statutes mention that women can be independent holders of rights and in this case were allocated a plot of land and a portion of straw (cf. Jörger 2006: 106).
6 The disputes seem to have been less virulent in the few pre-alpine and alpine regions, where the rights of access to highland pastures could be bought within a restricted circle of neighbouring communities as, for example, in the *Geteilschaften*/corporations of the Valais or the *Kapitalistenalpen* in Nidwalden.
7 The discussions sometimes lasted from the seventeenth century right up to the beginning of the twentieth century, as can be observed in the case of the corporations in Glarus, Nidwalden and the Grisons.
8 The total disappearance of common land before 1800 in the canton of Geneva was due to its annexation to France.

References

Bluntschli, J.C. 1839. *Staats- und Rechtsgeschichte der Stadt und Landschaft Zürich, 2. Theil: Die neuere Zeit*. Zürich: Orell Füssli.

Buchmann, W. 1997. *Schweizer Bürgergemeinden als Landschafts-Aktoren. Ihr Grundeigentum und dessen aktuelle Entwicklung am Beispiel von Bern, Chur und Solothurn*. Basel: Wepf.

Dönz, O. 1943. Die Churer Alpen, eine alpwirtschaftliche Skizze. *Alpwirtschaftliche Monatsblätter* 77 (11): 341–352.

Dosch, M. 2002. Alpweide als Übungshang: Alpwirtschaft und kommunale Organisationsverdichtung in Graubünden um 1500. *Bündner Monatsblatt* 2002 (5): 324–340.

Dubler, A.-M. and Siegrist, J.J. 1975. *Wohlen: Geschichte von Recht, Wirtschaft und Bevölkerung einer frühindustrialisierten Gemeinde im Aargau*. Aarau: Sauerländer.

Head-Konig, A.-L. 2016. Common land and collective property in pre-Alpine and Alpine Switzerland. Tensions regarding access to resources and their allocation (Middle Ages–twentieth century). In *Ländliche Gemeingüter. Kollektive Ressourcennutzung in der europäischen Agrarwirtschaft – Rural Commons. Collective use of resources in the European agrarian economy*, edited by N. Grüne, J. Hübner, and G. Siegl, 232–243. Innsbruck et al.: Studien Verlag.

Ineichen, A. 1996. *Innovative Bauern: Einhegungen, Bewässerung und Waldteilungen im Kanton Luzern im 16. und 17. Jahrhundert*. Luzern et al.: Rex Verlag.

Jäggi, L. 1953. Von der Rechtsame im Bucheggberg. *Jahrbuch für solothurnische Geschichte* 26: 281–286.

Jörger, A. 2006. *600 Jahre Genossame Lachen: Geschichte einer Allmeindgenossenschaft in der March im Kanton Schwyz*. Lachen: Gemeinde Lachen.

Jörimann, P. 1970. *Die Churer Alpen*. Arosa: Schweizerischer Verband der Bürgergemeinden (unpublished presentation).

Kläui, P. 1964. *Geschichte der Gemeinde Uster*. Uster: Gemeinderat.

Lemmenmeier, M. 1983. *Luzerns Landwirtschaft im Umbruch. Wirtschaftlicher, sozialer und politischer Wandel in der Agrargesellschaft des 19. Jahrhunderts*. Luzern et al.: Rex-Verlag.

Netting, R.M. 1972. Of Men and Meadows. Strategies of Alpine Land Use. *Anthropological Quarterly* 15 (3): 132–144.

Pfister, U. 1993. *Die Zürcher Fabriques. Protoindustrielles Wachstum vom 16. zum 18. Jahrhundert*. Zürich: Chronos.

Renaud, A. 1845. Die Gemeindenutzungen, insbesondere die Realgemeinderechte in ihrer geschichtlichen Entwicklung und rechtlichen Natur betrachtet, mit vorzüglicher Berücksichtigung der schweizerischen Verhältnisse. *Zeitschrift für deutsches Recht und deutsche Rechtswissenschaft* 9: 1–100.

Schüpbach, A. 2012. *Ökonomie in der Herrschaft Worb (1645–1850)*. Nordhausen: Verlag Traugott Bautz.

Stadler, A. 1982. *Geschichte der Genossame Goldingen. Ein Beitrag zur Siedlungs- und Wirtschaftsgeschichte des Kantons St. Gallen*. St. Gallen: Stiftsarchiv.

Tiefenbach, M. 2013. *Alpkorporationen – traditionelle Institutionen nachhaltiger Landschaftsentwicklung. Das Beispiel der Bergschaften Grindelwalds im Kontext aktueller gesellschaftlicher Herausforderungen*. Grindelwald: Sutter Druck.

Trevisan, P. 1983. Weidgangauflösung und Allmendverteilung in den solothurnischen Amteien Bucheggberg und Kriegstetten. *Jahrbuch für solothurnische Geschichte* 56: 61–107.

9 From natural supply to financial yields

The common fields of the Bernese Civic Corporation since the seventeenth century[1]

Martin Stuber and Sarah Baumgartner

Introduction

Today the Bernese Civic Corporation (*Burgergemeinde*) consists of about 18,000 burghers. The municipality (*Einwohnergemeinde*) of the city of Bern, on the other hand, has 140,000 inhabitants. If one considers the great share of burghers living outside the city boundaries, the actual share of the burghers must be estimated to be less than 5 per cent. Whereas, in terms of numbers of the Bernese Civic Corporation, the burghers' share of the total population has continuously gone down since the early nineteenth century, their economic potential has grown massively. At the end of the twentieth century, the official value of the burghers' landed property was estimated to be about 433 million Swiss francs (Rieder 2008; Stalder and Stuber 2015). Where do these great financial means of the Bernese Civic Corporation come from? In a 2009 information brochure, the corporation gives a threefold answer to this question (Burgergemeinde Bern 2009): First, the corporation is not entitled to collect taxes. Second, to fulfil its tasks the corporation depends on returns from its assets. Thirdly, the assets of the corporation consist most of all of landed property and woodlands.

Thus, the current financial structure of the Bernese Civic Corporation, which is fundamentally different from that of a municipality, is basically revealed: Whereas the municipality, like the Swiss cantons and the Swiss Federation, is financially based on tax revenues, the only financial source of the Civic Corporation is investment income. What stays invisible, however, are the sometimes dramatic historical developments and conflicts, which have led to this structure. The landed property of the burghers has its origins in the common land of the medieval city, which had an important function when it came to supplying the inhabitants of the city with milk, meat, grain, and firewood. The transformations of the collective resources since then have to be analysed in the context of changing energy ages. The 'agrarian society' was based on water and solar energy (until 1850); the storable energy carriers were primarily the food for humans and animals produced on the basis of photosynthesis, and firewood, which made the production of biomass land-bound. The 'industrial society' (1850–1950) was reliant on coal imports after the country had been connected to the railway network. This led not only to the development of industrial production sites, but also to fundamental environmental and historical changes in many other areas, such as mechanical agricultural modernisation, the extensive drainage of wetlands, and the expansion of cities. The 'consumption society' was characterised by mass consumption and mass motorisation based on the global availability of mineral oil after 1950, whereupon settlements grew in the areas, which had a favourable traffic connection (Pfister 1995; Pfister and Egli 1998). How was the common property of the burghers affected by the changing economic significance in the course of these three energy ages? How did resource users adapt the common property in order to react to these changes?

Additionally, we will also have to ask in which ways resource management was affected by the political dynamics during the analysed period. How did pre-modern institutions succeed with adjusting to the modern constitutional state? How did they adjust their collective use of resources to the changing political environment, or how – vice versa – did they influence it? And how are the political changes related to those of the energy system?

In the context of these political and energy transformations, three complexes of questions will be discussed: first, the question about the actual practice of land utilisation and thus the produced yields; second, the question of which groups of actors were participating; third, the question of how the Bernese Civic Corporation and the municipality were interacting (Barth et al. 2003). The focus will be on the fields.[2] The woods owned by the Civic Corporation, on the other hand, will only be addressed selectively, for a comparison (cf. Stuber 2018).

Finally, it has to be pointed out that such transformations did not all happen only in terms of economy or politics, but also concerning the social structure. By the end of the twentieth century, more than two-thirds of the burghers belonged to families, which had become burghers only after the end of the Ancien Régime. However, this is just background information for our article but will not be the topic itself (Arn 1999; Stalder 2015; Stalder and Stuber 2015).

Exclusive utilisation during the Ancien Régime

The common lands, consisting of fields and woods, can be traced back to the time when the city of Bern was founded in the Middle Ages (Messmer 1830: 4–7). Originally, each household was entitled to be supplied with a certain amount of wood from the forests each year, to grow grain on a part of the burghers' arable land and pastures (*Stadtfelder*), as well as to graze some sheep or one cow there. Soon, however, access to this kind of utilisation, to which originally all inhabitants of the city had been entitled, was more and more limited, as the available area per head became increasingly smaller because of the constantly growing population. Newcomers were granted the *Burgerrecht* (membership with the Civic Corporation), due to which one was entitled to utilisation only after some time or after having paid a fee. Therefore, a growing number of households had no access to these resources, whereas at the same time the often wealthy native families lost their interest in agricultural activities. This was the situation in the seventeenth century, the time of the oldest preserved sources.

Apart from allowing for a reconstruction of the legal situation, these sources also give detailed insights into the practices of utilisation.[3]

Whereas the woods were subject to the authority of the City Republic (*Schultheiss und Räte*) already at an early stage, the fields were administered relatively autonomously until the end of the eighteenth century by the yearly assemblies of the two communities into which the city was divided in those days. Participants were the heads of the burghers' (*burgerliche*) households, as being entitled to utilisation, representing slightly less than one-third of the city population. These communities each elected two *Vierer*, as the public servants in charge of supervision were called, as well as further staff, such as the herdsmen. Only individual decisions of greater concern, such as the cultivation or exchange of lands, had to be taken to the Councils (Baumgartner 2013: 17–26).

Thus, two-thirds of the inhabitants of Bern were excluded from any participation in the common land. Another factor producing severe differences among those entitled to utilisation was the strict division of city and fields. The "Münstergässchen" served as an administrative boundary between the Upper Community ("Gemeinde Obenaus") and the Lower Community ("Gemeinde Untenaus"); two-thirds and one-third of the inhabitants, respectively, were living in these two districts. The Upper Community owned the 165 hectares of the Obenaus fields, i.e. on the left bank of the River Aare, and the Lower Community owned the 370 hectares (that is, more than double the area) of the Untenaus fields, i.e. on the right bank of the river (Figure 9.1). This disparity of available area and burghers entitled to utilisation produced the result that the Upper Community was provided with theoretically only one-quarter of the area per household of what the Lower Community had. As a consequence, the methods of utilisation and the degree of access were quite different for the two halves of the city, despite similar regulations (Baumgartner 2013: 16–17).

The common fields served, first, for grazing cattle and, second, for the production of grain, meaning mostly spelt as the main diet of the era. Compared to the kind of agriculture common in the environment of the city, the meadows covered the bigger area; this was due to the necessity to produce animals closer to the central place of consumption, as they were less amenable than grain to store and transport. The fertility of the fields benefitted from the fact that there were more meadows than grain fields and that fertiliser was available, which became obvious from more than average yields. Overall, the grain produced in the fields covered slightly less than one-tenth of the caloric requirements of the city population, and in the summer months the milk of the cows grazing on the meadows covered one-third of the city's consumption. If we refer this to only those entitled to utilisation, the degree of self-subsistence was of course much higher, at least theoretically (Baumgartner 2013: 44–45, 60).

In the late eighteenth century, and under the effects of the famine of the early 1770s, it was allowed at first temporarily and then also permanently to additionally grow clover or potatoes on the fallow land (Baumgartner 2013: 49–50).[4] This more intensive type of utilisation was in line with the postulates of the agricultural reform movement, which was very active in Bern. Its representatives used to

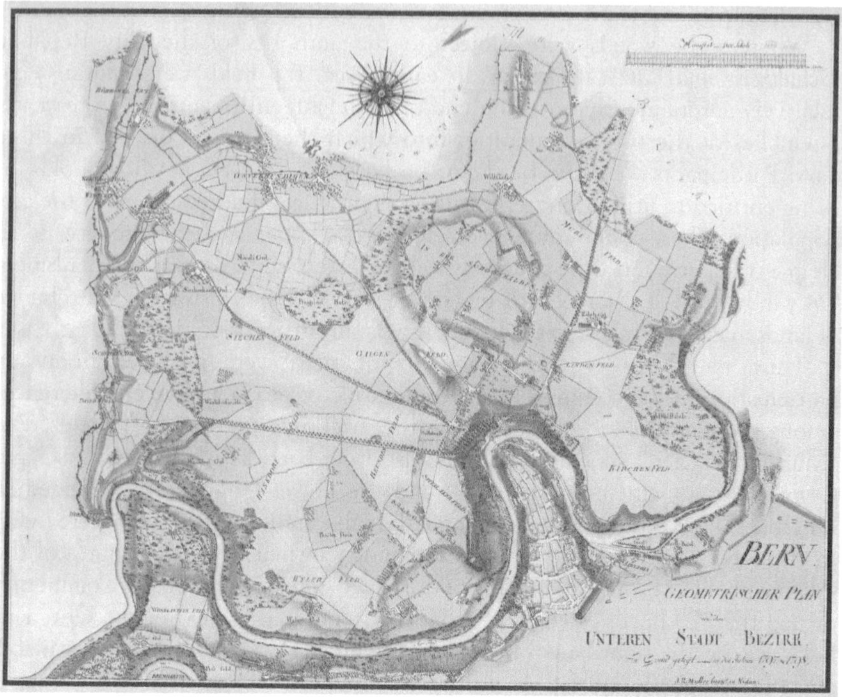

Figure 9.1 The common fields "Untenaus", map by J.G. Müller, 1797–1798.
Source: Stadtarchiv Bern.

meet in the Bernese Economic Society (*Oekonomische Gesellschaft*), founded in 1759 (Stuber and Wyss 2012: 157–181). However, its implementation had to be enforced against severe resistance. For example, the tithe lord, the *Grosse Spital* (the city's main hospital), feared negative effects on grain production and thus on the connected dues, and the butchers resisted this change of utilisation because they lost fallow lands they needed for grazing their animals to be slaughtered (Baumgartner 2013: 49–51).

The Obenaus fields were divided into three *Zelgen*; every second-year one of them was newly ploughed and winter grain (spelt) was sown there, whereas the other two were fallow and used as meadows. In the following year, summer grain (oat) was grown on the field, which stayed uncultivated in the third year, whereas a new, previously fallow, field was ploughed. The Untenaus fields, on the other hand, were divided into seven *Zelgen*. On one of them, people sowed summer grain each year, to serve for growing spelt in the following year. Thus, two *Zelgen* were always dedicated to grain, whereas the other five served as meadows. These utilisation systems showed principles of grain production according to three-field rotation, which was common all over Central Switzerland, but connected to a more or less prolonged fallow period during which these areas served as meadows. On the Untenaus fields, at least the availability of pastureland was never a problem. Households, which, according to their entitlement, wanted to graze a cow there, were usually not prevented from this, and about 80 per cent of the entitled households did so. In the Upper Town, the situation was less favourable: only one-tenth of those who were entitled could participate in utilising the pastureland (Baumgartner 2013: 43–45, 53–54).

Until the late eighteenth century, the scarcity of pastureland in Obenaus was partly compensated by utilising the much bigger woodlands there. In those days, the silvo-pastoral utilisation of forests was very common all over Europe; it provided an indispensable part of the early modern subsistence of peasants, although this affected the function of the forests as suppliers of wood (Stuber and Bürgi 2012: 25–32). Indeed, this was the reason why at many places the authorities attempted to prevent this practice which, in their eyes, was an 'abuse' – and also in Bern where the ban on using forests as pastures could be enforced only at the end of the eighteenth century, after decades-long quarrels between the Councils and the Upper Community (Baumgartner 2013: 54–56).[5]

In no part of the city, however, was the arable acreage sufficient to allow every burgher household the utilisation of 2 acres (approximately 0.66 hectares) in the Upper and 4 acres (approximately 1.32 hectares) in the Lower Community, to which they were theoretically entitled; that is why entitlement was decided by drawing lots. In the course of time, however, this distribution mode became uncommon and the fields were increasingly considered a kind of private property. Consequently, fields were bequeathed and even sold. This resulted in the concentration of land use in the hands of a comparably small number of burghers who practised agriculture not only for subsistence but also for the market (Messmer 1830: 34–36). Thus, by the end of the eighteenth century, the type of utilisation was considerably different from the original regulations. Now and then, there were protests against this unlawfulness, but they were too weak to change anything, even more so as in most cases they were based on less propertied circles. Suggestions to extend participation to wider circles were expressed on several occasions, but there was never any chance of realisation (Baumgartner 2013: 28–29; Messmer 1830: 43–46). Only the radical political upheaval at the end of the eighteenth century brought fundamental change.

Extension of utilisation – exclusivity of profits

In spring 1798, French troops occupied Bern as well as the entire Switzerland, which marked the end of the Ancien Régime and the beginning of a transition period, which would finally lead to the modern democratic state (Junker 1982, 1990). During this period, the political constitution was radically changed several times, which to a certain degree also affected the administration of the common fields and woods; however, the fundamental changes actually happened immediately after the old authorities stepped down. The new rulers installed new authorities, and the Bernese City Republic, which previously had ruled a large territory, lost its former status and was reduced to a common municipality (Künzler 2013; Stalder and Stuber 2015: 50–69). Like any other place in the Helvetic Republic, into which the French had transformed Switzerland, the city of Bern was changed into a *Munizipalität*, as this new communal institution was called. It was governed by a Council, which was elected by all adult males of the city, regardless of having the status of burgher or not. The French intended to extend their principle of legal equality also to the countries they conquered. Still they had to make concessions, and that was similar in all parts of Switzerland, as far as regulating access to common property was concerned. Even if now all inhabitants were formally citizens with equal rights, the utilisation rights to common woods and fields stayed restricted to those who had previously been entitled. The reason for this was the fierce resistance by influential old native families (Schläppi 2011: 47–95). These estates were made subject to a second municipal authority, the *Gemeindekammer*, which was exclusively elected by burghers and was mostly independent from the *Munizipalität*. In this way, the basis of municipal dualism (*Gemeindedualismus*) was created, which still exists today in many places in Switzerland through the parallel existence of a personal or burgher community (*Burgergemeinde*) and a territorial municipality (*Einwohnergemeinde*) (Leonhard and Mattmüller 2001; Sieber 2005). Soon thereafter, the Bernese *Gemeindekammer* appointed a field commission, which from that time onwards was in charge of administering the common fields instead of the old municipal communities (Untenaus and Obenaus) (Baumgartner 2013: 62–63, 71–72).

As already mentioned, many burgher families, many of whom belonged to the financially better-off classes of the city, had little interest in utilising a plot of the common fields, whereas among the huge segment of poorer non-burgher inhabitants of the city – who were not only numerous but constantly growing – there was definitely a demand for agricultural land for self-subsistence. That is why in the year 1800 the *Gemeindekammer* created a completely new mode of utilisation.[6] Participation in utilising the fields was now open to anybody, whereas the burghers' privilege was changed from a material to a financial privilege. In concrete terms, from then on the utilisation rights for each plot of agricultural land were distributed on publicly accessible auctions. Whoever made the best offer could then make use of his land mostly as he liked for a certain period of time, or he could even lease it to third parties. The choice of crops to be grown was no longer subject to any regulation; only the grazing of animals was prohibited. The idea behind this system was the maximisation of interest returns in the form of a yearly rent payment (the *Feldgeld* or 'field money')[7]; these intakes were distributed among the entitled burghers. In the following decades, the payments of *Feldgeld* grew continuously. Starting out from 22 Swiss francs and 50 rappens (nominal values) per entitled person and year (1802), it was raised to 56.3 (1833), 62 (1841), and 75 Swiss francs (1858). How could we estimate the paid *Feldgeld* in terms of contributing to a family's budget? Certainly it was not enough for a living. However, particularly for less well-off burgher families, the sums paid in the mid-nineteenth century, for which a craftsman had to work a whole month, may be supposed to have been a welcome benefit. Nothing emphasises the significance of the *Feldgeld* more than the fact that over the entire nineteenth century the political discourse on the common fields issue was determined by debate on the distribution of this money (Stuber 2015: 296–302).

For the time being, this rent system survived all radical political changes, although on several occasions there was protest from burgher circles, and reform projects were formulated. For example, the field commission had to defend itself against the accusation – for which there is no confirmation from the sources,[8] by the way – that there had been irregularities in the context of distributing the *Feldgeld*. Another point of issue was the question about the circle of beneficiaries, in particular whether the money should also be paid to unmarried female burghers living in their own households or to burghers not living in the city. In addition, the demand to sell the common land and to distribute the resulting intakes among the burghers could not be enforced (Baumgartner 2013: 116–119).[9]

Finally, in 1852 there was a definite separation of goods between the *Burgergemeinde* and the *Einwohnergemeinde*.[10] In this context there had been some disputes about several funds accumulated by the burghers; on the other hand, the claim to the woods had been comparably less controversial, and the distribution of the fields was almost undebated. Even concerning the spatial details, these decisions were based on the endowment document of 1803 (with the addition of lands purchased subsequently), which again referred to the regulations of the Ancien Régime. The impending rapid growth of the city, which resulted in a rapid increase of the value of the lands, could not be predicted at that time (von Werdt 2009; Stuber 2015: 295–296).

The "Burgersturm" and the abandoning of the burghers' exclusive financial gain

We have just demonstrated how, in the case of the common fields, already in the first half of the nineteenth century – that is, still at the time of an agrarian society – the objective of utilisation changed fundamentally, from agricultural to financial yields. However, the financial gain was still exclusively reserved to the burghers, in the form of yearly paid fixed deposits (*Feldgeld*). In the course of the nineteenth century, criticism of the burghers' exclusive right to utilisation grew ever louder, and from the 1860s onwards, in the context of the so-called 'Burgersturm', it became a decisive period for the Civic Corporation: here their further existence was put into question. In this context, the lines of conflict were indeed running even among the burghers themselves. For example, the leaders of the two opposite parties both came from the burgher patriciate. On the one hand, Rudolf Brunner (1827–1894), a radical member of the municipal and the national council, considered himself first of all not a burgher but a citizen and, for reasons of equal justice, wanted to abandon all privileges of the burghers. On the other hand, Alexander von Tavel (1827–1900), a conservative publicist and long-time clerk of the burghers, tried to maintain inequality, which in his eyes was historically valid, by help from the political means of modernity – public meetings, letters of protest, and media campaigns (Brunner 1884; von Tavel 1884; Stuber 2015: 302–310).

The situation escalated when in 1885 a referendum at the cantonal level about abandoning the civic corporations took place. The abandonment of all civic corporations in the whole canton of Bern could be prevented, however, but the extremely narrow result of the vote was a warning even for the conservative wing around Alexander von Tavel. In 1888, in the course of a memorable assembly, the Bernese Civic Corporation gave up on the burghers' exclusive right to utilisation (*Feldgeld, Burgerholz*). An important role in this process was played by the new prospects resulting from the transition from an agrarian society to an industrial society, which started to develop after Bern was connected to the railway system in 1858, and thus to the possible import of coal (Zeerleder 1886; Stuber 2015: 310–320).[11]

The abandonment of the burghers' exclusive rights of utilisation released the financial means, which only allowed for the comprehensive commitment of the burghers to general municipal purposes (culture, sciences, etc.). The timber from the civic forests was no longer delivered free of charge as *Burgerholz* but could now be sold on the open market. In addition, the rent income from the common fields was no longer paid to the bourgeois families as *Feldgeld*, but generated cash receipts. From these two revenues, the Civic Corporation financed the newly founded museums of natural history and of history as well as the library (the later Burgerbibliothek) (Antener 2013; Stuber 2015: 408–412, 428–429). The legitimation formula for the Bernese Civic Corporation – still valid today – was depicted by the structure of intakes and expenses. By abandoning their exclusive *Burgernutzen*, due to public pressure, the Bernese Civic Corporation indeed sacrificed its exclusive utilisation rights. In this way, however, it secured the further existence of its political power. And that the common fields, due to being located around the city, were connected to political power was a fact emphasised both by the burghers' representatives and their radical critics, precisely when it came to future planning in the context of the expansion of the city (Stuber 2015: 323).

The transformation of the common fields into building land

Still in the mid-nineteenth century, the territory of what is today called the old city ("Altstadt") stood out from the environment as a clearly visible and closed block. To the south, east, and north the deep "Aaregraben" was a natural obstacle for expansion. At the eastern entrance to the city, for a long time the "Untertorbrücke" formed a bottleneck, which was accessible only via steep approaches. In Bern, a general dynamic of development as it unfolded in many other Swiss cities in those days demanded a leap across the river Aare by means of high bridges (Grosjean 1973: 140; von Bergen 1998: 25–40). Concerning the history of the Bernese Civic Corporation, this general significance of high bridges is even more marked. At a stroke, the construction of a high bridge transformed the common fields on the right banks of the Aare, opposite the old part of the city, into potential building land, thus enormously increasing their value. It may thus not come as a surprise that the history of each of the four high bridges realised in the nineteenth century – the Nydeggbrücke, Rote Brücke, Kirchenfeldbrücke, and Kornhausbrücke – is closely connected to the Bernese Civic Corporation.

In the process of political decision making concerning the construction of the high bridges the Bernese Civic Corporation was an important actor. However, it could achieve its goals only in cooperation with the municipality and the canton. Tellingly, it was least successful in the case of the Nydeggbrücke, which was most clearly initiated and realised by the burghers. Just a short time after its opening in 1844 it was in competition with the Tiefenaubrücke, opened in 1850 and created by the canton, in terms of transport policy offside (Emch 2013; Stuber 2015: 325–330).

In the case of projects initiated by others, the Bernese Civic Corporation hooked up extremely skilfully, brought its financial power to bear, and successfully pushed through with its objectives. The Civic Corporation deliberately used this argument in the public discourse, which developed around different bridge variants. The Bernese Civic Corporation considered several projects to be incompatible with its interests, and refused to support them financially.

In the case of the Rote Brücke, the Civic Corporation explicitly linked its financial contribution to that of the municipality and kept back three-quarters of its contribution until the pillars had been constructed. The burghers' community secured its interests even more strongly in the cases of the Kirchenfeldbrücke and the Kornhausbrücke, by completely staying away from any financial contribution. In these two cases its contribution consisted of selling land to the building principals at a much reduced price *after* the realisation of the bridges, the reduction being calculated in comparison to the massive rise of the value of real estate to be expected from the building of the high bridge. Furthermore, the Civic Corporation connected this sale of building land at a reduced price to a way of planning quarters which would take the burghers' interest "in a bourgeois residential area" into consideration or to the choice of the location it favoured (Stuber 2015: 331–346).

In terms of policy and democracy, however, in this way the burghers' community was confronted with serious problems. During the voting campaign preceding the municipality's decision about the locations, the burghers were asked who was ruling Bern, after all, if the referendum by all inhabitants was only supposed to agree with the burghers' previous decision about the funding. That the Bernese Civic Corporation was aware of this problem becomes obvious from their relief when the referendum finally produced the result that the majority of the inhabitants agreed with the burghers.

The burghers' contribution to the high bridges was on the one hand guided by their readiness to contribute to the development of the city. Given the high costs, the citizens and the canton would not have been able to realise such a project without support from the burghers. Just as significant, however, seem to have been the latter's own interests, for the Bernese Civic Corporation connected its commitment with a certain high bridge to the necessary capital being safely and profitably invested and having a positive effect on the burghers' real estate. At the same time, the burghers' policy concerning the high bridges pursued two general objectives of urban development. First, the "Untere Altstadt", which was traditionally a burgher stronghold, was supposed to be kept from becoming economically second class, and second, the Bernese Civic Corporation wanted to prevent the share of the working class among the inhabitants from becoming too high. The burghers' council declared its explicit commitment to all these interests – not least because it always needed the agreement of the general assembly of the Bernese Civic Corporation.[12]

The exploitation of the building land

The common fields provided the Bernese Civic Corporation with the key to the urban development of the city. In what followed, the Civic Corporation indeed influenced the development of the region of the city of Bern very much (Egli 1998; Bähler 2003), and still today it is the biggest estate owner in the city. At the same time, since the beginning of the nineteenth century the revenues from real estate have provided the Bernese Civic Corporation with its main intakes. From this a double, sometimes even contradictory, goal of the burghers' land policy has resulted: on the one hand it is obliged to the common good, as the Civic Corporation is a public law institution, and on the other hand it strives for maximum, continuous profit and for long-term value increase like any other rationally acting private owner. In doing so, at most times in the public discourse it did not deny its own interests, such as profitability, self-representation, and maintaining its influence. The extent of its land-political influence is illustrated by a look at the extensive areas of the region of the city of Bern, which, in the course of the past 150 years, went through the hands of the Bernese Civic Corporation (Figure 9.2) (Stuber 2015: 282–283, 474–476).

Whereas the space of the burghers' woods had been absolutely unchanged since the beginning of the nineteenth century, great changes can be observed concerning real estate owned by the Bernese Civic Corporation. Due to a good situation regarding sources – the sales contracts have been preserved almost completely since the end of the nineteenth century, and for the time before they were copied into the so-called *Dokumentenbücher* – this dynamic can be reconstructed in great detail. From the original endowment document (1803), which exactly determined the extent of the burghers' common fields, 450 purchases and 1,350 sales were made until the end of the twentieth century. In the course of this, already before the definite separation of goods (1852), the original area was extended by about one third, mostly through the purchase of the "Oberes Galgenfeld" as well as by land purchases by the Burgerspital (Baumgartner 2015).[13]

The first phase, with massive shifts (1852–1914), was closely connected to the general dynamics of urban development at the beginning of the industrial age, when the city's population tripled from about 30,000 to 90,000 inhabitants

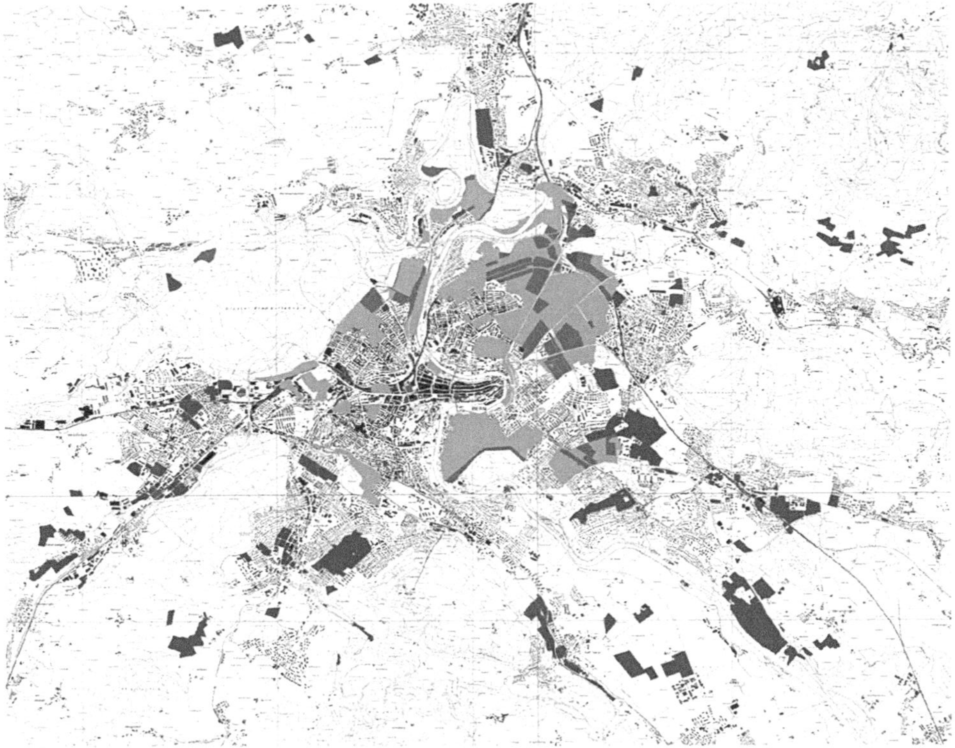

Figure 9.2 Real estate owned by the Bernese Civic Corporation since 1852: unchanged (medium grey)/purchase (dark grey)/sale (light grey).
Source: Map designed by Sarah Baumgartner (2018), Institute of History, University of Bern. Base map: UP5, Copyright: Amt für Geoinformation des Kantons Bern.

between 1850 and 1910. The Bernese Civic Corporation met the quickly growing demand for building land by way of extensive sales of land for residential quarters (Kirchenfeld, Spitalacker), urban infrastructure (schoolhouses, cemeteries, slaughterhouses), military facilities, and for the tracks and station buildings of the railway companies. To compensate for the reduced area of the burghers, the intakes from land purchases were continuously reinvested in new real estate. Like onion rings, the Bernese Civic Corporation sold land near the city centre as building land and purchased land situated farther away from the centre that was previously used for agricultural purposes as possible future building land. However, those purchases still took place almost exclusively within the city's boundaries, e.g. parts of the Tiefenaugut, Wankdorfgut, and Saaligut on the eastern fringe and Schwabgut, Ziegelacker, and Untermatte on the western fringe (Stuber 2015: 429–434).

In the second phase (1914–1957), in the later industrial age, this onion ring principle was extended beyond the boundaries of the city. At the end of this period, the burghers' real estate was still about as big as at the time of attribution by the separation contract. However, it was no longer at the same place; in the mid-twentieth century, the share within the city's boundaries made only about one-half of the burghers' entire real estate. During this phase, the purchase of parts of the burghers' real estate within the city's boundaries had been continued, not least as a consequence of negotiations with the municipality which, given its problems with finding new building land to solve the housing shortage, had asked for support. At the same time, there had been increasingly fewer opportunities for compensation purchases within the boundaries of the city, and the Civic Corporation had extended its purchases of plots to neighbouring municipalities, explicitly expecting a future increase in value. Now the burghers' real estate formed a ring around the closer region of the city, reaching from Köniz, Kehrsatz, Rubigen, Muri, Gümligen, and Bolligen as far as to Zollikofen (Stuber 2015: 434–438).

In its report of 1953, the administration of the burghers' domains stated concisely: "Today land has become a much demanded and scarce commodity. Those estates as being in the municipal area are slowly sucked in by the growing city."[14] Being aware of this and despite the great demand, the Bernese Civic Corporation did not sell its entire real estate within the city's boundaries but looked for other ways of making a profit out of it. For example, apart from the still practiced leasing out for agricultural purposes (*Pacht*), considerable parts of the burghers' domains were rented out for purposes of gaming and sports since the beginning of the twentieth century. When bigger building investments were planned, one did not sign lease contracts but agreements on building rights (*Baurecht*), which grants the utilisation rights of a plot for a certain time against payment of a certain interest rate while the estate owner retains ownership. Examples of this practice are two football stadiums at Wankdorf and Neufeld (1926) as well as the wave pool and the artificial ice rink called Ka-De-We (1933). From the 1940s onwards, increasingly the building right became the preferred legal form of the valorisation of the burghers' real estate, such as in the case of the huge residential blocks of Brückfeldstrasse, Mülinenstrasse, and Morgartenstrasse (Stuber 2015: 438–439).

During the third phase (1957–2012), both sales and purchases of plots within the boundaries of the city owned by the burghers practically came to a standstill. In the neighbouring municipalities, on the other hand, the Civic Corporation purchased land until the end of the 1960s earmarked for development, but no longer in the region immediately bordering the territory of the city of Bern like in the previous phase. In line with the transition to the consumption society and its basic features such as mass motorisation and agglomeration, now the burghers' onion rings covered municipalities that were a bit farther away, such as Belp, Kirchlindach, Schüpfen, and Worb. The planning and realisation of building projects on existing real estate of the burghers became more important than the purchase of new plots. Now real estate was usually not sold but left according to agreements on building rights (*Baurecht*). This resulted in continuous revenues, depending on the duration of the contract over 80 to 100 years. Typical for building projects by the burghers during this third phase are high-rise estates (Buchmann 1997: 44–46; Stuber 2015: 442–443, 453). Concerning the "Schwabgut", the Civic Corporation announced a project competition in 1957; in a referendum in 1961, the municipality of Bern agreed with the development plan resulting from the victorious project. In what followed, only six years after the start of the construction works, more than 1,000 dwelling units could be added to the housing space of the city of Bern. The agreement on building rights was of utmost significance for the execution of the development plan. First, it functioned as a set of regulations for the consortium of investors (rent control, prevention of speculation, cheap rents). Second, the Bernese Civic Corporation was able to provide the public with building land for housing purposes without having to reduce its real estate. Third, the calculated interest rates for the building rights, which were many times higher than the intakes from rents for the agricultural use of the land, helped maintain the financial basis of the Bernese Civic Corporation in the long run (Gutjahr 2018).

A counter-example to the successful Schwabgut is the lengthy planning history of "Wittigkofen", starting in 1961. Together with the "Schloss Wittigkofen" community of heirs, the Civic Corporation worked out a development plan providing for more than 5,500 flats for about 20,000 people, and also for four schools, two churches, one shopping centre, one cemetery, sports facilities, and administrative and commercial zones. After a run-up period including preliminary planning, first zones planning, redisposition, new planning, contract negotiations, and preliminary contracts, the voting inhabitants accepted the special building regulations in 1970. In the following year, the City Council agreed with the development plan for the "Thoracker" with a convention and shopping centre. In a referendum in 1972, however, the Bernese municipality rejected the zones planning both surprisingly and clearly. The opponents, consisting most of all of "Berner Heimatschutz" and the women, who had just recently been granted suffrage, accused the planners of technocratic thinking and one-sided profit orientation. The lost referendum marked the beginning of the end of further construction. In the end, only the first stage, with 1,258 flats for about 3,000 people, was realised. There were several reasons why Wittigkofen remained uncompleted. On the one hand, at that time the population growth had slowed down, and on the other hand the oil crisis of 1973 led to a general slump, which also reached the building sector. Crucial, however, was probably the progress-critical reassessment of big urban development projects from the 1960s (Schnell 2013; Stuber 2015: 444–447).

To be able to continue its policy of building rights agreements, the Civic Corporation reacted, on the one hand, by way of participation-oriented residential building projects such as "Baumgarten" (1990–2012), where in the course of three stages a total of 233 flats for families and 25 studios was realised. In 2002, the Great Burgher Council spoke out in support of the master plan for "Schönberg-Ost", where 54 multi-family houses with 411 flats for about 1,200 people were erected.

On the other hand, the portfolio was diversified and the share of commercial property was increased. Since the 1980s, for example, the Civic Corporation in cooperation with partners (COOP, Berner Kantonalbank, PTT) planned new shopping centres in several neighbouring communities, based on building rights agreements, which could, however, only be realised at Bolligen and Köniz (Stuber 2015: 456–461). At the turn of the millennium, the "Entwicklungsschwerpunkt ESP Wankdorf" became the Corporation's most important planning area. Like the canton and municipality of Bern as well as the Swiss Federal Railways, the Bernese Civic Corporation was also much involved. Being the biggest real estate owner, the Civic Corporation was striving for increased utilisation and new projects based on building rights agreements and took over a share of five million Swiss francs of the infrastructure costs, particularly near the new suburban train station. Consequently, this area, which had been re-designated from an industrial and commercial area into a services and commercial or business area, yielded much higher intakes for the Bernese Civic Corporation. To a certain degree, the Civic Corporation balanced the pursuit of its own interests by also serving public interests. A comparative political science study drew the conclusion that in the case of ESP Wankdorf the outspoken long-term prospects of the Civic Corporation had a price-dampening effect and increased the quality of urban development (Gerber et al. 2011).

Conclusion

During the Ancien Régime, the common fields were exclusively utilised by those burghers who were entitled to do so. The yields contributed considerably to the city's subsistence. In contrast to the common woods, where already in the late Middle Ages the Council, dominated by patricians, pushed through as a regulating authority against the Lower and the Upper Community, the autonomy concerning the common fields was hardly denied over centuries. This changed when, after the radical upheaval of 1798, the common fields were incorporated into the general burgher administration and at the same time were liberated from the many legal restrictions. Whereas the financial yield, in the form of the *Feldgeld*, was still exclusively reserved to the burghers, the circle of possible tenants was opened up beyond the burghers and even beyond the territory of the city, to achieve higher interests by way of rising rents as a result of competition. Now meant for maximum financial yield, the common fields developed into the crucial financial and power resource of the burghers. To legitimate this towards the *Einwohnergemeinde* and the canton, the centuries-old autonomous administration by the Lower and Upper Communities became significant and was used as a key argument.

During the general transition from an agrarian society to an industrial society in the second half of the nineteenth century, the Bernese Civic Corporation transformed its common fields into building land, by way of investing in traffic infrastructure. In the course of the burghers' complex process of self-constitution, which happened under strong pressure from the outside, the liberal-conservative position pushed through with abandoning the *Burgernutzen* (*Feldgeld*, *Burgerholz*) at the decisive meeting of the Bernese Civic Corporation in 1888. Now it used the rising intakes from agriculture and forestry as general funds for culture and sciences. In this way, the Bernese Civic Corporation gave up on its exclusive utilisation contract but maintained its political power. It only lost the exclusive right to the yields of the Burgergut, but not its decision-making power over their distribution. In the following years, according to the onion ring principle, the Civic Corporation sold estates close to the city as building land and used the intakes to buy estates farther away from the city as development land. In this way, it met the building land needs of the growing city while at the same time safeguarding sustainably growing intakes for itself.

During the age of the consumption society, the Bernese Civic Corporation reacted to the agglomeration, on the one hand, by extending its onion rings beyond the boundaries of the city, and by systematically purchasing development land in municipalities which were somewhat farther away. On the other hand, given the increasing scarcity of land, the Civic Corporation itself hardly sold any building land but utilised it in the form of building rights agreements (*Baurecht*). In this context, it planned most of all tower block projects and shopping centres, but as a reaction to progress-critical reassessments in the 1990s it increasingly pursued participative building projects.

The Bernese Civic Corporation reinvented itself again and again. Developing from the radical changes by the transition from the Ancien Régime to modernity in the late eighteenth and the nineteenth centuries and facing the challenges of political, social, and economic change, Switzerland's richest civic corporation was able to maintain its existence and keep its influential position. At the same time, it referred to traditions going back as far as to the founding of the city in the Middle Ages.

How can we understand the Bernese Civic Corporation in the Swiss context? Arnold Niederer (1991), an expert on Swiss municipal dualism, identified three main types within the broad range of Swiss civic corporations, based on the different ways of exploiting their real estate: first, rural collectives "for which the exploitation of common lands and woodlands has a function adding to private property"; second, "philanthropic and culture-oriented activities"; and third, civic corporations of a "joint stock company nature", which exploit their real estate commercially and in the context of tourism.

In the course of its history, the Bernese Civic Corporation covered the whole range. Until the abandonment of the yearly distribution of the *Burgerholz*, the municipal woodlands owned by the burghers of Bern had a function, which was adding to private property as it was the case with the common fields until they were transformed into building land. Consequently, the Civic Corporation, due to exploiting its lands commercially, took on the nature of a joint stock company. Based on this, it represents the classical civic corporation of the patrician kind by its activities oriented to culture funding.

Notes

1 The contribution presented here was written in the context of a research project funded by the Swiss National Science Foundation titled *SCALES Sustainable Commons Adaptations to Landscape Ecosystems in Switzerland* (contributors: Tobias Haller, Christian Rohr, Stéphane Nahrath, Jean-David Gerber, Stefan Mann). It is based on the empirical findings worked out in the context of the publication project *Geschichte der Burgergemeinde* (A History of the Civic Corporation), funded by the Bernese Civic Corporation (Stalder et al. 2015). We thank Mirko Wittwar (Morsbach) for the translation.

2 The following analysis is based on the two recent and more detailed studies (including detailed information about sources and further literature) by Baumgartner (2013) and Stuber (2015). Furthermore, we can reach back to a concise contemporary presentation by Messmer (1830).

3 Today the majority of these documents are preserved at the Burgerbibliothek Bern (BBB) but also at the Stadtarchiv Bern (SAB) and the Staatsarchiv des Kantons Bern (StAB). The oldest sources are the records of the municipal assemblies: SAB, A 087, Manual der Untern Gmeind (1683–1782); SAB, A 088, Manual der Untern Gmeind (1783–1798); SAB, A 089, Manual der Obern Gmeind (1622–1766); and SAB, A 090, Manual der Obern Gmeind (1767–1798). Somewhat more recent are other important sources from the Ancien Régime: the lists of leased out areas SAB, A 142, Acher Rodel Obere und Untere Gemeind und Vierferfeld (1734–1796); BBB, VA DV 81–87, Acker Rödel [Untere Gemeinde] (1715 (ca.)–1803), and of the cows listed for grazing, BBB, VA DV 89, as well as the regulations called *Viererbücher*: BBB, VA FB 2, Vierer-Büchlein Einer Ehrenden Oberen Gemeinde (1772); BBB, VA FB 4, Vierer-Büchlein Einer Ehrenden Unteren Gemeinde (until 1779).

4 SAB, A 087, Manual der Untern Gmeind, 1683–1782: Gemeindeversammlung vom 11.03.1771, Zedel an MehH Vierer Obmann Jenner: 277; SAB, A 090, Manual der Obern Gmeind, 1767–1798: Gemeindeversammlung vom 26.04.1770, Tr. 5: 14.

5 StAB, A I 470, Polizeibuch, 22.01.1771–18.04.1780, vol. 16: 23.06.1773, Decret wegen Einfristung und Verwahrung der Stadt Waldungen und derselben Befreyung von dem Weydgang: 214–216.

6 *Publikation der Gemeinds-Kammer von Bern, in Betreff einer bessern Benutzung der Stadtfelder*. Bern 1800.

7 BBB, VA BK 1690, Gutachten des Feldkassaverwalters Ulrich, 16.05.1808.

8 The intakes and expenses of the fields' administration were recorded in the *Feldrechnungen*, BBB, VA BK 1775–1857.

9 These disputes happened also via the press, i.e. the conservative *Allgemeine Schweizer Zeitung* and the liberal *Berner Volksfreund*. See *Allgemeine Schweizer-Zeitung*, 09.11.1833, No. 135: 723–724; *Allgemeine Schweizer-Zeitung*, 27.05.1834, No. 63: supplement; Wohlgemeinte und beherzigungswerthe Ansichten über die Theilung des hinter der burgerlichen Gemeindsverwaltung liegenden Burgerguts. *Berner Volksfreund*, 30.03.1834, No. 26: 204–206.

10 Ausscheidungsvertrag 1852. In *Verwaltungsbericht der Burgergemeinde Bern 1853–1862*. Bern: Stämpflische Buchdruckerei: 3–5; BBB, Mss. h. h. LI. 9. 4 (20) Urkunde der Aussteuerungen für die Stadt Bern. Dotationsurkunde ausgestellt durch die schweizerische Liquidationskommission (Kopie), 20.09.1803: 18–22.

11 *Vortrag des Verstärkten Burgerrathes der Stadt Bern an die Burgergemeinde. Betreffend die Reorganisation der Burgergemeinde*. Bern: K.J. Wyss 1886.

12 This public debate has been reconstructed from the *Intelligenzblatt für die Stadt Bern* and the numerous relevant printed publications, such as *Das Kirchenfeld-Unternemen und seine Ausführung die Berne-Land-Company. Bericht des Kirchenfeld-Comite*. Bern 1881; Comité für die Waisenhausbrücke (ed.). *Antwort auf die Schrift des Kornhausbrückencomités*. Bern 1891; *Überbrückung der Aare auf der Nordseite der Stadt Bern. Vorlagen der städtischen Bau- und Finanzdirektion und vorläufiger Beschluss des Gemeinderathes*. Bern 1891; and *Vortrag des Burgerrathes an die Burgergemeinde betreffend Abtretung des Spitalackerfeldes an die Einwohnergemeinde behufs Unterstützung des Baues der Kornhausbrücke*. Bern 1892.

13 BBB, VA FBB 133–153; BBB, VA DV 39–92.

14 *Verwaltungsbericht der Burgergemeinde Bern 1951–1953*. Bern: Buchdruckerei Rösch, Vogt & Co 1954: 39.

References

Antener, E. 2013. *Finanzgeschichte der Burgergemeinde Bern 1852–1914*, unpublished M.A. thesis. Bern: University of Bern.

Arn, K. 1999. *"Mehr Sein als Scheinen". Die Burgerschaft der Stadt Bern im 19. und 20. Jahrhundert*, unpublished M.A. thesis. Bern: University of Bern.

Bähler, A. 2003. Von der Altstadt in der Aareschlaufe zur Stadtregion. Stadtentwicklung, Wohnungsbau, städtische Versorgungsnetze und Verkehr. In *Bern – die Geschichte der Stadt im 19. und 20. Jahrhundert*, edited by R. Barth, E. Erne, and C. Lüthi, 231–293. Bern: Stämpfli.

Barth, R., Erne, E., and Lüthi, C. (eds.) 2003. *Bern – Die Geschichte der Stadt im 19. und 20. Jahrhundert. Stadtentwicklung, Gesellschaft, Wirtschaft, Politik, Kultur*. Bern: Stämpfli.

Baumgartner, S. 2013. *Die Burgerlichen Stadtfelder von Bern 1700–1852*, unpublished M.A. thesis. Bern: University of Bern.

Baumgartner, S. 2015. An "Aussichts- und zukunftsreicher Lage" – Kauf und Verkauf von burgerlichem Grundeigentum, 1800–1990. In *Von Bernern & Burgern. Tradition und Neuerfindung einer Burgergemeinde*, vol. 1, edited by B. Stalder, M. Stuber, S. Meyrat, A. Schnyder, and G. Kreis, 448–452. Baden: Hier+jetzt.

Brunner, R. 1884. *Das burgerliche Nutzungsgut*. Bern: Suter & Lierow.

Buchmann, W. 1997. *Schweizer Bürgergemeinden als Landschafts-Aktoren. Ihr Grundeigentum und dessen aktuelle Entwicklung am Beispiel von Bern, Chur und Solothurn.* Basel: Komm. Wepf.

Burgergemeinde Bern. 2009. *Rundgang durch die Burgergemeinde.* Bern: Burgergemeinde.

Egli, H.-R. 1998. Stadtentwicklung. In *Historisch-Statistischer Atlas des Kantons Bern 1750–1995. Umwelt, Bevölkerung, Wirtschaft, Politik*, edited by C. Pfister and H.-R. Egli, 88–89. Bern: Haupt.

Emch, U. 2013. *Die Berner Nydeggbrücke. Geschichte einer bautechnischen Pionierleistung.* Bern: Haupt.

Gerber, J.-D., Nahrath, S., and Knoepfel, P. 2011. The role of Swiss Civic Corporations in land-use planning. *Environment and Planning* A 43: 185–204.

Grosjean, G. 1973. Die Entwicklung des Berner Stadtbildes seit 1800. In *Bern, von der Naturlandschaft zur Stadtregion*, edited by K. Aerni, 135–166. Bern: Geographische Gesellschaft Bern.

Gutjahr, M. 2018. Die Überbauung des Schwabguts 1957–1971. Die Burgergemeinde Bern als Akteurin in der städtischen Wohnraumpolitik. *Berner Zeitschrift für Geschichte* 80 (3): 3–41.

Junker, B. 1982. *Geschichte des Kantons Bern seit 1798, vol. 1: Helvetik, Mediation, Restauration, 1798–1830.* Bern: Stämpfli.

Junker, B. 1990. *Geschichte des Kantons Bern seit 1798, vol. 2: Die Entstehung des demokratischen Volksstaates, 1831–1880.* Bern: Stämpfli.

Künzler, L. 2013. *"Stäts unserm Mutz getreu". Vom alten Patriziat zur Burgergemeinde: Kontinuitäten und Brüche einer Elitetransformation in Bern (1795–1852)*, unpublished M.A. thesis. Bern: University of Bern.

Leonhard, M. and Mattmüller, M. 2001. Allmend. In *Historisches Lexikon der Schweiz (HLS)*, online-version, www.hls-dhs-dss.ch/textes/d/D13704.php (last accessed 17 November 2018).

Messmer, B.L. 1830. *Die burgerlichen Stadtfelder von Bern.* Bern: Carl Stämpfli.

Niederer, A. 1991. Gemeindedualismus – ein schweizerisches Unikum. In *Alpine Alltagskultur zwischen Beharrung und Wandel. Ausgewählte Arbeiten aus den Jahren 1956 bis 1991*, edited by A. Niederer, 314–331. Bern: Haupt.

Pfister, C. 1995. *Im Strom der Modernisierung. Bevölkerung, Wirtschaft und Umwelt im Kanton Bern 1700–1914.* Bern: Stämpfli.

Pfister, C. and Egli, H.-R. 1998. *Historisch-Statistischer Atlas des Kantons Bern 1750–1995. Umwelt, Bevölkerung, Wirtschaft, Politik.* Bern: Haupt.

Rieder, K. 2008. *Netzwerke des Konservatismus. Berner Burgergemeinde und Patriziat im 19. und 20. Jahrhundert.* Zurich: Chronos.

Schläppi, D. 2011. Grenzen der Gleichheit. Wie und warum die helvetischen Regenten vor dem Gemeinbesitz von Korporationen kapitulierten. In *Grenzen des Zumutbaren. Erfahrungen mit der französischen und der Helvetischen Republik (1798–1803)*, edited by A. Würgler, 47–65. Basel: Schwabe.

Schnell, D. 2013. *Die Architekturkrise der 1970er-Jahre.* Baden: Hier+jetzt.

Sieber, B. 2005. Bürgergemeinde. In *Historisches Lexikon der Schweiz (HLS)*, online version, www.hls-dhs-dss.ch/textes/d/D26443.php (last accessed 17 November 2018).

Stalder, B. 2015. Die soziale Gruppe – Burgerrechtspolitik und Habituspflege. In *Von Bernern & Burgern. Tradition und Neuerfindung einer Burgergemeinde*, vol. 1, edited by B. Stalder, M. Stuber, S. Meyrat, A. Schnyder, and G. Kreis, 137–278. Baden: Hier+jetzt.

Stalder, B. and Stuber, M. 2015. Die Burgergemeinde als Körperschaft – ihre institutionelle Entwicklung. In *Von Bernern & Burgern. Tradition und Neuerfindung einer Burgergemeinde*, vol. 1, edited by B. Stalder, M. Stuber, S. Meyrat, A. Schnyder, and G. Kreis, 41–135. Baden: Hier+jetzt.

Stalder, B., Stuber, M., Meyrat, S., Schnyder, A., and Kreis, G. (eds) 2015. *Von Bernern & Burgern. Tradition und Neuerfindung einer Burgergemeinde*, 2 vols. Baden: Hier+jetzt.

Stuber, M. 2015. Vom Burgerholz zur Bodenpolitik – der burgerliche Grundbesitz in der Entwicklung der Stadtregion. In *Von Bernern & Burgern. Tradition und Neuerfindung einer Burgergemeinde*, vol. 1, edited by B. Stalder, M. Stuber, S. Meyrat, A. Schnyder, and G. Kreis, 281–493. Baden: Hier+jetzt.

Stuber, M. 2018. Kollektive Ressourcen und Nachhaltigkeit in der Longue durée. Die burgerlichen Wälder der Stadt Bern vom Ancien Régime bis in die Gegenwart. In *Von der Allmende zur Share Economy. Gemeinbesitz und kollektive Ressourcen in historischer und rechtlicher Perspektive*, edited by D. Schläppi and M.-C. Gruber, 173–188. Berlin: Berliner Wissenschafts-Verlag.

Stuber, M. and Bürgi, M. 2012. *Hüeterbueb und Heitisträhl. Traditionelle Formen der Waldnutzung in der Schweiz 1800 bis 2000*. Bern: Haupt.

Stuber, M. and Wyss, R. 2012. Paternalism and Agricultural Reform: The Economic Society of Bern in the Eighteenth Century. In *The Rise of Economic Societies in the Eighteenth Century*, edited by K. Stapelbroek and J. Marjanen, 157–181. Basingstoke: Palgrave Macmillan.

von Bergen, S. 1998. Brückenschläge in die Zukunft – die Schlüsselrolle der Hochbrücken in Berns Urbanisierung. In *Bern – eine Stadt bricht auf. Schauplätze und Geschichten der Berner Stadtentwicklung zwischen 1798 und 1998*, edited by C. Lüthi and B. Meier, 25–40. Bern: Haupt.

von Tavel, R. 1884. Eine Lösung der Burgergutsfrage. *Intelligenzblatt der Stadt Bern* (Separatabdruck). Bern.

von Werdt, C. 2009. Der Ausscheidungsvertrag zwischen Burger- und Einwohnergemeinde Bern von 1852. Quellenanalyse statt Verschwörungstheorie. *Berner Zeitschrift für Geschichte* 71 (3): 57–97.

Zeerleder, A. 1886. Zur Reorganisation der Burgergemeinde der Stadt Bern. Bemerkungen und Anträge zu den Vorschlägen des verstärkten Burgerrathes. *Intelligenzblatt der Stadt Bern* (Separatabdruck). Bern.

10 Universal values and the protection of commons

Fighting corruption with bottom-up process in Mallorca

Ramez Eid

Introduction

Many anthropologists and other social scientists have criticized weak participatory approaches to common-pool resources management (West, Igoe, & Brockington 2006; Susser 2016) as "Trojan horses" (Blaikie 2006). That is, that these schemes benefit powerful actors (external or internal) rather than "commoners" for whom the commons historically were managed by and for. David Harvey argues that common resources today risk "accumulation by dispossession" by corporate rapacity and neo-liberal economic policies around the world (Harvey 2012). Harvey and other critics have called for a better understanding of new bottom-up movements around the world, concerned with environmental or urban commons, who try to adapt to and offset the effects of the external forces of the modern world. Such understandings might help bring the power to change the status quo of ongoing elites (De Angelis and Stavrides 2010; Susser 2017).

In this article, I discuss some of my ethnographic research in the Serra de Tramuntana at the western side of the island of Mallorca, Spain. I will argue that local level active involvement in creating new rules and platforms for the democratic management of cultural and environmental resources in the Serra de Tramuntana has reversed some of the domination of "elite capture" related to environmental corruption issues and introduced new sustainable practices to the island's "development culture".

The history of the natural resources management (or better say the mismanagement) such as forests, beaches, protected lands and natural resources in Mallorca is with no doubt a case example of non-sustainable practices which lead to destructive environmental results. Historically, privatization policies and legal changes in the local Mallorcan government have transformed some of the most important resources of the island into commercial assets that were sold to powerful businessmen or multi-national firms. These include unlawful building permits for tourism development projects in naturally protected areas. As will be discussed in this paper, such corruption cases that involved major politicians were the main cause for environmental destruction in Mallorca, creating what might be described as an "open access" context. It is claimed here that under such conditions, the political game became a one-sided act where politicians have de facto served the economic elites and neglected all laws and policies that might have protected these environmental resources. Corruption in this case can be seen as a way to breach the legal framework which was modeled democratically and collectively by the regular legitimate political government in the island. From an anthropological viewpoint that is concerned with common-pool resources, when the laws are not implemented, an open access case will take place, and less powerful local groups might find themselves out of the game (Ostrom 1990; Ensminger 1992; Haller 2009). In this context of extreme corruption, the state is not only absent (both the Federal state of Spain and the local Mallorcan autonomous government) but takes part of a political culture that encourages the powerful players (with the biggest financial abilities) to take over the limited natural resources of the island. Kalb and Gabor (2011) argue that the 2008 global economic crisis combined with the increase in numbers of migrants and refugees in Europe was exploited by nationalist parties with anti-environmental agendas to gain political power in few states. This is also true in the local political game in Mallorca, where we see continuous efforts of the right-wing parties in the government to reverse environmental achievements while promoting pro-development policies.

Under such context, since 2005 there has been a growing bottom-up demand in Mallorca to adopt the UNESCO platform for the sustainable protection of the cultural and natural heritage of the island (under the international UNESCO World Heritage Sites program). Such call for a new direction in the management of the natural commons has gained more and more support of local stakeholders and activists, and from other political parties and community organizations in Mallorca. These stakeholders finally helped in creating an alternative sustainable understanding and new laws or local by-laws by promoting this international status and new open and democratic platform, and hence preventing future privatization development projects.

The article will discuss the emic perceptions of different local stakeholders who promoted and participated in the UNESCO process, which culminated into official declaration of the Serra de Tramuntana as a Cultural Landscape World Heritage Site (WHS). I argue that such bottom-up process has been able to reverse some of the effects of environmental corruption, by introducing a new culture that supports the sustainable management of natural resources.

Constitutionality as an analytical approach

In the past few years, Tobias Haller, Stephan Rist, and other researchers have introduced a new theoretical approach they call "Constitutionality" (Haller et al 2016; Haller et al 2018), according to which they analysed examples of successful local institution building (differing from top-down imposed participation) that highlights natural resource management initiatives, from a perspective that emphasizes community members' views and historical precedents on participation, and those leading to independent governance. It suggests conditions that have existed in contexts of successful bottom-up institution building. It evaluates local

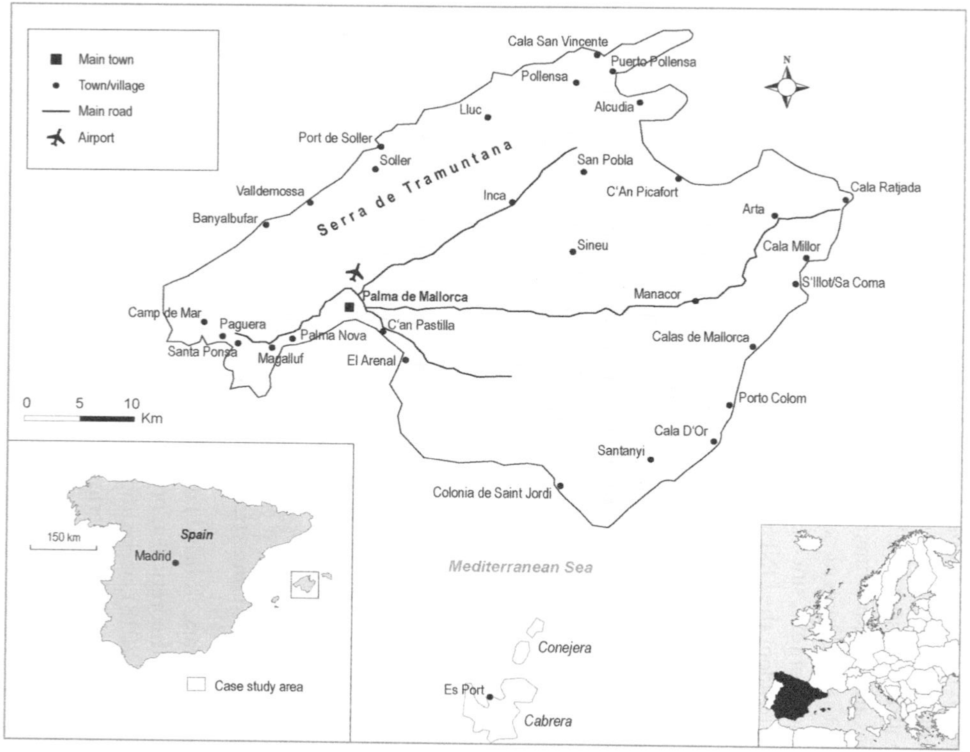

Figure 10.1 Map of Mallorca.

Source: Map compilation and design by Ulla Gämperli Krauer (2018), CDE, University of Bern, Bern, Switzerland. Geodata source: OSM data by OSMaxx (2018). Geometalab, HSR University of Applied Sciences Rapperswil, Switzerland.

agents' strategies in negotiating the development and use of such initiatives, and in so doing, illustrates the large extent of local agency and creativity in constructing novel and personally beneficial governance institutions for using and protecting environmental resources in their livelihoods.

Their analysis highlights six conditions of constitutionality:

1. Emic perceptions of the need for new institutions should be understood and analysed.
2. Participatory processes of negotiating power asymmetries, and the development of a sense of ownership among participants.
3. Preexisting institutions as a basis for newer institution building.
4. Outside catalyzing agents (fair platform) to promote new institutions.
5. Recognition of local knowledge by the state or higher authorities.
6. Higher level acknowledgment of new institutions.

It is important to mention that this approach pays attention to bargaining-power dynamics, since local stakeholders are usually heterogeneous in terms of internal power distribution and characterized by a relative lack of power in dealing with outside actors, such as the state or immigrants. Constitutionality illuminates how in particular places conflicts over resource management has led to institutional solutions formalized as compromises with a collective benefit for most or even all actors. Our Mallorca case study shows such characters, as outcomes often entailed a compromise due to the relative of bargaining power between actors (Haller & Merten 2008, Chabwela & Haller 2010).

This approach also focuses attention on how local levels are linked directly and indirectly to outside legal and political-economic contexts and frameworks (provincial, national and global levels). Such links are crucial not only to enable "institution shopping" (Haller 2013) but also for the processes of state recognition and legitimation required for the sustained operation of local frameworks achieved through constitutionality. Using constitutionality is crucial to emphasize the "pro-active and strategic participation in the design of local institutions negotiated in settings with heterogeneous stakeholders focused upon the use of natural resources" (Haller et al 2016: 2). Such strategic activity was also witnessed in Mallorca, with the adoption and creation of inter-related local and international process for the promotion of new sustainable framework under UNESCO status.

The long history of the weakened local community of the island took a different direction, due to these local bottom-up processes described in this paper. It is claimed that this process has developed a new sense of trust and ownership among different stakeholders, an important condition for developing constitutionality (Chabwela and Haller 2010; Haller 2013). It is also important to emphasize that the local government in Mallorca had a very weak reputation, thus a weak bargaining power as discussed below. This context has created an opportunity for local action, combined with the UNESCO label and the state's international obligation to allow the creation of local management institution for the protected area, leading to debates and measures by the local inhabitants. These different elements of the constitutionality approach can be combined to be a very useful tool to analyse such local level social dynamics and the change which followed the process.

Methods

The theoretical and empirical scope of this study required a mixed methods approach, involving both qualitative and quantitative tools, including ethnography as well as household questionnaires with closed ended questions. But since the focus was put on the local perceptions and strategies of local actors, participant observation in social anthropology was the key method. In addition, other research tools were used such as: open and structured ethnographic interviews, archival research, focus groups discussion and oral history. The research was conducted first in 2012, and followed up by another research period in 2014 as part of PhD research project at the University of Bern, Switzerland.

A brief history of the island

The Island of Mallorca, in the western side of the Mediterranean Sea, has a long history of wars and invasions by those seeking to colonize it, followed by hunger or prosperity. The island was ruled by the Muslims of Andalusia, as Spain has become the main "motherland" since the Middle Ages. These multi-cultural compositions of history are still today, as elements of Muslim, Spanish, Catalan and other European cultures can be traced in the somewhat cosmopolitan aura of the Mallorca, and as such, they are still very much related to the heterogenous imagined identities and politics of modern Mallorca.

The Muslim rulers brought great new advances in technology and agriculture, mainly the sophisticated water irrigation systems (some of which is still in use in Mallorca and other parts of Spain), in addition to the *alqueries* – the farmsteads system they established in the island. This lasted until the middle of the 13th century when Jaume I (1208–1276) king of Catalonia sent his Christian groups into the island, then he gave his lieutenants the farmsteads and villages of the Arabs (Shubert, 1990).

During the first decades of the 20th century, the politics of Mallorca, as much of the mainland Spanish history, revolved around political struggles between the Right (pro-mainland Spain) and the Left (pro-Catalan). As the Spanish general Francisco Franco led an uprising against the central government in Madrid, the island witnessed several battles between the Republican right-wing forces and the left forces backed by Catalan and Valencian troops. Franco finally won the civil war in Spain, and Mallorca followed the mainland. The result was the banning of the use of Catalan language in public, and the language was removed from the education system of the island. Up until 1976, with the end of the Franco regime, Mallorca was controlled by military personnel loyal to the Spanish dictator (Tusell 2007).

While the central Spanish state took control for centuries, the inhabitants of the three Balearic Islands (Mallorca, Menorca and Ibiza) never thought of themselves as composing a unique political and cultural unit, as each has a strong exhibition of local cultural malleability and openness. In Mallorca, for example, the local *Mallorqui* dialect, which emerged from Catalan since the conquest of Jaume I in the 13th century, has become a crucial part of their island identity (Waldren 1996). As the Franco regime was replaced by democracy in 1978, *Mallorqui* has witnessed a new revival period, supported by the Catalan nationalists in Barcelona, who would claim that Mallorca belongs to the historical Catalonia (Tusell 2007). This issue continues to be a very sensitive and important issue in the local politics of the island, which has also contributed to the dynamics of the process for the declaration of the Serra de Tramuntana as a UNESCO World Heritage Site.

History, language and the political game are so intertwined in Mallorca, that one wonders why people still mention events that happened 500 years ago. One such event is the yearly festival in Port de Soller of the pirates against the Christians, as almost 10,000 participants gather in a day-long simulated war, with the regular winner every time. Battles re-enacted and celebrated with enthusiasm and joy. This is probably one way that the local residents of the Tramuntana region are trying to keep and build (or rebuild) their tradition and their history. "Tradition" has gained much social value in the past years, as the identity of the region, and the island's common resources have been under perceived threat. Jaume Fortes, one of the residents from Soller in the Tramuntana region and a well-known activist in the city, reminded me:

> The people of the Tramuntana are the original people of the island. We are a strong group, and if we stop behaving like a strong group, we might lose our war against our heritage and language without even knowing it.
>
> (Interview with Jaume Fortes on 25 April, 2013)

When I asked him back: "but to whom you think you will lose the war these days? There are no more pirates, there are no more wars in Europe or in Spain?"

His answer is very decisive:

> There is still a war in the island. It is the invasive war of the big companies with the little dirty politicians against real Mallorca. Against the natural resources, and against the local Mallorqui heritage and language. There has been a constant war since the early days of Franco, that is almost 80 years long. He banned our language from schools and public offices, and he wanted to alter the culture of Mallorca to that of Spain. The Spanish government is still trying to impose Spanish identity here.

The tourist industry

Following a hundred years of global attention and fascination by the unique natural and cultural "resources" this island has to offer, tourism has become today the most, and probably the only, significant economic sector in the economy of Mallorca. The income from this industry amounts to over 80% of the local GDP (Perry 2004; Spilanis and Vayanni 2004; Pack 2006). Mallorca has provided many attractions even to the early visitors a century ago: the mountains of the Serra de Tramuntana, beaches, caves and the Mediterranean forests. All these were reachable in a one-day journey from any point in the island. If we add to this the mild Mediterranean climate, mainly during winter, greatly warmer than northern European areas, combined with a long heritage and archaeological sites and "attractions", the island was easily promoted as the playground for the rich and the famous, as well for the occasional travelers (Salva 2002; Lyth 2009; Royle 2009; Buswell 2011). Some researchers claim that the fact that a place is an island gives it a certain inherent fascination and intimacy to outsiders (Trauer & Ryan, 2005). While other islands of the Mediterranean can be considered as steps to somewhere else, as the famous French historian Braudel once described them (Braudel 1992:116), Mallorca has truly become through the 20th century "somewhere" on its own.

The real big "boom" in tourism started in the early 1960s, as Mallorca was one of the cheapest destinations for travelers in Europe. With the building of the new airport in Palma, the numbers multiplied soon: from less than one million tourists in 1962 to four million in 1969 and eight million in 1973. The big increase in tourist numbers must be understood in the context of the post-World War II period and its links to the economic and national politics in mainland Spain (Barke et al 1996). The fascist regime of General Franco, which ruled till 1975, has always feared the growing regional trends in some Spanish regions such as Mallorca. For Franco, it was important to implement a unified political system, in which only one Spanish cultural image and language (Castilian) must be in use. He often tried to suppress by force any expressions of regionalism, as in Catalonia and Mallorca, in order not to undermine the central authority of Madrid (Purdy 2012). Even though, the real problems of Spain in the 1950s forced the Madrid ruling man to allow certain tourist development projects to support the declining economy and get the international acceptance he was deprived of. The revenues from tourism would enable him to use it as collateral for getting international loans for larger projects. Tourism, then, was a political tool for Franco more than a mere administrative issue to arrange (Buswell, 2011; Purdy 2012). Tourism both supported the dictator and helped to "sugar a very bitter pill for much of the population and ultimately made it far easier for a nation to be reborn and accepted on the wider international stage when 'the tyrant' had gone" (Bull 1997; Purdy 2012).

Another important factor leading to the unique expanding of the development pressure of tourism was the growing local business innovation in the first half of the 20th century in the island, led by few super-rich residents. Buswell (2011: 39) emphasizes the new form of capitalism which spread across the island, dependent on the power of a banking and financial oligarchy of local Mallorcan families, who purchased large areas of land. During these decades, the banks owned and built huge tourist projects themselves, especially as some of these bank owners had large estates of agricultural lands which no longer provided enough cash flow as tourism would do.

This unique link which existed in Mallorca between (1) capital (strong banks); (2) land (large agricultural estates) both owned by the same old families of Mallorca; (3) outside demand for more hotels; (4) the national political and neo-liberal policies of the Franco regime, created an overlap of forces to push forward national development projects in the island, and establish the pressure that would destroy many natural commons all over the island (Buswell, 2011: 30–42). These same factors were considered a blessing to the economy at the time but will become the target of much of the arguments (by political parties, left/right wing activists, mayors, etc.) in the Serra de Tramuntana against further development of tourism in this mountainous region (Wheeler 1995; Bardolet 2001; Garcia and Servera 2003).

Corruption

Besides being famous for its natural resources such as forests and beaches, the island, unfortunately, has been the focus of the media coverage in Spain and in Europe, for its never-stopping series of corruption stories. It was described as "the most corrupted region" in Spain, for having a third of all corruption cases in the country (BBC 2013). These cases usually included the mismanagement of economic projects, the issue of illegal construction permits or even bribing top politicians and police officers.

An earlier example was the 1996 project of constructing a national highway (*Autopistas*), for which the government allocated over 900 million Pesetas, a staggering number by its own. Even though many called for a review of the plan, the actual spending topped the billion *pesetas* limit, which made many residents raise their eye-brows and distrust the government. In the decade that followed, corruption increased in numbers and forms, to the extent that scandals were given names, such as Gurtel, Barcenas, Andratx and Palma Arena, mentioning just a few. The most important of these were the cases of illegal construction of tourist development projects, the gloomy sign from the historic real estate boom in pre-2008 Spain. These were usually big projects, owned by business owners who pushed to get the permits to build a new hotel or entertainment projects in protected areas. Beaches have been disappearing, coral reefs destroyed and forests were harmed. Tenure arrangements such as commonly owned forested areas, centuries-old water management customs were also severely changed or eliminated (Dodds 2007). Political power, personal connections and money have been mixed too often to create an environment of corruption. Public resentment and protests grew more and more during those years (Ulrich 2007; BBC 2013, 2014). Many resedints claim that they started to by take new actions. For example, Maria Sorroco, who lives in the Port de Soller, was very involved since the early 2000s in the local organization of residents who met and discussed the UNESCO proposal:

> These were the years of the traumatic effects of the corruption period. The image of the island was being destroyed by greedy people in power. Do you know how many times I was ashamed to read the newspaper and know all about this selling of our land and heritage? Do you know how many tourists who dined in our restaurant asked me: "what is going on with Mallorca"? These were horrible times, one corruption story followed the next, and no end was seen on the horizon since the people in power started behaving as if this island is their own grandfather's property. Therefore I think so many people supported the idea of fighting back with all means. I started only then to be very interested in politics and to be active with some friends who invited me to attend meetings to discuss ways of fighting back to take the island out of their hands. This is how I first heard about the UNESCO initiative, and I was fully supportive and active with it since then.
>
> (Interview with Maria Sorroco on 30 March, 2012)

Other residents of the Tramuntana frequently mention this state of external threat not only by the foreign companies but foreign European migrants ("*wealthy expats*") as well, who found their place here in great numbers in the past three decades. With their deep pockets and the ability to compete with the residents over land and resources, many of my informants have claimed that prices of land have been continuously rising for many years. It became very tempting to sell one's grandparents' house if it is empty or no one lives in it anymore. Stories were told about many young persons who got over two or three million Euros for their nice property in villages. In Deia, a tiny but famous village on the slopes of the mountains, some of those locals who sold their house to some Russian or German multi-millionaires were looked upon in disdain by many villagers who expected them to give a notice, or to try to keep these centuries-old houses within the control of the original families, but the sudden selling was almost considered in some cases as a stab in the back:

> Why do we need empty houses, nobody living in it for most of the year? They come and spend two or three months during summertime, and then we never get to meet them again. They do not participate in the music or the parties or the little mixing in the neighborhoods that regular neighbors usually do in Deia.
>
> (Focus group, Deia, held on 2 May, 2013)

This resentment to those who change their social boundaries, and alien themselves too close to the *Estrangers*, grew more and more with the influx of more outside buyers. The result has been greater heterogeneity in the area: these new-comers acted to promote their own interests, as with the language problem would illustrate: should Catalan or Spanish take the lead in the public, educational and social life of the island? And should more English lessons be introduced into the educational system?

These conflicts stand at the center of the different political and cultural streams in Mallorca: That is, the question whether Mallorquin (or Catalan) should be the main official language of the island. While trying to fit in and thus become a legitimate partner to "enjoy" the resources and life of Mallorca, the outside residents took contradictory stances towards that effort. It was a puzzling dilemma: keeping the "authenticity" of the Serra de Tramuntana and at the same time maintaining their original identities.

Jakau Charlston, who came to Mallorca from Britain with his wife and brought up his kids there for the last 14 years, was able to buy a very old building on the top of a mountain in the Tramuntana. He was very proud about it:

> This building was built in the middle of the 19th century, and was kept in the same family for 145 years. Then it was neglected by the owners for many years until we arrived here and bought it from the lady. So this is an original Mallorcan architecture and heritage, and we are the second owners. We took care of it, and rebuilt it to its original function of the 19th century, as a service point for the travelers who take this long and uneasy road between the two parts of the island.
>
> (Interview with Jakau Charlston on 7 June, 2012)

Jakua and his wife were very pleased to show me the progress their children made in the local school, as they were taught many lessons on issues related to the local environment and culture of the Tramuntana. But the conservation issue became more complicated when I asked about language:

> This year, in September, the official language in schools will become Castellan, as the new government decided. We think this is not that bad, since if you want to find jobs in mainland Spain, or even South America, you need to speak it very well.

When I asked them about the perception of the local Mallorquins about this matter, Jakau answered:

> Oh well, most of them don't like it. Especially since it is part of how they see themselves, their identity.

The couple seemed very happy about the new change, mainly since their own language, English, will also get more teaching time, while Mallorquin will get less. But on the other hand, I could feel their great difficulty in expressing this position. After all, they told me they have so many good Mallorcan friends who opposed it vehemently. For this reason, they usually avoided discussing the subject with them.

This silence might have caused some of the residents to accuse them as holding "double standards", and that they have other interests than the "real locals", as Maria Sorroco told me:

> If they want to stay here and live forever, why don't they learn Catalan, and why don't their children learn Catalan as well? Only people who come here for economic reason, such as making good investment in real estate, are the ones who do not care about the real Mallorca and the real Tramuntana. So they want more Spanish to keep their options open … they might move on again.
> (Interview with Maria Sorroco on 25 June, 2012)

Such opposing views did create new "cultural boundaries" among the two sides and added more heterogeneity onto the dynamics of the area. But these differences were discussed and resolved among participants of the UNESCO platform. It was the idea of "universal value" of the Serra de Tramuntana that created a feeling of collectiveness for all the residents of the area.

It is worth bringing at this point the case example of the local forests as a common-pool resource in the Tramuntana, and to analyse their change in value due to the constitutionality process, as external pressure is crucial when analyzing such process. During the past century, the island witnessed massive deforestation process: the number of residents increased, and machinery helped to clear the land for more agricultural uses. Near all villages and settlements forests were usually heavily exploited, using woods to construct houses and furniture. Other important industries heavily affecting the forests were the charcoal and lime industries, as coal was the main heating method (Gil & Diaz-Fernández 2003). The tourist boom since the 1960s brought a further tremendous change to the environment of Mallorca, first with the increased pressure on some areas due to development projects, forested areas were cleared and destroyed. Second, the traditional economic uses of the forest became less profitable, much like agriculture, and with new employment opportunities in the rising tourist market, the rural areas were depopulated. This last trend might have caused a disruption in the traditional forest management methods (Ibid). Between 1965 and 2005 it was estimated that over 90,000 hectares of land and forested areas were neglected and permanently abandoned due to the tourist boom (CAIB, 2011). Researchers have suggested that all these factors combined have caused the following effects to the forests (Rosello 2012):

1. **The economic values of trees were undervalued:** in market terms, the trees became less valued during the last 30 or 40 years, as they were not needed in the same quantities as in the first half of the century (Sureda, 2011).
2. **Trends of putting the forest in the market for new uses:** many landowners tried to transform large areas of private forests and fields into more profitable urban lands.

3. **Forest fires increased:** the neglected forests and the cease of all activities inside their territories have helped to expand undergrowth. After few years of such growth, fires were easier to spread, especially as more dry wood accumulated. (CAIB, 2011; Rosello 2012).
4. **New understanding for the values of forest emerged:** many of this research local ethnographic sources claim that both the increase in the number of accidental fires, and the continuous growing pressure of the development projects and urbanization have led to public concern and interest in protecting the natural resources of the island.
5. **New sustainable uses and more integrated protection measures for the forests:** with the declaration of UNESCO, the importance of the forests in the eyes of the public has changed even more, as issues of how to protect them while keeping them open for more public uses and tourist activities became very popular in the media and the political arena. New public policies encouraged the creating of new forests for touristic and educational purposes. Biking was also popular in the eastern part of Mallorca, but now it spread into the Tramuntana mountains and forests (mountain biking in some reserved areas). New hotels specializing in such new "alternative" tourism were built in the last ten years, which joined the market for these new uses of the forests. Also, some of the large rural estates have built walking and biking paths into their private lands and allowed visitors to pass through them for a fee. The UNESCO declaration accelerated more educational activities, as compulsory environmental classes in schools were offered for all levels of classes, and the students started visiting the forests and studying the old traditional practices of the islanders (Rosello, 2012).

It is within these backgrounds of changes to the local economy of Mallorca that we need to understand the new values and uses of the forests in the Tramuntana. External factors of pressure, mainly the increasing numbers of fires and the market pressure caused by a pro-development, and corrupted, economic system, have caused new public debates and created a process of re-valuing of the forest commons.

Constitutionality and the UNESCO platform: a discussion

Data collected through interviews or focus groups show that the UNESCO platform was conceived from the beginning as an international "*legitimizing tool*" for the stakeholders' quests, fears and ambitions, forming one one of constitutionality elements. This is how they hoped that "their Tramuntana" will get the recognition it deserved, "*before it is changed or altered forever*". As has been discussed, foreign outsiders were forming a contrast image to the local identity, but as the bottom-up process proceeded, these migrants had a new opportunity to fit in, and find their interests and plans, and even their new identities, recognized by the locals.

With the election of the Socialist Party in 2007, the new local government decided that Mallorquin will be officially introduced into public offices and the educational system. During these years, different local NGOs, activists, politicians and community leaders were meeting to discuss new possibilities of action to protect the western side of the island. The main person to lead the four-year campaign was the Councilor for Territorial Planning, Maria-Lluïsa Dubon, who successfully advocated the case for a UNESCO candidacy in 2008 up till its declaration in 2011. When I met with her in Palma in May 2012 she mentioned that for her, Mallorca is the local expressions of culture, music and food, and of course it is the local Mallorquin language that she cares about the most:

> We wanted to preserve many things in the Serra de Tramuntana, from the first stages. As you might already know, the island is divided among those who support the nationalist politics, the pro-Madrid people, and those who support the local identity and culture of Mallorca, who are usually inclined towards Catalonia. But what we thought to do, after several meetings and discussions with many stakeholders in the island, that we should find the basic common story for all the people who live here, including the non-citizens. That was "our" UNESCO idea. Here you have a platform, which is based on something called "universal" values, like culture, language and advanced levels of human and environmental principles. This became much more suitable for Mallorca because these values can be incorporated into the most important industry of the island, for which everybody has his unconditioned support: tourism. So, by this step, we thought we will get three important birds: we unite all stakeholders under an international umbrella, since there was no way to do it under the local politics of the island. Second, we get to promote tourism in a new nondestructive sustainable way in the Tramuntana. Third, we will be able to promote, or even force a protection for the important cultural elements of our heritage and nature by putting it up on the international and prestigious list of the World Heritage Sites. So people can be really proud about it. For all these reasons, the UNESCO was the perfect platform for us, and it really worked.

Dubon also described how at the beginning she worked with few local activists, who formed the core group. They slowly but steadily organized public meetings for the discussion of the idea. They met with local municipal council members, mayors and professionals. Within a year the group grew to include business persons, academics, prominent land owners, European residents and migrants and other stakeholders. Well into the second year the platform was also part of the public media coverage of the island.

Analyzing some of the actors' emic perceptions emphasized their initial need to create new constitutional agreements in order to stop "the destruction of the island caused by corruption, uncontrolled tourist development, and fires of the forests" (Marcelo, a landowner living in northern Mallorca 11 May, 2013). While finding the middle way was a big challenge during the meetings, especially with the introduction of new environmental concepts such as "sustainable development", which was met with suspicious reactions by local municipal leaders.

Jaume Mateu was one of the main coordinators of the public meetings. He told me that his six years' experience into the process has given him "real life experience to bring different people together". He thinks that this platform has created a new shared story. Foreign residents were "finally able to meet with locals on an equal status", and deal with their differences:

> In the morning of one of the earlier meetings, which included about 35 participants, both locals and foreign residents, a new corruption story was uncovered. An angry discussion developed that day, focusing on the eminent threat of corruption to the environmental resources of the island. This shared anger has brought them to neglect their initial differences, and trust in their mission increased. Finally, all agreed to support the efforts of change.
>
> (Interview with Jaume Mateu on 12 May, 2013)

Some Mallorcan academics think that the UNESCO process has promoted a compromise within the corruption aura. "Corruption after all is the dominance of the elites over the little citizens," tells Jaume Deyà, a well-known archaeologist from Soller who directed the excavations in Almallutx, the last Islamic stronghold in the heart of the Serra de Tramuntana, and participated in the heritage meetings:

> Powerful persons or big companies were controlling us, by their own set of rules. The UNESCO declaration forced the government to stop such practices of destruction and to promote an equal participation processes for the local inhabitants. Now there is a lawful commitment to the international body to give these local people the power to operate the new World Heritage Site.
>
> (Interview with Jaume Deyà on 20 May, 2013)

Many locals' perceptions presented in this paper show a clear case of constitutionality in the formation of new institution of the UNESCO World Heritage, which has a long-term potential of sustainability. Native Mallorcans did use an ideology of local identity connected to the Serra de Tramuntana area, coupled with "the need to protect the Mallorquin language and heritage" in order to achieve higher level of participation in decision-making vis-à-vis the historical attempts by the central Madrid government to impose its own policies on the islanders. On the other hand, foreign residents used the UNESCO platform to create legitimacy for their recent arrival to the island, and thus they usually enthusiastically supported the principles of sustainability developed by UNESCO. This empowerment to engage in the institution building process given to the local actors in the area, coupled with the forming of heterogeneous and open groups of actors which were able to discuss and form the core principles of the new institution has finally succeeded to create a real grass-roots process for the sustainable management of common-pool resources (land, forests, beaches, natural habitats, historical sites) in western Mallorca. The corruption cases coupled with the continuous destruction of these resources did invoke a feeling of emergency among all actors, who tried to prevent their elite capture.

Conclusions

The case presented in this chapter, using constitutionality as the main analysis framework, shows a successful process of a pro-active and strategic participation in the protection of the local natural and cultural commons in the Serra de Tramuntana in the western side of the island of Mallorca. The platform created through adoption of UNESCO participatory rules was to a large extent a successful bottom-up institution building. Constitutionality emphasizes the importance of negotiating power asymmetries, and here we see that the incorporation of immigrants and new residents in this process was able to bridge the gaps with the local Mallorcans on various aspects by taking part and developing new norms under the UNESCO definitions of "sustainability". Within this coordination with the locals, we also see a clear case for institution shopping. Each group have tried to increase their bargaining power, and the new alliances that were created through the process helped them to legitimize their own aspirations, while formalizing compromises with collective benefits.

New local political leaders who adopted the international UNESCO principles, such as Maria-Lluïsa Dubon, rose to power and were able to legitimize their roles with downward accountability. Those leaders have positioned themselves in contrast to the corrupted political agenda of Mallorca and the weak state thereafter. It is important to emphasize here that research in similar areas with contexts of weakened state show that such context tends to facilitate local level actions and successful institution building. For example, the case in Mount Carmel Biosphere Reserve on the eastern side of the Mediterranean. There, the local Druze-Arab minority was able to challenge the fortress approach of managing the Carmel forest, shortly after a horrendous fire caused mainly due to the state's misconducts in the region and the public anger directed towards government ministers and state officials (Eid 2018).

Another important element of constitutionality is the needed recognition of local knowledge by the state. This bottom-up process in Mallorca was based on a reenacted local identity through years of debates and has managed to affect the highest levels of the political game, as the local government adopted many of these ideologies and argued for a change in the policies towards the forests and the environment. These new narratives were used and embedded in the plan to nominate the area under the UNESCO international platform, resulting in real change in the emic perceptions of many stakeholders in the island.

While conservation and protective considerations where some of the main driving incentives for many actors, the issue of how to combine conservation and development schemes brought into the system new questions that needed new debates. The actors' testimonies show concerns related to economic policies and environmental and heritage protection, such as the way to further promote tourism in the Serra de Tramuntana without hurting the natural landscape. Their credence in their bargaining powers, and in new strategies to create constitutional agreements based on more equal statuses, was a very important reason they wanted to use the UNESCO World Heritage Sites platform.

This UNESCO umbrella was generally believed by most stakeholders to be a fair and neutral platform, concerned mainly with sustainable universal principles, created by social and natural scientists. This helped to foster a feeling of trust and ownership towards the new institutions among the heterogenous participants. Thanks to this process, the level of natural degradation has been greatly decreased, and new by-laws that were created following the UNESCO process were able to protect the local heritage and language.

References

Bardolet, E. (2001) 'The path towards sustainability in the Balearic Islands', in Ioannides, D., Apostolopoulos, Y., and Sonmez, S. (eds), *Mediterranean Islands and Sustainable Tourism Development: Practices Management and Policies* (pp 193–213). London: Continuum.

Barke, M., Towner, J., and Newton, M. (eds) (1996) *Tourism in Spain: Critical Issues*. Wallingford: CABI.

BBC (2013) 'Corruption Casts Shadow on Spain's Balearic Islands', www.bbc.com/news/world-europe-21538593, accessed 11 November 2017.

BBC (2014) 'Spanish duke Inaki Urdangarin to appear before judge', www.bbc.co.uk/news/world-europe-16352703, accessed 11 November 2017.

Blaikie, P. (2006) 'Is Small Really Beautiful? Community-based Natural Resource Management in Malawi and Botswana', *World Development*, vol 34, no 11, pp 1942–1957.

Braudel, F. (1992) *The Mediterranean and the Mediterranean World in the Age of Philip II*. London: HarperCollins.

Bull, P. (1997) 'Mass tourism in the Balearic Islands: an example of concentrated Dependence', in Lockhart, D. and Drakakis-Smith, D. (eds), *Island Tourism: Trends and Prospects* (pp 137–151). London: Pinter.

Buswell, R.J. (2011) *Mallorca and Tourism: History, Economy and Environment*. Bristol: Channel View Publication.

CAIB (Gestió Forestal) (2011) *Conselleria d'Agricultura, Medi Ambient i Territori*. Palma.

Chabwela, H. and Haller, T. (2010) 'Governance issues, potentials and failures of participatory collective action in the Kafue Flats, Zambia', *International Journal of the Commons*, vol 4, no 2, pp 621–642.

De Angelis, M. and Stavros S. (2010) 'On the commons: a public interview with Massimo De Angelis and Stavros Stavrides', *e-flux*, vol 6, no 17, pp 1–17.

Dodds, R. (2007) 'Sustainable tourism and policy implementation: Lessons from the case of Calvia', Spain', *Current Issues in Tourism*, vol 10, no 4, pp 296–322.

Eid, R. (2018) 'Burning forests, rising power: towards a constitutionality process in Mount Carmel Biosphere Reserve', *Human Ecology*, vol 46, no 1, pp 41–50.

Ensminger, J. (1992) 'A proper marriage: new institutional economic anthropology' in Ensminger, J. (ed.), *Making a Market* (pp 1–32). Cambridge: Cambridge University Press.

Garcia, C. and Servera, J. (2003) 'Impacts of tourism development on water demand and beach degradation on the island of Mallorca (Spain)', *Geografiska Annaler: Series A, Physical Geography*, vol 85, no 3–4, pp 287–300.

Gil, M. and Diaz-Fernández, P. (2003) *La transformación histórica del paisaje forestal en las Islas Baleares*. Madrid: Ministerio de Medio Ambiente.

Haller, T. (2013) *The contested floodplains: institutional change of the commons in the Kafue Flats, Zambia.* Lanham, MD: Lexington/Rowman & Littlefield.

Haller, T., Acciaioli, G., and Rist, S. (2016) 'Constitutionality: conditions for crafting local ownership of institution-building processes', *Society & Natural Resources*, vol 1, pp 1–20.

Haller, T., Belsky, J., and Rist, S. (2018) 'The constitutionality approach: conditions, opportunities, and challenges for bottom-up institution building', *Human Ecology*, vol 46, no 1, pp 1–2.

Haller, T. and Chabwela, H. (2009) 'Managing common pool resources in the Kafue Flats, Zambia: from common property to open access and privatization', *Development Southern Africa*, vol 26, no 4, pp 555–567.

Haller, T. and Merten, S. (2008) '"We are Zambians – don't tell us how to fish!" institutional change, power relations and conflicts in the Kafue flats fisheries in Zambia', *Human Ecology*, vol 36, no 5, pp 699–715.

Harvey, D. (2012) *Rebel Cities: From the Right to the City to the Urban Revolution.* London: Verso.

Kalb, D. and Gabor, H. (2011) *Headlines of Nation, Subtexts of Class: Working-Class Populism and the Return of the Repressed in Neoliberal Europe.* Oxford: Berghahn Books.

Lyth, P. (2009) 'Flying visits: the growth of British air package tours', in Segreto, L., Manera, C., and Pohl, M. (eds), *Europe at the seaside; the economic history of mass tourism in the Mediterranean* (pp 11–30). Oxford: Berghahn Books.

Ostrom, E. (1990) *Governing the Commons: The Evolution of Institutions for Collective Action.* Cambridge: Cambridge University Press.

Pack, S. (2006) *Tourism and Dictatorship: Europe's Peaceful Invasion of Franco's Spain.* New York: Palgrave.

Perry, A. (2004) 'The Mediterranean: how can the world's most popular and successful tourist destination adapt to a changing climate?', in Hall, C.M. and Higham, J. (eds), *Tourism, Recreation and Climate Change* (pp 86–96). Clevedon: Channel View Publications.

Purdy, M. (2012) 'Tourism and tyranny in Franco's Spain', *Diffusion*, vol 3, no 2.

Rosello, J. (2012) 'Is forest use in the Balearic Islands the result of an environmental conscience?' *Earthzine*, 2 July 2012, http://earthzine.org/2012/07/02/is-forest-use-in-the-balearic-islands-the-result-of-an-environmental-conscience/, accessed 6 November 2018.

Royle, S. (2009) 'Tourism changes on a Mediterranean Island: experiences from Mallorca', *Island Studies Journal*, vol 4, no 2, pp 225–240.

Salva, T.P. (2002) 'Foreign immigration and tourism development in Spain's Balearic Islands', in Hall, C. and Williams, A. (eds), *Tourism and Migration: New Relationships between Production and* Consumption (pp 119–134). Berlin: Springer.

Segreto, L., Manera, C., and Pohl, M. (Eds.) (2009) *Europe at the Seaside: The Economic History of Mass Tourism in the Mediterranean.* Oxford: Berghahn Books.

Shubert, A. (1990) *A Social History of Modern Spain.* London and New York: Routledge.

Spilanis, I. and Vayanni, H. (2004) 'Sustainable tourism: Utopia or necessity? The role of new forms of tourism in the Aegean Islands', in Bramwell, B. (ed.), *Coastal Mass Tourism: Diversification and Sustainable Development in Southern Europe* (pp 269–291). Clevedon: Channel View Publications.

Sureda, J. (2011) 'Perception of pine trees among citizens of the Balearic Islands: Analysis and description of some mistaken ideas', *Applied Environmental Education & Communication*, vol 10, no 1, pp 31–42.

Susser, I. (2016) 'Considering the urban commons: anthropological approaches to social movements', *Dialectical Anthropology*, vol 40, no 3, pp 183–198.

Susser, I. (2017) 'Introduction: for or against commoning?' *Focaal – Journal of Global and Historical Anthropology*, vol 79, pp 1–5.

Trauer, B. and Ryan, C. (2005) 'Destination image, romance and place experience – an application of intimacy theory in tourism', *Tourism Managmanet*, vol 26, no 4, pp 481–491.

Tusell, J. (2007) *Spain, from Dictatorship to Democracy: 1939 to the Present*. Oxford: Blackwell publishing.

Ulrich, A. (2007) 'Corruption in Spain: Shady deal in the Sun'. Speigel Online magazine', www.spiegel.de/international/spiegel/corruption-in-spain-shady-deals-in-the-sun-a-469272.html, accessed 11 November 2017.

Waldren, J. (1996) *Insiders and Outsiders: Paradise and Reality in Mallorca*. Oxford: Berg.

West, P., Igoe, J., and Brockington, D. (2006) 'Parks and peoples: the social impact of protected areas', *Annual Review of Anthropology*, vol 35, pp 251–277.

Wheeler, D. (1995) 'Majorca's water shortages arouse Spanish passions', *Geography*, vol 80, no 348, pp 283–286.

11 Constitutionality and identity

Bottom-up institution building and identity among Coastal Sami in Northern Norway

Angelika Lätsch

Introduction

> The Coastal Sami population has always lived in the shadow of the reindeer herd.
>
> (H. Johansen, personal communication November 2011)

The citation above indicates that 'being Sami' in the sense of identity performance has been relatively easy to Sami reindeer herders, whereas Sami who settled along the coast and fjords of Northern Norway and who live as fishermen and small-scale farmers, collectively known as the *Coastal Sami*, have long remained invisible both on the political agenda and within the cultural Sami landscape. Owing to political experiences such as an enduring assimilation policy conducted by the Norwegian state, Coastal Sami have come to accept their identity as fishermen while concealing their stigmatised ethnic identity. However, in recent decades this situation has changed, as Coastal Sami have undergone a process of ethnic revitalisation, refusing to accept the marginalisation of their culture and economic adaptation. This article explores the factors behind this situation, arguing that institutional changes have resulted in reduced access to local commons – the fisheries – in the process triggering a shift in local people's perceptions of historically hidden Sami ethnic identity. Fishing quotas for small-scale fisherman have become more constrained by the state, stimulating the Sami community's strategic legitimisation of ideological claims with reference to being just as indigenous as their Sami reindeer counterparts. Therefore, re-enacting a Sami identity appears to have represented the fundamental strategy. However, this process is much more profound than mere contestation over institutional arrangements. Indeed, alongside this discourse of being Sami, local people's identities have become reshaped, with increased engagement with the past and a movement to acquire new legislative rights, as well as developing both an old-new identity and self-consciousness as people with a Sami background.

The Sami are an indigenous people in Scandinavia and the Kola-Peninsula in north-western Russia, and who have prehistoric roots in large parts of the region. They have traditionally lived within their own small-scale societies, with diverse (but not mutually exclusive) livelihoods of fishing, hunting, trapping, animal husbandry, farming and reindeer-herding, creating a significant degree of heterogeneity within contemporary Sami culture. Today, it is formally accepted that the Norwegian state was established on the territory of two peoples. However, this shared history of Sami and Norwegians reveals that co-existence has not always been idyllic. From the mid-nineteenth century, the Norwegian state followed a consistent policy of assimilation (Norwegianisation) towards the Sami in order to establish supposedly fully fledged Norwegian citizens. The assimilation policy had far-reaching consequences, especially for the Sami people living in coastal areas, such as a loss of language skills in their mother tongue as well as damage to their self-esteem and identity as a distinct group (see Brantenberg 1991; Drivenes et al 1994; Minde 2005; among others). The development of the Norwegian welfare state after World War II went hand-in-hand with national standardisation, and in a Sami context this was characterised as 'passive Norwegianisation', as social aspects including accommodation, education, nutrition and health care became tied to the government's notion of equality, in which every citizen, whether Sami or Norwegian, should have the same autonomy to participate in the development of the Norwegian welfare state (Thuen 1995; Nilsen 2003). In addition, newly established economic structures gradually led to the integration of the Coastal Sami into the national economy, thereby instigating their long-term invisibility as a distinct group.

The conflict over the damming of the Alta-Kautokeino watercourse during the 1980s[1] can be interpreted as the turning point in the relationship between the Sami people and the Norwegian government, resulting in the so-called Sami Paragraph of the Norwegian Constitution (1987), which guarantees the Sami in Norway the status of an indigenous people by ratifying International Labor Organization (ILO) Convention no. 169 as well as the establishment of the Sami Assembly (1989). Whereas the reindeer-herding Sami were able to (involuntarily) dominate public discourses as the 'real Sami', traditional Sami fishermen remained marginalised and mostly invisible as an indigenous group, given that they were not really distinctive in terms of economic adaptation and language from other Norwegian small-scale fishermen.

This paper describes how Coastal Sami regained their ethnic identity via conflicts over reduced access to fisheries. This common-pool resource, which was previously managed as part of a common property regime in the fjords and at the coast, was later considered state property for commercial use. In introducing technical changes and promoting modernisation and industrialisation, the Norwegian authorities contributed to a process of overcapitalisation and significant overuse of fish stocks (Andersen 1997). The state reacted to this depletion with the introduction of an individual quota system to cod fisheries in 1990 in order to relieve the pressure (Hersoug 2005). The quota system led to fishing rights becoming concentrated in fewer hands, from which the large-scale fishing industry profited at the expense of Coastal Sami and other small-scale fishermen, who in contrast faced a 'tragedy of the commoner'. Certainly, they lost most of their access to their previous fishery ground, even though they were not primarily responsible for the decline in this common-pool resource. In response, the newly established Sami Parliament (1989) immediately entered the debate regarding fishing rights, albeit with little immediate visible success. Subsequently, some fishermen of Sami origin began to act collectively in partly self-established local organisations, using channels of the Sami Parliament in order to boost their bargaining power and thereby (re-)incorporate their ethnic identity. This illustrates three new insights to constitutionality. First, it was perceived as far easier to achieve success regarding higher quotas once an indigenous political identity based on an historic notion of Saminess became incorporated into the debate. Indeed, this notion provided justification for access to the fisheries, as access rights to resources had long formed a central aspect of indigenous identity for Sami reindeer-herders. Second, bottom-up institution building was stimulated, as Sami were able to discuss certain issues regarding fisheries in their fjordal fishing areas such as regulations. Third, a basis for a re-established Sami identity that was different from other small-scale fishermen was generated, reflecting their past. This represents a novel aspect in the discussion of constitutionality, as this process does not merely consider local participation based on a sense of ownership of the institution-building process, but additionally incorporates the role played by the reproduction and revitalisation of identity in the management of local resources.

Theoretical considerations: the role of identity in constitutionality

In the management of the commons, the approaches of political ecology and new institutionalism have focused on the ways in which external pressures and power relations have shaped the management and outcomes of common-pool resources (see Blaikie and Brookfield 1987; Ensminger 1992; Haller 2017). I use constitutionality as an analytical approach to argue, based on empirical case studies, that in successful bottom-up and participatory institution building, a process by which a conscious sense of ownership in the crafting of new rules for the governance of common-pool resources and based on a platform where all actors can participate occurs. As a critique on the state-driven crating of environmental subjects coined by Agrawal (2005) as environmentality, constitutionality instead refers to the participatory bottom-up process, in which local actors create their own rules and come to embody them as their own. Empirical studies have shown that this can occur where local knowledge is accepted and where the heterogeneity of power constellations, creativity and recognition of previous local institutions are all recognised (see Haller et al 2018). The cases highlight how local actors with different levels of bargaining power and reflexive processes of collective action craft their own institutions based on six principles: 1) internal or emic perception for the need to create new institutions; 2) capacity to stimulate participatory processes that address rather than ignore power asymmetries; 3) availability of a pre-institutional setting; 4) external catalysing agents help provide an equal platform; 5) recognition of local knowledge; and 6) acknowledgement of higher-level recognition and support (Haller et al 2016). More recent studies refer to the building of identity as a new element of the concept (Eid and Haller 2018; Gambon and Rist 2018). All of these studies have demonstrated that if a process of constitutionality takes place, local users can attain greater bargaining power relative to other users, as well as negotiate the internal heterogeneity of the actors involved. In this chapter I discuss the role played by identity and identity transformation in the management of the commons, whereby resource conflicts trigger issues of identity, again instigating legal claims regarding resource allocation and resource management. The role of identity has rarely been discussed in the commons literature and in newer publications on constitutionality. I will demonstrate how discussing identity as a fluid concept of emic and etic perspectives on who might be considered a group (and related group identity processes) helps shape demands and legal claims on the commons. The chapter also contributes to the discussion on forum and institution shopping (Haller 2016) and highlights how actors strategically choose legal and institutional as well as ideological elements that are both nationally and internationally available for their interests. The case of the Sami fishermen who claim their rights as indigenous people offers not just empirical but theoretical insights on the creation of institutions and the management of the commons. I will therefore also refer to the literature on ethnicity (see Barth 1998 [1969] and others) to show which elements of identity building correspond with the process of constitutionality.

Barth (1994) refers to the establishment of ethnic boundaries, which emerge when in constant contact with other groups. The characteristics of self-ascription and ascription by others become significant in rendering membership of a group a reality (Barth 1994). Thus, Barth states that it is not actual differences that create ethnicity, but rather that cultural differences are made relevant in the interaction between two or more groups. The Norwegianisation process was especially impactful in Coastal Sami areas, as individuals here had greater contact with other Norwegians. The Sami culture was perceived as 'inferior' and 'primitive', and thus many families assimilated within Norwegian culture and customs. Parents would not tell their children about their Sami background for fear that they would have to experience the same discrimination (Dahl 1970). Thus, many Coastal Sami redefined themselves as solely Norwegian (Bjørklund et al 2002). The disappearance of the Coastal Sami from the public sphere has been well-documented and conceptualised in different studies, the most prominent being Eidheim's (1998) paper 'When ethnic identity is a social stigma'. Through reading Eidheim's (1998) study, it becomes evident that the hidden and stigmatised Coastal Sami identity commanded limited potential for organisation and resistance during the assimilation process. Nilsen (1998a) has analysed particular problems pertaining to the Coastal Sami experience in the political and organisational field that owed to ethnic and economic suppression. An important question is how the Coastal Sami managed to revitalise their culture and enter public discourses on indigenous rights and marine resources, or how these ethnic boundaries were 'reconstructed'. I see (ethnic) identity as an important precondition for the increased use of the powerful concept of 'indigenous' in the process of recognition, empowerment and entitlement (see also Minde et al 2007). In line with new institutionalism approaches in social anthropology (see Ensminger 1992; Haller 2010), identity is part of ideology, containing discourses and narratives of what is right or wrong and thus what may act as a resource of legitimacy. The narrative of being of indigenous origin and the discourse of the rights of indigenous peoples thus in theory boosts local people's bargaining power and can be used in empowerment strategies. In this context, I also discuss the strategies used by the actors to influence the political process within the institutional conditions of the nation-state. In remaining aligned with the anthropological approach of adopting a bottom-up perspective, I pursue an analysis of processes taking place at the local level and how these interact with national and international settings of laws and politics.

The setting: the Sami and fisheries in Northern Norway

The Coastal Sami settlement is, to a large extent, located in the two northernmost counties of Norway (Finnmark and Troms), including the northern part of the county of Nordland. Sami participation in saltwater fisheries can be traced back more than 1,000 years (Hersoug 2005). The largely ice-free coastal waters (due to the Gulf Stream) have attracted people to northern parts of Norway for a long time and have rendered settlement here possible. In this context, Nilsen (2003) has stated that cod has historically enjoyed the same significance as reindeer for Sami living in the inner parts of Finnmark. The fjordscape is unique in a European context as it provides shelter as well as abundant fish that can be caught with relatively little effort (Nilsen 1998b). The Coastal Sami traditionally lived in the inner parts of the fjords, whereas the outer fjord and coastal areas came to be populated by Norwegians, in the process limiting the area of resource exploitation of the Coastal Sami (Nilsen 2003). The Sami's economic adaptation has varied regionally due to ecological and topographic features and available resources in the locality in question (Ween and Lien 2012). The key words for describing traditional Coastal Sami livelihoods include combination, versatility and locality (Eyþórsson 2008). Early Coastal Sami adaptation was characterised by seasonal transhumance: the exploitation occurred over a wide ecological range, among other things sea fishing, sea hunting, freshwater fishing and land hunting, including the exploitation of forested areas in order to obtain fuel and timber for boat construction (Paine 1957). Farming and animal husbandry became important, alongside fisheries; this mixed subsistence economy is generally described as *fiskarbonden* (fisherman-farmer) (Nilsen 1998a). From the latter part of the nineteenth century, Coastal Sami settled in one place rather than having different seasonal households in their area of exploitation; nevertheless, they were not completely sedentary. The different households in the villages in the inner parts of the fjords managed the area commonly in social units, known as *siidas*. *Siida* is a term commonly used to describe the traditional social organisation of Sami communities, their economic adaptation to the natural environment and use of natural resources (Grydeland 1997). With the exception of the winter period, when collective forms of hunting took place, household units were

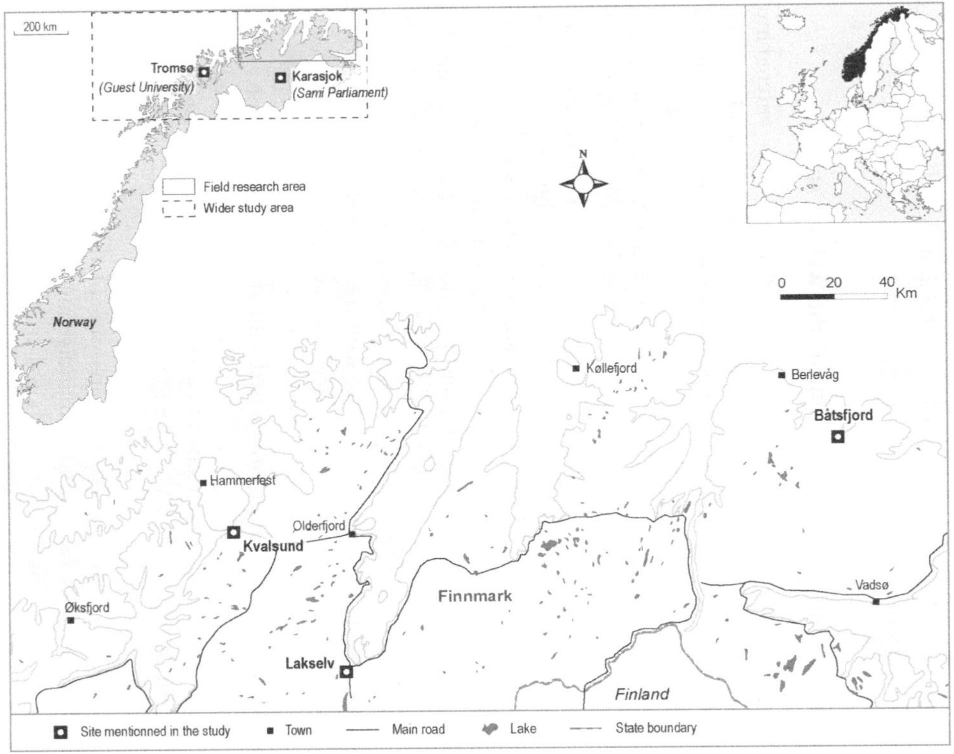

Figure 11.1 Map of (wider) field research area in Northern Norway.

Source: Map compilation and design by Ulla Gämperli Krauer (2018), CDE, University of Bern, Bern, Switzerland. Geodata source: OSM data by OSMaxx (2018). Geometalab, HSR University of Applied Sciences Rapperswil, Switzerland. Further geoinformation provided by the author of this article.

based in different locations where they were separately able to exploit available resources (Hansen and Olsen 2004).

Contemporary Sami fisheries are typically small-scale, part-time and seasonal, but traditional livelihoods combine fishing with other natural resource extraction activities such as small-scale farming and hunting or some form of wage labour. Contemporary Coastal Sami fishing is mainly based on 'home-fishing' (fishing in waters close to their homes, in the fjord), and has some distinctive features, including the use of conventional fishing gear, small vessels, little capital investment and fishermen with an extensive knowledge of the local environment that has been transferred from generation to generation (Einarsbøl 2006). In recent decades, and especially since 1990, the socio-cultural landscape along the Northern Norwegian coast has been changing. Traditional fisheries have faced different challenges due to increased actors with powerful interests and a tendency towards privatisation. In this context, the transferability of fishing rights can be observed, as I will explain further below (Søreng 2007).

Norwegianisation and the development of the welfare state

Norway became internally independent from Denmark in 1814, but entered into union with Sweden until 1905, when Norway became an independent state. The subsequent construction of a homogeneous state as well as the assimilation policy practised in Norway from the 1850s until after the Second World War resulted in the apparent loss of Sami language skills and a distinct Sami ethnic identity in most Sami areas (Bull and Gaski 1994; Minde 2005). The systematic 'scorched-earth policy' of the retreating German army in 1944 destroyed a huge portion of Northern Norway, and erased nearly all visible ethnic markers in the landscape, such as traditional boats, fishing equipment and coastal settlements (Bjørklund 2000). The reconstruction of Northern Norway after the Second World War did not correspond to earlier and traditional settlement forms, but instead signalled a new era where all ethnic groups should have the same social and cultural possibilities. The idea of the central Norwegian administration was to rebuild the northern region in accordance with the requirements of a modern economy, and was therefore defined in Norwegian terms (Thuen 1980). However, this led to the further reinforcement of the assimilation policy, as the population became socially and economically integrated into and thus dependent on the greater Norwegian society (Eidheim 1971). Modernisation, the construction of the Norwegian welfare state and assimilation were thus heavily intertwined and had a significant impact on local communities, especially the ethnic identity of coastal Sami.

Today, formally all Sami in Norway, in spite of different specific socio-economic adaptations, are recognised as an indigenous people by the Norwegian state. The Norwegian Sami policy is rooted in international law, the most important legal sources being the International Covenant on Civil and Political

Rights (ICCPR) of 1966 (mainly article 27) and the International Labor Organization (ILO) Convention no. 169 concerning Indigenous and Tribal Peoples in Independent Countries of 1989 (Bjerkli and Selle 2003). However, political legal claims are not easily transferred into the context of the nation-state, but are succeeded by struggles embedded in different ideologies. Even though the Norwegian state has been an active promoter of international law and has made a significant shift in its policy towards the Sami, state actors follow an ambiguous strategy. Thus, although the state affirms the indigenous status of the Sami, they hesitate to secure for the Coastal Sami special access to fjordal and marine-related resources. The issue is important because it contributes to the debate regarding the ways in which identity and political strife within a state are linked to a resource base and conflicts about this base. It concerns a process of institutions that are chosen and shaped in order to trigger institutional change. In reviewing the Coastal Sami struggle for obtaining rights to natural resources, I argue that identity and ethnicity play a crucial role in crafting institutions from below, as they can both facilitate and constrain collective action, and provide an important source of legitimation when it comes to resource rights and increasing one's bargaining power.

Fisheries and the fishing industry have constituted the most important economic activities for the purpose of maintaining coastal Sami and Norwegian settlements. A special feature of fisheries in the North is the high proportion of fisheries conducted as home fishing with relatively small vessels. In recent decades, Norwegian fisheries have developed from being open access fishing to a regulated system with quotas and concessions. The open fishery commons have increasingly become closed, privatised and centrally state-governed industries (Skogvang 2010). In this section I will develop how fisheries in Norwegian coastal waters underwent a process from local common property of fish in fjords to state property and state management (Hersoug 2005).

Fjord fishing was an exclusive right for people living in the fjords at the beginning of the 1900s (Skogvang 2010). Between 1687 and 1778, the right to fish of fishermen living outside of Finnmark was continuously restricted and limited to sea fishing. The aim was to reserve the local and regional resources for those who lived in Finnmark. This measure should at the same time be a pull factor to render migration to Finnmark more attractive. The law of the fisheries in Finnmark (1830) was the formal legal frame that laid the basis for substantial changes in Finnmark. The law introduced the principle that all Norwegian citizens should be treated equally in relation to access to and the performance of fishing. The fisheries law from 1830 broke with the relatively old and formalised principle about the first or priority rights of Finnmark's population to exploit local fish resources in the county. Therefore, in many places the new law was not accepted and the old system continued to be practised for a period of time. Given the appeal of rich natural resources in the sea as well as beneficial trade conditions to the south, the population in Northern Norway increased substantially. The 1800s came to be referred to as the second big Norwegian colonialisation wave, and was closely connected with possibilities for commercial fishing. The time around 1900 marked a new era inside the fisheries in Finnmark. Among other things, the types and sizes of the boats as well as the fishing equipment used diversified, and motors became increasingly common. Such modernisation was supported by the Norwegian authorities through the introduction of several programmes as well as a loan system to finance larger vessels and motors. In connection with the new technologies and types of equipment available, resource conflicts developed to a greater extent, and claims of protection of the resources (especially those located in the inner parts of the fjords) emerged from local (Sami) fishermen (Eyþórsson 2008).

A combination of technological, structural and regulatory changes has thus led to a greatly reduced participation in fishing in Coastal Sami areas (West and Hovelsrud 2010). First, the development of modern and effective fishing techniques (especially trawlers, seines and echo sounders) following the Second World War led to overfishing out at sea, resulting in fewer fish migrating into the fjords. It was no longer easy for small-scale fishermen to catch a sufficient amount of fish in the fjords, as these areas had also been intruded by trawlers and purse seines with active types of fishing equipment. Second, the ongoing assimilation and stigmatisation of the Coastal Sami diminished possibilities to promote their own fishing interests. Third, following the war, the state focused on modernisation and specialisation and thus further marginalised the Coastal Sami as it became much more difficult to obtain loans and subsidies for small-scale fishers who did not merely concentrate on fishing but followed a combination of different livelihoods. Finally, the general rationalisation of fishing has also affected Coastal Sami fishermen, drastically reducing their numbers and the likelihood that new individuals will become involved in this sector (Nilsen 2003). These processes provide further evidence that the Coastal Sami, with their traditional knowledge, traditions and culture, have long remained invisible as a distinct group.

Studies by Eidheim (1971) and Bjørklund (1985) thus demonstrate that following the Second World War, most Coastal Sami came to define themselves as Norwegian, leading to a 'sudden' disappearance of the Coastal Sami as a distinct group (Eyþórsson 2003). Most local fishermen (including Coastal Sami) entered into local branches of the Norwegian Fishermen's Union (Norges Fiskarlag) in order to obtain the same rights and benefits as the rest of the Norwegian population. The Fishermen's Union is the biggest national fishery organisation in Norway and aims to represent all fishermen in the country. In other words, membership was defined through occupation and position, and pays no attention to ethnic minority identity. In contrast, reindeer-herding Sami have always been alone in their economic branch and have been perceived as exotic and 'indigenous' from other populations. Consequently, they have been able to attract greater attention as regards their economic and political situation. Given these developments, the question arises as to why or how the Sami appeared in the public sphere, and how the coastal Sami re-constructed themselves as one group inside the Sami 'nation'.

Identity politics in the context of resource conflicts

The situation of the Sami and the possibilities of a Sami movement changed after the Second World War, in a context of an increasingly liberal political attitude towards minorities and the greater value of human rights at the international level (Eidheim 1997). At the same time, local and national cultural and interest Sami organisations began to flourish, with an initial peak in the 1970s, and since then, Sami culture has seen an active revitalisation of ethnic identity, cultural activities and language (Bjerkli and Selle 2003; Brattland 2010). The conflict over the damming of the Alta-Kautokeino watercourse in the Sami reindeer heartland in Finnmark during the 1980s is often considered the turning point of the political and legal relationship between the Sami people and the Norwegian authorities (Thuen 1980). The conflict attracts national and international attention, with the result that the Sami became compared and connected to other indigenous peoples. The emergence of an international indigenous movement, as well as the fact that the Sami were actively participating within the movement to develop an indigenous people's network, helped alter perceptions of the Norwegian authorities on matters pertaining to the situation of their Sami populations (Minde 2003). Even though the Sami still lost out on all essential issues disputed, the emerging public debate had decisive effects on the political work of the Norwegian authorities towards minorities in their country, as well as revealing power structures and fundamentally changing the status of the Norwegian Sami. On a national level, the Norwegian state established two official committees: the Sami Rights Committee and the Sami Culture Committee, which launched a process that ultimately resulted in the so-called Sami Paragraph of the Norwegian Constitution (§110a), conferring upon the state the responsibility of guaranteeing that they are able to preserve and develop their culture and society. A 'Sami Act' (Sameloven) was enacted in 1987, legalising among other things the organisation of an assembly democratically elected by the entire Sami population living in Norway: the Sami Assembly (or Parliament), which began operating in 1989. In addition, Norway had been a main participant in the drafting of the ILO Convention no. 169 and was the first state to ratify the ILO Convention (Bjerkli and Selle 2003). In the aftermath of the Alta conflict, the Sami issue has developed into a public discussion regarding the scope and content of Sami rights in relation to Norwegian legislation and management systems (Thuen 1980). At a different level, another important result of the Alta conflict was discussion regarding the definition of Saminess itself, and the development and revitalisation of Sami self-understanding and self-esteem as a people, thus providing a basis for Sami mobilisation through the constant use of ethnic identity in order to realise political goals, and reinforcing Sami ethnicity in general.

Having outlined the more general economic and political developments of Norwegian Sami and the changes in the institutional setting, I continue by analysing the development of claims that have been articulated by different actors since the twentieth century and that can be seen as both actions and reactions towards transforming a social, political and legal landscape.

Trigger for local identity development: the fishery crisis

Local fishermen had for many decades attempted to warn state authorities of the negative developments in the fjords, especially the negative consequences of large and modern vessels entering these areas, destroying both local stocks and important spawning sites. The state was of the opinion that it was not possible to empty the sea and constantly ignored local fishermen's claims, which were mostly handled by local branches of the Norwegian Fishers' Union or the respective municipalities (Andersen 1999). The observations of local fishermen were generally overlooked as they did not regard traditional local knowledge to be important or 'scientific' (Andersen 2011, personal communication; see Berkes 1999 for similar issues). In addition, the Norwegian Fishers' Union, including the associations of the larger vessels, had attained such influence in the Norwegian Fisheries Management that short-term economic interests prevailed. In this context, small-scale fishermen were deemed less relevant and were thus often neglected. This situation also owed to the fact that many active fishermen from the large-scale fisheries at the coast comprised leaders of the union and were uninterested in 'closing the fjords'.

For many years, Coastal Sami and local fishermen had observed how their fjords were being intruded by fishermen using efficient fishing methods such as purse seining, and considered themselves increasingly displaced from their traditional fishing grounds.

> One could walk over the fjord without getting wet feet. No one believed the local population when telling that this would go wrong.
>
> (Frode Gulbjørnsen about the purse seine in 2008, DVD 'Hva nå')

This statement is just one of many illustrative statements revealing the feelings of local fishermen when the fjords became invaded by large vessels, a situation they considered themselves helpless to prevent. This problem was exacerbated during the reconstruction process when fisheries were modernised with support from the state, and competition in the fisheries was won using economies of scale, thereby increasing the pressure on fisheries and leading to the overuse of the common-pool resource due to overcapitalisation.

Reading the detailed documentation of these conflicts in Einar Eyþórsson's (2008) book *Sjøsamene og kampen om fjordressursene* (The Coastal Sami and their fight for fjordal resources), it becomes evident that the local communities had few if any instruments or strategies guaranteeing success in their claims against powerful interest organisations and the state. People felt completely powerless, neglected as citizens and devoid of a resource base. Thus, for more than 100 years, (Sami) fjord fishermen have conducted an almost hopeless resource fight against the Norwegian fishery authorities, trying to obtain protection for the fjords and their fishing areas. 'They have tried to warn about overexploitation and asked repeatedly for limitations. And they have lost. They have lost in organisational life, they have lost in the management, and they have lost the whole way,' expressed I.A. Eriksen (fishery expert of the Sami Parliament) at a public meeting in Olderfjord. At the same meeting, B. Wilhelmsen stated to the *Fiskeribladet* (a journal on fishery issues) that 'finally we are listened to, finally we are taken seriously' (in Nordlys, 2 November, 2006). These two statements express the poor treatment experienced by Coastal Sami and other fjord fishers for several generations.

In April 1989, the concerns of local fishermen became a reality when the Directorate of Fisheries stopped cod fishing with immediate effect in the middle of the most important cod fishery season due to a dramatic decline in cod stocks (Nilsen 2003). In order to ensure that this situation would not happen again, the state in 1990 introduced a new vessel quota system in cod fisheries to limit access to fisheries. This was supposed to represent a temporary instrument for meeting the cod crisis, but nevertheless 1990 must be considered a turning point in Norwegian fisheries policy, as coastal and fjord fishery became subjected to individual quota management (Davis and Jentoft 2001). The open access principle for that particular fishery, which had constituted an important factor in the original settlement and development of Northern Norway, was abolished. The regulation linked quotas to vessels and not to fishermen, being free of cost to those able to show a sufficient cod dependency in the form of catches. The objective of distributing the quotas was to prioritise the most cod-dependent fishermen by assigning some larger quotas than others. The cod-dependency of fishermen had been measured in the three preceding years. Given resource decline, the invasion of seals into the fjords during these years and the combination of livelihoods, the amount of catches necessary to obtaining a quota was impossible for fjord fishermen (Pedersen 2012). Thus, specialised full-time fishermen were perceived as being more dependent on cod than part-time Coastal Sami who fished as one of several economic activities. Consequently, Coastal Sami and other small-scale fishers were practically excluded from fisheries, or, in other words, their rights to fish were removed and they were only able to fish under the maximum quota arrangement, known as the 'competition quota' or 'Olympic quota' (Nilsen 2003). This was based on a small-group quota where the fishery was closed when the total allowable catch (TAC) had been reached, independently of whether an individual fisher had caught his or her individual maximum quota (Hersoug 2005; Einarsbøl 2006). Thus, this system instigated competition between fishermen to meet the quota as quickly as possible. The formerly flexible fisheries of Coastal Sami were thus considerably constrained by increased regulation by the state, as they for example cannot respond to variations in the fish stocks and must fish in a certain season in order to receive a share of the maximum quota. The state-determined management of individual fish stocks does not take into account the factors that influence the conditions under which fishing can take place, such as weather conditions. Another consequence of the limited access to fisheries was that fishermen often started to fish alone instead of having a crew, as the costs of fishing gear, reduced quotas and technology all increased (West and Hovelsrud 2010).

Furthermore, the authorities allowed individual quotas to attach to vessels at the point of sale. Thus, a market was established where the price of the vessel was much higher than its value given the attached quota. Fishermen who were allocated a cod quota were able to earn a considerable share of money. However, especially for newcomers it was almost impossible to access the fishing industry owing to the requirement for capital, and so recruitment and therefore also the transfer of local knowledge from generation to generation has been negatively affected.

Ole Henrik Magga, former president of the Sami Parliament, stated that the loss of fishing rights of the Coastal Sami represented the biggest abuse against the Sami people in modern times (NRK-Brennpunkt, 5 October, 2009). Even though the cod stocks have recovered in recent decades, most Coastal Sami still lack a sufficient quota to make a living. Where they have obtained such a quota, they have been obligated to pay for it (Pedersen 2012).

> In spite of Sami people from the Sami areas having fished cod for hundreds and thousands of years it was the missing 'historical catch' in the years 1987/88/89 that was used as an argument for denying people the right to live from what had always been a pillar of settlement and culture. It can as a matter of fact, be said that the authorities used a natural catastrophe, like the seal invasion in the 1980s, and the overfishing of Arctic cod in the same period, to get rid of a large part of the small-scale fishers in the Sami areas.
>
> (Sami Parliament 2004: 16 in Hersoug 2005: 197, translation by Hersoug)

Losing the commons: legal demands and outcomes in the fisheries

As explained above, the quota system has gradually been consolidated and the quasi-privatisation of harvesting rights to marine resources has continued. With the introduction of the new management system in the cod fisheries, the claims of local fishermen and different organisations appeared on the political and public agenda. From this very moment, the content of claims was no longer rooted in resource protection, but rather to reversing the new regulation policy and to reclaim local fishing rights. Not only was the content of the claims affected by institutional change in the fisheries, but the period from 1990 onwards marked a new stage in the organisation of the claims due to the establishment of the Sami Assembly in 1989, which immediately entered into the articulation of claims, strengthening the Sami voice and their minority situation in public discourse regarding fishing rights, albeit with little visible success in the beginning.

The Sami Parliament entered into a long-lasting evaluation process whereby different experts and committees were appointed by both the Sami Parliament (the so-called Storslett-report from 1995) and the Norwegian government (Smith Report 1990; Sami Fisheries Committee 1997) to assess and examine the rights that Coastal Sami might enjoy in fisheries. However, most reports were confined to a drawer without any significant consequences. During this 'phase of reports', it seems that the local population temporarily stopped mobilising and actively articulating its claims (Lätsch 2012). I argue that through the institutionalisation of a new Sami policy, the articulation of claims and the negotiation of those claims was channelled to the Sami Parliament, which thereby adopted the role of a 'pressure group'.

Through the passage of the Finnmark Act in 2005[2] following a long-lasting political process, Norway's Parliament recognised the right of people living in Finnmark to manage the land in the northernmost county of Norway. As the act stopped at the shoreline and left the rights to marine resources unsettled, the government was again asked to clarify this issue. Yet another commission was thus established in June 2006 by the Department of Fisheries and Coastal Affairs, the Coast Fishing Commission, to investigate the rights of the Sami and other groups to fish in the sea off Finnmark (Norges Offentlige Utredninger (NOU) 2008). The work of the commission was comprehensive and different strategies were used. The most important of these was the arrangement of public meetings in all 17 coastal municipalities in Finnmark to include the claims, opinions, local knowledge and observations and legal perceptions of the local population. The effect was that people actively reflected on their use of the fjords, their local history and the problems they had encountered in the past. For the first time they were actively asked to contribute to the discussion with traditional local knowledge. It was the first time that the local population had been included in the preparation of a report and was able to influence its drafting. Interestingly, the concerns of the locals were more or less the same throughout all 17 coastal municipalities. The locals who participated at the public meetings expressed their perspectives on questions of rights, concerns over the use of active gear in the fjords, the fish farming industry that had been developed in many fjords, and the quota distribution system, among other issues (see NOU 2008 for detailed meeting minutes). People often asked why the commission had been established or what fishing rights meant in a situation where there were no fish. This shows that the awareness of the problem of overfishing and the experience of the fishermen's concerns being overlooked remained. According to my observations, these public meetings arranged by the Coast Fishing Commission did however set in motion a stronger revitalisation process by reinforcing the awareness of the local people that they are living in a traditional Coastal Sami area. Or, as Persen (2011: 21) states:

> It [the report] gave recognition to a history and a necessary proposal for action as a consequence of this history.

The findings of the Commission's work, published in a White Paper (NOU 2008), resulted in a series of proposals to strengthen and safeguard coastal Sami culture. The most important findings of the Commission in this context are the affirmation of the Sami having a historical right to fishing and thus being entitled to prior access to fisheries based on international human law. Moreover, regarding management, the Commission proposed the establishment of a Finnmark Fishing Administration (in Norwegian) that should have the authority and capacity to regulate local fjord fishery four nautical miles out to sea as regards size of vessels, gear usage and quotas. In June 2012, following a year of consultations with the Sami Parliament on the issue, the Norwegian Parliament clearly voted against the proposals from the Coast Fishing Commission, rejecting the notion that the Sami have a unique historical right to marine fishing in Finnmark compared to other fishermen in the country, which would call for a separate law and would empower the Coastal Sami in fisheries governance. Nevertheless, they agreed to certain points, the most important of which are as follows (Andersen and Eyþórsson 2015: 161):

a. The state commits itself to consult with the Sami Parliament on issues of particular concern to Sami fisheries.
b. A new paragraph in the Marine Resource Act will emphasise the need to consider Sami resource use and the impact on Sami local communities, when quotas are allocated.
c. Identification and recognition of local fishing rights: claims can be forwarded to the Finnmark Commission, which already treats land disputes.
d. A regional management body with advisory function shall be established.

In general, the agreement pertains to a surprisingly limited extent of the proposals of the Coast Fishing Commission. To many Coastal Sami, this result was very disappointing, especially given that the government did not recognise their historical rights to fishing. To Coastal Sami, the recognition of historical rights was not only about the right to fish, but also recognition of the Coastal Sami as a distinctive group and identity. However, the agreement shows that Coastal Sami have managed to become relevant stakeholders in fisheries, and the establishment of a regional management body – even though the questions about representation and tasks were not defined in the agreement – show that the state is today more willing to pay attention to people's local knowledge and is open to a limited degree of influence in the management of marine resources.

Creation and revitalisation of Coastal Sami identity

> Events may provide symbolic material which invite interpretations that subsequently change concept about social identities and relationships.
>
> (Thuen 1995: 201)

This quote is taken from the chapter *Communicating the Challenge* in Thuens' book *Quest for Equity*. In this chapter, he demonstrates how indigenous policies use symbolic actions to challenge dominant perceptions of minority issues. I have chosen this quote to show the importance of 'events' or external influences such as the Alta conflict and the introduction of the 1990 vessel quota system in order to mobilise people through a shared experience, triggering a common identity. Even though the implementation of the new vessel quota system in 1990 did not have the same vast political and social consequences as the Alta conflict, as

I discuss later in this chapter, it nonetheless marks an important turning point because it revitalised the questions regarding rights to fishing and the distribution of fish resources in a Sami context.

Throughout the 1990s, the Sami rights struggle entered into the international global discourse on indigenous people's rights. It follows that arguments based on the concept of indigeneity appeared on the political agenda, further legitimising claims for rights. The first substantial change of Norway's Sami policy consisted of the ratification of the Finnmark Act and the signing of the Consultation Agreement in 2005. The Consultation Agreement had a major impact on the Sami's political organisation to obtain fishing rights as they now enjoyed greater opportunities to express their opinions in all matters concerning Sami interests, including fisheries (Brattland 2010).

The exclusion of the Coastal Sami fishing rights from the Finnmark Act in 2005 not only led to the appointment of the Coast Fishing Commission but also crucially stimulated new forms of mobilisation and cooperation at the local level. Since 2005 new local Sami fishery organisations have been established and pre-existing fishermen's unions also entered the discussions. Among those were the first Coastal Sami fishery organisation *Bivdi* (2005) and the *Fjord Fishermen's Association* (2008). Both organisations worked hard for their goal of securing local fjord fishermen's access to fishing and protecting the fjords from vessels with active fishing equipment or from the negative impacts of the aquaculture industry. Backed by claims that had already existed for decades, the claims for greater local participation inside the resource management system and other decision-making institutions became increasingly important. This shift towards more precise claims and increased participation in the management system must be seen in relation to the publication of the report of the Coast Fishing Commission (NOU 2008) and its inclusion of international law as well as the growing cooperation and mobilisation of indigenous peoples at an international scale. I argue that the bargaining power of local actors by increasingly using the notion of indigeneity and by referring to international law has been significantly enhanced during this period, and has become one of their most important strategies. This is evident when looking at the strategies and claims of local organisations. Indeed, from the beginning, Bivdi based most of its argumentation on locals' ethnic backgrounds of being part of the Sami people who are protected by law, whilst the Fjord Fishermen's Association until recently argued for special measures and rights for the entire population being Kven, Sami or Norwegian in order to safeguard the coastal culture of Finnmark in general. When I talked to the leader of the Fjord Fishermen's Association during my field research, it became clear that he had changed his strategy. He argued that he enjoyed greater opportunities to be listened to by introducing an indigenous perspective to his work. His change in strategy is now highlighted by the fact that he succeeded in being elected to the Sami Parliament in 2017, following several years of experience in other Sami parties and organisations.

With the ratification of the Consultation Agreement in 2005,[3] the different local organisations seized the opportunity to collaborate with the Sami Parliament, with the aim of directly transmitting their concerns and interests towards decision-makers. The cooperation between the different local organisations and the Sami Parliament has become an important element in their fight for resource rights and a source of both identity and legitimacy. Traditional media such as local and regional newspapers and the use of social media have become important instruments of communication between organisations, representatives of the Sami Parliament and locals, as well as providing a forum for discussion and debate (Lätsch 2012).

Sigvald Persen and Jon Egil Nilsen (former leader of Bivdi) explained the relationship to the Sami Parliament and its benefits as follows:

> What we have discussed with the Sami Parliament is that they also have needs for grassroots organisations which can tell them what the opinions and understandings are, what the claims are; thus, the argumentations become thin if they do not have a population to refer to as 'here are the claims from them', so I also perceive our role to be really important. Thus, we are likely to be the legitimacy – and this the Sami Parliament told us clearly – yes they did, we have discussed this, that one needs to understand what happens in the local environment – opinions, interpretations – because it is really important. This reciprocal aspect is really important here.
>
> (Personal communication 2010)

The citation clearly shows that Sami fishermen representatives are labelling themselves as central to the legitimacy of the Sami Parliament and that through the discourse of knowing the local environment, increase the value of local views, which is emically described as being reciprocal. However, this is only possible through recognition as a central basis for creating bottom-up legitimacy, as desired by the Sami Parliament. Thus, the objective is not merely that rights become recognised by the state and the fishery authorities; the Sami fishing community also wants to be recognised in the democratic organisation of the Sami, which the Coastal Sami perceive as being dominated by the reindeer Sami.

Discussion and conclusion: constitutionality and identity

In the presented case of the Coastal Sami, bottom-up institution-building processes are linked with issues of identity and ethnicity to a greater extent than in the case studies proposed by Haller et al (2016). At the beginning of the overall mobilisation process in the aftermath of the Alta conflict, the issue of identity could be seen as an obstacle to collective action and institution-building in Coastal Sami areas, as the Coastal Sami were yet to be revitalised and re-established as a distinct group. There are two main reasons for this. First, in their fight for recognition as an indigenous people, the Sami movement used strong traditional symbols from the reindeer-herding Sami, who had obtained a sort of cultural predominance, resulting in the increased although unintentional marginalisation of the Coastal Sami, who could not identify themselves with the same symbols. Second, in comparison to reindeer-herding, fishing was never perceived as an exclusive traditional Sami mode of production, but rather as a crucial part of Norwegian culture and identity. Furthermore, given their combination of different livelihoods, their interests were never represented in national or specialised interest organisations in the same way as reindeer herders, and so they had no platform to promote their interests. Nevertheless, reflections on the past and the whole process of articulating different claims over time have contributed to the cultural revitalisation and political consciousness of the Coastal Sami, resulting in the (re)incorporation of ethnic identity. This was especially true in the aftermath of the Alta conflict through discussion of different notions of Saminess.

The exclusion of many Coastal Sami from making a living from cod fishing in 1990 has been described as the biggest abuse against the Sami population in modern times, and triggered a process that we call constitutionality. What does the presented case tell us regarding the concept of constitutionality as described by Haller and others? Of the *six principles*, four most clearly apply, as discussed below.

The case shows how fish resources may increase in value due to external factors and that they can be captured by more powerful actors in an unsustainable way. For a long time, Coastal Sami had warned of the overuse of the resource basis. This awareness and recognition by a larger set of local actors then provides the basis for constitutionality. Resource overuse as well as exclusion from sufficient cod quotas resulted in inferior resources being made available for Coastal Sami. Thus, there was a clear *emic perception* that there is a need to change the situation in fisheries. Having promoted claims of resource conservation for numerous decades, the main goal from 1990 onwards was to (re-)obtain rights to fishing. The Coastal Sami then entered into a *participatory process* even though participation varied, as I have shown elsewhere (see Lätsch 2012). In this process, the relative weakness of the Coastal Sami was addressed and challenged, and their views came to be adopted and recognised by the Sami community, the Sami Parliament, and in 2008 by the Coast Fishing Commission. In this process, the Sami Parliament and the Coast Fishing Commission functioned as *catalysing outside agents* and provided a platform that helped welcome Sami fishermen. However, Sami fishermen also had to strive for otherness as a distinctive group. In the preparation of the report, the Coast Fishing Commission conducted public meetings in all coastal municipalities and gathered local knowledge, observations, local perceptions on fishing rights and schemes of management. Through the published report (NOU 2008) it thus becomes evident that *local knowledge was clearly recognised* by the Commission and was given special attention. The same is true for the Norwegian government, as they have agreed to the establishment of a regional management body, the Finnmark Fishery Administration. It has, however, yet to be seen how the state is going to include the issue of local knowledge into a concrete setting. It is unclear how the subject of old fishing institutions were used but it seems that this aspect was integrated by conducting public meetings arranged by the Coast Fishing Commission and finally by analysing the legal history in respect to customary usage or rights, respectively.

Identity seems to be the main driving force in this case, triggering the process of constitutionality as well as politically induced resource scarcity. In their long fight to obtain resource conservation and to attain fishing rights, specific identities were formed. For example, in their respective villages they had experience of being constantly ignored in matters pertaining to ecological change inside the fjord system. Andersen (2006) has observed that at the same time as ecological changes were being experienced in a fjord, a revitalisation of Sami culture, livelihood and interest could be seen in the local villages. The reflection of the changes, consequences for livelihoods and society resulting from the changes can be thought of as influencing and giving content to an ethnic self-understanding at a local level. The awareness of the ecological changes in this context may then be seen as an expression and evidence of local affiliation and an important background for customary use. It refers to continuous settlement and can be used as a framing perspective in the fight to obtain historical rights to fish resources by establishing a sense of ownership. In order to become recognised as a stakeholder group, it is thus crucial for the Coastal Sami to establish a distinct identity, and more importantly, to present themselves as a conscious group to the public. In recent decades, the Coastal Sami have assumed an active role as participants, challengers and important drivers in the process of clarifying the relationship between the state and the Coastal Sami as an indigenous people of Norway as regards fishery issues. This corresponds closely with two central elements that include strategies of institution shopping. Indeed, conflicts over the fishery common-pool resource triggered legal claims by local people making strategic use of state and global legal institutions related to human and indigenous rights. Succeeding in contestation over a higher quota was perceived as being much easier by reinstalling a new indigenous political identity that provided justified access to indigenous rights in the fisheries. Active involvement in the process and the strategic use of the Sami label have contributed to an altered image of the Sami in general, being 'more than only reindeer herders', in turn triggering a revitalisation process in Coastal Sami areas. In sum, the entire process of articulating different claims over time has contributed to the cultural revitalisation and political consciousness of the Coastal Sami, forming specific identities within the overall identity of being Sami. These in turn boost local people's bargaining power as the established identities serve as resources of legitimacy vis-à-vis the state, among other agents.

Thus, small-scale fishermen of Sami origin have rediscovered their identity via legal options to reclaiming access to the commons; moreover, legal frameworks have contributed to bottom-up institution-building, as the Coastal Sami were able to discuss and participate in issues of regulation, resulting in a re-established Sami identity that was different from other small-scale fishermen as a reflection of their past. In this way, the Coastal Sami were able to offer their perspectives, knowledge and creativity in order to attain a sense of ownership in the management of the fishery commons. This process is thus not only about local involvement of institutions based on a sense of ownership of the institution-building process, but goes further by potentially triggering debates on how identity is linked with the management of common pool resources.

Notes

1 The Norwegian Water Resources and Energy Administration (NVE) followed the plans to develop a huge hydro-electric power station on the Finnmark plateau by damming the Alta-Kautokeino River in the Sami reindeer heartland (Anderson 1987). The consequences were huge because it implied the flooding of a vast area important to reindeer-herding Sami in terms of settlement, pasture and migration. The announcement of these plans provoked huge forms of civil disobedience and spectacular demonstrations and was covered by a massive international media presence. Even though the Sami lost the fight and the dam was built (with some modifications), the Alta conflict made power asymmetries visible and fundamentally changed the status of Norwegian Sami (Minde 2003).

2 After a long political process the Finnmark Act was finally adopted by the Norwegian Parliament [Storting] in 2005. By adopting the Act, approximately 95 per cent of the land in Finnmark (previously managed by the state-owned enterprise Statskog) was transferred to local ownership via Finnmarkseiendommen [Finnmark Estate]. The Finnmarkseiendommen is a private landowner and is governed by a board consisting of six members, three of whom are appointed by the Sami Parliament and the remaining three by the Finnmark County Council. In addition, the Finnmark Act authorised the Finnmark Commission 'to investigate the scope and content of the rights held by the Sami and other people living in Finnmark on the basis of prescription or immemorial usage on the former state lands of Finnmark' (Ravna 2013: 443).

3 The Consultation Treaty was a direct outcome of the process that led to the Finnmark Act. The Consultation Treaty set the Sami Parliament in a new formal position vis-à-vis the state, as they have the right to be consulted in matters that may affect them directly. Thus, the Sami Parliament appears not only as an advisory body or an interest representative towards the state, but as an informed formal participant in decision-making processes, increasing the influence of the Sami Parliament in negotiations regarding laws and measures important to the Sami community (Bjerkli and Selle 2015).

References

Agrawal, A. (2005) *Environmentality, Technologies of Government, and the Making of Subjects*. Durham: Duke University Press.

Anderson, M. (1987) 'Transformations of centre and periphery for the Sami in Norway', *Antropologica*, vol 29, no 2, pp 109–130.

Andersen, S. (1997) 'Ressurskamp og møter med offentlige instanser – noen utfordringer i en norsk og samisk utkant', Hovedsfagoppgave i sosialantropologi, Universitetet i Tromsø.

Andersen, S. (1999) 'Samiske fjordfiskere og offentlige myndigheter – et eksempel fra Vest-Finnmark', in Bjørklund, I. (ed.), *Norsk ressursforvaltning og samiske rettighetsforhold. Om statlig styring, allmenningens tragedie og lokale sedvaner i Sápmi* (pp 109–129). Oslo: Ad Notam Gyldendal.

Andersen, S. (2006) 'Landskap, forvaltning og identitet', *Diedut*, no 3, pp 12–34.

Andersen, S. and Eyþórsson, E. (2015) 'Kystfiskeutvalget – drømmen som brast?', in Bjerkli, B. and Selle P. (eds), *Samepolitikkens utvikling* (pp 148–170). Oslo: Gyldendal Akademisk Forlag.

Barth, F. (ed.) (1969 [1998]) *Ethnic Groups and Boundaries. The Social Organization of Culture Difference*. Bergen: Universitetsforlaget.

Barth, F. (1994) *Manifestasjon og prosess*. Oslo: Universitetsforlaget.

Berkes, F. (1999) *Sacred Ecology. Traditional Ecological Knowledge and Resource Management*. Philadelphia, PA: Taylor & Francis.

Bjerkli, B. and Selle, P. (eds) (2003) *Samer, makt og demokrati. Sametinget og den nye samiske offentligheten*. Oslo: Gyldendal Akademisk Forlag.

Bjørklund, I. (2000) *Sápmi – en nasjon blir til*. Tromsø: Tromsø museum.

Bjørklund, I. (1985) *Fjordfolket i Kvænangen*. Tromsø/Bergen/Oslo: Universitetsforlaget.

Bjørklund, I., Brantenberg, T., Eidheim, H., Kalstad, J.A., and Storm, D. (2002) 'Sapmi – becoming a nation: the emergence of a Sami national community', *Australian Indigenous Law Reporter*, vol 7, no 1, pp 1–14.

Blaikie, P. and Brookfield, H. (1987) *Land Degradation and Society*. London: Methuen.

Brantenberg, T. (1991) 'Norway: Constructing indigenous self-government in a nation-state', in Jull, P. and Roberts S. (eds), *The Challenge of Northern Regions* (pp 66–128). Cauarina: Australian National University.

Brattland, C. (2010) 'Mapping rights in coastal Sami seascapes', *Arctic Review on Law and Politics*, no 1, pp 28–53.

Bull, T. and Gaski, H. (1994) 'Fornorskningspolitikk og språkundertrykkelse', in Drivenes, E.-A., Hauan M.A., and Wold, H.A. (eds), *Nordnorsk kulturhistorie II. Det mangfoldige folket* (pp 246–258). Oslo: Gyldendal Norsk Forlag.

Dahl, T.E. (1970) *Samene i dag og i morgen. En rapport*. Oslo: Gyldendal norsk forlag.

Davis, A. and Jentoft, S. (2001) 'The challenge and promise of indigenous peoples' fishing rights – from dependency to agency', *Marine Policy*, no 25, pp 223–237.

Drivenes, E.-A.. Hauan, M.A., and Wold, H.A. (1994) *Nordnorsk Kulturhistorie II. Det mangfoldige folket*. Oslo: Gyldendal Norsk Forlag.

Eid, R. and Haller, T. (2018) 'Burning forests, rising power: towards a constitutionality process in Mount Carmel Biosphere Reserve', *Human Ecology*, no 46, pp 41–50.

Eidheim, H. (1971) *Aspects of the Lappish minority situation*. Oslo/Bergen: Universitetsforlaget.

Eidheim, H. (1997) 'Ethnopolitical development among the Sami after World War II', in Gaski, H. (ed.), *Sami Culture in a New Era* (pp 29–61). Karasjok: Davvi Girji OS.

Eidheim, H. (1998 [1969]) 'When ethnic identity is a social stigma', in Barth, F. (ed.), *Ethnic Groups and Boundaries. The Social Organization of Cultural Difference* (pp 39–57). Bergen: Universitetsforlaget.

Einarsbøl, E. (2006) 'Noen juridiske betraktninger vedrørende samiske rettigheter i saltvann. Eiendomsrettens utstrekning i saltvannsområder etter norsk rett og folkretten', *Gáldu Čála, Tidskrift for urfolks rettigheter*, no 1, http://galdu.custompublish.com/getfile.php/3309440.2388.ytesxdtafq/sjosamisk_norsk.pdf, accessed 8 October, 2018.

Ensminger, J. (1992) *Making a Market. The Institutional Transformation of an African Society*. Cambridge: Cambridge University Press.

Eyþórsson, E. (2003) 'The Coastal Sami: a "Pariah Caste" of the Norwegian Fisheries? A reflection on ethnicity and power in Norwegian resource management', in Jentoft, S., Minde, H., and Nilsen, R. (eds), *Indigenous Peoples. Resource Management and Global Rights* (pp 149–162). Delft: Eburon Academic Publishers.

Eyþórsson, E. (2008) *Sjøsamene og Kampen om Fjordressursene*. Karasjok: ČálliidLágádus.

Gambon, H.A. and Rist, S. (2018) 'Moving territories: Strategic selection of boundary concepts by indigenous people in the Bolivion Amazon – an element of Constitutionality?', *Human Ecology*, no 46, pp 27–40.

Grydeland, S.E. (1997) 'Women's position in former Saami society. Some reflections on demographic changes in Kvænangen, Northern Norway', *Acta Borealia*, no14, pp 3–32.

Haller, T. (2010) *Disputing the Floodplains: Institutional Change and the Politics of Resource Management in African Floodplains*. Leiden and Boston: Brill.

Haller, T. (2013) *The Contested Floodplain. Institutional Change of the Commons in the Kafue Flats, Zambia*. Lanham, MD: Lexington Books.

Haller, T. (2016) 'Managing the commons with floods: the role of institutions and power relations for water governance and food resilience in African Floodplains', in Ostegard, T. (ed.), *Water and Food – Africa in a Global Context* (pp 369–397). Uppsala: The Nordic African Institute. I.B. Tauris.

Haller, T. (2017) 'Perceptions and control of assemblage in a "Glocal World". Dialogues in human geography', *Sage*, no 7, pp 207–211.

Haller, T., Acciaioli, G., and Rist, S. (2016) 'Constitutionality: conditions for crafting local ownership of institution-building processes', *Society & Natural Resources*, 29(1):68–87.

Haller, T., Belsky, J.M., and Rist, S. (2018) 'The constitutionality approach: conditions, opportunities, and challenges for bottom-up institution building' *Human Ecology*, no 46, pp 1–2.

Hansen, L.I. and Olsen, B. (2004) *Samenes historie fram til 1750*. Oslo: Cappelen Akademisk Forlag.

Hersoug, B. (2005) *Closing the Commons: Norwegian Fisheries from Open Access to Private Property*. Delft: Eburon.

Lätsch, A. (2012) 'Coastal Sami revitalization and right claims in Finnmark – two aspects of one issue? Preliminary observations from the field', in Andersen, S. (ed.), *Fávllis. Innblikk i et forskningsprojsekt om local fjordkunnskap*. Tromsø: Senter for samiske studier. http://septentrio.uit.no/index.php/samskrift/article/view/2356, accessed on 8 October, 2018.

Minde, H. (2003) 'Urfolksoffensiv, folkerettsfokus og styringskrise: kampen for en ny samepolitikk 1960–1990', in Bjerkli, B. and Selle, P. (eds), *Samer, makt og demokrati. Sametinget og den nye samiske offentligheten* (pp 87–123). Oslo: Gyldendal Akademisk.

Minde, H. (2005) 'Fornoskninga av samene – hvorfor, hvordan og hvilke følger?', *Gáldu Čála, tidskrift for urfolks rettigheter*, no. 3, http://galdu.custompublish.com/getfile.php/3131397.2388.ebafttewvb/3_2005_fornorskning_norsk.pdf, accessed 8 October, 2018.

Minde, H., Eide, A., and Åhren, M. (2007) 'The UN Declaration on the rights of indigenous peoples. What made it possible? The work and process beyond the final adoption', *Gáldu Čála, tidskrift for urfolks rettigheter*, no 4, http://galdu.custompublish.com/getfile.php/3131390.2388.pufwaqcwus/4_2007_un_declaration_eng.pdf, accessed 8 October, 2018.

Nilsen, R. (1998a) *Fjordfiskere og ressursbruk i nord*. Oslo: Ad Notam Gyldendal.

Nilsen, R. (1998b) 'The coastal survivors – industrialization, local adaptations and resource management in the north Norwegian fisheries', in Jentoft, S. (ed.), *Commons in a Cold Climate* (vol. 22, pp 83–96). Man and the Biosphere Series.

Nilsen, R. (2003) 'From Norwegianization to Coastal Sami uprising', in Jentoft, S., Minde, H., and Nilsen, R. (eds), *Indigenous Peoples. Resource Management and Global Rights* (pp 163–184). Delft: Eburon Academic Publishers.

Norges Offentlige Utredninger (NOU) (2008: 5) *Retten til fiske i havet utenfor Finnmark*. Oslo: Norges offentlige utredninger.

Pedersen, S. (2012) 'The Coastal Sámi of Norway and their rights to traditional marine livelihood', *Arctic Review on Law and Politics*, vol 3, no 1, pp 51–80.

Paine, R. (1957) *Coast Lapp Society I. A study of neighborhood in Revsbotn Fjord*. Tromsø Museums Skrifter Vol. IV. Tromsø: Tromsø Museum.

Pedersen, S. (2012) 'The Coastal Sami of Norway and their rights to traditional marine livelihood', *Arctic Review on Law and Politics*, vol 3, no 1, pp 51–80.

Persen, S. (2011) 'Lokalhistorien, kunnskapen og demokratiet' in Andersen, S. and Persen S. (eds), *Den gang var det jo rikelig med fisk* (pp 15–21). Indre Billefjord: Sjøsamisk kompetansesenter.

Ravna, Ø. (2013) 'The first investigation report of the Norwegian Finnmark Commission', *International Journal on Minority and Group Rights*, vol 20, no 3, pp 443–457.

Skogvang, S.F. (2010) 'Innføringen av fartøykvoter for torsk i kystfiskeflåten i 1990', in Bull, K.S. (ed.), *natur, rett, historie* (pp 211–245). Oslo: Akademisk Publisering AS.

Søreng, S.U. (2007) 'Fishing rights struggles in Norway: Political or legal strategies?' *Journal of legal pluralis*, no 55, pp 187–210.

Thuen, T. (1980) *Samene: urbefolkning og minoritet*. Tromsø/Bergen/Oslo: Universitetsforlaget.

Thuen, T. (1995) *Quest for Equity. Norway and the Sami challenge*. Newfoundland: Institute of social and economic research.

Ween, G.B. and Lien, M. (2012) 'Decolonization of the Arctic? Nature practices and land rights in the Norwegian High North', *Journal of Rural and Community Development*, vol 7, no 1, pp 93–109.

West, J. and Hovelsrud, G.K. (2010) 'Cross-scale adaptation in the coastal fisheries: findings from Lebesby, Northern Norway', *Arctic*, vol 63, no 3, pp 338–354.

12 Swiss alpine pastures as common property

A success story of bottom-up institution-building in Sumvitg, Canton of Grisons, Switzerland

Gabriela Landolt

Introduction

The institutions regulating access and use of common-pool resources (CPRs), such as alpine pastures, in the Swiss mountain village of Sumvitg date back to late Medieval Times. The creative force behind those norms, rules and regulations (institutions) is the farming population organized in corporations (*Korporation* or *Genossenschaft*), which are recognized under civil law and guarantee access and user rights (see Landolt and Haller 2015, p. 103). From the written regulations of 1805, it is clear that the farmers were willing and able to adjust their resource management to social, economic, legal and political changes in order to maintain the primary aim of the institution: to enlarge the fodder base on the privately-owned meadows in the valley by the collective use of pastures at the alpine level. Looking at the maintenance of the natural resources (pastures) as well as their management as a CPR over a long period of time, this case study is predestined for common property research. An earlier publication by the author (Landolt and Haller 2015) was based on a New Institutionalism approach (see Ensminger 1992; Haller 2013) about external and internal factors influencing institutional robustness and the ability of common property regimes to adapt to changing conditions (Dietz et al. 2003; Folke et al. 2007; Baur and Binder 2013). Jean Ensminger's (1992) model of institutional change and Elinor Ostrom's (1990) eight design principles were applied, revealing external and internal variables explaining the successful adaptation of CPR institutions and the organizational structure of the farmers as property rights holders (common property regime) to policy changes, market pressure and structural adjustments. External factors such as the legal recognition of the local regime, the high degree of autonomy of the commune as lowest governmental entity (local municipality) and direct payments have been identified as favouring conditions for collective action and bottom-up institution-building processes. However, those preconditions did not necessarily lead to sustainable institutional developments and the fulfilment of the eight design principles: internal factors and dynamic processes related to power and social relations, history of interaction, ideology, leadership and bargaining power to reduce transaction costs for collective action defined the direction of institutional change. The findings of the earlier publication lead to further questions about how – meaning the process – and why compromises are reached that are perceived as legitimate and fair by local actors (referred to as emic perception). Where (platforms), how (processes) and by whom (agency) is institutional change discussed and how do the local institutional outcomes relate to higher governmental level expectations (nesting)? A promising approach to look at those questions in more detail is the concept of "constitutionality" introduced by Haller et al. in 2016 (see also Haller et al. 2018), who define constitutionality as "pro-active and strategic participation in the design of local institutions negotiated in settings with heterogeneous stakeholders focused upon the use of natural resources" (Haller et al. 2016, p. 69). What makes this approach relevant to the study presented in this publication is the focus on local agency, the different roles, engagement and perceptions of the stakeholders in institution-building processes, including emic views in regard to fairness, equity and sense of ownership as well as catalysing forces such as legitimate leadership facilitating discussion platforms. Based on the description of the history of institutional change in the mountain village of Sumvitg the preconditions for constitutionality processes identified by Haller et al. (2016) are discussed and insights revealed for answering the research questions outlined above.

Theoretical framework

According to Jean Ensminger (1992), institutional change occurs if the bargaining power between individuals or interest groups shifts. Ideologies (world views, values, beliefs), knowledge, communication and leadership skills may be used as power assets influencing the outcomes. These assets may be transformed to bargaining power to bring about institutional change, but may also legitimize the persistence of existing rules. Ideology, including notions about fairness and justice, enhance the respect and trust towards leaders, the acceptance of collective decisions taken and present institutional settings (Ensminger 1992, p. 141). Ensminger further points out that the institutional outcome from the interplay between changes in relative prices, power shifts, ideological legitimacy and distribution are complex and cannot be generalized. I argue that this complexity stems from the relevance of local agency in institution-building processes and I argue that understanding local agency is the key for understanding what is happening in Ensminger's "internal box" of institutions, organization, ideology and bargaining power. Local agency as referred to in the constitutionality approach by Haller et al. (2016) requires understanding the different views, motivations and power assets the actors bring to the negotiation table and how discussions are guided and decisions taken. Based on four case studies describing the histories of change of CPR institutions, Haller et al. (2016) identified six preconditions favouring constitutionality processes, i.e. the formation of institutions that facilitate a sustainable use of the CPRs: 1) emic perception of need of new institutions; 2) participatory processes addressing power asymmetries; 3) pre-existing institutions; 4) outside catalysing agents (fair platform); 5) recognition of local knowledge; and 6) higher-level state recognition. Constitutionality looks at institution-building processes for CPR management from a bottom-up perspective. Constitutionality reflects the ability of local communities to craft their own rules and regulations, to monitor and enforce them and to mitigate conflicts and broadly accepted institutional changes.

Setting and characteristics of the case study

Sumvitg is located in the south-eastern part of Switzerland, in the valley of the Vorderrhein that belongs to the Surselva region encompassing the valley of the Vorderrhein and all its side valleys, from the source near the Oberalppass down to Reichenau. Sumvitg comprises four villages, so called fractions: Rabius, Surrein, Cumpadials and Sumvitg itself. Sumvitg is located at around 1,000 metres above sea level, and some 49 per cent of the total territory of 10,258 hectares comprises pastures.

The population of Sumvitg has remained stable, at around 1,300 residents, since the first population census in 1656. Unlike the general population of Sumvitg, the farming population decreased rapidly. Until the late 19th century almost all households were farmers. However, already in 1920, when the first data on the employment structure in Sumvitg were made available by the Federal Population Census (1920), industrialization had left its mark: 64 per cent of the working population was engaged in the agricultural sector, but already 26 per cent found a living in the nearby textile industry, in craft, trade or the transportation sector. In 1970, tourism centres close to Sumvitg began to develop creating further job opportunities. Employment in the primary sector decreased to 30 per cent (Federal Population Census 1970) and continued to decline to 7.2 per cent in 2015 (communal statistics). The number of farms reduced accordingly after the First World War: From 213 farms in 1939 (Maissen 2000) the number went down steadily to 38 farms remaining in 2014.

The number of cattle has decreased by almost 30 per cent since 1906 (Table 12.1) while the number of cattle per farm has increased from 9.5 in 1906 (Federal Livestock Census 1906) to 37.6 in 2011 (FSO 2011). Although the number of cattle per farm has increased, the paid staff per farm has decreased from an average of 2.7 persons in 1939 to 1.4 persons in 1985 (Federal Agricultural Business Census 1939, 1985) and beyond. The decrease in dairy cows and young cattle has been partially compensated for by an increasing number of suckler cows (meat cattle breeds) and newly introduced animals such as horses and lamas.

Methodology

The data on which this article is based are derived from a triangulation of methods: in order to observe behaviour and to get to know the local context in which collective decision-making takes place, I carried out participant observation

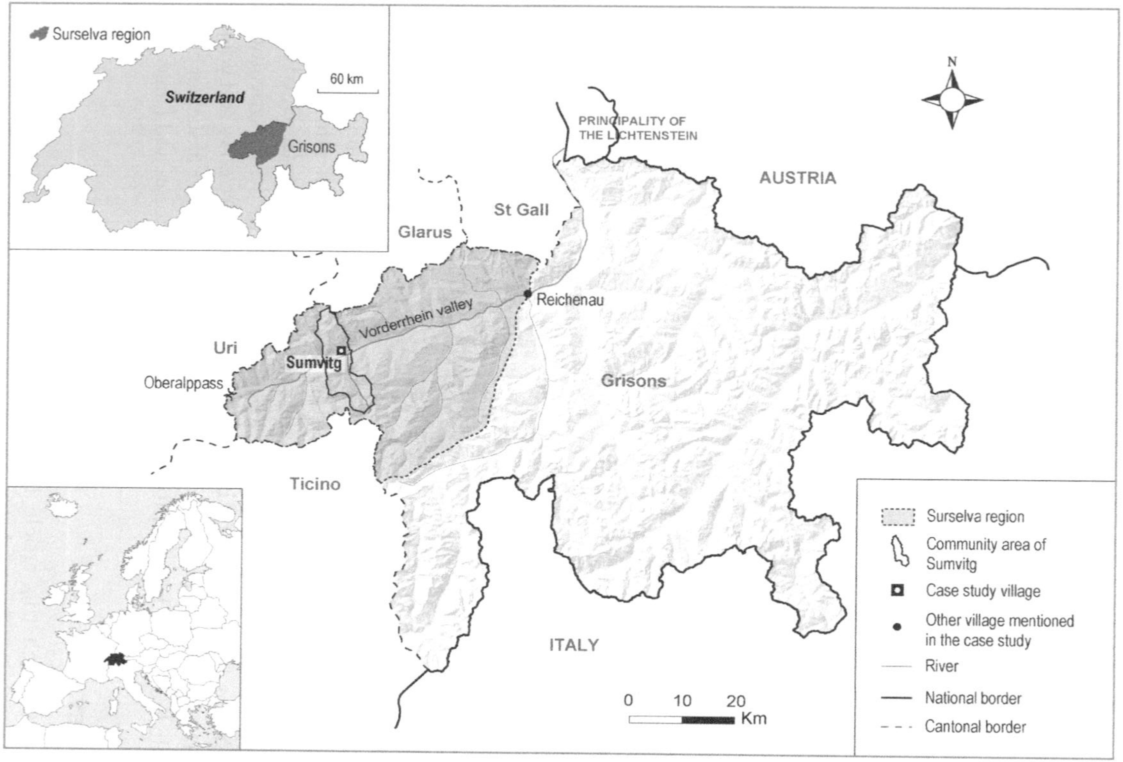

Figure 12.1 Map of the case study.

Source: Map compilation and design by Ulla Gämperli Krauer (2018), CDE, University of Bern, Bern, Switzerland. Geodata source: OSM data by OSMaxx (2018). Geometalab, HSR University of Applied Sciences Rapperswil, Switzerland. Hillshade: Swiss Hillshade by Juerg Krauer (2017), CDE, University of Bern, Bern, Switzerland.

Table 12.1 Development of animal husbandry in Sumvitg

	1906	*1926*	*1946*	*1966*	*1983*	*1990*	*1999*	*2003*	*2007*	*2011*
Cattle	1,567	1,486	1,186	1,527	1,373	1,424	1,198	1,281	1,208	1,090
Dairy cows	530	534	445	418	365	360	346	317	276	239
Young cattle	1,037	952	741	1,109	1,008	994	624	791	658	559
Suckler cows and calves						70	228	173	274	292
Sheep	2,085	1,427	1,022	1,578	1,553	1,554	1,793	1,322	1,224	1,477
Goats	1,402	1,178	1,191	334	190	234	136	103	70	51
Animals total	5,054	4,091	3,399	3,439	3,116	3,212	3,127	2,706	2,502	2,618

Source: Federal Statistical Office

during the alp season of 2010. Half of the alp season (1.5 months) I worked on the summer dairy farm Alp Naustgel owned by the commune Sumvitg. In the following winter of 2010/2011 I interviewed ten out of 12 farmers of the alp corporation (*Alpgenossenschaft*) of Alp Naustgel as well as the local authorities (president, head of the agricultural office) and of the alp corporations (board members, alp masters). During the summer of 2011, the objective was to observe the activities on the private lands in the valley in order to get a more holistic picture of the agricultural cycle. On four farms I participated in harvesting activities on the private meadows (two days per farm). In the winter months of 2011/2012 I carried out oral history interviews with the oldest and retired farmers of the village (N=5) and a quantitative survey on the significance of communal work and summer pasturing (N=17). To go further back in time and to complement oral with written data I analysed the regulations from 1805, 1890, 1940, 1985 and 2000, highlighting the institutional changes that occurred.

Alpine pastures as a common-pool resource

Traditional services of the alpine pastures

The only way to adapt to the seasonally variable, hardly predictable availability of fodder and the difficult accessibility of the vast, higher altitude alpine pastures was to use those natural resources collectively (Netting 1981, p. 144). The rationale to use the alpine pastures on altitudes of 1,600 to 2,000 metres above sea level was to enlarge the fodder base during the summer month so that the farmers could harvest the private meadows for hay. This way, the famers were able to keep more livestock than only using their private land. In spring, the common pastures in the valley, the so-called home pastures (*Heimweiden*), as well as private meadows were opened to all small livestock of the commune. This practice is called communal grazing (*Gemeinatzung*) and benefited the poorest households that did not have sufficient private meadows. Around a month before the ascent to the alpine pastures, the animals moved to the *Maiensäss* level for grazing. A *Maiensäss* is a smaller private barn including surrounding land at an altitude of around 1,200–1,600 metres. The poorer households without a *Maiensäss* were allowed to use the home pastures. During the summer months, usually from mid-June until mid-September, the animals grazed on the alpine pastures under the supervision of employed herders and cheese makers (in the case of dairy cow alps), while on the valley floor and the *Maiensäss* the private meadows were harvested for hay. In mid-September the animals returned from the alps to the *Maiensäss* or the home pastures. The communal grazing was repeated in September/October, after the autumn harvest (Weiss 1941; Netting 1981; Stevenson 1991).

The state's demand for "new" institutional services

In 1980, the federal state started to subsidize summer pasturing. The introduction of those direct payments was essential for the CPR institutions in order to cope with the increasing labour costs and the decreasing economic value of agrarian products in the course of market globalization and structural adjustments (Landolt and Haller 2015, p. 104). In 1992, direct payments became bound to ecological services such as the maintenance of cultural landscapes. The alpine pastures were identified as an essential part of the Swiss cultural landscape and hence, the local CPR institutions benefitted from steadily increasing direct payments, which facilitated their economic survival. In 1999, the former direct payment system was revised and replaced by the ordinance on summer pasturing payments (*Sömmerungsbeitragsverordnung* (SöBV)). The SöBV also include sustainability criteria regarding the maintenance of the pastures: stocking limits per alp, protection against scrub and weed, biodiversity enhancing fertilization, and the maintenance of alp buildings and infrastructure. Neither stocking limits nor activities to maintain the pastures were new to the local CPR institutions as those duties have been institutionalized in the "communal work", the obligation of each farmer grazing animals on the alps to contribute a certain amount of working hours to maintain the pastures and related infrastructure. However, the shifting of those rules to a higher institutional level becomes relevant in the light of the difficulty to control them (Schulz 2011).

The written regulations and the oral history in Sumvitg presented in the next section provide insights into the political, economic, social and institutional processes since the early 19th century and will provide the bases for discussing the concept of constitutionality regarding the locally driven institutional changes in the management of the CPRs.

The local CPR governance system in Sumvitg: a history of institutional change

The first written regulations in Sumvitg date back to 1805. It was a time right after the integration of the federal state *Freistaat Gemeiner III Bünde* that later developed into the Canton of Grisons, and a time characterized by accelerated local, regional and federal formalization processes (Mathieu 1992, p. 183).

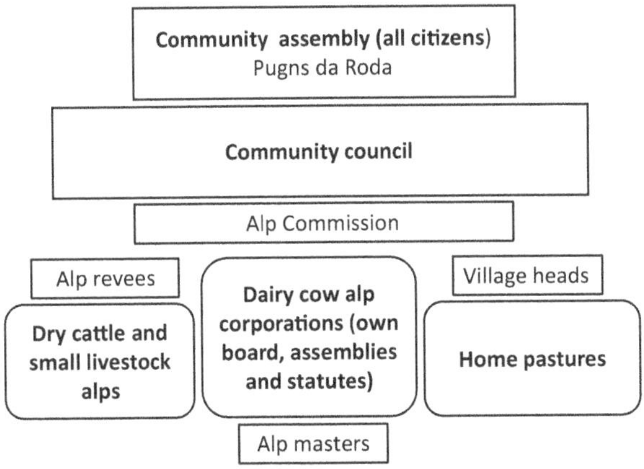

Figure 12.2 CPR government system 1805.
Source: Drawn by the author.

Sumvitg was still organized as a neighbourhood that later developed into a commune, with the primary aim to manage the CPRs. The CPR regulations were called *Pugns da Roda*. *Pugns* stands for the rules and the *Roda* for the usufruct of the CPR during a ten-year period. The *Pugns da Roda* from 1805 reflect the executive structure developed in the High Middle Ages in different parts of Switzerland (Weiss 1941; Netting 1981; Stevenson 1991; Mathieu 1992; Head 1995): The community assembly emerged as the central organ to manage the CPRs in which a two-third majority takes decisions. Each household head and citizen (from original families) has user as well as voting rights. Residents (later settlers) or foreigners have no decision-making power in CPR affairs. The community assembly elects the community council, the alp reeves, to manage the cattle alps on a yearly basis and a commission to revise the regulation every ten years. For the organization and monitoring of the grazing on the home pastures and the communal grazing, village heads are nominated. While the use of the dry cattle alps, the small livestock alps, and the home pastures are under close supervision of the commune, the dairy cow alps have been managed by alp corporations in a largely autonomous way: They rent the alp from the commune for the duration of the *Roda*, establish their own statutes, take decisions on operational issues during their general assembly (by a two-third majority) and elect their own board members, alp reeves and alp masters (see Figure 12.2).

The alp master is in charge of employing the alp staff, organizing the communal work, taking care of the inventory and dividing the costs and benefits proportionate to the animals per user. The rules were monitored by those holding positions. For every breach, a fine was defined proportional to its severity. Besides the grazing taxes, Matthieu (2011, pp. 35–47) argues that the fines have been an important contribution to the communes' income in the Canton of Grisons and that the yearly public reading of the regulations, also mentioned in the *Pugns da Roda* 1805, was part of raising awareness among the community members of the consequences of breaking the rules. The terms for all executive positions were limited to one year and had to rotate among the members. Overstocking was prevented by applying a rotational system (by lot) to balance the stocking of the alps and the restriction that only animals wintered in the commune, fed with own fodder, were allowed to be stocked. Every ten years the animals were redistributed among the alps. The right to use the common pastures was bound to the duty of each farmer to do one day of communal work on each common pasture used and another day for infrastructure maintenance. The farmers were obliged to share the costs arising from the use of the common pastures proportionally to the number of cattle owned. An exception was made for poor households that did not own a cow: in order to equally benefit from the returns of the alp (butter and cheese), they were allowed to rent a cow during summer and to put more pigs on the alps. In order to assure that the small livestock finds sufficient fodder during the communal grazing (private meadows becoming common property) there are rules restricting the individual use of private meadows in spring and autumn.

As part of the formalization process, the cantonal government assigned more and more social and political duties to the neighbourhoods increasing their expenditures. The CPRs (alps, forests, pastures) were the only sources of income for Sumvitg and also had to cover non-CPR related expenditures. Financial hardship drove the neighbourhoods to allow citizenship for money, to sell alps and forests, and to increase taxes (Liver 1968, p. 207). To stop tendencies toward privatization a federal law released in 1848 prohibited the use of communal and cooperative goods for private purposes (Liver 1968, p. 207). With the constitution of 1854, the *Freistaat Gemeiner III Bünde* became the Canton of Grisons of today. The communes that emerged from the reform period have remained identical with the former neighbourhoods and kept autonomy in regard to managing the CPRs and their internal affairs (Barber 1974, p. 208f). In 1890, the *Pugns da Roda* reflect the structural reform, and the increased political and social responsibilities the communes had to take over. New offices were created such as the commune council (*Gemeinderat*), the school council (*Schulrat*) or the poverty relief commission (*Armenkommission*). The taxes mentioned in the *Pugns da Roda* of 1890 reflect the need for income. A cantonal law that directly influenced the local CPR management was the law on settlement (*Niederlassungsgesetz*) released in 1874, guaranteeing residents access rights to CPRs for taxes. This new law, combined with the pressure to generate more income, opened access to the CPRs for residents, led to the introduction of grazing taxes, and the abolishment of the rule that only allows animals wintered in the commune to graze on the common pastures in the *Pugns da Roda* in 1890.

The first article of the *Pugns da Roda* 1940 notes that the income of the commune is not generated solely by the communal utilities anymore but also through a direct tax (*Haushaltssteuer*) per household. It is safe to assume that the direct taxes were a reaction to an increasing number of households no longer dependent on the CPRs, leading to a reduction in income from grazing taxes. The national pressure to intensify production and to modernize alp and farm infrastructure are reflected in the rules on the fertilizing of pastures and the creation of a fund to meliorate the alp infrastructures. The main duties of the alp masters and the reeves have not changed, but article 20 indicates that is has become difficult to fill the positions: It states that every farmer using the alps is obliged to accept the position at least once when he is elected by the community assembly. The duties within the communal work were reduced but the compensation per working hour increased. The communal grazing is not mentioned in the *Pugns da Roda* 1940 although it was not abolished. In Sumvitg the communal grazing was only discussed in 1947 when communal grazing became a source of conflict due to the patchwork of redeemed land, which was difficult to protect. The last regulation on the communal grazing in Sumvitg dates from 1978. In the light of increasing forest protection policies and the various authors, dating the abolishment of the communal grazing to earlier times (Weiss 1941; Barber 1974; Mathieu 1992), the survival of this institution in Sumvitg indicates the high autonomy of the commune and the resistance to change in this matter. The main changes in the *Pugns da Roda* 1985 were the approval of longer-term positions and the further reduction of the communal work.

In 1995, the period of enforcement of the *Pugns da Roda* was reduced from ten to five years in order to prepare for a more substantial revision of the regulations.

In the year 2000, the farmers created the cooperative named Cooperaziun d'Alps Sumvitg. The cooperative leases the common pastures from the commune for a yearly rent. Besides the general obligation of the cooperative to use the pastures in compliance with cantonal and federal laws, the commune withdrew from regulating any constitutional and operational issues. The cooperative established statutes and regulations and so did all the subordinate alp corporations to cover membership and operational issues. Two new rules are worth mentioning: 1) the rule raising the costs for members that graze their animals on foreign alps was abolished and consequently, the farmers were free to choose an alp outside the commune without punishment; and 2) every three years an alp master is elected for a six-year term on every alp, which means that two farmers are sharing this position.

The *Pugns da Roda* are written testimonies of institutional change: Between 1805 and 1890 we see the formalization of the commune's legal, economic, political and social function, the monetization of the governance system (introduction of monetary fines, compensations, taxes) and the release of rules limiting access. Newer changes relate to structural changes triggered by industrialization, technological development and national agrarian policies. Nevertheless, it also indicates that the commoners themselves tried to adapt to these changes by adjusting the use and membership of the common pastures as

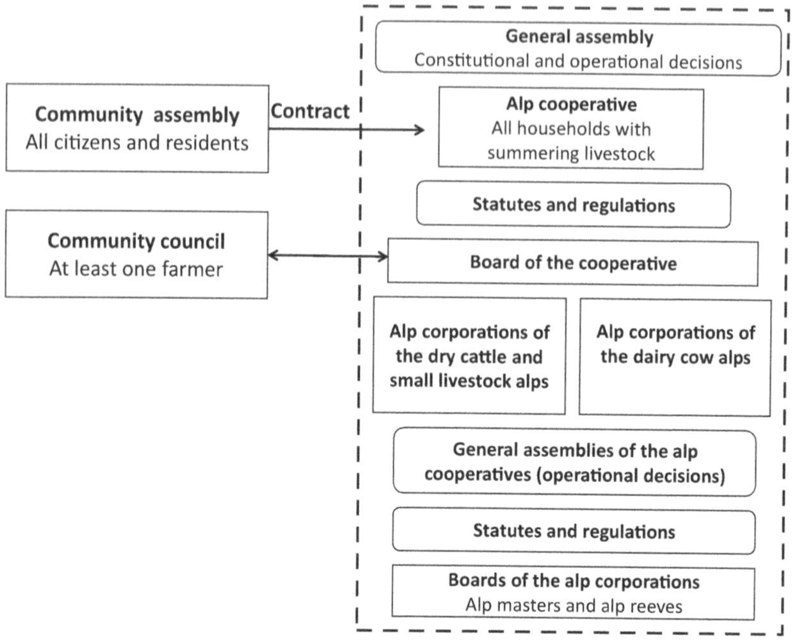

Figure 12.3 CPR government system since 2000.
Source: Drawn by the author.

well as trying to reduce the burden on the people in the commons governance administration.

Emic perceptions explaining the institutional changes

However, how were these changes discussed and what contributed to a common process of transformations from below? These questions, central to the constitutionality framework, can only be addressed historically via looking for information on emic perceptions through oral history since the 1950s onwards. This section of the chapter provides information as reported by retired and practicing farmers of Sumvitg on the following institutional changes:

a) Abolition of the communal grazing practice called *Gemeinatzung*.
b) Shift from yearly rotating positions to long term positions.
c) Decrease of working hours within the communal work and reactions.
d) Structural changes: Decrease of farming households, livestock, and specialization of production.
e) Creation of the Cooperaziun d'Alps in 2000.

Abolition of the communal grazing practice on private land and a primary service of the traditional CPR institution

The *raison d'être* of the temporary communal grazing on private land in spring and autumn, was poverty alleviation, a central task of the early CPR management system. This practice was not questioned as long as private meadows could not easily be enclosed. The introduction of the electric fence, however, changed this situation:

> The abolition of the *Gemeinatzung* was highly politicized. Decisive was the introduction of the electric fence: In 1953 we were the first household to own one. In the seventies, the farmers able to protect their private meadows with electric fences reached a majority and were able to outvote the others. Without electric fences available, it would not have been possible to abolish this practice.
>
> (Personal communication by a retired farmer age 82, 19 October, 2012)

The clear presence of this institutional change in the memory of the old farmer shows how emotional and debated it must have been. The communal grazing was also a symbol for solidarity with the poorest farming households. On the other hand, it became outdated due to the separation of the management of the CPRs from social services, the reduced interest due to structural changes and the scarcity of herders.

Shift from yearly rotating positions to long term positions: the development of good leadership

The rotational system in holding positions is rooted in the distrust in any kind of authority and was based on the possibility of mutual control within the community. In the times before the 1980s legitimacy was not derived from trusting an individual leader but from the decision-making power of the collective. The decreasing number of farmers, however, necessitated longer terms but also led to the insight that it is more sustainable not to force people into positions but to allow motivated farmers to prove their leadership skills and to gain trust and reputation during a longer term. This insight has been highlighted by one of the first long term alp master of one of the dairy cow alps:

> I became alp master in 1990. In the first year, my work was closely monitored and criticised. But when you are successful and lucky [referring to unpredictable factors such as weather, quality of alp staff, and cheese quality] during the next few years, you gain trust and authority.
>
> (Personal communication, 15 December, 2011)

Because the decreasing pool of possible alp masters forced the alp corporations to allow longer terms, this institutional change was vital for building trust in leaders that were able to achieve broadly supported compromises despite increasing individualism, divergent interests and unequal power-relations.

Structural changes: diversification of interests and ideologies

The eldest (retired) farmers in the village estimate that until the beginning of the 1970s almost every household kept some livestock, being full-time farmers or not, to sustain their livelihoods. In the context of specialization and adaptation to the market, agrarian policies, and laws, decisions had to be taken by the farmers regarding the branch of production. With specialization, interest groups developed according to their particular needs for livestock production. Besides a general reduction of livestock (Table 12.1) resulting in less alpine pastures used, the introduction of suckler cow farming (for beef production, the calves remain with the mother and suckle the milk from their mothers instead of milking the mother cow) in the 1990s challenged the CPR institution:

> Suckler cow farming became an option because we had no possibilities to increase production and because the milk price dropped. The suckler cows provided a way to stay a farmer. The meat price was more stable than the milk price and the extensive production allowed having a part-time job.
>
> (Personal communication, one of the first suckler cow farmers in Sumvitg, 13 March, 2011)

The president of the alp cooperative explained (personal communication 2 March, 2011) that the ideological resentment towards part-time farming and extensive production was substantial, meaning that conflicts arose regarding the integration of the suckler cows in the alpine pasture management system. Only the intervention by the commune leaders referring to the equal and indisputable right of every farming household to graze their livestock on the common pastures made the dairy farmers tolerate the suckler cows.

Besides low prices for milk, direct payments had a decisive influence on individual livelihood strategies affecting collective decision-making:

> Individual strategies depend highly on the agrarian policies. If subsidies are mainly bound to agricultural land, then the incentives to put the animals on the alps are low. The farmers will harvest the land in the plain of the valley once and put the animals on the steeper meadows. It does not matter anymore if a farmer harvests once, twice or three times. And, in case that the policies support an extensive agriculture with low production of fodder and less animals, what would this mean for the investments made in the last years in order to prepare for a rising production and cattle number? It's also an ideological shift: for me the animal serves for producing food and nothing else. If that changes, the whole ideology of farming changes.
>
> (Personal communication, 65-year-old dairy cow farmer, 8 January, 2012)

This old farmer was referring to the discourses in the preparation of the AP 2014–2017 that announced a further shift from rewarding production (number of livestock) to rewarding the maintenance of the land(scape). He explains how the direct payments pushed the farmers to adopt practices contradicting the wide spread self-perception of being a producer rather than an "eco-service provider". The statement of the old farmer expresses a relevant issue, namely the discrepancies between the traditional self-perception as producer versus the image of the farmer as landscape manager produced by the public. Such ideologies are brought to the assemblies, creating emotional discussions and heterogeneities that raise transaction costs for finding compromises.

Communal work: an indicator for ecological outcomes

The reduction of farmers and changes in livelihood practices leading to less livestock, and particularly to fewer small livestock using the summer pastures has increased the need for communal work in order to maintain the cultural landscape. At the same time, efforts have decreased over time due to the high workload on the farms and high labour costs, as one of the long-term alp masters observed:

> When I was a child we spread the dung manually and we cut weeds. But when I grew older the investment in the communal work decreased. 20 years ago we did not even cut back the young trees anymore. Around eight years ago a new regulation by the canton assigned to segregate the pastures from the forest. Those maps made us aware of how much pasture we lost to the forest and since then the young trees are cut back rigorously.
>
> (Personal communication, alp master of alp Naustgel, 10 December, 2010)

The statement refers to the neglect of ecological aspects in the alpine pasture use since the 1970s and 1980s when the alp infrastructure was still rudimentary and decentralized demanding high labour input while the costs were already increasing and direct payments not yet introduced.

The visualization of what has been lost created awareness and motivation to combat further forestation. Investments in the maintenance of alpine pastures and alp infrastructure further increased with the establishment of the cooperative in the year 2000. A beef cattle farmer pointed out:

> Renting the common pastures from the commune created a sense of ownership. All that we do and invest is now for our own good.
>
> (Personal communication, 37-year-old suckler cow farmer, 14 December, 2010)

The statement reflects the autonomy gained from the commune, and the increased commitment of the farmers to invest time and money in something that is now felt to be their own. This marks a central point of the constitutionality framework actions taken collectively create a sense of ownership of creating one's own rules. The results of the quantitative survey carried out in March 2011 (N=17, ten dairy cow farmers, seven suckler cow farmers) further supports the assumption that the relevance of the common pastures for the survival of the farm businesses (100 per cent), the importance of the summer pasturing as part of the alpine farming culture (75 per cent) and the social aspect of the communal work (65 per cent) benefit the maintenance of the communal work and the ecological outcomes (see also Landolt and Haller 2015).

Bottom-up institution-building: creation of the Cooperaziun d'Alps in 2000

All these institutional changes are a response to new challenges leading to bottom-up processes culminating in new organizational structures: In 1995, Glivers – the second dairy cow alp – lost the permission to produce cheese because the old production structures did not comply with federal and cantonal food hygiene guidelines. For Alp Glivers, a decision had to be made: either the infrastructure to produce cheese had to be renovated or the dairy cow alp was transformed to a suckler cow alp. This is a basic institutional change as grazing dairy cattle needs more work for milking and milk processing, while keeping cows with their calves (suckling) and using them for beef production is less labour demanding. The decision-making process was characterized by ideology rather than rationality: The dairy farmers emphasized the traditional value of Alp Glivers as a dairy cow alp and ignored the fact that the number of dairy cows has decreased substantially. A former farmer that attended the decision-making meetings explained the behaviour of the dairy cow farmers with the following words:

> At that time, the dairy farmers still had a slight majority over the beef farmers. The dairy farmers were aware of the high probability that soon they would not have enough cows anymore to entirely stock the alp. They ignored this obvious trend and emphasized ideological and cultural aspects in order to follow their own interests. They took their last chance to decide in their favour.
>
> (Personal communication, retired dairy cow farmer, 2 March, 2011)

The milk-producing farmers used ideology to legitimize their decision and took advantage of their remaining power to outvote the others. The renovation of the alp infrastructure took place in 1999. In the year 2000, the alp commission was asked to revise the alp and pasture law of the commune. The commune asked three farmers in the commission who proved to be good mediators and leaders in the past. Aiming at operational independence in order to react more rapidly to change, particularly to the continuing trend to shift from dairy cows to suckler cows, the alp commission elaborated the idea to unify all six alps on an alp cooperative. The creation of the alp cooperative was a way to restore the principle of democratic decision making within the CPR institution, to decrease the transaction costs for collective action and to gain institutional flexibility in order to react more rapidly to change:

> Before [the creation of the cooperative], the community assembly decided over the main alp affairs, the farmers did not have much decision making power. That changed with the alp cooperative. Imagine discussing the stocking of the alps every year in the community assembly; that would be much more difficult than reaching agreements only among the farmers.
>
> (Personal communication, board member of the alp cooperative, 1 March, 2011)

Although there were many good reasons for creating the institutional structure and the voices of the farmers were very positive after the transformation was accomplished, the process to secure agreement from all farmers in order to get a majority in the community was only possible due to the leadership skills of the commission members in combination with a participative bottom-up process (see also Landolt and Haller 2015).

However, despite the success of creating a new institutional structure, there is a continued need for negotiation and reaching compromises:

> The assemblies are always emotional. We as the board have to prepare in order to be able to find good compromises. A challenge is the unequal power distribution between interest groups and particularly between suckler cow farmers and dairy cow farmers. We need to find fair compromises for all as we fear that otherwise farmers move to other alps and do not come back.
>
> (Personal communication, president of the alp cooperative, 2 March, 2011)

The leaders are aware of the importance of satisfying all members of the cooperative to support decisions particularly considering the newly allowed freedom of the farmers to stock their livestock on foreign alps. So far, there is no member drain to observe and the board members are convinced that the incentives are still high to stock on the closest alps and to maintain social ties with the community.

Discussion

The concept of constitutionality by Haller et al. (2016) assumes that a certain decision-making environment has to be given or created in order to facilitate bottom-up institution-building to deliver sustainable ecological outcomes. When analyzing the situation in Sumvitg the following preconditions are found.

Emic perception of need of new institutions

Although institutional changes have occurred every ten years since 1805, economic and policy changes on the national level gained momentum in the 20th century. The traditional CPR system reached a threshold due to the ten-year cycle of adapting the regulations and redistributing the livestock on the alps that was perceived as too long in the light of the fast-changing environment. Furthermore, the marginalization of the farming population in the community assembly, where major decisions about investments and distribution of livestock were taken, was a clearly perceived threat to their decision-making power in regard to the alpine pasture management. Nevertheless, it was not easy for the farmers elected into the commission to unite the different interest groups (old and young generation, beef and dairy cow farmers) driven by divergent worldviews and ideologies, to create the perception of a need for institutional change in order to speak as one voice in the community assembly. The creation of the cooperative per se was an act of regaining ownership in regard to the farming population "owning" the future development of the CPR institution and the CPR.

Participatory processes addressing power asymmetries

The democratic institution of the assemblies is deeply rooted in the Swiss and particularly in the political culture of the communes of the Canton of Grisons. Although participatory and democratic, it does not necessarily lead to decisions perceived as fair by everyone as there are winners and losers. With an increasing non-farming population, the community assembly became a heterogeneous crowd of individual decision makers. To address power asymmetries in such a decision-making environment is difficult. The farmers had to join their forces in order to reach a two-third majority in the community assembly for creating the cooperative, to become independent from the commune and to regain a sense of ownership in the institution-building process. The success of the leaders to raise high expectations of benefits within the farming population made it possible to overcome heterogeneity and to win the vote (see Landolt and Haller 2015, p. 109). Speaking with one voice became possible thanks to the equal participation rights of all interest groups during discussion and decision-making processes, which led to legitimate outcomes.

Trust and good reputation paired with a good educational background and communication skills have enabled the "new generation" of leaders to bridge divergent worldviews and self-perceptions and to reach compromises about the distribution of costs and benefits. The successful integration of the suckler cows in the CPR management system and the creation of the cooperative is based on the commitment of the leaders to share information, to debate in a participative way and to reach broadly accepted decisions that bridge power asymmetries and increase the sense of ownership.

Pre-existing institutions on which to build

There is a strong institutional foundation on which the CPR management of Sumvitg is based. Part of that foundation is the principle of political equality and economic proportionality, and the democratic institution of the assembly to legitimize decisions and to assure equal rights.

Those principles have guided the decisions taken by the members of the CPR institutions and have been referred to by leaders in order to mediate conflicts. Only the principle of economic proportionality was altered when the suckler cow farmers had to be convinced to pay proportionally more than their real costs produced (see Landolt and Haller 2015, p. 107). Although the long history of institutional adaptation in Sumvitg has certainly facilitated the process of bottom-up institution-building in the year 2000, traditional rules such as economic proportionality can turn into a challenge if there is no good leadership that is able to reach a compromise perceived as fair, despite economic concessions made. Furthermore, the pre-existing CPR institution to manage common pastures is labour intensive in regard to the maintenance of its organizational structure and in regard to the maintenance of the natural resource and hence requires a sufficient pool of human capital to fill positions and accomplish the communal work. Longer term positions, less hours of communal work, and also the creation of the alp cooperative have been reactions to a decreasing number of farmers allowing the CPR management system to survive.

Outside catalysing agents creating neutral platforms

Although there is no physical outside catalysing agent, such as tourism or political actors, there is the federal state influencing the local CPR management by setting the legal and economic framework. The communes have developed CPR institutions independently and without direct interference from the Swiss state over centuries (see also Landolt and Haller 2015). A strong institutional foundation was already present when the Swiss state increased its influence by releasing political (law on settlement or law on forest protection) and economic policies (subsidies and direct payments) triggering reactions at the local government level. Direct payments became a precondition for the economic survival of the CPR institutions but at the same time it catalyses changes in individual livelihood strategies challenging the maintenance of the natural resources as well as the CPR institutions. Nevertheless, local leaders acted as catalysing agents. This paper contributes to the discussion that if recognized, also locally operating leaders can facilitate the creation of a neutral platform.

Recognizing local knowledge and innovations

The creation and the development of the CPR institution in Sumvitg is based on local knowledge. The comparatively few changes in the main principles, the user rights and regulations prove that the knowledge and the perception of the ancestors are still respected and valid until today. Nevertheless, new forms of management had to be found in order to react to the irreversible consequences of structural change: the power loss of the farming population within the commune on the one hand weakening the decision-making power of the CPR institution (and hence ownership), and the labour disinvestment on the other hand leading to pasture loss and transformation of the cultural landscape in the area. The recognition of local knowledge became apparent again in 2000 with the rule determining overlapping terms of two alp masters sharing one position in order to reduce the workload and to guarantee the transfer of local knowledge from one alp master to the next. In the light of the slow extinction of alp masters with their own working experiences on the alps, this is a forward-looking strategy to sustain local knowledge. The state does recognize institutional innovation and relies on the survival of the local CPR institutions in order to reach its goal to maintain cultural landscapes and to provide ecosystem services. At the same time, the state directs livelihood strategies through direct payments, which do not necessarily strengthen local knowledge and innovation (see above).

Higher-level state recognition

The federal state and the cantons recognize the commune's autonomy and their local knowledge in the CPR management. In fact, they rely on them in regard to the provision of the goods the public expects to be produced (ecosystem services, leisure areas, cultural landscape) and provide financial subsidies to strengthen the sustainability of the CPR institutions. The recognition of local knowledge can play a role in institution-building processes on the level of the commune particularly if there are resentments within the non-farming population against the farmers due to the financial aid of the federal state: if the farmers are not respected, local knowledge of the farmers is not respected either and decision making in the community assembly becomes an act of frustration (see also Landolt and Haller 2015). The farmers of Sumvitg were aware of the increasing difficulty to discuss farming issues with the whole community and it was one of the important reasons to create the cooperative.

Without a trusted leader who was able to gain broad support within the farming community, the window of opportunity in the year 2000 to reorganize the CPR management would have most likely remained unused despite the favourable preconditions. I argue that theoretical models not addressing local agency are substantially limited in explaining the success or failure of institution-building. I further argue that leadership is key not only for the sustainability of the primary function of the CPR institution (the provision of the common property good) but also for sustainable ecological outcomes through the maintenance of the communal work.

Conclusion

Written and oral history proved to be key for explaining institutional change by: a) detecting external drivers of change within the economic and political environment (including technological developments); and b) by revealing internal drivers of change through uncovering social and power relations among the actors taking decisions in regard to the direction of institutional change.

The constitutionality approach proved to be suitable to address the nesting of local CPR institutions in multilevel legal-constitutional and political-economic frameworks characterizing the Swiss decentralized governance system while at the same time recognizing the high autonomy of Swiss communes. This was achieved by looking at local institution-building considering negotiation processes characterized by power relations and agency, and participation processes guided by good leadership leading to a sense of ownership of the results achieved. Trusted and broadly legitimized leadership creating a fair platform for discussions played an important role in building up new institutions for the management of the common pastures which re-established a sense of ownership by gaining autonomy from the community, allowing operational independence by each alp nested under the umbrella of the alp cooperative, strengthening the communal work to maintain the pastures, and an independent financial administration allowing their own investments and economic independence.

The federal state and the commune remain crucial actors in creating an enabling environment for bottom-up institution-building processes and the provision of incentives to direct institutional change towards sustainable ecological outcomes. Economic factors such as the changes in relative prices that would have led to the demise of the CPR institutions in the 1980s have been compensated for by increasing direct payments disarming this powerful factor when it comes to the financial survival of the CPR institutions. Despite financial support, the structural changes remain a fundamental challenge for all local CPR systems: the decreasing farming population ultimately leads to a decreasing workforce and availability of good leaders. What we can learn from the farmers of Sumvitg is that CPR institutions managing alpine pastures need to react to the structural changes before they entirely depend on the benevolence of the commune, the federal state or other external actors losing their right of self-determination. Furthermore, Sumvitg is a role model in regard to how constitutionality can be achieved: by involving all relevant actors in decision-making processes and by creating a sense of ownership in the institution-building process. The future will reveal whether the trend in the CPR management of community alps will develop towards sustainable institutional changes or towards increasing institutional failure (see also Landolt and Haller 2015) leading the way to alternative property rights arrangements, private or public.

The case study of Sumvitg shows clearly how the interaction of different institutional levels determine the margin for bottom-up institution-building and how bargaining power, agency and leadership finally determine if, and how, this margin is used for adapting the CPR institution and to reach sustainable institutional and ecological outcomes.

This case study will remain interesting to follow in the context of further structural changes. Is there a threshold in regard to the group size for managing the alps? Further institutional changes will most likely be needed in order to react to the scarcity of labour for maintaining the institution and the pastures and hence, to maintain the CPR management system.

References

Barber, B.R. (1974) *The Death of Communal Liberty. A History of Freedom in a Swiss Mountain Canton*. Princeton, NJ: Princeton University Press.

Baur, I. and Binder, C.R. (2013) Adapting to socioeconomic developments by changing rules in the governance of common property pastures in the Swiss Alps. *Ecology and Society*, vol 18, no 4, p. 60.

Bundesamt für Statistik (1970) *Eidgenössische Volkszählung 1970*. Bern: Bundesamt für Statistik.

Bundesamt für Statistik (1985) *Eidgenössische Betriebszählung 1985*. Bern: Bundesamt für Statistik.

Bundi, M. and Rathgeb, Ch. (2003) *Die Staatsverfassung Graubündens. Zur Entwicklung der Verfassung im Freistaat der Drei Bünde und im Kanton Graubünden*. Chur: Verlag Rüegger.

Dietz, T., Ostrom, E., and Stern, P.C. (2003) The struggle to govern the commons. *Science*, vol 302, pp. 1907–1912.

Eidgenössisches Statistisches Amt (1920) *Eidgenössische Volkszählung 1920*. Bern: Eidgenössisches Statistisches Amt.

Eidgenössisches Statistisches Amt (1945) *Eidgenössische Betriebszählung 1939*. Bern: Eidgenössisches Statistisches Amt.

Ensminger, J. (1992) *Making a Market. The Institutional Transformation of an African Society*. Cambridge: Cambridge University Press.

Fleischman, F.D., Boenning, K., Garcia-Lopez, G.A., Mincey, S., Schmitt-Harsh, M., Daedlow, K., Lopez, M., Basurto, X., Fischer, B., and Ostrom, E. (2010) Disturbance, response, and persistence in self-organized forested communities: analysis of robustness and resilience in five communities in southern Indiana. *Ecology and Society*, vol 15, no 4, p. 9.

Folke, C., Pritchard, L., Berkes, F., Colding, J., and Svedin, U. (2007) The problem of fit between ecosystems and institutions: ten years later. *Ecology and Society*, vol 12, no 1, p. 30.

FSO (2011) Farming businesses and livestock in Sumvitg 2011. www.bfs.admin.ch, accessed 5 September, 2012.

FSO (2013) *Arealstatistik 2013*. Neuchâtel: FSO.

Haller, T. (ed.) (2010) *Disputing the Floodplains: Institutional Change and the Politics of Resource Management in African Wetlands*. African Social Studies Series. Leiden: Brill.

Haller, T. (2013) *The Contested Floodplain. Institutional Change of the Commons in the Kafue Flats, Zambia*. Lanham, MD: Lexingthon/Rowman & Littlefield.

Haller, T., Acciaioli G., and Rist, St. (2016) Constitutionality: conditions for crafting local ownership of institution-building processes. *Society & Natural Resources*, vol 29, pp. 68–87.

Haller, T., Belsky, J.M., and Rist, S. (2018) The constitutionality approach: conditions, opportunities, and challenges for bottom-up institution-building. *Human Ecology*, vol 46, no 1, pp. 1–2.

Head, R.C. (1995) *Early Modern Democracy in the Grisons. Social Order and Political Language in a Swiss Mountain Canton, 1470–1620*. Cambridge: Cambridge University Press.

Landolt, G. and Haller, T. (2015) Alpine common property institutions under change: conditions for successful and unsuccessful collective action by alpine farmers in the Canton of Grisons, Switzerland. *Human Organization*, vol 74, no 1, pp. 100–111.

[LBBZ] Plantahof (2007) *Situationsbericht Alpwirtschaft im Kanton Graubünden 2001–2005*. Im Auftrag des Amtes für Landwirtschaft und Geoinformation. Fact Sheet Kanton. Juni 2007.

Liver, P. (1968) Das Eigentum an den Liegenschaften des Nutzungsvermögens der Gemeinde in Graubünden. Verwaltungspraxis 7/8, pp. 206–210.

Maissen, A. (2000) *Somvix: eine kulturhistorische Darstellung Gemeinde Sumvitg*. Sumvitg: Gemeinde Sumvitg.

Mathieu, J. (1992) *Eine Agrargeschichte der inneren Alpen. Canton of Grisons, Tessin, Wallis 1500–1800*. Zürich: Chronos Verlag.

Mathieu, J. (2011) *Gemeinde als sozialer Prozess: Der Freistaat der Drei Bünde 1500–1800*, in Hitz, F., Rathgeb, C., and Risi, M. (eds), *Gemeinden und Verfassung. Bündner Politik und Gebietsstruktur gestern, heute, morgen* (S. 35–47). Chur.

Netting, R. (1981) *Balancing on an Alp: Ecological Change and Continuity in a Swiss Mountain Community*. Cambridge: Cambridge University Press.

Ostrom, E. (1990) *Governing the Commons*. Cambridge: Cambridge University Press.

Schulz, T. (2011) Politikanalyse-Evaluation bestehender und alternativer Steuerungsinstrumente für das Sömmerungsgebiet. www.wsl.ch/publikationen/pdf/10914.pdf, accessed 23 January, 2012.

Statistisches Bureau des eidg. Departements des Innern. (1906) *Viehzählung 1906*. Bern: Kommissionsverlag A. Francke.

Stevenson, G.G. (1991) *Common Property Economics. A General Theory and Land Use Applications*. Cambridge: Cambridge University Press.

Weiss, R. (1941) *Das Alpwesen im Kanton Graubünden. Wirtschaft, Sachkultur, Recht, Älplerarbeit und Älplerleben*. Reprinted version 1992. Chur: Octopus-Verlag.

Werthemann, A. and Imboden, A. (1982) *Die Alp- und Weidewirtschaft in der Schweiz*. Bern: Bundesamt für Landwirtschaft, BLW.

Part III

Features and effects of global (e.g. European) investments on commons in the world

Top left: Local women in a large-scale rice plantation in Ghana. Picture by Divine Harrison. For more details on the impacts of a 'best-practice' large-scale land acquisition on access to commons in the Volta Region of Ghana, see Chapter 14.

Top right: Slag deposit next to neighbourhood at Mopani Copper Mines, Zambia. Picture by Meinrad Schade. For an analysis of impacts of mining by transnational corporations on access to commons, see Chapter 17.

Bottom left: Solar panels on common land by the Solar Project Noor II in Ouarzazate, Morocco. Picture by Tobias Haller. For further details on impacts of a green energy investment on access to commons in Ouarzazate, Morocco, see Chapter 18.

Bottom right: Looking down on lost land in Iringa District, Tanzania. Picture by Désirée Gmür. For a detailed analysis of impacts of a forest plantation on women's access to commons, see Chapter 15.

13 Impacts of large-scale land acquisitions on common-pool resources

Evidence from the Land Matrix

Markus Giger, Kerstin Nolte, Ward Anseeuw, Thomas Breu, Wytske Chamberlain, Peter Messerli, Christoph Oberlack and Tobias Haller

Introduction

When prices for agricultural commodities reached record highs in 2008 and again in 2011, reports about a rush for land made headlines in news media around the world. Observers and analysts explained this phenomenon – which we refer to as large-scale land acquisition (LSLA) – by pointing to factors like the expectation of continued high commodity prices, fear of food shortages in the face of an increasing (and increasingly better-off) global population, agrofuel production, speculation, and the quest for finding new investment opportunities. Reports about large land-based deals made in secrecy and without involving affected land users raised concerns about a wave of "land grabbing" happening to the detriment of smallholders, pastoralists, forest dwellers, and other poor land users with low land tenure security and little political voice.

At the same time, the debate about the best model for agricultural development is still far from resolved (Deininger and Byerlee, 2011a; Byerlee, 2014). For instance, the World Bank Report 2008 (Bank, 2007) as well as IAASTD 2009 (McIntyre, 2009) argued for smallholder-based development as the right model for rural development, based on both theoretical and empirical evidence. However, others highlighted the limited success of promoting small-scale agriculture in Sub-Saharan Africa (Collier and Dercon, 2014) and argued in favour of large-scale, modern and resource-intensive agriculture. The perceived urgency to achieve food security lead both important donors and governments in the South to opt for large scale agricultural production in the global South (Deininger and Byerlee, 2011b, Fouilleux et al., 2017).

This debate is of high relevance also for the study of the commons. We posit that expansions of large-scale agriculture often affect areas formerly used as common-pool resources (CPRs), for instance, by smallholders and pastoralists. CPRs such as water, pasture, fisheries, wildlife, forests, and veldt products are resources linked and related to land and are central for food security and sustainability as an extensive body of research shows (see McKean, 2000; Haller, 2007; Haller et al., 2013 for summaries). The question whether or how CPRs are impacted by LSLA is of high interest in this article, as they have formerly been neglected in the debate on land investments (Adams et al., 2019).

CPRs are defined by a number of characteristics, in particular the difficulty to exclude users and the subtractability[1] of their use (Haller et al., 2013). They may comprise a wide range of natural resources, including forested land, grassland, so-called "wasteland", wetlands, and water bodies. Such resources often require flexible and seasonally adapted use, for example in the form of pastoralist migration between lower and higher altitudes. CPRs provide a wide range of goods and services, such as wood for construction and fuel, edible and medicinal plants, fodder, bush meat and fisheries products and many more. In so-called mosaic landscapes dominated by cropland, access to commons can be crucial in helping local people meet their food and income needs, which they might not be able to cover fully based on privately owned land. Social anthropology research shows that CPRs not only play an important role in ensuring food security but also provide cash resources relevant for securing livelihoods, especially among marginalized groups and women (Haller, 2010b, 2010c, 2013).

CPRs can exist in the context of common property, private property, or state property regimes, or a mixture of these (Netting, 1993; Haller, 2010b). In many cases they have been – and continue to be – managed under customary rules that define a common property regime. Many such systems have been significantly weakened or transformed, with negative effects; reasons for this include contradictions between government regulations and customary rules (Stücklin and Frei, 2010, Haller et al., 2016). Fox et al. (2009) listed several factors contributing to a decline in CPR-based livelihoods, among them the rise of conservation, commoditization of land, and promotion of industrial agriculture. Customary management systems are frequently overlooked, or their benefits for users are underestimated or considered outdated and irrational (Lavers, 2012, Haller et al., 2016).

Land titling has been proposed as a strategy addressing the existing or perceived weaknesses of CPRs, creating more land tenure security and paving the way for increased productivity and incomes by creating better incentives for investments in developing countries (Lawry et al., 2017). However, these outcomes are not certain (Sitko et al., 2014) or can vary widely depending on the context (Chimhowu, 2019). Furthermore, an important question is whether land titling of customary land is a way to prevent negative consequences of LSLAs also because CPRs often cut across land boundaries (Haller, 2010b). For instance, in Cambodia this is not the case (Dwyer, 2015).

Two recent meta-analyses of the scientific literature indicate that LSLAs frequently impact on CPRs and common property regimes. Oberlack et al. (2016) investigated the livelihood impacts reported in scientific case studies of LSLAs and found that common property regimes were affected in 38 out of the 44 cases in the sample. Loss of access to land and natural resources was the most frequent of all adverse livelihood impacts. In more than 70 per cent of the cases investigated, the land acquired had previously been held as common property or as a mosaic of common and private property. A small number of archetypical processes explained the adverse impacts: enclosure of livelihood assets based on privatization of land rights; elite capture by local and state elites; selective marginalization of people already living in difficult conditions; and polarization of development discourses. Dell'Angelo et al. (2017) systematically reviewed case studies of LSLAs and found that these took place in the context of common-property regimes in 55 per cent of the 56 cases investigated.

In this chapter, we analyse the database of the Land Matrix (LM) to find out more about the impacts of LSLAs on CPRs and common property regimes. The LM database covers a comparably large number of cases and contains information not only from scientific studies but also from other sources such as media, governments, the private sector, and NGOs. Although this information is often incomplete and partly derives from secondary sources, it nonetheless provides an overview of patterns and trends characterizing the recent wave of land appropriation. We exploit the LM data in light of what they tell us about impacts on CPRs and address the following questions:

- What general patterns and processes characterize LSLA in the global South?
- How do LSLAs impact CPRs and common property regimes?
- To what extent are CPRs affected by LSLAs, and what impacts have been observed?

In the following section, we present the LM database and describe what information it contains about recent trends in LSLA in the global South. We then discuss how LSLAs impact CPRs and common property regimes. Further, we analyse empirical evidence of these impacts in the LM database and illustrate them with brief descriptions of individual cases. We conclude with a discussion of our overall findings.

What general patterns and processes characterize LSLA in the global South?

Our source of information: the Land Matrix database

As the phenomenon of LSLA grew, several initiatives started to collect data on as many cases as possible. Among them was the Land Matrix Initiative (LMI), which is widely considered to maintain the most comprehensive database on LSLAs in the global South. An international partnership of research organizations and regional land-focused organizations, the LMI continues to collect data on international LSLAs in low- and middle-income countries and provides open access to these data. The aim is to increase transparency in the context of LSLAs and contribute to more balanced and equitable decision-making on land.

The LMI collects data from a wide range of sources, including governments, the private sector, academia, civil society, and the media. All contents of the LM database are open access and are frequently used in research, media coverage, lobbying, and policymaking. Besides collecting data and making them widely accessible, the LMI also engages in research and analysis, with the aim of supporting multi-stakeholder processes and enabling informed participation in decision-making around land governance. The fact that all data are accessible through the database or via the primary source facilitates research into individual cases.

The LM database focuses on international LSLAs that entail "a transfer of rights to use, control or ownership of land through sale, lease or concession" (Anseeuw et al., 2012). Other criteria for inclusion of cases are a minimum area of 200 hectares, conclusion of the deal in 2000 or later, and implication of potential conversion of land from smallholder production, local community use, or important ecosystem service provision to commercial use. The focus is mainly on LSLAs by international investors for the purpose of agricultural production, although the database is also beginning to include domestic LSLAs and LSLAs for non-agricultural purposes (timber extraction, carbon trading, industry, renewable energy production, conservation, and tourism). Data collection is restricted to deals in low- and middle-income countries.

The LMI acknowledges that LM data are incomplete and subject to potential biases (Nolte et al., 2016). Governments often choose not to document deals publicly, and official figures are sometimes contradictory. Where possible, data on an individual deal are collected from multiple sources, enabling triangulation of information. Currently almost 80 per cent of cases are based on two or more sources, and 40 per cent on three to seven sources (Nolte et al., 2016). Biases can result from different effects. The level of presence of land-focused media coverage, NGO activities, and researcher networks may differ from country to country, which can lead to under- or over-reporting in certain countries. In addition, the LMI network is not equally strong in each country. Finally, media attention may be focusing on certain sectors (e.g. agrofuels), regions (e.g. Africa), or types of investors (e.g. state companies from emerging countries) (Nolte et al., 2016). But although the data in the LM database are neither complete nor free of errors and omissions, they nonetheless offer valuable insights into current trends in international LSLA.

Insights into broad patterns and frequent processes

The analyses presented in this chapter refer to the data set used in the LMI's second analytical report (Nolte et al., 2016), which contained 1,004 concluded deals for agricultural purposes covering 26.7 million hectares.

These deals focus mainly on the production of food crops (38 per cent of total area), unspecified agricultural products and non-food commodities (32 per cent), agrofuels (21 per cent), and livestock (9 per cent). The most important crop types across all continents are oilseed, including oil palm and jatropha (44 per cent), cereals (20 per cent), and sugar crops (10 per cent).

Investors come from all regions of the world. However, investors from Western Europe (the UK, the Netherlands, France, Jersey, Cyprus, and others) are involved in 315 concluded deals, which makes this the biggest investor region. The second most important investor region is Southeast Asia. The top individual investor countries are Malaysia, the USA, the UK, Singapore, and Saudi Arabia. Non-listed and listed private companies account for more than 70 per cent of deals. Investment funds and state-owned entities together are involved in as little as 15 per cent of deals.

Taken together, these findings nuance and contradict widely held perceptions by the media that state investors from emerging countries (Gulf States, China) are the main investor category. On the contrary: private-sector actors from developed countries in the global North – especially the USA and Europe – are key players behind LSLA worldwide.

Figure 13.1 shows a global heat map of LSLA target areas based on the LM data set. The higher the density of deals affecting a region, the darker the shading.

Africa is the most important target region of deals in the LM data set. LSLAs occur in many countries across the continent. Overall, 422 concluded agricultural deals (42 per cent of all deals) covering 10 million hectares (37 per cent of total area) targeted land in Africa. Many LSLAs are found along major rivers and in Eastern Africa. Eastern Europe is the second most targeted region in terms of area, and accounts for 96 concluded deals on 5.1 million hectares. Comparison of data from 2012 and 2016 shows that although a number of deals failed, the overall trend points towards increased implementation.

Nolte et al. (2016) analysed the location of acquired land with regard to climatic zones and found tropical savannahs (38 per cent of all land deals) and tropical rainforest (18 per cent) to be the two most targeted zones. In Eastern Africa, temperate zones in highland areas are targeted as well. Land in arid areas is less frequently acquired, except along major rivers and waterbodies. This points to the importance of water availability for agricultural production (Breu et al., 2016) and to potential externalities.

Looking at how strongly LSLA affects individual countries, Indonesia, Ukraine, Russia, Papua New Guinea, and Brazil emerge as the five most targeted states in the LM dataset. In the context of CPR, it is important to note that a number of major LSLA target countries also appear to have weak land tenure security (e.g. Cambodia, Ethiopia, Madagascar, Laos, or Ghana). Some studies suggest a strong correlation between poor tenure security and a high occurrence of LSLAs (Deininger, 2013; Nolte et al., 2016) – although a more recent study confirms this relationship only with regard to smaller-sized LSLAs (Lay and Nolte, 2018).

How do LSLAs impact on CPRs and common property regimes?

Having outlined the overall patterns in the geographical distribution of LSLA target areas and investor origins, in this section we proceed to investigate how LSLAs affect CPRs and common property regimes. Any discussion of impacts of recent LSLAs on commons needs to consider the historical background of commons appropriation by settlers and colonial states. In this process, much of the land traditionally managed as common property was placed under the ownership of private settlers or of the state (Peluso and Lund, 2011; Alden Wily, 2012; Haller et al., 2016). In the postcolonial era, common property regimes were additionally

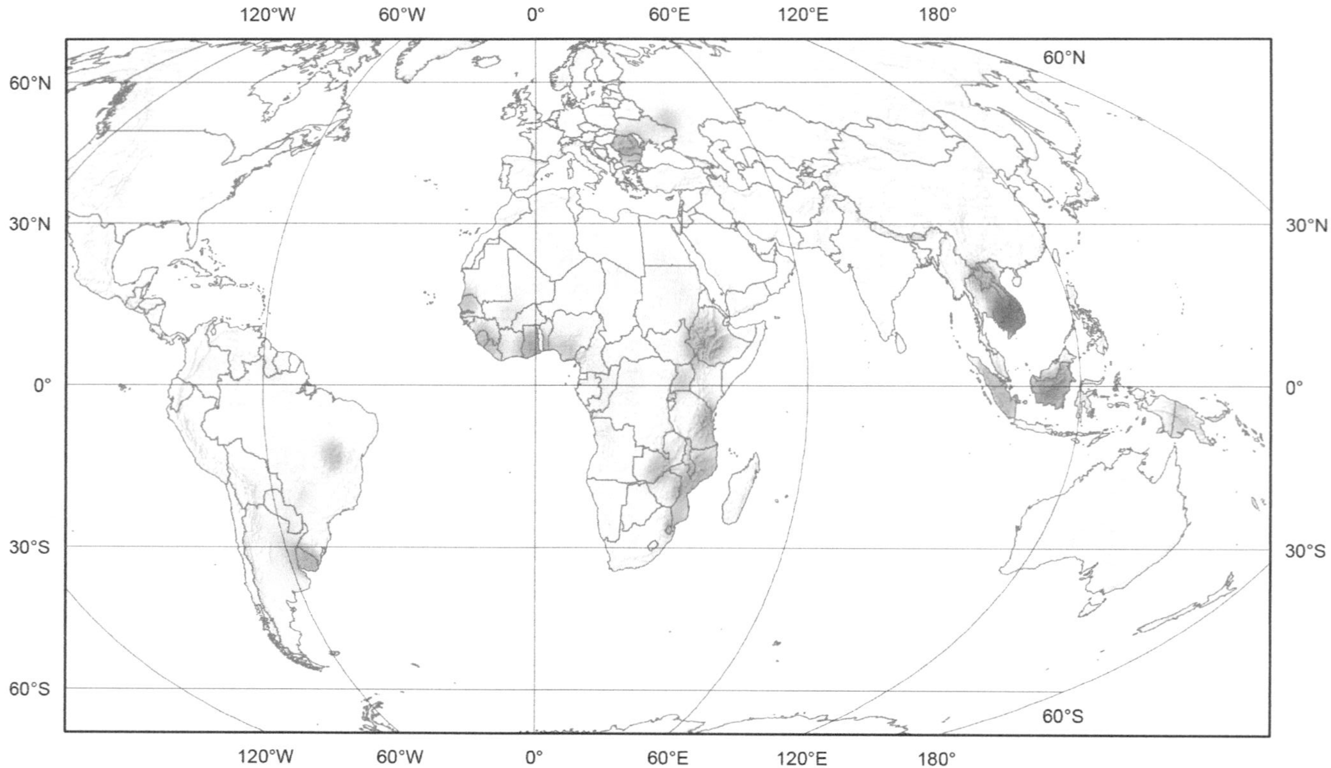

Figure 13.1 Global heat map of LSLA target areas (Nolte et al., 2016).
Source: Land Matrix, 2016.

weakened by the establishment of nature reserves and increasing division and fragmentation of land into land designated for forestry, agriculture, or rangeland (Alden Wily, 2011, 2012). Most states created separate management entities for specific types of resources (e.g. animal husbandry, forestry, or fisheries), further weakening and sometimes even destroying common property management systems by separating responsibilities for different components of interlinked socio-ecological systems (Haller et al., 2013, 2016).

Common property regimes in formerly colonized countries are particularly susceptible to continued destruction in the contemporary global land rush, as these countries disproportionately attract private investors from their former colonial powers (Arezki et al., 2013). Furthermore, postcolonial state structures facilitate the expropriation of customary land users (Alden Wily, 2011; German et al., 2013).

LSLAs may thus be a further step in the history of the progressive weakening of such customary practices (White et al., 2012). If corporate actors acquire large tracts of land for commercial use, this can damage CPRs and common property regimes in various ways. At least four mechanisms are crucial to understanding how LSLAs affect CPRs and common property regimes.

First, the rationale behind such deals is often to establish "modern", highly productive agricultural systems based on intensive, often year-round cultivation and high usage of water and chemicals compared to more traditional production systems. Such plantation style production systems were already established during colonial times, due to perceived and partly real economic advantages for certain crops, but also ideologies and beliefs on the part of the colonial powers (Byerlee, 2014). The current wave of large-scale investments could be a resurgence of this tendency, fuelled by high commodity prices and easy access to land (Byerlee, 2014). The high degree of mechanization, new agricultural technologies and the large amounts of inputs required make such intensive systems capital-intensive to establish. Accordingly, the premises are frequently fenced off to protect expensive equipment. However, this also cuts off access to CPRs on the premises for other users. This can cause locals to be displaced from the land on which they have built their house and grow their crops, or which they use collectively for herding, gathering firewood, and other activities. Land rights under common property regimes often allow powerful traditional or government leaders to reallocate rights from community members to investors (German et al., 2013; Nolte and Väth, 2015).

Second, establishment of a mechanized and rationalized production system often comes with externalities that affect CPRs. For example, adjacent water bodies may suffer from increased abstraction of water, contamination, or other off-site effects, such as the loss of important landscape elements like hedges, small streams and ponds, trees, and bushes (Tejada and Rist, 2017; Lanz et al., 2018). Such biotopes frequently generate products that local users consider to be common property and that constitute important elements of subsistence and household resilience. This type of environmental impact has been reported for LSLAs in Kenya that involved substantial intensification of horticulture and floriculture production (Zaehringer et al., 2018). Furthermore, CPRs may also be overexploited and degraded, when communities lose access to land and subsequently resource users concentrate more on the remaining land (Haller, 2010b).

Third, CPRs are frequently overlooked because their use may constitute a secondary use of land – even if this secondary use is also very important. For example, even if an area's land use is indicated as smallholder farming, in addition to growing crops, smallholder farmers and other inhabitants may also be using that land to collect and hunt a range of products which are communally managed (Haller, 2010b; Lavers, 2012; Marfurt et al., 2016). Such secondary uses are affected, for example, when an LSLA cuts off access to water points or blocks pastoralists' migration routes. This situation is particularly frequent in areas characterized by seasonal water shortages and droughts, such as floodplains or other types of pastoral lands, where herders frequently move to wetlands in the dry season, and drylands in the rainy season. Moreover, drylands are not only used by pastoralists, but often also serve for grazing poor farming households' small ruminants, collecting firewood, and hunting and gathering. When such land is transformed for large-scale commercial production, it is frequently enclosed (Oberlack et al., 2016), depriving farmers of access to its CPRs. Loss of such resources is not readily visible to outsiders. Often it takes an in-depth anthropological study of the socio-ecological systems involved to detect it.

Fourth, water use rights usually form an integral part of investment contracts (Breu et al., 2016). The implementation of these contracts often comes with increased water resource appropriation, altering the local availability of freshwater and thereby directly affecting local livelihoods and CPRs in downstream areas outside the perimeter of acquired land (Breu et al., 2016). Thus, water resource appropriation by investors can impact on ecological and hydrological cycles in a larger area, for instance by cutting off seasonal river flows through dams (Haller, 2010c; Haller et al., 2016; Kamski, 2016). It can reduce the practicability of traditional irrigation patterns or make them entirely obsolete (Tejada and Rist, 2017), negatively affect local fisheries (Haller and Merten, 2008) and wildlife (Chabwela and Haller, 2010), impede seasonal migration by pastoralist groups (Haller et al., 2013, 2016), raise the groundwater table, damaging activities such as brick production (Tejada and Rist, 2017), or restrict access to swamps used by women for off-season vegetable production (Marfurt et al., 2016).

To what extent are CPRs affected by LSLA, and what impacts have been observed?

Evidence in the Land Matrix data that points to adverse impacts on CPRs

In this section, we review available data on the context in which LSLAs occur and present information on previous uses, covers, and ownership of the acquired land. This gives an indication of what types of land are targeted; and it enables conclusions regarding the potential extent of LSLA impacts on CPRs.

Research based on the LM data (Messerli et al., 2014) showed that the local surroundings of these LSLAs had a considerable average population density of 81 people per square kilometre; and most cases were located in relatively accessible areas with fairly well-developed infrastructure. This resonates with concerns about competition for scarce land, other natural resources, and infrastructure between investors and local communities (Anseeuw et al., 2012). However, these observations did not apply to all cases; many LSLAs targeted areas that had a low population density (47 per cent of the deals were in areas with a population density below 25 persons per square kilometre) and were relatively remote (43 per cent of the deals were in areas with more than six hours travelling time to the nearest city with 50,000 or more inhabitants). The land cover types of LSLA target areas ranged broadly: cropland mosaics, forested landscapes, as well as grass, shrubs, and other cover types were targeted by around one third of deals each. Looking at these characteristics together, three distinct socio-ecological contexts (Messerli et al., 2014) were identified:

1 **Densely populated and easily accessible areas with cropland mosaics.** These areas are well suited for agriculture, and strong competition for land is to be expected.
2 **Moderately accessible and moderately populated shrub- or grassland.** This type of land corresponds broadly to what the LM database refers to as "marginal land".[2] However, such land may be highly important to local land users and can be very productive, especially when it includes wetlands (Haller, 2010b).
3 **Largely remote and sparsely populated forestland.** Forest-related land use systems are often targeted by LSLAs intended for producing crops like oil palm, rubber, soy, and many others (Nolte et al., 2016). Large areas of the forest/agriculture frontier are still partially or entirely used for shifting cultivation (Van Vliet et al., 2012, Heinimann et al., 2017).

Each context pose different challenges for local populations and CPR management. In the next sessions we conduct an analysis of more data contained in the LM to further investigate the context in which LSLA take place.

Type of land acquired: previous landowners, land covers, and land uses

Three variables describe the types of land that investors are acquiring: previous landownership, previous land cover, and previous land use. Data on these variables can be used to examine the links between CPRs and LSLAs.

Previous landownership

Communal landownership is found in many parts of Africa, Asia, and Latin America. Under communal landownership, a traditional authority regulates and manages individual and communal land use rights. In many regions state ownership coexists with customary land tenure, which may be either individual or communal. Historical research has shown that, in many cases, land and corresponding CPRs have been transferred to state property (Haller et al., 2013). Allocation of such land to large-scale investors will further dispossess local populations of these resources, to which they may still have had access, albeit with limited control (Haller et al., 2013). Therefore, we investigate whether LSLA actually targets such land within state property or land held by communities.

The LM distinguishes between community-owned, state-owned, and privately owned (by large landowners or smallholders) land. Nolte et al. (2016) report that previous ownership of the land acquired in the 336 deals for which information on this variable was available[3] was attributed to communities (32 per cent of total acquired area), the state (27 per cent), private smallholders (13 per cent), and private large-scale farmers (28 per cent) (Nolte et al., 2016).

Together, community and state ownership indeed account for more than half (59 per cent) of the acquired land. This strongly suggests that CPRs are widely affected by LSLA. Of course, communal ownership can take different forms, and we cannot assume that all of this land is managed by common property institutions. Communal ownership can also entail that members of the local community de facto hold the land as private, though in many cases they will face restrictions on sale, lease, or inheritance.

Previous land cover

All land cover categories may be important in the context of CPRs, although each may concern different types of CPRs and present different challenges. Of high interest to local communities are rangelands in pastoral systems, which are often managed as a CPR (Haller et al., 2013; Messerli et al., 2014). Haller et al. (2013) discuss this with respect to African pastoralist regimes and conclude that common property regimes are still important in many regions, despite having been strongly affected by both colonial and postcolonial state interventions. Floodplains in seasonally flooded areas, often categorized as "marginal" land, frequently contain CPRs (e.g. water, dry-season pasture, fisheries) that are managed by locally developed common property institutions (Haller, 2010b). Forested landscapes are likewise used and managed as CPRs by people living near or in the forests (Agrawal and Gibson, 1999; Nagendra and Ostrom, 2012). But common property institutions may also apply on croplands in traditional land-use systems, particularly when it comes to rules on the right of way for humans, grazing of cattle and small ruminants during off-season periods, gathering and hunting, migration of herds, gathering of firewood, and other activities.

In the LM data, cropland proved to be the most frequently reported land cover type found on acquired land prior to its acquisition, with almost half of all deals (49 per cent) implemented (at least partially) on former cropland. Another 32 per cent of deals targeted forestland, 18 per cent "marginal" land (a category which is not clearly defined, but most likely is often used to describe pastoral land and also floodplains[4]), and 2 per cent land reported as "grass and shrubs". As noted earlier, the two categories of "marginal" land and "grass and shrubs" largely overlap and relate to similar underlying perceptions of land less suitable for intensive production.

Previous land use

The previous use of acquired land reveals information about former livelihood activities. Smallholder agriculture and pastoralism are particularly important land-use categories with regard to CPRs, but forestry and conservation may also point to former uses of CPRs.

Among the deals in the LM data set, the most dominant previous land use is agriculture: 43 per cent of the total area acquired was used for commercial agriculture and another 31 per cent for smallholder agriculture. The high percentage of commercial agriculture requires an explanation. Although such land may have been used commercially prior to the most recent acquisition, it was also used by local communities,[5] for example if the former owner at some point neglected a plantation and local communities replanted it. A detailed analysis from Mozambique shows that a number of these cases concern land that was at one time colonized by European settlers, then nationalized, and subsequently sold or leased to private investors (Adalima, 2017). However, this land was also used by local people, and the recent LSLA led to the dispossession and eviction of local land users, creating land tenure insecurity and conflict (Reys et al., 2018). Similar cases have been recorded in the LM database for Ethiopia and Tanzania, among others. Cases with previous private ownership are not included in our further analysis, however, as they do not directly relate to land as a CPR, despite their potentially negative impact on local communities.

Smallholders previously used 31 per cent of the total area acquired for agriculture, that is about 8.3 million hectares. If we assume an average farm size of 2 hectares (Lowder et al., 2016), it becomes evident that LSLA may have affected a very substantial number of smallholders. Further land uses were less targeted: 16 per cent of the total acquired area was previously used for forestry and 5 per cent each for pastoralism and conservation.

Combining information about previous land use and landownership and previous land use and land cover

In a next step of analysis, we compared the available data on previous land use and previous landownership. Our assumption is that the combination of state or community landownership with land uses that are likely to occur under CPR regulations will provide a better indication of how frequently LSLAs risk affecting CPRs. In Table 13.1, we display these combinations that potentially reflect an impact of LSLAs on CPRs. We therefore do not show the combinations

Table 13.1 Comparison of previous land use and previous land landownership for areas potentially managed communally

		Previous landowner					
		State		*Community*		*Total*	
		Number of deals	*Size in 1,000 ha*	*Number of deals*	*Size in 1,000ha*	*Number of deals*	*Size in 1,000 ha*
Previous land use	Smallholder agriculture	67	578	93	957	160	1,535
	Pastoralism	11	43	13	71	24	114
	Forestry	8	433	42	1,107	50	1,541
	Conservation	9	269	22	105	31	375
	Total	**95**	**1,324**	**170**	**2,241**	**265**	**3,566**

Source: LM database, 25 April 2016; analysis by the authors.

Note: LM records may indicate more than one previous land use and more than one previous landowner, but the LM does not provide any information on the shares of area relating to each former land use or landowner. For this analysis, we have we have attributed equal shares of the area under contract to each land use or landowner and divided the deals in order to count each combination separately. As a result, the original sample of 297 records that include information on both the previous land use and the previous landowner increased to 537 cases. Of these, 265 fall into the categories taken into account in this table. The numbers of deals indicated show in how many deals a specific combination occurs, but it may concur with other combinations in the same deal.

for previous private land owners (smallholders or large scale) with the different previous land uses – albeit these may also include CPR regulated land.

The data clearly show that most of the land previously owned by the state or by communities was used for smallholder agriculture and forestry (1.5 million hectares each). However, some of it was also used for pastoralism and for conservation.

In 67 cases, land previously used for smallholder agriculture was owned by the state. Most of these cases are located in Southeast Asia and Eastern Africa. In another 93 cases where the land acquired was previously used by smallholders, it was community-owned. Most of these are located in Eastern or Western Africa and in Southeast Asia. Pastoralists who used land previously owned by the state (11 deals) or communities (13 deals) are reported to have lost access to land in Eastern and Western Africa and in Southeast Asia.

In all of these regions, governments legally own most of the land. In Cambodia, for example, the government grants land titles to individuals and in some cases to local communities, but it also grants concessions on state land (Dwyer, 2015). This is a case of a legal pluralism, which is observed in many other countries as well (Haller, 2010a, 2010c).

Finally, we compared data on previous land use and previous land cover (Table 13.2). We did so based on the assumption that the land cover of land potentially used under common property regimes can give an indication of what types of CPRs are affected, and where.

Table 13.2 Comparison of previous land use and previous land cover for areas potentially managed communally

		Previous land cover							
		Cropland		*Forest*		*Shrubland and marginal land*		*Total*	
		Number of deals	*Size in 1,000 ha*	*Number of deals*	*Size in 1,000 ha*	*Number of deals*	*Size in 1,000 ha*	*Number of deals*	*Size in 1,000 ha*
Previous land use	Smallholder agriculture	173	1,713	30	174	64	769	**267**	**2,657**
	Pastoralism	9	17	29	67	31	359	**69**	**44**
	Forestry	11	141	60	1,652	20	24	**91**	**1,818**
	Conservation	5	24	27	383	50	30	**37**	**438**
	Total	**198**	**1,897**	**146**	**2,277**	**120**	**1,184**	**464**	**5,359**

Source: LM database, 25 April 2016; analysis by the authors.

Note: LM records may indicate more than one previous land use and more than one previous land cover, but the LM does not provide any information on the shares of area relating to each former land use or land cover. For this analysis, we have we have attributed equal shares of the area under contract to each land use or land cover and divided the deals in order to count each combination separately. As a result, the original sample of 277 records that include information on both the previous land use and the previous land cover increased to 612 cases. Of these, 464 fall into the categories taken into account in this table. The numbers of deals indicated show in how many deals a specific combination occurs, but it may concur with other combinations in the same deal.

Table 13.2 gives a nuanced view of how previous land covers relate to previous land uses. In terms of area, former cropland was clearly most frequently used for smallholder agriculture, and former forests were most frequently used for forestry. However, this provides no additional indication of whether acquisition of this land by investors may have affected CPRs. It is important to recall that the information in the LM regarding previous land cover and land use is often incomplete and not fully able to capture the details of how smallholders use resources in a highly diverse manner. Nevertheless, it is interesting to note that, in many deals, previous forests, shrubland, or marginal land had been used by smallholders and pastoralists. The likelihood that this land was used as CPRs is high. For example, in cases where the previous land cover is indicated as forest and the previous land use as smallholder farming, the land may have been used for shifting cultivation or for collecting forest products. It is important to mention that even if forest is indicated as the previous land cover and forest as the previous land use, CPRs may still have been affected; in addition to forestry, the land may have also been used for various purposes by people living in or around the forest. The same is true of areas previously used for conservation. They may still have been used by local people, even though they might already have lost the right to manage these resources. Loss of access to such land particularly affects marginalized groups and women, as they typically base their livelihoods on CPRs (Johnson, 2004).

Insights into potential LSLA impacts from case studies reported in the LM database

In this section, we present four short case studies of LSLA impacts on CPRs. All four cases are included in the LM data set analysed above. However, they provide deeper insights into the impacts of LSLA on CPRs than analyses of entire data sets can. The cases are located in Senegal, Sierra Leone, Ethiopia, and Kenya and are related to European or North American investments and actors. We would like to underline that these cases are not representative of the general impact of LSLA on CPRs; we selected them because they are relatively well-documented. For each case, we briefly describe its main features, the CPRs concerned, and how these are affected by the LSLA.

Senhuile in Senegal (Case ID in LM database: #3433)

The case shows how land used for grazing, growing crops and collecting timber, as well as access to water collection points was lost – all of it relevant for CPRs. The main mechanism for the loss of these CPRs was the loss of access to land.

This deal covers 10,000 hectares in the Ndiael nature reserve and was undertaken by private investors, including one from Italy. Initially, the Senegalese government allocated land to this company in the area around Fanaye. Strong community protests against the project forced the government to find alternative land. A first lease of 20,000 hectares was granted in 2012 for a period of 50 years. This land, owned by communities and the state, was previously used by local communities for smallholder agriculture. Since the first lease was signed, 10,000 hectares have been revoked by the government as the lessor. A subsequent lease for an additional 5,000 hectares has also been revoked.

The project has affected a total of around 9,000 community members from 40 villages. None have been displaced, but people have lost access to land. The community was only consulted during the social impact assessment in 2013, when the land had already been acquired. Villagers living in close proximity to the project are constantly threatened with eviction by company representatives and the local police.

In an effort to improve working relations between the company and the community, an agreement was signed in 2014 outlining the compensation and benefits that the communities would receive. This stipulates that Senhuile will provide 0.3 hectares of land per family for grazing and cultivation; so far, 189 hectares have been allocated to families. The memorandum also promises other community benefits, such as the construction of classrooms and the creation of community gardens for women, but these have not yet materialized. However, Senhuile has delivered fodder to affected community members to compensate them for the loss of grazing land (Franchi and Manes, 2012; GRAIN, 2013; Action Aid, 2014a, 2014b, 2016; Harding, 2016).

Addax Bioenergy Ltd in Sierra Leone (Case ID in LM database: #1798)

The case shows the loss of land managed under a traditional community system – to which not all community members have equal access – impacting disproportionally families not originating from the region and women who depend on swamps for their specific crops. The main mechanisms are the loss of land, externalities created by the LSLA (abstraction of water) and the loss of secondary uses (such as swamps for horticulture production in off-season).

This project was initiated in 2008 by a Swiss-based company to produce sugar cane in Sierra Leone. The lease originally covered 54,000 hectares, but was later downscaled to about 23,000 hectares. Partly as a result of the Ebola crisis, implementation of the project came to a stop in 2015. The project was then sold to another company, Sunbird Bioenergy Africa Ltd, and production started again. The case was the subject of detailed anthropological field research and a large socio-economic survey, whose findings inform this case study.

The research showed that the project's large-scale monoculture has destroyed a highly diverse cultural landscape, significantly changing the quality of and access to land, water, and grassland products, especially for the more marginalized groups such as women, youth, tenants, and migrants, and reducing the resilience of local livelihoods to external shocks. Many land users have been excluded from accessing CPRs (land for production, swamps) and have thus lost previous access rights based on common property institutions. The average area of land a family uses for agriculture is much smaller within the project area than outside the project area (2.53 hectares versus 9.16 hectares). Those who have no land of their own – because they are originally not from the area – are more seriously affected by this land scarcity than landowners.

Payments for the land lease are low and have been made only to landowners, who make up about 50 per cent of the people living on agriculture. The compensation scheme has exacerbated existing tendencies towards elite capture of the project's economic benefits, further intensifying tensions and conflicts among different groups within the *Temne* society.

In the beginning, local people welcomed the project as they anticipated it would bring development and salaried work to the area. Failure of these expectations to materialize triggered various responses. Local elites as well as the younger generation have activated both old and new ways of resistance, resorting to old institutions of resistance (secret societies) and to a combination of old and new tenure institutions and international legal rights. With the help of a local NGO they are aiming to win back control over the commons (Bottazzi et al., 2016; Marfurt et al., 2016; Nolte et al., 2016; SiLNoRF and BFA, 2017).

Kuraz Sugar Development Project in Ethiopia (Case ID in LM database: #4623)

The Kuraz Sugar Development Project exemplifies how the use of, and access to, water can be jeopardized by an LSLA and pasture managed as CPRs can be destroyed by these changes in the hydrological cycle. It provides insight into the cumulative effects of hydropower infrastructure and land acquisition on local people, which creates huge externalities in terms of changes in water and pastoral resources.

This case is located in the lower Omo Valley in Ethiopia. The Gibe III dam was inaugurated 2016 and constitutes the biggest hydroelectric dam in the country. Construction of the dam is intrinsically linked to a huge irrigation scheme which was meant to enable ESC (Ethiopia Sugar Corporation) to irrigate 175,000 hectares of land allocated to them by the Ethiopian government. Plans included five factories and several sugar cane estates and the creation of 700,000 jobs as well as five urban centres and 42 villages; however, these plans have recently been downscaled.

NGOs and UNESCO (2011) raised concerns about the project's environmental and social impacts on people living downstream of the dam. Their livelihoods rely on flood-recession agriculture and pastoralist activities and will likely be severely damaged or even destroyed. The pastures of the Mursi and Bodi – two of the ethnic groups affected (Stevenson and Buffavand, 2018) – are managed as common property and controlled by local groups (Turton, 1995). But "the permanent alteration of the hydrological cycle has also ruled out the continuation of flood-recession agriculture, which is the most reliable component of local agro-pastoralist livelihoods" (Kamski, 2016). Government plans foresee the villagization of 45,000 people. But while the dam has been constructed, the establishment of the sugar cane estates has been severely lagging. Kamski (2016) reports that mitigating measures such as watering points for livestock have been implemented. But in February 2016, sugar cane had been planted only on about 10,600 hectares. As a result, people are deprived of their traditional livelihoods with no alternative income sources available, and their traditional way of life cannot continue in the new settlements (The Land Matrix Global Observatory, Salini Impregilo, Stevenson and Buffavand, 2018).

Dominion Farms Ltd in Kenya (Case ID in LM database: #1374)

This case shows the impact of a LSLA through blocking the access to land. In this case, land had multiple uses including fishing, collecting wild products, grazing and small-scale agricultural production. These uses are of importance for a large number of people under CPR regimes.

The Yala Swamp in Kenya is a wetland region of more than 200 km^2 in Kenya. Kenya's legal pluralism, dating back to colonial times, provided a legal basis for a US investor to lease 6,900 hectares of swampland, primarily to produce rice. The lease was agreed with local county councils in the name of development, and as such was welcomed by political leaders. However, reclamation of the swamp affects the resilience of local communities by blocking access to CPRs. It is estimated that indirectly many more people benefitted from the swamp's resources than those who benefit directly from employment created by the investment. The loss of a major livelihood basis combined with a lack of employment opportunities limits the diversification of livelihood strategies.

An area of 182 hectares was allocated as compensation for the loss of resources. But after Dominion Farms Ltd drained the land, local institutions were ignored and no further steps were taken to distribute the land to local people. Some wealthy community members, using paid labour, rushed to clear this land in order to lease it to others themselves. Vulnerable people, including elderly women and poor peasant farmers, were unable to continue using the land as they had before.

Currently, local newspapers report that the investor is preparing to leave the country. Whether the project will be continued by a new investor remains unclear (Schubiger, 2015; Schubinger and von Sury, 2016; von Sury, 2016; Kamau, 2017; Odhiambo, 2017).

Conclusions

This contribution discussed impacts of LSLAs on CPRs and common property regimes. We considered multiple ways in which land deals may affect CPRs. First, capital-intensive and rationalized agricultural production systems are generally fenced off, blocking access, and eliminating important components of CPR users' livelihoods. Second, impacts might also be related to indirect externalities, as the change in landownership often comes with changes in the production mode and techniques – affecting off-site CPRs like adjacent water bodies, biodiversity, landscape components, and others. Third, impacts can affect secondary land uses. Many forms of CPR use, such as collecting and hunting communally managed products, are often "hidden" and therefore frequently overlooked when assessing potential impacts. LSLAs nonetheless restrict these uses. Lastly, LSLAs can directly affect natural resources on- and off-site: for example, water use rights are often part of investment contracts, affecting downstream areas and users.

We assess the extent to which LSLA may be impacting CPRs and provide deeper insights into the nature of these impacts. Our review of LM data indicates that LSLA may be impacting widely on CPRs. As LSLAs target a wide variety of socio-ecological contexts, potential impacts will likewise range widely, and challenges will differ by context. Based on a subset of LM data we show that acquired land was previously owned by communities or the state in 59 per cent of cases. The importance of this finding becomes apparent when we consider that in Africa, and in other regions of the global South, land officially owned by the government is very often de facto used by local land users. Against this background, the high percentage of previously community- or state-owned land affected by LSLAs indicates that land appropriation by commercial users may be having profound effects on local populations in more than half of the cases contained in the LM database.

We further describe four case studies that provide deeper insights into how LSLAs can affect CPRs. In the four cases, CPR users indeed lose access to grazing land, water, and timber as a result of LSLA. In this context it is important to consider that this will likely lead to exclusion of weaker groups within the local setting, such as immigrants, women, or poor households.

We have not investigated to what extent new options and income opportunities arising from LSLA may counterbalance losses due to enclosure and reduced availability of, and access to, CPRs with other benefits, or how these benefits are distributed. However, the fact that elite capture, selective marginalization, and discourse polarization are widely observed in the context of LSLA (Oberlack et al., 2016) raises serious doubts about the prospects LSLA providing sizable benefits for marginalized groups.

Customary institutions for the management of commons very often include rules and mechanisms that lead to a certain level of balance between powerful and marginalized groups. Institutions for customary management of CPRs also often include ways and means to mitigate risks of climate variability and extreme events. Common property regimes have been weakened throughout colonial history, and LSLAs are a further step in the marginalization of such customary practices. Existing power imbalances in traditional systems (for instance between gender, or with regard to actors not originating from the area or same ethnic group) can be exacerbated under the impact of LSLA and the loss of resources they entail. New mechanisms that could compensate for this loss are often absent or not sufficiently implemented when commercial actors start using the land. The impact of LSLA on CPRs may therefore be even higher than the mere quantity of land affected suggests.

We would like to conclude this chapter by discussing how the relevance of LM data for the study of the commons might be increased. We have shown that LSLAs have tremendous impacts on the commons; however, LM data only indirectly cover the topic of CPRs and common property regimes. In the upcoming process of decentralizing the LM database, a greater focus of the LMI on these aspects would therefore be welcomed. Great potential lies in the establishment of national land observatories. In regions where CPRs are important, these observatories could focus on impacts of LSLAs on the commons. Ideally, researchers and practitioners investigating these impacts would share their insights with the LMI in general and with relevant national land observatories in particular. The LMI is currently also reaching out to other data initiatives, including LandMark,[6] which focuses on mapping lands that are collectively held and used by indigenous peoples and local communities. The national land observatories, data sharing, and new alliances could advance research and practice alike and ultimately benefit the communities depending on CPRs.

Notes

1 Subtractability means that CPR users compete for CPRs; if one user benefits from use of a CPR, this reduces other users' potential benefits from that CPR. See chapter by Haller et al. in first section of this book.

2 Note that the land cover typology (GlobCover 2009) used in the study by Messerli et al. does not contain a category of "marginal land"; instead, it contains land cover types such as "closed to open shrublands", "grasslands", and "sparse vegetation".

3 Individual deals list up to three different previous owners. The LM does not provide any information on the share of area owned by each previous owner; for the present analysis, we have attributed equal shares of the area under contract to each previous landowner and divided the deals in order to count each combination separately. N (deals) = 336; N (previous landowners) = 386.

4 According the LM guidelines "land of poor quality with regard to agricultural use, and unsuitable for housing and other uses. OECD 2001: Glossary of statistical terms" LMI Draft 2015.

5 Although the LM's criteria for inclusion of deals in the database exclude transfers from commercial large-scale owners to other large-scale owners, some deals include land that was originally owned by communities or smallholders and acquired by a large-scale investor, but has been resold from one private actor to another after it was first recorded in the database. In such cases, the previous land use is listed as large-scale agriculture, but the deal is retained in the LM database, as the land was originally lost by the local community (Nolte, et al., 2016).

6 www.landmarkmap.org, accessed 4 April, 2018.

References

Action Aid (2014a) *The Great Land Heist. How the World is Paving the Way for Corporate Land Grabs*. Johannesburg: Action Aid.

Action Aid (2014b) *L'investissement Senhuile-Senethanol à Ndiael, Sénégal: Quel avenir sans ma terre? Des communautés mobilisées pour récupérer leur terre*. Johannesburg: Action Aid.

Action Aid (2016) '#LANDfor SENEGAL: una buona notizia!', https://petizioni.actionaid.it/news/landfor-senegal-buona-notizia/, accessed 31 March, 2016.

Adalima, J.L. (2017) *Workpakage 2 Report*. Afgroland Project.

Adams, T., Gerber, J.D., and Haller, T. (2019) 'Who Gains from Contract Farming? Dependencies, Power Relations, and Institutional Change' (accepted for publication).

Agrawal, A. and Gibson, C.C. (1999) 'Enchantment and disenchantment: the role of community in natural resource conservation', *World Development*, vol 27, pp 629–649.

Alden Wily, L. (2011) '"The law is to blame": the vulnerable status of common property rights in Sub-Saharan Africa', *Development and Change*, vol 42, pp 733–757.

Alden Wily, L. (2012) 'Looking back to see forward: the legal niceties of land theft in land rushes', *The Journal of Peasant Studies*, vol 39, pp 751–775.

Anseeuw, W., Boche, M., Breu, T., Giger, M., Lay, J., and Messerli, P. (2012) *Transnational Land Deals for Agriculture in the Global South. Analytical Report based on the Land Matrix Database*. Bern, Montpellier, Hamburg: Centre for Development and Environment, Centre de coopération internationale en recherche agronomique pour le développement, German Institute of Global and Area Studies.

Arezki, R., Deininger, K., and Selod, H. (2013) 'What drives the global "land rush"?', *The World Bank Economic Review*, vol 29, pp 207–233.

Bank, W. (2007) *Agriculture for Development: World Development Report 2008*. Washington, DC: World Bank.

Bottazzi, P., Goguen A., and Rist, S. (2016) 'Conflicts of customary land tenure in rural Africa: is large-scale land acquisition a driver of "institutional innovation"?', *The Journal of Peasant Studies*, vol 43, pp 971–988.

Breu, T., Bader, C., Messerli, P., Heinimann, A., Rist, S., and Eckert, S. (2016) *Large-scale land acquisition and its effects on the water balance in investor and host countries*. PLoS One, 11, e0150901.

Byerlee, D. (2014) 'The fall and rise again of plantations in tropical Asia: history repeated?', *Land*, vol 3, pp 574–597.

Chabwela, H. and Haller, T. (2010) 'Governance issues, potentials and failures of participative collective action in the Kafue Flats, Zambia', *International Journal of the Commons*, vol 4.

Chimhowu, A. (2019) 'The "new" African customary land tenure. Characteristic, features and policy implications of a new paradigm', *Land Use Policy*, vol 81, pp 897–903.

Collier, P. and Dercon, S. (2014) 'African agriculture in 50 years: smallholders in a rapidly changing world?', *World Development*, vol 63, pp 92–101.

Deininger, K. (2013) 'Global land investments in the bio-economy: evidence and policy implications', *Agricultural Economics*, vol 44, no 1, pp 115–127.

Deininger, K. and Byerlee, D. (2011a) *The Rise of Large Farms in Land Abundant Countries: Do They Have a Future?* Washington, DC: The World Bank.

Deininger, K. and Byerlee, D. (2011b) *Rising Global Interest in Farmland: Can it Yield Sustainable and Equitable Benefits?* Washington, DC: World Bank Publications.

Dell'Angelo, J., D'Odorico, P., Rulli, M.C., and Marchand, P. (2017) 'The tragedy of the grabbed commons: coercion and dispossession in the global land rush', *World Development*, vol 92, pp 1–12.

Dwyer, M.B. (2015) 'The formalization fix? Land titling, land concessions and the politics of spatial transparency in Cambodia', *The Journal of Peasant Studies*, vol 42, pp 903–928.

Fouilleux, E., Bricas, N., and Alpha, A. (2017) '"Feeding 9 billion people": global food security debates and the productionist trap', *Journal of European Public Policy*, vol 24, pp 1658–1677.

Fox, J., Fujita, Y., Ngidang, D., Peluso, N., Potter, L., Sakuntaladewi, N., Sturgeon, J., and Thomas, D. (2009) 'Policies, political-economy, and swidden in Southeast Asia', *Human Ecology*, vol 37, pp 305–322.

Franchi, G. and Manes, L. (2012) *Land Grabber. Italy's involvement in the Great Land Grab.* Rome: Common.

German, L., Schoneveld, G., and Mwangi, E. (2013) 'Contemporary processes of large-scale land acquisition in Sub-Saharan Africa: legal deficiency or elite capture of the rule of law?', *World Development*, vol 48, pp 1–18.

Grain (2013) 'Who is Behind Senhuile-Senethanol?', www.grain.org/article/entries/4815-who-is-behind-senhuile-senethanol, accessed 30 January, 2017.

Haller, T. (2007) *The Contested Floodplain: Institutional change of Common-Pool Resources Management and conflicts among the Ila, Tonga and Batwa, Kafue Flats (Southern Province) Zambia. Department of Social Anthropology*. Switzerland: University of Zurich.

Haller, T. (2010a) 'Common Pool Resources, Legal Pluralism and Governance from a New Institutionalist Perspective: Lessons from the African Floodplain Wetlands Research Project (AFWeP)', in Eguavoen, I., and Laube, W. (eds), *Negotiating Local Governance. Natural Resource Management at the Interface of Communities and the State* (pp. 137–180). Berlin: LIT Verlag Münster.

Haller, T. (ed.) (2010b) *Disputing the Floodplains: Institutional Change and the Politics of Resource Management in African Floodplains*. Leiden: Brill.

Haller, T. (2010c) 'Institutional Change, Power and Conflicts in the Management of Common-Pool Resources in African Floodplain Ecosystems: An Introduction', in Haller, T. (ed.) *Disputing the Floodplains: Institutional Change and the Politics of Resource Management in African Floodplains*. Leiden: Brill.

Haller, T., Fokou, G., Mbeyale, G., and Meroka, P. (2013) 'How fit turns into misfit and back: institutional transformations of pastoral commons in African floodplains', *Ecology and Society*, vol 18.

Haller, T. and Merten, S. (2008) '"We are Zambians – don't tell us how to fish!" institutional change, power relations and conflicts in the Kafue flats fisheries in Zambia', *Human Ecology*, vol 36, pp 699–715.

Haller, T., Van Dijk, H., Bollig, M., Greiner, C., Schareika, N., and Gabbert, C. (2016) 'Conflicts, security and marginalisation: institutional change of the pastoral commons in a "glocal" world', *Revue scientifique et technique (International Office of Epizootics)*, vol 35, pp 405–416.

Harding, A. (2016) 'Box 10: resistance to land deals – the case of Senhuile in Senegal', in Nolte, K., Chamberlain, W., and Giger, M. (eds), *International Land Deals for Agriculture. Fresh insights from the Land Matrix: Analytical Report II*. Bern, Montpellier, Hamburg, Pretoria: Bern Open Publishing.

Heinmann, A., Mertz, O., Frolking, S., Christensen, A.E., Hurni, K., Sedano, F., Chini, L.P., Sahajpal, R., Hansen, M., and Hurtt, G. (2017) 'A global view of shifting cultivation: recent, current, and future extent', *PloS one*, vol 12, e0184479.

Johnson, C. (2004) 'Uncommon ground: the "poverty of history" in common property discourse', *Development and Change*, vol 35, pp 407–434.

Kamau, J. (2017) 'Rumours, truths and lies on Dominion Farm', www.nation.co.ke/news/Rumours-truths-lies-Dominion-Farm/1056-3827670-3n14phz/index.html, accessed 5 March, 2017.

Kamski, B. (2016) 'The Kuraz Sugar Development Project (KSDP) in Ethiopia: between "sweet visions" and mounting challenges', *Journal of Eastern African Studies*, vol 10, pp 568–580.

Lanz, K., Gerber J.D., and Haller, T. (2018) 'Land grabbing, the state and chiefs: The politics of extending commercial agriculture in Ghana', *Development and Change*, vol 49, no 6, pp 1526–1552.

Lavers, T. (2012) 'Patterns of agrarian transformation in Ethiopia: State-mediated commercialisation and the "land grab"', *The Journal of Peasant Studies*, vol 39, pp 795–822.

Lawry, S., Samii, C., Hall, R., Leopold, A., Hornby, D., and Mtero, F. (2017) 'The impact of land property rights interventions on investment and agricultural productivity in developing countries: a systematic review', *Journal of Development Effectiveness*, vol 9, no 61–81.

Lay, J. and Nolte, K. (2018) 'Determinants of foreign land acquisitions in low- and middle-income countries', *Journal of Economic Geography*, vol 18, pp 59–86.

Lowder, S.K., Skoet, J., and Raney, T. (2016) 'The number, size, and distribution of farms, smallholder farms, and family farms worldwide', *World Development*, vol 87, pp 16–29.

Marfurt, F., Käser, F., and Lustenberger, S. (2016) 'Local perceptions and vertical perspectives of a large scale land acquisition project in Northern Sierra Leone', *Homo Oeconomicus*, vol 33, pp 261–279.

Mcintyre, B.D. (2009) *International Assessment of Agricultural Knowledge, Science and Technology for Development (IAASTD): Global Report*. Washington, DC: Island Press.

Mckean, M.A. (2000) 'Common property: what is it, what is it good for, and what makes it work'. *People and Forests: Communities, Institutions, and Governance*, pp 27–55.

Messerli, P., Giger, M., Dwyer, M. B., Breu, T., and Eckert, S. (2014) 'The geography of large-scale land acquisitions: analysing socio-ecological patterns of target contexts in the global South', *Applied Geography*, vol 53, pp 449–459.

Nagendra, H. and Ostrom, E. (2012) 'Polycentric governance of multifunctional forested landscapes', *International Journal of the Commons*, vol 6.

Netting, R.M. (1993) *Smallholders, Householders: The Ecology of Small Scale, Sustainable Agriculture*. Stanford: Stanford University Press.

Nolte, K., Chamberlain, W., and Giger, M. (2016) *International Land Deals for Agriculture. Fresh insights from the Land Matrix: Analytical Report II*. Bern, Montpellier, Hamburg, Pretoria: Centre for Development and Environment, Centre de coopération internationale en recherche agronomique pour le développement, German Institute of Global and Area Studies, University of Pretoria.

Nolte, K. and Väth, S.J. (2015) 'Interplay of land governance and large-scale agricultural investment: evidence from Ghana and Kenya', *The Journal of Modern African Studies*, vol 53, pp 69–92.

Oberlack, C., Tejada, L., Messerli, P., Rist, S., and Giger, M. (2016) 'Sustainable livelihoods in the global land rush? Archetypes of livelihood vulnerability and sustainability potentials', *Global Environmental Change*, vol 41, pp 153–171.

Odhiambo, O. (2017) 'Dominion farms in new round of trouble', *The Standard Digital*, www.standardmedia.co.ke/article/2001262339/dominion-farms-in-new-round-of-trouble, accessed 5 March, 2018.

Peluso, N.L. and Lund, C. (2011) 'New frontiers of land control: Introduction', *Journal of Peasant Studies*, vol 38, pp 667–681.

Reys, A., Mercandalli, S., Anseeuw, W., Burnod, P., and Giger, M. (2018) *Large-scale Agriculture Investments in Northern Mozambique: Impacts on Households' Livelihoods and Rural Dynamics*. Afgroland WP4 Households and Livelihoods Research Report. Not published. Bern, Montpellier, Hamburg, Pretoria: CDE, CIRAD, University of Pretoria.

Salini Impreglio 'Global Player in Large, ComplexInfrastructure Projects', www.salini-impregilo.com/en/, accessed 27 January, 2018.

Schubiger, E. (2015) *Ethnography of a Large-scale Land Acquisition in West Kenya: 'No Matter How LOUD the Frogs Croak, the Cow Still Drinks the Water'*. Bern: Universität Bern, Institut für Sozialanthropologie.

Schubiger, E. and von Sury, A. (2016) 'Box 13: rice project falls short of development potential', in Nolte, K., Chamberlain, W., and Giger, M. (eds), *International Land Deals for Agriculture. Fresh insights from the Land Matrix: Analytical Report II*. Bern, Montpellier, Hamburg, Pretoria: Bern Open Publishing.

Silnorf and BFA (2017) *Monitoring Report On the Operations of Addax/Sunbird Bioenergy Mabilafu Project Sierra Leone (Period July 2016 – August 2017)*. Sierra Leone Network on the Right of Food and Bread For All.

Sitko, N.J., Chamberlin, J., and Hichaambwa, M. (2014) 'Does smallholder land titling facilitate agricultural growth?: An analysis of the determinants and effects of smallholder land titling in Zambia', *World Development*, vol 64, pp 791–802.

Stevenson, E. and Buffavand, L. (2018) '"Do our bodies know their ways?" Villagization, food insecurity, and ill-being in Ethiopia's Lower Omo Valley', *African Studies Review*, vol 61, no 1, pp 109–133.

Stücklin, S.B. and Frei, K. (2010) 'Between water spirits and market forces: institutional changes in the Niger Inland Delta fisheries among the Somono and Bozo fishermen of Wandiaka and Daga-Womina (Mali)', in Haller, T. (ed.), *Disputing the Floodplains: Institutional Change and the Politics of Resource Management in African Floodplains*. Leiden: Brill.

Tejada, L. and Rist, S. (2017) 'Seeing land deals through the lens of the "land–water nexus": the case of biofuel production in Piura, Peru', *The Journal of Peasant Studies*, pp 1–24.

The Land Matrix Global Observatory. 'The Online Public Database on Land Deals', International Land Coalition (ILC), Centre de Coopération Internationale en Recherche Agronomique pour le Développement (CIRAD), Centre for Development and Environment (CDE), German Institute of Global and Area Studies (GIGA) and Deutsche Gesellschaft für Internationale Zusammenarbeit (GIZ), www.landmatrix.org/en/, accessed 27 January, 2018.

Turton, D. (1995) *Pastoral Livelihoods in Danger: Cattle Disease, Drought, and Wildlife Conservation in Mursiland, South-Western Ethiopia*. Oxford: Oxfam.

UNESCO. (2011) *Decision of UNESCO's World Heritage Committee at its 35th session (June 2011) on the Lake Turkana National Parks World Heritage Site (Kenya)*.

van Vliet, N., Mertz, O., Heinimann, A., Langanke, T., Pascual, U., Schmook, B., Adams, C., Schmidt-Vogt, D., Messerli, P., and Leisz, S. (2012) 'Trends, drivers and impacts of changes in swidden cultivation in tropical forest-agriculture frontiers: a global assessment', *Global Environmental Change*, vol 22, pp 418–429.

von Sury, A. (2016) *Ethnography of Land Deals: 'Dominion Has the Biggest Shamba Now': Local Perceptions of the Dominion Farms Project in Western Kenya*. Bern: Universität Bern, Institut für Sozialanthropologie.

White, B., Borras, Jr, S.M., Hall, R., Scoones, I., and Wolford, W. (2012) 'The new enclosures: critical perspectives on corporate land deals', *The Journal of Peasant Studies*, vol 39, pp 619–647.

Zaehringer, J.G., Wambugu, G., Kiteme, B., and Eckert, S. (2018) 'How do large-scale agricultural investments affect land use and the environment on the western slopes of Mount Kenya? Empirical evidence based on small-scale farmers' perceptions and remote sensing', *Journal of Environmental Management*, no 213, pp 79–89.

14 "They said they were bringing a development project"

'Best-practice' large-scale land acquisition or 'commons grabbing' in Ghana's Volta Region?

Kristina Lanz

Introduction

The world is experiencing an unprecedented land rush, as national elites and international investors are acquiring vast amounts of land, particularly in poorer, seemingly 'land abundant' countries (Deininger and Byerlee 2011; Anseeuw et al. 2012). As McMichael has argued, "the conjunction of food, energy and financial crises, has resulted in international capital markets gravitating towards agriculture as a relatively safe investment haven for the relatively long-term" (2012: 690).

Sparked by a critical publication of the NGO *Genetic Resources Action International* (GRAIN[1]) in 2008, a huge amount of academic and policy publications has dealt with the size of the current land rush, the motivations of the involved actors and the impacts these land deals have on local populations – a fact that Oya (2013) ironically called the 'land grab' literature 'rush'. Early studies on the impacts of specific large-scale land acquisitions (LSLAs) were criticized for using inadequate (mostly short-term) research methodologies and for overemphasizing the negative effects on subjectively perceived homogenous 'local populations' (Cotula 2013; Oya 2013; Scoones et al. 2013). A number of more recent case studies have thus highlighted how LSLAs create both winners and losers, and how the effects of LSLAs are shaped by various factors, such as class, gender, age or migration background (Boamah and Overå 2015; Hall et al. 2015; Bottazzi et al. 2016; Nyantaki Frimpong and Bezner-Kerr 2017). Several authors have also emphasized the prominent role of state actors, customary authorities and other elites in reaping most of the benefits from large-scale land investments (Fairbarn 2013; Haller 2013; Wolford et al. 2013; Lanz et al. 2018) and in legitimizing the land deals over affected land users (Levien 2013).

Analyzing the role of the state, Levien argues that states usually try to justify dispossession with "an ideological (and legal) claim to be serving the 'common good' or a 'public purpose' – typically cloaked in the language of 'development'" (2013: 402). The labelling of vast tracts of land under common property regimes as 'unused' can also be seen as a means to legitimize LSLAs (Alden-Wily 2011;

Cotula 2013). A prominent World Bank study found that in Sub-Saharan Africa, huge tracts of 'unused' land are available, resulting in a 'yield gap', which can be bridged by converting the land to commercial agriculture (Deininger and Byerlee 2011). This assessment has been questioned by various authors, who have accentuated that "the real issue is not that land is not used, but that the ways in which local people use the land may be treated as unproductive and backward" (Cotula 2013: 86). In fact, various African case studies collected by Hall et al. (2015) outline the importance of pasture, water and other common-pool resources to local livelihoods, highlighting that "livelihood strategies involve multiple activities dependent on social and economic relationships and ecosystems and that, once disrupted by land deals, cannot be revived and are seldom if ever adequately compensated for" (ibid.: 15). Various authors have furthermore shown that marginalized groups, such as pastoralists, landless people and women, mostly rely on common-pool resources for their livelihoods and that these resource users are often left out of compensation measures in the case of LSLAs (Daley and Pallas 2014; Tsikata and Yaro 2014; Marfurt et al. 2016). In this broad array of literature no detailed studies exist to analyse the trade-off between CPRs and new resources created by investments, which are supposed to benefit the whole community, i.e. infrastructural developments or other corporate social responsibility measures. Therefore, my case study focuses on a 'best-practice' large-scale land acquisition in Ghana's Volta Region, and asks: How are benefits and losses regarding the enclosure of commons and the creation of new communal resources distributed and can these newly created resources ultimately offset the loss of common-pool resources (CPRs)?

In order to analyse this question, I use a new institutionalism in social anthropology framework, which I will outline below. The article then proceeds with a description of the methodology used for this case study. Before outlining the process of privatization of the commons and the accompanying loss of various CPRs instituted by the LSLA, the research results section shows how the commons were transformed even prior to the large-scale land acquisition. I will then discuss the benefits of the new communal resources that were created by the company and highlight the perceptions of various local actors with regards to this 'local development'. The conclusion discusses how the process of institutional change from common to private property is structured by local power relations and can also be seen as a process of 'resilience grabbing' of the most vulnerable members of society.

Theoretical perspective – institutional change from common to private property

Common property regimes are often confused with open access regimes, where no rules or regulations exist to protect resources from over-exploitation. As a consequence, it is argued that land that is not managed by the state or held privately is bound to be overexploited by rational individuals seeking to maximize their own gain, thus leading to a 'tragedy of the commons' (Hardin

1968). Various authors have however highlighted that these types of open access situations are very rare and that most so-called 'open access' resources belong to specified local groups (villages, lineages or kinship groups) and that there are indeed rules and regulations governing the use of these resources (Acheson 1990; McCabe 1990). Under common property, the same plot of land is often used for different purposes by different user groups, giving rise to the notion that these groups have different bundles of rights (see Meinzen-Dick and Mwangi 2009). Who has which rights and which decision-making power in a given common property system is based on gender, ethnicity, religion or wealth and thus also on power relations (Agrawal 2003).

While Elinor Ostrom's seminal work in particular has shown how these locally devised rules and regulations are in many cases more appropriate and effective at protecting resources from overuse than state or private property rules and how they indeed contribute to local sustainability and resilience (Ostrom, 1990; Becker und Ostrom 1995), the 'tragedy of the commons' is still evoked by advocates of privatization. Private property is also seen as a means of cutting transaction costs, as the use of the land is decided by the owner without need for complex negotiations amongst different resource users (Acheson 1989). This is also highly relevant in the current global context of LSLA, where large tracts of land previously held under common property are effectively privatized.

New institutionalism in social anthropology helps us to understand how the transformation from common to private property occurs (Ensminger 1996; Haller 2010, 2013). Ensminger's (1996) framework of institutional change highlights how rising prices of land locally lead to institutional change. Her framework is particularly useful to show how such change is shaped by the local political system, whereby heterogeneous local actors with different bargaining power evoke different rules and regulations (institutions) and ideologies to protect their own interests in the context of institutional change. Unlike neoclassical economic understanding that privatization will always lead to the most efficient outcome, Ensminger's framework predicts that while "such change may be in the direction of increasing the economic benefit of the powerful parties who initiate the change, there is no reason to expect that these interests will be consistent with increasing economic output for society as a whole" (ibid.: 166).

Study design: a 'best-practice' large-scale land acquisition

To answer my research questions, I analysed a large-scale land acquisition for rice production in the Fievie Traditional Area in the South Tongu District of Ghana's Volta Region. Tongu is an Ewe word meaning "along the river", as the District is stretched out along both sides of the Volta River, making it ideal for water-intensive rice production (even though rice is not customarily produced or eaten in the area).

The suitability of the area was determined by the Global Agricultural Development Company (GADCO), a company made up of a multi-national team of investors (UK, Nigeria and India), which is registered in the

Netherlands and has received funding from various sources.[2] The investment has variously been described as a 'best-practice' example of a LSLA in international and national newspapers, as well as in a UNDP Report (see Darko Osei 2012; Wan 2013). The positive appraisals were mainly the result of GADCO's innovative 'community-private partnership' (CPP) agreement, which regulated that in exchange for the land, the company paid 2.5 per cent of their sales revenue into a local 'community development account'. The investors enclosed both individually used farmland, as well as communally used land – the largest parts having previously been in the latter category. GADCO thus prided itself to have put large tracts of, as the discourse goes, 'unused land' to productive use, neglecting that this land had been used extensively and contained various important CPRs. Despite its proclaimed commitment to sustainability and long-term engagement, GADCO went bankrupt shortly after the first field research was conducted in 2014. It was bought up by the Swiss company RMG Concept in 2015 and continues its operations still using the name of GADCO.

To analyse the LSLA, data was collected between March and July 2014 and between April and July 2016.[3] Through a first phase of participant observation, countless informal discussions and various transect walks, a general impression of the LSLA was gained and four villages were chosen for further investigation, based on the make-up of their respective populations and the different effects of the investment on these villages (see Figure 14.1).

Fievie-Dugame is a village of indigenes that hosts the seat of the chieftaincy; Kpodzi and Kpevikpo are two cattle-rearing communities made up of indigenes and settlers. These communities have been most affected by the loss of commons, specifically grazing land and water ponds. Bakpa Adzani was created after the Akosombo Dam was built in the 1960s, and two Bakpa communities from the North Tongu District were resettled to the Fievie Traditional Area. Approximately 100 hectares of farmland used by this community were enclosed.

Thirty semi-structured household interviews were conducted with differently affected people (outgrowers, employees, people who lost land) from these villages. Furthermore, 18 biographical interviews, representing differently affected women and men of different age groups and social status in the four focus villages, and 27 expert interviews with traditional leaders, governmental representatives at district and national level, as well as company representatives (including company managers) were carried out. The researcher also participated in various community meetings and events, and five gender-segregated focus group discussions were organized to explore contentious issues. All data (including observations captured in a research diary) was thematically coded, using a grounded theory approach, which included both inductive and deductive coding.

Some of the data gathered in the field could also be triangulated through various written documents, which were provided to the researcher.[4] Moreover, two community workshops were hosted at the end of each research period, which brought together women and men from the different communities, chiefs,

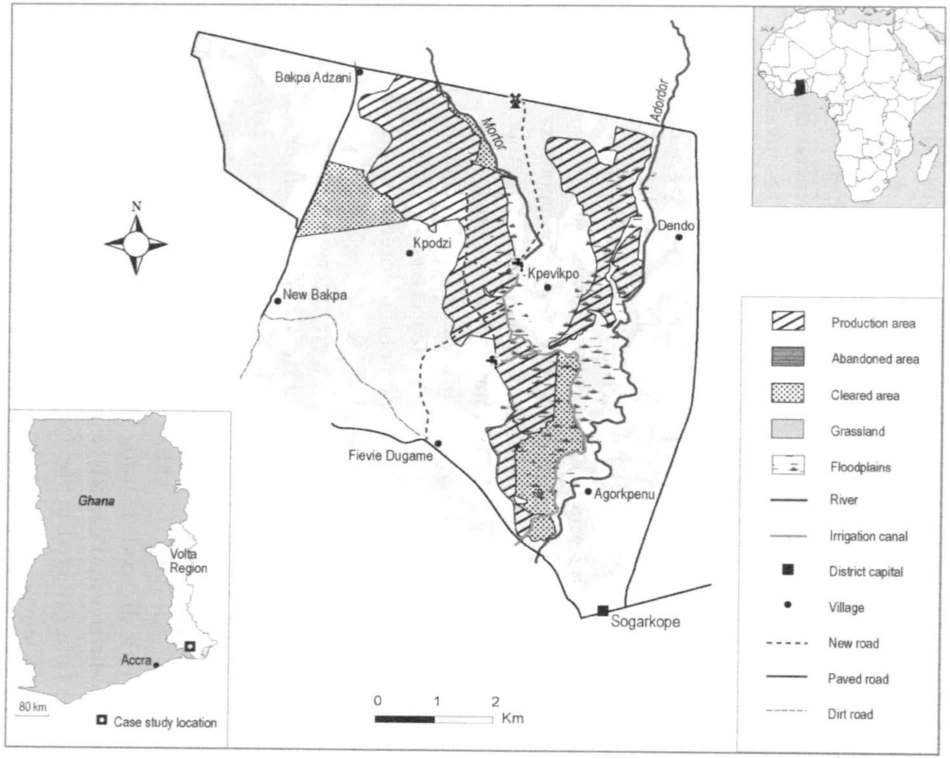

Figure 14.1 Study Location and GADCO Production Area in 2015.
Source: Map compilation and design by Ulla Gämperli Krauer (2018), CDE, University of Bern, Switzerland. Map base: Dominic Schuppli (2016). *The Impact of Large-Scale Land Acquisitions on Land Use and Local Actor's Access to Land. A Case Study of Southern Ghana*, pp. 66 and 74. GIUB, University of Bern, Switzerland.

government representatives and company representatives to discuss and validate the preliminary research results.

Changing livelihoods in the study area before the GADCO acquisition

Being riparian communities, people in the South Tongu District – including the study villages – used to depend on the Volta River and its various tributaries for their livelihoods. According to Tsikata, the inhabitants of the Lower Volta "had developed a successful system of livelihoods organised around the seasonal flooding of the Volta" (2012: 85), which included fishing and clam harvesting (the latter activity mainly carried out by women) as well as farming.

While the Lower Volta communities were known to be quite prosperous, especially as a result of the clam harvesting, their fortunes changed drastically with the building of the Akosombo Dam in the late 1960s. Resulting from the need to keep enough water upstream to run the turbines, the regular flow of the Volta River was impeded, which not only prevented fish from flowing downstream, but also led to the growth of various weeds and water hyacinth, making fishing very difficult. Clams were now buried deep in the sand and mud, making them unreachable to the female clam harvesters. Both the clam and fishing industries went into decline, throwing many households into poverty and leading to large scale out-migration (ibid., 2012).

The study communities were also vastly affected by this development. Bakpa Adzani for example was previously located in the North Tongu District and as a result of the construction of the Akosombo Dam was completely flooded in 1963. Together with another Bakpa community, the village was resettled to the South Tongu District. While various people decided to migrate North to continue with their fishing activities, others had to adapt to their new circumstances and focus their livelihoods on farming and, in the case of women, trading. The general effect of the building of the Akosombo Dam is well summarized by the following quote:

> We, the people of Tongu, were fishermen and when the dam was constructed it blocked the river and weed took over so we didn't get fish again. We the Ewe people got money from clams to build our houses, look after our children in school and everything but now the Akosombo made weed come on the water and we didn't get fish to catch and no clam to extract as well. Though it is good because it provides us with electricity, what we lost through it is greater than the electricity we got. (…) If it were not the illegal mining [galamsey] that came to exist, Tongu towns would have gone extinct because that is the work that the youths have resorted to. For the Akosombo Dam, it only came to destroy Tongu.
>
> (Expert Interview, Paramount Queen Mother, Fievie-Dugame, 1 May, 16)

In recent years, livelihoods have additionally been affected by situations probably related to climate change manifesting in instable weather conditions such as increasing droughts or large rainfall that have made farming more unreliable, thus putting growing strains on the once prosperous communities. All of the study communities still relied on the rain-fed flooding of the two creeks Mortor and Adordor running through the study area in order to collect fish from the various fishponds located in the floodplains of the creeks. The floodplains furthermore continued to offer other important CPRs (see also Haller ed. 2010 for other floodplain cases and commons), as will be outlined below.

The continued importance of common-pool resources under common property regimes

While many young men had migrated to other areas of Ghana to attempt to earn money in illegal gold mining, the majority of the population in the four study communities was still engaged in rain-fed small-scale agriculture, growing traditional subsistence crops, such as maize and cassava. Most households also produced cash crops, mainly pepper and groundnuts and some wealthier families operated medium-scale mango, cashew and woodlot plantations. Most farm plots were located near the settlements and depended solely on rainfall for irrigation. People living near the floodplains of the two creeks also used the seasonally flooded areas for agriculture during the dry season, as the greater soil moisture in these areas enabled farmers to cultivate crops with higher water requirements, such as okra and tomatoes. Figure 14.1 highlights land use and land cover in the study area prior to the arrival of GADCO.

The vast areas of grassland surrounding the settlements of Kpodzi and Kpevikpo were used for grazing cattle and had been reserved for this activity over a century ago. There were approximately 100 cattle owners within the Fievie Traditional Area, mostly owning between 50 and 300 cows. However, many of these cattle owners also looked after other people's livestock, which made some herd sizes exceed 1,000 cows. During the dry season, cattle were moved to the floodplains, where the greater soil moisture allowed grasses to sprout when there was no more fodder available on the general grazing areas. While the area was mainly reserved for cattle rearing from within the Fievie Traditional Area, good relations among cattle rearers from different Traditional Areas meant that in case of droughts access was also frequently granted to pastoralists from adjoining Traditional Areas.

Apart from pastoral and seasonal agriculture, the floodplains also provided many other important CPRs. The many scattered trees and bushes as well as small forests served as the main source of firewood to local communities. The collection and sale of firewood (and in some cases of charcoal made from tree stumps) was a major income-generating activity for many women and was also seen as a 'buffer' activity in times of distress, i.e. when quick money was needed to pay for hospital bills, buy food or provide for other emergencies, thus playing an important role for households' resilience. As one research partner explained:

> Sometimes when the harvest season doesn't come, it becomes very difficult for some few days but we usually manage to get some food and wait for the season. (…) Sometimes we have to go to the bush and cut down some trees to make charcoal to sell and get some money for food.
>
> (Elderly man, Bakpa Adzani, 18 June, 16)

In addition, Neem trees were used by local women for various medicinal purposes and in the two Bakpa village the fruits from the Baobab trees served them to make a local porridge, which was sold by the road-side. Other wild fruits, such as wild mangoes, as well as thatch for roofing and reeds used for weaving mats were collected in the floodplains and adjoining grassland.

The two creeks, the various water ponds and small dams provided drinking water to all four communities. While most of the ponds dried up during the dry season, the largest ones contained water the whole year round and were of crucial importance to pastoralists for the watering of cattle (Schuppli 2016). The numerous man-made fishponds in the floodplains also constituted an important resource to local livelihoods. Even though the ponds were owned by individual families, fish could be caught by anybody living in the area, when the creeks overflooded their banks. During the dry season, when the water receded, water was retained in the ponds and owners harvested fish. Most of the fish was dried and smoked after harvest and then kept for home consumption. Surplus fish was also sold to derive an income. The importance of fish to local diets cannot be overstated, as people ate fish with every single meal, and it was often mentioned that they would rather not eat than eat a meal without fish. As one interviewee said:

> Fish is important for us, because without fish we cannot eat.
>
> (Household Interview, fishpond owner, Fievie-Dugame, 3 March, 2015, *DS)

In some cases, the elders also ordered community members to harvest and sell some fish in order to pay for the hospital bills of a villager who had fallen ill. Thus, these fishponds also had an added social security function (focus group discussion, differently affected women, 4 May, 2014).

The various CPR were thus subjected to a complex common property regime. While cattle were moved into the floodplains during the dry season, several parts of the floodplains were also used privately during this time for seasonal farming activities and harvesting fish from the many man-made fishponds. During the rainy season, when the creeks overflooded, cattle were moved out and the whole area reverted to common property, as all members of the Fievie Traditional Area could now enter and catch fish in the floodplain areas. Fuel wood, on the other hand, was harvested by women from within the Traditional Area the whole year round, whereby the only rules were not to fell whole trees and not to use young trees.

However, despite their various uses and functions, the commons were threatened even prior to the arrival of GADCO; as a result of rising relative prices of land and increased control over land by customary authorities, the institutional context slowly shifted from common to private property, as will be highlighted below.

Institutional change in the Fievie Traditional Area prior to the GADCO investment

While Ghana is a decentralized democratic republic, customary authorities remain a powerful political force – mainly a result of their control over land, which was established during colonialism when boundaries of traditional areas were surveyed and recorded (Boni 2008). It is estimated that customary authorities control approximately 80 per cent of land in Ghana, with the remaining 20 per cent being public land (Kasanga and Kotey 2001). The system of chieftaincy was heavily formalized during colonialism (when Paramount Chiefs, Divisional and Sub-divisional Chiefs were established), and it continues to be heavily protected by the state (Lanz et al. 2018). The actual content of customary laws and land tenure regulations, however, varies across different regions in Ghana. In the Volta Region, where the case study area is located, chiefs are seen as the functional and political leaders of their communities with important roles in dispute settlement. However, their land management functions are not as far-reaching as in other areas, as most land is held by families or clans who are represented by their family/clan heads (Kasanga and Kotey 2001).

During colonial times, the system of chieftaincy was also formalized in the Fievie Traditional Area. At the time of research, the Fievie Traditional Area was headed by a Paramount Chief and his Paramount Stoolfather (in charge of enstooling all new chiefs in the traditional area), as well as by a Paramount Queen Mother. Most of the 57 villages that made up the Traditional Area were also headed by a village chief or village head (sub-divisional chiefs). These chiefs were in charge of organizing village life, settling minor land and other disputes within the village, as well as fostering general developments at the village level. Traditionally, customary authorities or clan heads could not give out land without consulting the relevant family heads.

In line with the general land tenure system in the Volta Region, all land in the Fievie Traditional Area was vested in the four clans and 'owned' by families within these clans. Family land was under the control of the male family heads (usually the oldest man in the family), who allocated land to individual family members and could make broader land management decisions. They could also provide any unused family land to other villagers as gifts, leases or for sharecropping. Individual family members held only use rights over their piece of land. If they stopped farming their land, it could be re-allocated to another person or be designed for a different use by the family head. Even though there were no obvious boundary demarcations, individuals generally knew which land belonged to which family. Women could access land either through the family head or upon marriage through their husbands, whereby many women lost access to their family land once they got married. Upon divorce or widowhood, their land rights were further threatened, as in many cases, their land was claimed by the husband's family. Due to their lack of secure land rights, many women – and especially widows – depended on CPRs, such as fuelwood, to sustain their livelihoods.

Large tracts of land in the Fievie Traditional Area were not under permanent cultivation and were used under a common property regime. Most of these commons had been allocated decades ago to settlers from within and outside the traditional area for the purpose of cattle rearing and various settlements had been established on them (including Kpodzi and Kpevikpo). Apart from cattle grazing, the commons were also used for various other purposes, as outlined above. Rules and regulations for the use of pasture and other CPRs (fishponds, firewood) were devised by the traditional authorities and infringements were reported to the traditional authorities and subjected to customary dispute resolution. While the settler communities living on these lands also used them for agricultural purposes, their land rights under customary land tenure were very vulnerable, since they could not trace their ancestry back to the four clans making up the traditional area (see also Stacey 2015; Lanz et al. 2018).

Rising relative prices of land and increased formalization of customary land tenure

The last decade has seen a considerable change in the structure of land ownership in the Fievie Traditional Area – a change directly related to rising relative prices of land and increased investor interest. As part of the Land Administration Project (LAP) – a national multi-donor funded initiative aimed at formalizing the customary land sector to enable investment – a Customary Land Secretariat (CLS) was established in the Fievie Traditional Area in 2008. Under LAP, control over land management, registration and dispute settlement is officially vested into CLSs, which are headed by the respective customary authorities. Land boundaries of traditional areas, as well as individual land rights within the traditional areas, are supposed to be mapped and registered at the Regional Lands Commission (RLC). While the Fievie customary authorities had registered the land boundaries of the Fievie State at the RLC, at the time of research, they had not registered and mapped family and individual land rights. Rather, it appeared that chiefs and clan heads were using the CLS to shift control away from the family heads towards a more centralized land tenure, as it is common in other areas of Ghana. Members of the CLS on various occasions mentioned that families were mere 'caretakers' of the land and that the allodial title to the land rested with themselves, who also had the power to divest large tracts of land to third parties and to make other major decisions regarding land use (Lanz et al. 2018).

All investments were now channelled through the CLS. According to some of the CLS members, several investors had already approached the secretariat for land since its establishment and lease contracts were generally negotiated by the Land Management Committee of the CLS, in theory in consultation with family heads, in practice however often without these consultations. This statutory intervention into the customary land tenure system thus considerably strengthened the bargaining power of chiefs to legitimately negotiate large-scale land acquisitions, including the GADCO deal, while at the same time weakening the land rights of land users, especially those with weak customary claims to land, such as settlers and people reliant on the commons (see Lanz et al. 2018).

Investment process and the enclosure of commons

In 2011, GADCO representatives approached the Paramount Chief of the Fievie Traditional Area for the acquisition of land to be converted into a rice farm. An agreement was negotiated whereby 1,000 hectares of land were leased to GADCO for 50 years. This lease was expanded in 2013 to cover 2,500 hectares. According to a profit-sharing agreement, the company was to pay 2.5 per cent of the sales revenue of rice and/or other cereals harvested and milled for the first five years. After five years, this amount would go up to 5 per cent. According to the investors, the money thus generated was paid into a 'community development fund', administered by a committee made up of chiefs and other prominent individuals. This committee was also to act as a connection between GADCO and the community. Both the GADCO manager and the customary authorities spoke of each other as business partners. According to the manager, the 2.5 per cent deal meant that the customary authorities, as business partners, would make sure to protect the investment. All 'community issues', such as organizing community consultation meetings, choosing outgrowers, dealing with compensation claims and resistance were also left to the customary authorities to deal with (Lanz et al. 2018).

The company initially started to establish its plantation in the Southern end of the research area, in the floodplains of the two creeks Mortor and Adordor, however the area was abandoned, as it was too difficult to cultivate. Continuous expansions saw the plantation moving upwards towards Kpodzi, Kpevikpo and Bakpa Adzani, and in the process, enclosing vast areas of communal land.[5] In order to irrigate the rice fields, the Mortor creek was redirected and the company dug a major canal following its former current to the Volta River. The course of the Adordor creek was also slightly altered and it was dredged and deepened to increase its velocity and prevent flooding of the company's fields (Schuppli 2016). Two pumping stations were built, as well as 10 smaller irrigation/drainage canals. Water was thus pumped from the Volta River to the highest elevated point and from there channeled through the smaller canals to the four different fields. In addition, dams were surrounding almost the entire GADCO plantation to prevent external water infiltration into the rice farm. Figure 14.1 highlights the changes brought about by GADCO.

In the process of transforming the commons into a rice plantation, more and more grazing land was made inaccessible to cattle rearers. In addition, the movement between the different grazing fields was curtailed by the rice fields and the various irrigation canals. Furthermore, many ponds, creeks and dams used for drinking water were destroyed, and some of the remaining water sources were reportedly poisoned by the run-off of fertilizers and pesticides. Kpodzi and Kpevikpo were completely cut off from their major water supply, as their main water pond – previously used for both human and animal consumption – was destroyed. While GADCO dug a new pond for the area, the villagers frequently mentioned that it was not well built, as it was small, muddy and the run-off from the farm entered the water, making it unusable. A walk through the area, confirmed these statements. Women, who were generally in charge of collecting water for the household, were mostly affected by the loss of water resources and had to find alternative means, as the following statement shows:

> They have also destroyed our source of drinking water. There was a pond that we used to fetch water from, it has been destroyed. They have dug some for us but it is not up to the standard that we can get potable water from, we are still fetching water from there to use and sometimes we drink it.
>
> (Elderly widow, Kpevikpo, 24 June, 2016)

The vast majority of the trees used to make firewood and charcoal for sale, as well as the Baobab trees, which grew near the Bakpa communities and whose fruits were used by Bakpa women to make local porridge, were uprooted and removed. The loss of these resources thus predominantly affected women, as the following quotes show:

> Women go to farm, they burn charcoal, and go to the market to sell charcoal and buy food stuff. The trees we used for the charcoal are those that have been cleared so we don't have much there to do now.
>
> (Middle-aged woman, Kpevikpo, 1 May, 2016)

> We were cutting woods to produce charcoal but now they have cleared all the places and now we have to struggle before we get something to eat.
>
> (Middle-aged woman, Bakpa Adzani, 21 May 2016)

Further resources that were largely lost because of the cutting down of trees and destruction of other plants included thatch and reeds. These resources were, however, mentioned much less frequently by respondents – an indication that they were less important to local livelihoods than other CPRs, such as water and firewood.

The company also destroyed many local fishponds. Additionally, the construction of the irrigation canal, as well as the dam along the creek Adordor, meant that fishponds outside the company's concession area were no longer flooded. This led to a general reduction in the availability of fish, which as mentioned before, was an essential part of local people's diets. Many villagers reported having to buy fish on the market as a result.

> We used to catch fish and get firewood from there. Our forefathers dug ponds in the water logs and fishes remain in them when the water dries up but all these things were destroyed and now we buy even firewood for cooking.
>
> (Elderly widow, Fievie-Dugame, 4 June, 2014)

While some people were compensated for the loss of farmland (Lanz et al. 2018), the loss of access to CPRs which resulted from the privatization of land was not compensated. However, the company and chiefs mentioned many general benefits that the investment was supposed to bring. This case study reveals that the much-praised improvements like employment creation and the outgrower scheme mainly favoured some selected individuals and their families (see Lanz et al. 2018 for a discussion of these benefits). The following section will look at those new resources created that were supposed to explicitly benefit the community as a whole – i.e. the community development fund, infrastructure and the increased availability of rice.

New 'communal' resources?

The investment is argued to provide several benefits which can be labelled as being new commons supposed to compensate the community as a whole for the loss of old commons. These consist of a community development fund, new infrastructure and increased availability of rice. In this section I explore who benefited from these resources and how these new commons are perceived by local people, elites and the company.

The community development fund

The GADCO 'community-private partnership' has received much positive media attention, as highlighted below in the Guardian:

> Unlike a lease for a set fee, this arrangement tied the fortunes of community and company together; greater profits for the farm would mean greater profits for the Fievie. All costs and expenditures were to be made publicly available, and the money paid to the community would be deposited into a special account to be used exclusively on local development projects.
>
> (Wan 2013)

Initial expert interviews with several chiefs and clan heads also conveyed the impression that the fund was used for various community development purposes, such as refurbishing a school house and building a library. According to one of the clan heads, the use of money from the account was generally discussed in community meetings. However, the majority of the researcher's respondents had no idea what the money had been used for and all respondents mentioned that the use of the money had never been made public.

> Many people are not happy because not many people have benefitted from the farm and the revenue being generated too is not benefitting anybody in the town. So many people are not happy for that fact.
>
> (Female outgrower representative, Fievie-Dugame, 1 May, 2016)

> I don't know what they are using the money for, because up till now we don't even have good drinking water in the town.
>
> (Young woman, Kpodzi, 20 May, 2014)

Many people were suspicious of the chiefs – several of whom were building new houses at the time of research. One of the sub-chiefs mentioned during a phone call:

> If I want to talk about what the chiefs are doing with that money, I have to call my Gods. Bad things are going on there and everybody just wants to chop.[6]
>
> (Research Diary, 13 June, 2014)

While it was obvious that only a very limited amount of the GADCO rent had been spent on community development purposes, both the former GADCO manager, as well as the new RMG Concept Manager, mentioned that they did not care what the money was used for.

> What they use the money for is none of our business. If you do business with me, it is not up to you to tell me to share the money with my wife. We have our business structures and they have their own. We don't meddle with their way of doing things.
>
> (GADCO Manager, Sogakope, 6 July, 2016)

Infrastructure creation

While the community development fund thus seemed to benefit mainly the chiefly elite, the infrastructure created by the company on the other hand primarily served the company's own interests, rather than those of community members. While GADCO's Environmental Impact Statement (GADCO EIS 2011: 31) claimed that "canals and water infrastructure developed will bring irrigation infrastructure to parts of the land that has never been irrigated. This offers the opportunity for local farmers to work with GADCO to support local farming activities through this shared infrastructure", both during the 2014 and the 2016 research, irrigation infrastructure was not shared, but exclusively used to water the company's fields. This discrepancy was all the more striking in 2016, as a drought had dried up many fields and shrivelled up plants could be seen right next to the lush irrigated rice fields of the company.

The dams, which were built to prevent flooding of the rice fields were a particular nuisance to local people, as during the rainy season their roads were frequently flooded, since the water was prevented from draining into the company's fields. In the case of Kpevikpo this was made worse by the fact that GADCO had built a canal right outside the village (see Figure 14.1). Even though the company had expanded the small footpath that previously linked Kpevikpo to other villages and had turned it into a rough road, thus making the village more easily accessible for cars, the canal heavily constrained accessibility during the rainy season, as well as during irrigation times. Whenever the company irrigated its fields or when there were heavy rains, people from Kpevikpo could only leave their village by wading through the chest-deep water of the canal. The bridge the company claimed to have built consisted simply of a few low-lying bricks inside the canal and did not improve the situation. The quote below encapsulates the sentiment of many villagers from Kpevikpo.

> How the road was before they came, we could move on it freely without any worries, but now the construction of the canal, with a little mistake, the water can carry you away so that is why we blame them. Our lives are now threatened. (…) I have not identified any positive benefit we derived from the coming of GADCO. They only destroyed our properties. The road, if it had rained, you would not be able to come here. We kept asking them to make a small bridge over the canal too, they refused to do it.
>
> (Middle-aged woman, Kpevikpo, 1 May, 2016)

Increased availability of rice

Finally, one much-praised aspect of the rice farm, was the fact that women were allowed to collect the leftover rice grains from the company's fields after every harvest. At the time of research, hundreds of women from many different villages could be seen on the rice fields gleaning rice. Most women, who engaged in this activity, used the rice both for home consumption, as well as to generate an income. Those who reported making an income from selling rice mentioned that they mainly used it to buy food for the family. The reported incomes from this activity ranged from 100 GHS to 400 GHS[7] per month, depending on the number of days and hours spent on the fields.

> My sister goes to glean rice anytime they harvest and when she gets time to go. It helps the small ones in the house, because now they don't suffer for rice and they also sell small, small. She can get about 200 GHS in one month because they gather the rice for several days.
>
> (Male electrician, Fievie-Dugame, 8 June, 2014)

All of the women interviewed agreed that the gleaning of rice had positive consequences for their household, especially in terms of availability of food. As a result of the practice of rice gleaning, the general availability of rice in the local area had greatly increased and rice had become more affordable to many families. A clear change in consumption patterns was visible, as many villagers reported eating less of their traditional food and eating more rice (which was previously only eaten on festive occasions).

> Sometimes we would not like to eat rice but we don't have the food we want, so we would just take the rice like that.
>
> (Male electrician, Fievie-Dugame, 8 June, 2014)

While certainly helping to offset part of the losses for many families, the gleaning of rice at the same time increased people's dependency on the company for their livelihoods and their food security. This dependency became very visible in the period between the bankruptcy of GADCO and the take-over of its successor RMG Concept, as the following quote illustrates:

> Since they [the company] have destroyed all these things and we have used the gleaning of the rice to compensate those losses, we were happy about it. But just that it did not keep long when they stopped farming on the field and now it has caused a lot of burdens on us again. At first, the gleaning of rice compared to the losses, it was somehow better but now that they have stopped, the burden is felt again. (…) They have destroyed everything and uprooted the trees with their stumps so the trees are not able to grow again. Even as they have stopped their operations it is not possible to get firewood or other resources from the land again.
>
> (Focus Group, rice pickers, Fievie-Dugame, 16 March, 2015, *ES)

Contested development

As has been discussed by Lanz et al. (2018), the investment – especially the loss of pasture and fishponds – created a lot of opposition in the affected communities, ranging from individual acts of sabotage (stealing, destroying machinery) to violence and organized forms of resistance. Especially cattle rearers, whose livelihoods were threatened by the enclosure of vast areas of pasture land engaged in pronounced resistance. At the time of the first research in 2014, the company was planning to expand its plantation by an additional 120 hectares, which would have made cattle rearing almost impossible. Community meetings relating to the expansion, many of which the researcher attended, were a platform during which grievances were voiced, while at the same time chiefs had to legitimate their actions in front of the community. Strong discourses of 'modernization' and 'development' clashed with claims based on discourses of 'tradition'.

Formally, the meetings had an aura of democracy where women and men could freely voice their concerns and common solutions were found to solve problems. But it quickly became clear that decisions had already been taken beforehand and that the expansion would go ahead anyway. Chiefs basically used these meetings to emphasize that the rice farm was in the interest of everybody–discourses relating to 'the greater common good' were frequently used. At the same time, the backwardness of previous activities carried out on the commons, such as cattle-rearing and the collection of firewood were underscored.

> So, before GADCO came, people were cutting firewood, but now they no longer do it. People were selling firewood. Back then if you wake up early in the morning and go to Fievie-Dugame, you won't like it, you see children carrying firewood to come and sell. But now that thing has stopped. (…) The cattle too, a whole year, you may not even sell 1 cow. They have that attachment to it, because they are cattle owners, but they are suffering. But let's say if everybody has a hectare of rice, in 5 years when you come, you will see that their life will transform tremendously.
>
> (Paramount Stoolfather, Sogakope, 24 May, 2014)

Thus, the discourse of the chiefs left little room to doubt that the commons were considered a thing of the past and that 'modernization' in the form of profit-oriented commercial agriculture was the way forward. While initially many villagers believed in the promises of 'development', from the beginning some vocally contested the common notion of development that was portrayed by the chiefs, as the following quote shows:

> I believe in destiny. Some people rear chicken to develop, some rear cattle to develop. Togbui (Chief) is now telling us that we should all farm rice to develop, but that is not our development. You should not suffocate us.
>
> (Elderly man at community meeting, Fievie-Dugame, 18 May, 2014)

The fact that the much-praised development was only beneficial for some people was also frequently emphasized by those most affected by the loss of CPRs:

> I don't know if I have even one cattle left. If your neighbour's beard is burning, you should go and fetch water to put it out. This is the 10th time that we meet on this project. The land we are talking about now is the major grazing field left for our animals, they go there from morning till evening. We have always gone there. The small portion that would be left, we can't even go there. And there is only one pond left, even we, Fievie-Dugame depend on that pond. (…) If we all go there to work and the project fails, we will be hungry. Only one person in another community is for this project, everyone else here is against it. Some people should not be crying, so that others will be laughing.
>
> (Elder at a community meeting, Fievie-Dugame, 18 May, 2014)

> We don't want people to claim Fievie is developing when some people are suffering.
>
> (Young man at a community meeting, Fievie-Dugame, 18 May, 2014)

It was clear that even though the discourses of the chiefs emphasized the benefits for all and the 'backwardness' of the previous uses of the commons, these had actually provided irreplaceable resources – especially to the most vulnerable members of society, who arguably became less resilient in the face of ongoing climatic changes, while the benefits of the rice farm were confined mainly to a few wealthy and influential members of society. The only new resource that was truly shared by the whole community was rice, which had become generally available as a result of the backbreaking work of gleaning of left-over rice from the company's fields. While the increased availability of rice was appreciated by most people, it led to clear change in consumption patterns, alongside a huge dependency on the company.

Conclusion

This chapter has described how livelihoods in the study area have been transformed several times in the past. Everyday life was strongly disrupted by the building of the Akosombo dam, as well as by climatic changes and increasing commercialization of land. In this changing context, the floodplains and adjoining pasture areas continued to offer important resources to local people's livelihoods, including fish, pasture and firewood. In particular, women, settlers and poorer community members relied on these resources. Prior to GADCO's arrival, these lands were managed under a common property regime, which was however threatened by the establishment of a Customary Land Secretariat in 2008. The chiefs strategically used this institution to increase their own control over the land. Recently, through the GADCO investment most of the commons were privatized.

The case study has shown that the people whose rights were already most vulnerable under customary law and who were most dependent on CPRs, i.e. women and members of settler communities, were losing out most from the company's operations, while the newly created 'communal' resources mainly benefited the elite. Furthermore, many people became dependent on the company to sustain their livelihoods through the gleaning of rice from the companies' fields. While their income from these activities in some cases was greater than the income generated from the land and resources they were using before, the legitimate fear that the company could leave at any moment made them vulnerable. This was very visible in the time between GADCO's bankruptcy and the takeover by RMG Concept when many women complained that there was no more rice available and they could not go back to cutting the trees for firewood as the trees were gone.

In line with Ensminger's framework of institutional change, I have shown that the shift from common to private property was filtered through the local political system – whereby powerful actors managed to benefit from the privatization of the commons, the least powerful actors have lost out the most, as their traditional access to the commons is considered 'backwards' and a thing of the past. It is furthermore likely that the loss of CPRs and the increased dependency on the company will reduce the resilience of those most vulnerable members of society in the face of possible future shocks. While the commons previously provided different resources, many of which could be used as fall-back livelihood options in times of personal problems or climate-related issues (i.e. the sale of fuelwood or fish harvested from the floodplains), these different resources have now been replaced by rice only –the production of which is heavily dependent on the company's continued operations. It is thus feasible to think of the company's operations not only in terms of 'land grabbing', but also in terms of 'resilience grabbing' (see Haller in this volume), as – in a volatile and changing livelihood context – livelihood options have been further reduced.

Notes

Research for this paper was financed by the Swiss National Science Foundation, Grant: 10001A_152773, by the Swiss Network for International Studies (SNIS) and by the Swiss Programme for Research on Global Issues for Development (R4D).

1 GRAIN (2008). *Seized! The 2008 land grab for food and financial security*. GRAIN Briefing. Barcelona: GRAIN.

2 Investors include the *Syngenta Foundation for Sustainable Agriculture* (SFSA), the *Alliance for a Green Revolution in Africa* (AGRA), the *Agricultural Development Company* (AgDevCo), *Acumen Fund, Africa Agriculture and Trade Investment Fund* (Aatif) and more recently the *Ghana Commercial Agriculture Project* (GCAP).

3 This article furthermore partly relies on qualitative, as well as remote sensing data collected by the two MA students, Eva Schober (interview quotes marked as *ES in the article) and Dominic Schuppli (all maps, interview quotes marked as *DS) from February to May 2015.

4 These included an Environmental Impact Statement from GADCO and an Environmental Management Plan 2015 to 2018 by RMG Ltd, both conducted as part of the state required Environmental Impact Assessment, various letters written by GADCO to the District Assembly, as well as a letter of complaint written by community members.

5 According to Schuppli's study on land use and land cover change in the affected communities, in 2015 GADCO had enclosed approximately 931 ha of previously communally used land (Schuppli (2016: 78).

6 "Chop" is a Ghanaian word, meaning to eat. However, it is frequently used with regards to corruption, i.e politicians are asking for "chop money", they just want to "chop".

7 100 GHS was equivalent to 31 USD (www.oanda.com).

References

Acheson, J. 1990. Management of common-property resources, in Platteau, J. (ed.), *Economic Anthropology*. Stanford: Stanford University Press.

Alden Wily, L. 2011. The law is to blame. The vulnerable status of common property rights in sub-Saharan Africa. *Development and Change* 42(3): 357–379.

Anseeuw, W. et al. 2012. *Transnational Land Deals for Agriculture in the Global South*. Bern/Montpellier/Hamburg: CDE/CIRAD/GIGA.

Agrawal, A. 2003. Sustainable governance of common-pool resources: context, methods, and politics. *Annual Review of Anthropology* 32: 243–262.

Becker, C.D. and Ostrom, E. 1995. Human ecology and resource sustainability: The importance of institutional diversity. *Annual Review of Ecology and Systematics* 26: 113–133.

Boamah, F. and Overå, R. 2015. Rethinking livelihood impacts of biofuel land deals in Ghana. *Development and Change* 47(1): 98–129.

Boni, S. 2008. Traditional ambiguities and authoritarian interpretations in Sefwi land disputes, in J.M. Ubink and K.S. Amanor (eds), *Contesting Land and Custom in Ghana: State, Chief and Citizen*. Leiden: Leiden University Press.

Bottazzi, P., Goguen, A., and Rist, S. 2016. Conflicts of customary land tenure in rural Africa: is large-scale land acquisitions a driver of 'institutional innovation'? *Journal of Peasant Studies* 43(5): 971–988.

Cotula, L. 2013. *The Great African Land Grab? Agricultural Investments and the Global Food System*. London: Zed Books.

Daley, E. and Pallas, S. 2014. Women and land deals in Africa and Asia. *Feminist Economics* 20(1): 1–24.

Darko Osei, R. 2012. *GADCO – A Holistic Approach to Tackling Low Agricultural Incomes*. UNDP Case Study. New York: United Nations Development Programme (UNDP).

Deininger, K. and Byerlee, D. 2011 *Rising Global Interest in Farmland: Can it Yield Sustainable and Equitable Benefits?* Washington, DC: World Bank.

Ensminger, J. 1996. *Making a Market. The Institutional Transformation of an African Society*. Cambridge: Cambridge University Press.

Fairbairn, M. 2013. Indirect dispossession: domestic power imbalances and foreign access to land in Mozambique. *Development and Change* 44(2): 335–356.

GADCO. 2011. *Draft Environmental Impact Statement*. GADCO (internal document).

GRAIN. 2008. *Seized! The 2008 Land Grab for Food and Financial Security. GRAIN Briefing*. Barcelona: GRAIN.

Hall, R., Scoones, I., and Tsikata, D. (eds) (2015). *Africa's Land Rush. Rural Livelihoods and Agrarian Change*. Woolbridge: James Currey.

Haller, T. (ed) 2010. *Disputing the Floodplains*. Brill: The Netherlands.

Haller, T. 2013. *The Contested Floodplains: Institutional Change of the Commons in the Kafue Flats, Zambia*. Lanham, MD: Lexington/Rowman & Littlefield.

Hardin, G. 1968. The tragedy of the commons. *Science* 162: 1243–1248.

Kasanga, R.K. and Kotey, N.A. 2001. *Land Management in Ghana: Building on Tradition and Modernity*. London: IIED.

Lanz, K., Gerber, J.D., and Haller, T. 2018. Land grabbing, the state and chiefs: the politics of extending commercial agriculture in Ghana. *Development and Change* 49(6): 1526–1552.

Levien, M. 2013. Regimes of dispossession: from steel towns to special economic zones. *Development and Change* 44(2): 381–407.

Marfurt, F., Käser, F., and S. Lustenberger 2016. Local perceptions and vertical perspectives of a large-scale land acquisition project in Northern Sierra Leone. *Homo Oeconomicus* 33(3): 261–279.

McCabe, T.J. 1990. Turkana Pastoralism: a case against the tragedy of the commons. *Human Ecology* 18(1): 81–103.

McMichael, P. 2012. The land grab and corporate food regime restructuring. *Journal of Peasant Studies* 39(3–4): 681–701.

Meinzen-Dick, R. and Mwangi, E. 2009. Cutting the web of interests. *Land Use Policy* 26: 36–43.

Nyantaki-Frimpong, H. and Bezner-Kerr, R. 2017. Land grabbing, social differentiation, intensified migration and food security in Northern Ghana. *Journal of Peasant Studies* 44(2): 421–444.

Ostrom, E. 1990. *Governing the Commons: The Evolution of Institutions for Collective Action*. Cambridge: Cambridge University Press.

Oya, C. 2013. Methodological reflections on 'land grab' databases and the 'land grab' literature 'rush'. *Journal of Peasant Studies* 40(3): 502–520.

Schuppli, D. 2016. *The Impact of Large-Scale Land Acquisitions on Land Use and Local Actor's Access to Land*. MA Thesis, University of Berne.

Scoones, I., et al. 2013. The politics of evidence: methodologies for understanding the global land rush. *Journal of Peasant Studies* 40(3): 469–483.

Stacey, P. 2015. Political structure and the limits of recognition and representation in Ghana. *Development and Change* 46(1): 25–47.

Tsikata, D. 2012. *Living in the Shadow of the Large Dam*. Accra: Woeli Publishing Services.

Tsikata, D. and Yaro, J.A. 2014. When a good business model is not enough. *Feminist Economics* 20(1): 202–226.

Wan, J. 2013. Ghanaian rice growers cultivate a food security solution. *The Guardian*. November 22, www.theguardian.com/global-development-professionals-network/2013/nov/22/ghanaian-rice-growers-cultivate-food-security-solution.

Wolford, W. et al. 2013. Governing global land deals: the role of the state in the rush for land. *Development and Change* 44(2): 189–210.

15 Grabbing the female commons

Large-scale land acquisitions for forest plantations and impacts on gender relations in Kilolo district, Iringa Region, Tanzania

Désirée Gmür

Introduction

Even though much has been written on the impacts of large-scale land acquisitions (LSLA) in Tanzania (Locher and Sulle 2013, 2014) and Africa in general (Cotula et al. 2009; Toulmin 2008), there is a notorious blindness regarding the impact on the commons and issues of resilience[1] regarding gender relations (see also Doss et al. 2014). Specifically, long-term social anthropological research on the gendered impacts of LSLAs is lacking. Scant literature deals with the broader institutional changes in gender relations due to LSLA and the strategies women adopt to cope with these changes. In this chapter, I discuss data from my research on large-scale land acquisitions for forest plantations made by the British-based investor New Forests Company (NFC) in the Kilolo district of the Iringa region. The chapter focuses on the impacts of the ways in which large-scale land acquisitions consolidate commonly owned land (i.e. create commons enclosures) that affects women's resilience differently than men. In particular, women's ability to fulfil their care work is hampered as they lose access to land and related common-pool resources (CPRs) (e.g. fruit trees, grasses, agricultural land) for which mostly only men are compensated. I argue that the commons enclosure through LSLAs mainly has negative impacts on women, involving increased workloads. It also increases wives' dependency on their husbands because land and related resources like fruit trees, previously an important source of cash controlled by women, have decreased or are no longer available and land is mostly controlled by elderly men. I further argue that resilience – in terms of food security[2] – is therefore negatively impacted, creating an imbalance between gender and generation based on this power relation.

In Tanzania, large-scale land acquisitions already began to take place in colonial times. However, they have mainly occurred since the mid-1980s following the political and economic shift to neo-liberalism after nearly 20 years of socialism, and especially in more recent times triggered by the so-called Triple-F-Crisis (food, finance and fuel) (Borras et al. 2010; de Schutter 2011). Influential players like the World Bank have long purported a win–win situation, as well as a green scenario for LSLAs (ibids). But on the other hand, the Tanzanian government has been actively attracting foreign investors under the current neo-liberal order. This is reflected institutionally in various laws, policies and initiatives as well as the establishment of a National Land Bank leading to an estimated area of around 200,000 to 1,000,000 hectares of leased land (Locher and Müller-Böker 2014). The more recent investment undertakings focus on food and forestry production and impact the use of former CPRs on the land, especially agricultural land, water, pasture and forests for subsistence and cash. This is done in the context of village common land that should legally protect common property based on the Village Land Act of 1999. But as I will show, formal legal institutions do not provide security for women. On the contrary, they can be shopped by men, the central government down to district government level as well as by the investor to grab women's rights to CPRs and thus reduce their resilience.

Theoretical and methodological background

This chapter uses a new institutionalism (NI) perspective in social anthropology (see Ensminger 1992; Haller ed. 2010). This approach is actor-oriented and examines how external effects shape the internal bargaining power of actors and gender ideologies shaping the institutional choice of, and distributional effects for, different actors. With regard to LSLA and gender this means discussing the context of external changes (LSLA) that raise land value and consequently influence bargaining power in gender relations, the ideologies that justify these relations, and the choice of gender-specific institutions related to resource governance. The NI perspective is about institutional transformations driven by change in the relative price of land due to the LSLA (changes in the value of land, rise of the relative price of land) that leads powerful actors to select among a plurality of institutions. These institutions can include rules and laws like land acts, property rights and regulations (see Ensminger 1992; Haller et al. 2010), as for example the 1999 Land Act, the 1999 Village Land Act, and the 2009 Water Resources Management Act, as well as local customary land rights, among others. The institutions employ ideologies through discourses and narratives that seek to legitimize land acquisitions in specific contexts. A new institutionalism perspective also looks at the impacts that the institution shopping process (Toulmin 2009; Haller 2010, 2013) has on people with less power in negotiation, which often are women. Additionally, it shows how new norms and institutions emerge and are selected by powerful actors. This process has implications for women's and men's previous access to CPRs. I combine this approach with theoretical insights on the governance of CPRs (see Ostrom 1990) that emphasize the characteristics of these resources and the way common property institutions are structured to manage land and related CPRs in a sustainable and robust way. By combining the two I not only show how institutions change but also how these CPRs were used and governed before the investment and are now being replaced by new forms of governance at the expense of women's access to vital resources. Furthermore, neo-Marxist approaches that focus on the household mode of production, the reproduction of the workforce, and gender relations of exploitation (see Claude Meillassoux 1981) indicate that exploitation of women is increasing under conditions of commons grabbing.

The data for this contribution was collected during several months of fieldwork from March–December 2015, April–December 2016 and February–October 2017. The scientific approach taken by this project is based on inter- and transdisciplinary cooperation between Social Anthropology, Human Geography and Gender Studies. I mainly used the following mixed research methods, and data was cross verified through data triangulation (Burke Johnson et al. 2007): empirical data collection from participant observation, followed by open and semi-structured interviews that were conducted with members of different interest groups as well as experts from the government and NGOs. Futhermore, biographies, narrative interviews and focus group discussions focusing on the life histories of local actors provided a basis for obtaining emic views on land use, the land investment process and its consequences for gender relations and institutional change (also regarding gender relations). Furthermore, value chain analysis and household questionnaires were used to obtain economic (household budgets, incomes, expenses) and demographic data.

National historical, socio-economic and political-legal background leading to institutional change before the investment

In pre-colonial times Tanzania was politically structured and organized into different chiefdoms comprised of diverse ethnic groups. Resource rights and responsibilities regarding land and related CPR like wildlife, water, forests, pastures and fisheries were in the hands of local groups, chiefdoms, ethnic groups, village groups, lineages, or kinship groups who held/owned and managed these resources under complex systems of rules and norms (institutions) within their territories, usually under a common property regime, which was based on the connectedness of different resources (Haller ed. 2010). Accessibility to these resources depended on the season and could vary from year to year based on an adaptation to the changing environment (ibid.). The chiefs, heads of the lineage, the clan or ethnic group were empowered to allocate community land (Rwegasira 2012). Usually, women had access to land and related CPRs like forests, fisheries, water, wildllife and non-timber forest products based on belonging to these groups through kinship ties, marital status and personal relations, as well as gendered roles and responsibilites in reproduction and production (Bryceson 1995; Suneri 1997; Benschop 2002).

In 1895, Tanzania was colonized by the Germans and later, after World War I, by the British (USAID 2011; Rwegasira 2012). The British introduced an export-oriented, cash-crop-based agricultural economy. Thus, during this time the management of land and related CPRs became increasingly controlled by the colonial state (Nelson 2007; Haller ed. 2010; Rwegasira 2012). Under the German administration all land in German East Africa became un-owned Crown land vested in the German Empire (Isinika and Mutabazi 2010; LNRTAP 2010; Rwegasira 2012). Tanzanians only had land use rights for land they occupied (ibid.). Under British rule title was still vested in the State (Isinika and Mutabazi 2010; Rwegasira 2012). Tanzanians were provided with deemed rights (no documentary evidence) of occupancy with land held in trust by The Governor (ibid.). Foreigners were provided with granted rights of occupancy (documentary evidence) that were superior to the customary land rights of the Tanzanians (ibid.). In 1928, the Land Ordinance was amended to formally recognize customary law (ibid.). Despite this, a dualistic system of land governance emerged with rights granted by the state still being functionally superior to customary rights in land (ibid.). Colonial rule and indirect rule affected women's access to, and control over, CPRs as it lowered their negotiation power as well as fostered men's authority over women (Bryceson 1995). While several anthropological studies argue that customary land tenure provided some significant interests to women, these started to erode under colonial processes (Tsikata 2003, 2004). The identity of men also changed as they were seen to be responsible for tax payment and thus the notion of the male as "breadwinner", and the female as dependent housewife was reinforced (Bryceson 1995). The cash men earned largely lowered womens' bargaining power in the household, as well as men's respect for women's work for the household as women did not have the same possibilities as men to earn cash (ibid.). At the same time, the out-migration of men also upset established gender spheres of authority and power in the community and within households because women had assumed greater control over fields that previously were the domain of men (Sunseri 1997). Therefore, both colonial government and male elders had a strong interest in circumscribing women's occupational choices and mobility, i.e. in controlling women, also administratively (Bryceson 1995).

Tanzania became independent in 1961 and The United Republic of Tanzania was established in 1964 (USAID 2011). At that time, a kind of African Socialism (Ujamaa) was installed (Mbeyale 2010). Two thirds of the rural population from different "ethnic" and local origins were mixed up and (re)settled in uniformly structured villages (Abdallah et al. 2014). The aim of this project, Operation Vijiji, or Villagization, was to modernize agriculture, and thus to increase agricultural productivity, as well as access to education, water and health services for everybody (LNRTAP 2010; Abdallah et al. 2014). Communal farms were cultivated by groups of village families (Havnevik and Isinika 2010). Traditional leadership was abolished, and instead new village governments were introduced (ibid.). However, the government executed a resource management similar to colonial times (Haller ed. 2010; Mbeyale 2010). Common property institutional regimes to manage CPRs were "legislated out of existence" and transformed by the government to conform to socialist policies and ideologies (ibid.). All resources of Tanzania, including land, were declared as belonging to all Tanzanians; land was redistributed, and pre-independence land rights were abolished (Tsikata 2003; Haller ed. 2010; Mbeyale 2010). Common property resources were managed and controlled as state property by centralized state management agencies (ibid.). In practice, this fostered boundaries of cross-cutting ecosystems which built the basis of indigenous institutions (Haller ed. 2010). It further diminished resource access for local people. The village government, i.e. village councils, received political (not legal) power over the allocation of land and the management of related resources within the village boundaries (Isinika and Mutabazi 2010; LNRTAP 2010). On the pretext of equality, customary tenure systems were generally disregarded (ibid.), and thus, also the rights women and men held under these regimes. Women did not have equal access to land, because normally the head of a family, which mostly was a man, was allocated one piece of land for the family (Donner-Reichle 1988).

In the mid 1970s, Tanzania faced an economic, political and social crisis, and in the mid 1980s the government signed economic stabilization agreements and reform programmes with the IMF and the World Bank with the aim of reducing the role of the state in the economy and opening up more space for the private sector (Abdallah et al. 2014). However, post-socialist governments have maintained ownership and control of most natural resources. Encouragement of private investment and property rights, including promotion of foreign investment, began in the era of President Mwinyi (1985–1995). This led to land grabbing by political elites in mainly rural areas, such as lands valued for wildlife tourism or agriculture (see Brockington and Igoe 1999). From the latter half of the 1990s, the Tanzanian state began preparations for improving the investment climate, including the adaptation of the institutional and legal frameworks and in 1997 the Tanzanian Investment Act was ratified. It established the Tanzania Investment Centre (TIC) that has a mandate to grant derivative land rights to foreign investors, to provide a Certificate of Incentive, which is a basic document for land processing and land allocation, and to establish a land bank (URT 1997). A new land policy was presented to, and accepted by, the Parliament in 1995 based on which two new land laws were enacted in 1999, namely the Land Act and the Village Land Act (URT 1999a, 1999b), which provide for the formal procedures for the land application.

These land laws, which actually present a new legal pluralistic setting, divide land into three categories: reserved, village and general land (Abdallah et al. 2014).[3] The Village Land Act respects customary land rights and legally empowers village governments, i.e. village councils, to manage village land under customary law on behalf of the village assemblies (URT 1999b). Thus, both customary and statutory land tenure systems are officially recognized. However, despite the villagers' possibility to have de facto ownership of land (they own rights over land, i.e. rights to occupy and use land, which may be bought or sold and inherited) they can be described more as landholders or managers than as common or private land owners, due to the fact that legally all land in Tanzania is public land vested in the president, who owns it on behalf of the whole nation and has final decision power (URT 1999a; Locher and Müller-Böker 2014). This limits women's access to land and land related resources and provides a huge challenge when the relative price of land changes because men – as well as government actors – are trying to sell land as it is getting interesting for investors. The idea to accommodate foreign investors and thus increase the price of land locally was further emphasized by the next president, Jakaya Kikwete, in 2005. He underlined the need to accommodate foreign investors. The Kilimo Kwanza (Agriculture First, Transforming Agriculture) Initiative, launched in 2009, promotes large-scale commercialization bringing legal and institutional change in order to prepare the ground for foreign investment in large-scale agricultural ventures, emphasizing, for example, the need to amend the Village Land Act of 1999 (Abdallah et al. 2014; Ngaiza 2012). The current President Magufuli, who is in power since October 2015, has on first sight adopted a harsh rhetoric against foreign investors, threatening license revocations where projects have failed to generate revenue for the government by taxation of investments. However, once investors pay their taxes, the environment is still investor friendly.

The case study area: local ethnographic, economic and ecological background before the investment

The forestry plantation of the New Forests Company (NFC) is located in the Kilolo district, Iringa region in the Southern Highlands of Tanzania. It was installed on land that NFC acquired in the nine villages of Kidabaga, Idete, Isele, Kising'a, Ukwega, Ipalamwa, Magome, Kiwalamo and Makungu, which are located within the three wards of Dabaga, Idete and Ukwega. These villages lie in highland zone (with an altitude that ranges between 1600–2700 metres above sea level) with a moderate climate and mean temperature of 15°C (URT 2007; Locher 2011). Rainfall ranges between 1000–1600mm per year (URT 2007). However, long-term climate records for the Southern Highlands of Tanzania show that the climate of the region has been changing with a steady increase of temperatures over the last 40–50 years, a higher frequency of droughts during the last few decades as well as a large inter-annual variability in rainfall (Kangalawe 2012; Msalilwa 2013).

Most of the people in the area belong to the Wahehe, which perceive themselves as indigenous to the region, and to the immigrated Wabena ethnic groups, who originate from Njombe region (Odgaard 2002; URT 2007; Locher and Müller-Böker 2014). Both groups are patrilineally and patrilocally organized and mostly Christian (ibid.), but polygamy is common in the area (DANIDA 1999). However, due to high immigration rates, the population in Iringa generally is now ethnically very mixed (Odgaard 2002) and is home to some minorities of other ethnic origins like Wakinga, Waudzungwa and Wasagara in the villages.

Small-scale agriculture is the most important activity in the area (Odgaard 2002), although there is also some livestock-keeping, mainly chicken and some pigs, goats, sheep, very few cows and some people keep fish in small fish ponds in the rivers down in the valleys. In 2011, Kilolo district was home to 54,400 agricultural households with an average of 4.5 people (URT 2011). Farm sizes are on average are 1.6 hectares per household (URT 2007). Kilolo is a very hilly district, cultivation is practised on valley bottom farms known as "vinyungu" (a Kihehe word for valley and/or garden) in the dry season, and on up-hill slope farms in the rainy season (DANIDA 1999; own research between March and

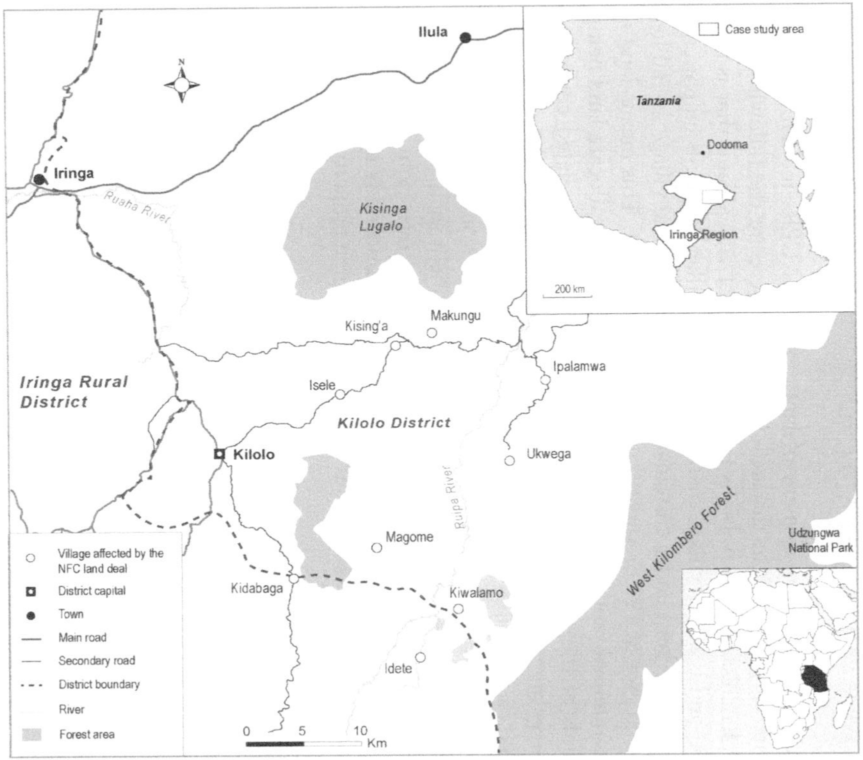

Figure 15.1 Map of the study area.

Source: Map compilation and design by Ulla Gämperli Krauer (2018), CDE, University of Bern, Bern, Switzerland. Geodata source: OSM data by OSMaxx (2018). Geometalab, HSR University of Applied Sciences Rapperswil, Switzerland. Further geoinformation provided by the author of this article.

July 2015). The *vinyungu* are the more productive areas as they are permeated by different sizes of streams and the fertile soils are not flushed out like those on the hills and are therefore preferred for agriculture (ibid., author's fieldnotes June 2015). Soils are red/yellow, well drained and highly weathered and leached clay soils (ibid.). Farmers practice crop rotation and shifting cultivation. They first slash, then burn and thus fertilize, and then till the plots. Agricultural work is done by household labour, normally involving both men and women. In between or on the agricultural plots there are often small areas with communal and private natural as well as planted forests with trees producing timber and non-timber forest products (NTFPs) such as fruits or leaves used as food or medicine but people also plant different kinds of fruit trees there. Some trees are left for water conservation purposes. In the context of climate change people are increasingly relying on NTFPs like wild mushrooms, wild fruits and wild vegetables for food, health and income security (see also DANIDA 1999; Msalilwa 2013), but also the planted fruit trees produce NTFPs. Many households are also involved in the planting of timber as savings. There is also temporary out-migration, especially involving young men looking for employment in forestry plantations in other regions.

Institutions for the management of land and other CPRs including tenure practices are very complex in the area. Generally, there are the following land tenure tendencies: legally, land falls in the category of village land that is held in common and is under village management with village councils empowered to manage village land under customary law on behalf of the village assemblies (Village Land Act of 1999 respects customary land rights). This means land is mostly a communal property of the clan, with the power to allocate land being in the hands of the clan head. Thus, much land and land-related common-pool resources are managed under elders and clan heads who distribute land use rights, some of them more communal but also some of them individual (land for agriculture). Among the Wahehe and the Wabena, women have the right to acquire land in their paternal home area, and to inherit such land from their fathers. They are therefore entitled to property in their own right and inherit such property from their fathers. Thus, people have considerably individualized rights but these are embedded in clan and lineage common property institutions.

New Forests Company's investment in Kilolo district and its impacts

Nowadays, the forestry sector plays an important role in Tanzania's economy contributing between 2.3 per cent and 10 per cent to the country's exports. It is estimated that the total gross area of forest plantations is around 250,000 hectares, with around 85,000 hectares in the hands of the government, 40,000 hectares privately owned, and between 80,000 and 140,000 hectares in out-grower schemes and woodlots. Pines (*Pinus patula*, *P. elliottii* and *P. caribaea*), cypress, eucalyptus and teak are the most important industrial plantation species. The leading export product is sawn timber (rough sawn), which is mostly (67 per cent) exported to Kenya to supply the high demand needed for construction in fast-growing cities. (Ngaga 2011.)

The New Forests Company (NFC) is UK-based and has six plantations and three sawmills in Uganda, Tanzania, Rwanda and Mozambique, and a head office in Johannesburg (NFC 2014a, 2014b). The NFC mixes commercial plantation forestry with protection and regeneration of indigenous tree species (ibid.). It produces wooden feed material from hardwood species of eucalyptus and pine trees for sawmills, board factories and pole treatment plants, as well as energy operations. It focuses on local and regional export markets (ibid.). In 2009, the NFC established a tree nursery with high-quality seedlings from South Africa and began operating the Lukosi plantation with pine and eucalyptus in Kilolo district. By early 2013, it had acquired land from seven villages. Wanting more connected pieces of land, the company is still looking for more land in neighbouring villages by asking new landholders in those villages where the company already had acquired land (Locher 2011; Locher and Müller-Böker 2014; NFC 2014a; author's fieldnotes May 2015). Since 2010, NFC has planted a total of 2,125,671 hectares of eucalyptus and pine trees (Kilolo District Council Strategic Plan) and has acquired more than 6,300 hectares of land (Locher 2011).

The following section demonstrates how the land was acquired in order to understand its impact on local women and men and their resilience capacities and position. It will also explain which institutions are used by the investor (institution shopping) and with which ideologies, discourses and narratives[4] (i.e. modernity, development, poverty) investments are legitimated. Also, of central importance is the quality of land taken from the people and the type of land and CPRs lost and by whom.

Since most of Tanzanian land is under village management, investors need access to village land for their investments. Land within village boundaries is managed by village authorities on behalf of the government and the Land Commissioner. According to the Tanzanian land laws of 1999, foreign investors can access village land only after it has been transferred into general land status, which is under the jurisdiction of the Ministry of Lands and Human Settlements Development. This process requires the village assembly's agreement, documented by meeting minutes. This has to be approved by the Land Commissioner on behalf of the President. Further, the involved parties have to agree on the compensation. Compensation should be based on market value, estimated by a government expert, and pay should include both the price of the land and unexhausted improvements, such as crops and/or trees. Finally, after several steps, the Ministry of Lands, on behalf of the President, enacts the land transfer. Once it has been transferred, it will not revert to village land again (URT 1999a, 1999b; Locher and Müller-Böker 2014). Under the 1999 Land Act, a foreign investor may occupy land through: 1) obtaining derivative rights from the Tanzania Investment Centre (TIC); 2) the rights of occupancy and derivative rights and leases ranging from periods of five to 99 years, which are renewable, but not beyond 99 years.

In the villages where I conducted research, namely Makungu and Ukwega, the New Forests Company acquired land on uphill as well as valley bottom areas. It belongs to a category of land I call "individual" but embedded in common property institutions since 2013. "Individual" land is land that an individual occupies as a member of a family or clan and is managed under customary law, mostly inherited land. However, it also involves land that individuals purchased or rented from other villagers, or that was allocated to them by the village council due to their right as Tanzanian citizens. In the villages I studied, many people have sold land, be it clan, family or individual land, to the investor since 2013 as NFC management was looking for more land after their first round approaching villages as a whole and targeting mainly village reserve land started in 2009. This means that the NFC has acquired granted land rights of occupancy for 99 years from villagers that usually have a Customary Right of Occupancy. However, none of them had acquired a certificate of Customary Right of Occupancy (CRO).

As outlined above, the legal pluralistic setting of the area regarding land and land-related resources contributed to the fact that the land deals are hampered by many conflicts. Different discourses and institutions are used by different actors to legitimize the transfer of land to the company. Depending on the perspective, position and interest of the involved actors in the land deal, the land acquisition can roughly be divided into "voluntary" and "involuntary" sale, in other words, loss of, or forced eviction from, the land. As the value of land was rising, stemming from the interest of the company and the land scarcity this created, more powerful actors, especially men as fathers and brothers, as well as the company itself, profited from their higher power to sell and buy land, which belonged to larger groups and to which women also had a customary entitlement. This has resulted in a series of conflicts.

In several cases, members of different clans are complaining that the clan or family head, sometimes in cooperation with the eldest son or just the sons, sold the land without involving the rest of the clan. Very often these deals left out female members in the sale decisions. In other cases, the sons decided on behalf of old fathers without consulting them. In addition, prior contexts, as well as Ujamaa, created tensions: using a discourse of "wasteland" or "unused land", many families and/or certain members thereof claim to not use land lost during Ujamaa times anymore because it is far away from their settlement area. As prices were low, this land became reserve land within a family or clan context. But this does not mean that clan members were not interested since many still claim to have used it or wanted to keep it for their children and/or future generations. Especially women have been among these clan members, who claimed a genuine interest in that lost land. Thus, when the investor came, land rights to land commonly owned before Ujamaa started to be contested, with different people, from different clans and within clans, claiming rights to the same land because they wanted to sell it. They often legitimate exclusion of the sisters, wives, daughters and mothers based on arguments like: it is the clan land of the husbands' clan, so the wives or sisters have no say in family land matters as they live with the families of their husbands anyway and now belong to the clan of the husband.

In several cases, brothers sold the land left behind by sisters when they moved to their husbands' place and left their land in the care of the brothers in their home areas. Brothers legitimated the sale by challenging, i.e. denying, the customary rule of Wahehe and Wabena daughters as having the right to inherit land from their fathers. These rights from fathers and/or parents that give women security, resilience and bargaining power in their marriage when they move to marriage are therefore increasingly undermined by the investment.

But most importantly for women are the cases of "fertile bottom valley land grabbing" (vinyungu land). Based on the company's interest in that land, it was largely the district government legitimizing evictions with the presence of the investor coming and planting trees. The government largely legitimized the acquisition by using the 1999 Land Act and the 1999 Village Land Act that actually exists to respect "customary land rights". It shopped on the institutional element of The 1999 Land Act that in the end all land in Tanzania is public land vested in the President having final decision power. In addition, the company itself was active in this process of institution shopping: making reference to, and activating, Article 34 of the 2009 Water Resources Management Act, the management of the NFC stipulated prohibition of human activities near water resources, which primarily includes vinyungu land. Even though this law has been enforced since 2009 the district did not enforce it up until the investor started being a powerful actor and good tax payer in the district. At the beginning, the NFC just bought upland areas but then started prohibiting villagers from using land in the vinyungus bordering the purchased upland areas. So technically/"legally" the land still belongs to the villagers as part of their customary rights but in practice the investor overtook it and did not even compensate the people as legally stipulated by the Water Resources Act, a policy backed by the district.

Discussion: impacts of the investment on local women and men

The land acquisition made by the NFC had several impacts regarding access to CPRs. Mostly, people lost access to a lot of agricultural land whose produce they used for subsistence and cash generation. In particular, the *vinyugunus* are the more productive areas where beans are usually planted. Beans are the primary cash crop that brings the most cash income for women, as the following citation illustrates:

> Income from the vinyungu is gone, this means … so the community depended on these vinyungus to get extra income to get money to educate our children but now we … and also that most of this lands in the vinyungus has been purchased and the investor hasn't refunded us, the money we used to purchase these vinyungus … even the present village government didn't help … yes the village government didn't help us in demanding our rights.
>
> (Female farmer, Ukwega, 5 October, 2015)

Many people depended on these vinyungus and their loss means the loss of possibilities to earn extra income to pay school fees, to buy extra food in cases of food scarcity when the harvests on subsistence plots are bad, and to buy fertilizers. Due to the loss of the vinyungus, food security is now low. This food shortage affects the entire family, but mainly the mothers, the wives, as it is mostly them who cut on meals when there are food shortages because they would rather starve than let their children eat less. Mothers often skip lunches or ration food. They especially starve between November and December. Usually they harvest maize in June which lasts them up until November, and then in November/December they normally harvest the beans in the vinyungus, sell part of it and buy maize because the next maize harvest is not until March. Thus, since for many people the income from the vinyungus is gone, they face food security problems during these months.

Generally, people say they don't have enough land anymore and they cannot produce enough food that will last them the whole seasonal cycle. Many people cope by renting land from others, but usually the new land is less in acreage, is not enough and is less productive land. Land rents have also increased due to the investment. People now pay around 30,000 to 40,000Tsh[5] per season per acre, whereas a few years back, it was half the price. Further, many had to seek land in other villages that are far away and this increases their workload. The large-scale land acquisition also led to lowering maize production in the villages, which has the effect of raising the price of maize within the villages, being especially a problem for poorer families with fewer acreages of land, as it is more expensive for them to buy maize. Some years ago, they used to buy 18–20kg of maize at 7,000Tsh. But it's now 10,000–13,000Tsh, which people, especially most women, cannot afford. As one farmer stated:

> We don't expect maize this year because even if those people from Iringa or Ilula bring maize, we don't have the capital from the beans to buy the maize because the beans were lost through drought in those vinyungus we still have, and because the other vinyungus are gone because of the investor.
>
> (Female farmer Makungu, 10 September 2015)

Women were especially also concerned of losing fruit trees like lime, avocado, banana, pear and Mikuti fruit, which is also a medicinal plant used against fever. Income from selling fruit often is controlled by women or at least used to buy things for the household and the children. Fruit are also important for nutrition, being a source of vitamins in an otherwise very much starch (mainly maize) based food system. The lost access to these trees increases wives' dependency on their husbands because the small money they had controlled by selling fruit is gone and they need to ask their husbands more for money when they want to buy goods for the household. Men generally are also more involved in other jobs like carpentry or seasonal work in other areas, including other forest plantations, for example, in Mufindi. Women, however, have more care responsibilities than men. Another loss is the access to two kind of grasses, namely a reed-like grass called *Mlulu (Sg.) (Milulu) (Pl.))*, which is used in making mats and baskets and is a bit thinner and *Malolo (Sg.) (Lilolp) (Pl.))*, which is thicker and mainly used for thatching roofs. Especially Milulu was a source of income for women, because it is mainly women who are making baskets and mats to sell.

These losses are hardly compensated for as the benefits proposed by the company (jobs, social services, schools, etc.) did not materialize or do not cover the losses of fertile land and related CPRs.

Conclusion

This land deal shows that impacts for men and women are different and also affect their resilience differently. In particular, women's ability to fulfil their care work is negatively affected as they lose access to land and related CPRs (e.g, fruit trees, *Milolo* grass) for which – if ever – mostly only men are compensated by the investor in the form of payments for the acquired land. Food security is therefore negatively impacted, creating inter-generational imbalances and imbalances between men and women. Especially mothers are affected as these losses mean lower food production and production of beans for cash generation (such as cultivation of beans and sale of CPRs and products made by women such as fruit and Milolo grass mats and baskets) by which access to food, health and education services could be bought before the investment.

The study reveals that this is due to an extreme case of women's resource grabbing for which no compensation is paid for valley bottom land and CPRs. The investment changes relative prices for land previously used by women, which leads fathers and brothers to sell off the land without consulting the clan or other family members. Especially women, wives, daughters and sisters, are not involved in the land deals. Legal pluralistic issues stemming from pre-Ujamaa times add to institution shopping of more powerful local – mostly male – actors. Furthermore, there is a lack of transparency in pay-out schemes and institution shopping on the part of the investor itself with the help of the district: land can be grabbed with no compensation payments making use of parts of the legal formal framework such as the Village Land Act and the Water Protection Act to legitimate land and commons grabbing in the very narrow sense of the term.

At the same time, jobs from the company are minimal and generally not available for local people and especially not for women. This leads to further negative impacts on women, which is predicted by neo-Marxist approaches (see Meillassoux 1981) involving increased workload. As access to food and cash generation options from the commons are reduced, women's time for reproductive work is increased as they have to travel further and work harder to compensate for food and cash losses – time that would be normally used for reproductive work. In addition, this LSLA also increases wives' dependency on their husbands as their resource base that gave them certain freedom in action has been reduced, leaving them completely reliant on the husbands' relatives' property, which is mostly controlled by elderly men. Furthermore, as predicted by the new institutional approach, land is getting scarce and increases in relative price due to the investment having triggered more and more land acquisition by domestic investors – on a smaller scale, but nevertheless adding up to land scarcity. This adds up to the marginalization of women who find it harder than ever to get access – even to less fertile land.

In conclusion, the institutional change in the area stems from several historical changes of the pre-colonial institutional setting, giving women rights to the clan commons in land and land-related resources to the colonial situation in which first losses were created. However, since independence creating legal pluralistic settings between so-called customary law and state laws and changes from socialism (ujamaa) to neo-liberalism (investment laws) as well as conservation legislation (protection of water and forests) lead to a situation in which the investment has created scarcity of land and commons grabbing constellations, which are legitimated by the more powerful on all levels: on the national and district level by the company and the district officers pursuing institution shopping to legitimate the grabbing process, down to the clan level by more powerful elders or influential men and even further down to the household level by men selling-off land and related CPRs. Women are mostly on the lowest scales of that process and have very little bargaining power to change that, primarily because their economic cash basis is removed from them and this has negative impacts on food security and resilience. It rests to be seen how women will react in the future to this challenge.

Notes

Research for this paper was financed by the Swiss National Science Foundation, Grant: 10001A_152773.

1 Resilience is the ability of a person and/or a household to restore basic livelihood capaciticies after a shock. Such capacities need to be available over time and remain high for the unit (household, community) to be resilient.

2 Regarding food security I adopt the definition of the FAO: "Food security exists when all people, at all times, have physical and economic access to sufficient safe and nutritious food that meets their dietary needs and food preferences for an active and healthy life." (FAO 1996). The FAO also speaks of four Dimensions of food security: Physical AVAILABILITY of food, Economic and physical ACCESS to food, Food UTILIZATION, and STABILITY of the other three dimensions over time (FAO 1996). During my research I also looked at the emic perspective of food security.

3 Village land can be further sub-divided into three categories: *Communal village land* is to be used for public purposes, such as schools and public markets or grazing areas (URT 1999b). *Individual land* is occupied or used by an individual or family or group of persons under customary law (ibid.). The third category is *spare land* for future communal or individual use and is sometimes referred to as reserve land (ibid.).

4 "Ideologies are considered here as worldviews that give major orientation and explain how the world is perceived. From a Marxist perspective, the term is of importance

in analysing the legitimacy of actions taken in both global and local contexts. Ideologies also embody discourses, referring to how meaning and orientation are produced in a coherent way in spoken or written language, as well as narratives, referring to how a specific situation is logically described and explained" (Haller and Galvin 2008: 14).

5 10,000Tsh are equivalent to around 4Euro.

References

Abdallah et al. (2014) Large-scale land acquisitions in Tanzania: a critical analysis of practices and dynamics. In: Kaag, M. and Zoomers, A. (eds), *The Global Land Grab. Beyond the Hype*. London/New York: Zed Books Ltd, pp 36–53.

Benschop, M. (2002) *Rights and Reality. Are Women's Equal Rights to Land, Housing and Property Implemented in East Africa?* United Nations Human Settlements Programme, UN-HABITAT.

Borras, M.S., McMichael, P., Scoones, I. (2010) The politics of biofuels, land and agrarian change: editors' introduction. *Journal of Peasant Studies*, 37(4): 575–592.

Brockington, D. and Igoe, J. (ed.) (1999) *Pastoral Land Tenure and Community Conservation: A Case Study from North-East Tanzania*. London: Pastoral Land Tenure Series No 11, IIED.

Bryceson, D. (1995) Gender relations in rural Tanzania: power politics or cultural consensus? In: Creighton, C. and Omari, C.K. (eds), *Gender, Family and Household in Tanzania*. Aldershot/Brookfield: Avebury, pp 37–69.

Burke Johnson, R. et al. (2007) Toward a definition of mixed methods research. *Journal of Mixed Methods Research*, 1(2): 112–133.

Cotula, L., Vermeulen, S., Leonard, R., and Keeley, J. (2009) *Land Grab or Development Opportunity? Agricultural Investment and International Land Deals in Africa*. www.ifad.org/pub/land/land_grab.pdf.

DANIDA (1999) MEMA, *Natural Woodland Management Project/Udzungwa Mountains Forest Management Project. Socio-economic Baseline Study*. Copenhagen: DANIDA.

De Schutter, O. (2011) How not to think of land grabbing: three critiques of large scale investments in farmland. *Journal of Peasant Studies*, 38(2): 249–280.

Donner-Reichle, C. (1988) Ujamaadörfer in Tanzania. Politik und Reaktionen der Bäuerinnen. Arbeiten aus dem Institut für Afrika-Kunde 58. Hamburg: Institut für Afrika-Kunde.

Doss, C. et al. (2014) Land, gender, and food security. *Feminist Economics*, 20(1): 1–23.

Ensminger, J. (1992) *Making a Market. The Institutional Transformation of an African Society*. Cambridge: Cambridge University Press.

Food and Agriculture Organization (FAO) (2010) *Global Forest Resources Assessment 2010. Main Report*. FAO Forestry Paper 163. Rome: FAO, www.fao.org/docrep/013/i1757e/i1757e.pdf (accessed 2 December, 2014).

Haller, T. (ed.) (2010) Disputing the floodplains. Institutional change and the politics of resource management in African wetlands. *African Social Studies Series*, **22**. Leiden/Boston: Brill.

Haller, T. (2013) *The Contested Floodplain*. Lanham: Lexington.

Haller, T., Gerber, J.D., Gmür, D., Ryser, S., and Lanz, K. (2013) *Large Scale Land Acquisitions and Gender in Africa*. SNF Research proposal. Bern: University of Bern.

Havnevik, K. and Isinika, A. (eds) (2010) *Tanzania in Transition. From Nyerere to Mkapa*. Dar es Salaam: Mkuki na Nyota.

Igoe, J. (2004) *Conservation and Globalization: A Study of National Parks and Indigenous Communities from East Africa to South Dakota*. Wadsworth, Belmont, CA: Young, J.A.

Isinika, A.C. and Muatabazi, K. (2010) Gender dimensions of land conflict. Examples from Njombe and Maswa Districts in Tanzania. In: Havnevik, K.J. and Isinika, A.C. (eds), *Tanzania in Transition: From Nyerere to Mkapa*. Dar es Salaam: Mkuki na Nyota Publishers Ltd, pp 131–158.

Kangalawe, R.Y.M. (2012) *Food security and health in the southern highlands of Tanzania: A multidisciplinary approach to evaluate the impact of climate change and other stress factors*. Full Length Research Paper, University of Dar es Salaam: Institute of Resource Assessment.

Land and Natural Resource Tenure in Africa Program (LNRTAP) (2010) *Focus on Land in Africa Brief*: Tanzania. Lesson 2 Village land.

Locher, M. (2011) *"How come that others are selling our land?" – Customary Land Rights, Rural Livelihoods and Foreign Land Acquisition in the Case of a UK-based Forestry Company in Tanzania*. Draft March 2011, Paper presented at the International Conference Global Land Grabbing, University of Sussex, 6–8 April, 2011.

Locher, M. and Müller-Böker, U. (2014) "Investors are good, if they follow the rules" – power relations and local perceptions in the case of two European forestry companies in Tanzania. *Geographica Helvetica*, 69: 249–258.

Locher, M. and Sulle, E. (2013) *Foreign Land Deals in Tanzania. An Update and a Critical View on the Challenges of Data (Re)production*. LDPI Working Paper 31, Bellville: The Land Deal Politics Initiative.

Locher, M. and Sulle, E. (2014) Challenges and methodolical flaws in reporting the global land rush: observations from Tanzania. *Journal of Peasant Studies*, 41(4): 569–592.

Mbeyale, G. (2010) From integrated slope management to fragemented use: common-pool resources, unstitutional change, and conflicts in Pangani River Basin, of Same district (Tanzania). In: Haller, T. (ed.), *Disputing the Floodplains. Institutional Change and the Politics of Resource Management in African Wetlands*. Leiden: Brill, pp 195–242.

Meillassoux, C. (1981) *Maiden, Meals and Money*. Cambridge. Cambridge University Press.

Msalilwa, U. (2013) *Role of non-timber forest products in climate change adaptation by forest dependen communities in Kilolo district, Iringa, Tanzania*. A dissertation in partial fulfilment of the requirements for the degree of Masters of Science in forestry of Sokoine Univeristy of Agriculture. Morogoro, Tanzania.

Nelson, F. (2007) *Emergent or illusory? Community wildlife management in Tanzania*. IIED, Drylands Issue Paper No. 146.

New Forests Company (NFC) (2014a) *Website*. www.newforests.net/ (accessed 10 October, 2014).

New Forests Company (NFC) (2014b) *Sustainability Report FY 14*. www.newforests.net/wp-content/uploads/pdf_docs/New%20Forests%20Co%20Sustainability%20Report_FY2014.pdf (accessed 10 January, 2015).

Ngaga, Y.M. (2011) *Forest Plantations and Woodlots in Tanzania*. African Forest Forum Working Paper Series Vol. 1, Issue 16, Nairobi: African Forest Forum.

Ngaiza, R.S. (2012) *Kilimo Kwanza*. Presentation at FAO-University of Nairobi –Regional Workshop on an integrated Policy Approach to Commercializing Smallholder Maiza Production at Norfolk hotel Nairobi Kenya. Revelian S. Ngaiza, Head Unit of Investment Policies and Private Sector, Development Department of Policy and Planning, Dar Es Salaam Tanzania. www.fao.org/fileadmin/templates/esa/Workshop_reports/Smallholders_2012/Presentations_1/Ngaiza_Kilimo_Kwanza_Tanzania.pdf (accessed 22 October, 2014).

Odgaard, R. (2002) Scrambling for land in Tanzania: processes of formalisation and legitimisation of land rights. *The European Journal of Development Research*, 14(2): 71–88.

Ostrom, E (1990) *Governing the Commons: The Evolution of Institutions for Collective Action.* Cambridge: Cambridge University Press.

Rwegasira, A. (2012) *Land as a Human Right. A History of Land Law and Practice in Tanzania.* Dar es Salaam: Mkuki na Nyota Publishers.

Sunseri, T. (1997) Famine and wild pigs: gender struggles and the outbreak of the Maji Maji War in Uzaramo (Tanzania). *The Journal of African History*, 38(2): 235–259.

Toulmin, C. (2008) Securing land and property rights in Sub-Sahara Africa: the role of local institutions. *Land Use Policy*, 26(2008): 10–19.

Tsikata, D. (2003) Securing Women's Interests within Land Tenure Reforms: Recent Debates in Tanzania. In: Razavi, S. (ed.), *Agrarian Change, Gender and Land Rights.* Oxford: Blackwell Publishing Ltd, pp 149–183.

Tsikata, D. (2004) Gender, land rights and inheritance. Securing women's land rights: Approaches, prospects and challenges. In: Quan, J. et al. (eds), *Land in Africa. Market asset or secure livelihood? Proceedings and summary of conclusions from the Land in Africa Conference.* London:Royal African Society, pp 89–97.

United Republic of Tanzania (URT) (1977) *The Constitution of the United Republic of Tanzania.* Dar es Salaam: URT.

United Republic of Tanzania (URT) (1997) *The Tanzania Investment Act of 1997.* Dar es Salaam: URT.

United Republic of Tanzania (URT) (1999a) *Land Act (No. 4 of 1999).* Dar es Salaam: URT.

United Republic of Tanzania (URT) (1999b) *Village Land Act (No. 5 of 1999).* Dar es Salaam: URT.

United Republic of Tanzania (URT) (2007) *Iringa Region – Socioeconomic profile.* 2nd edition. Ministry of Planning, Economy and Empowerment, the National Bureau of Statistics and the Iringa Regional Commissioner's Office.

United Republic of Tanzania (URT) (2011) *Tanzania Bread-Basket Transformation Project. Annex A: District Profiles.* Ministry of Agriculture, Food Security and Cooperatives.

USAID (2011) *USAID Country Profile: Property Rights and Resource Governance.*

16 Gendered impacts and coping strategies in the case of a Swiss bioenergy project in Sierra Leone

Franziska Marfurt

Introduction

The food, finance and fuel crisis of 2007–2008 triggered a sharp increase in land investment in the Global South, and in the process sparked the interest of journalists, academia and the wider public alike. Although research on different aspects of large-scale land acquisitions (LSLA) increased, data regarding concrete processes at the community level have remained sparse. This paper builds on ethnographic fieldwork and addresses the local impacts of a large bioenergy project in Sierra Leone. The investor is a Swiss-based bioenergy company that leased 57,000 hectares of land for the production and export of biofuel to the European market. The company's use of the biofuel discourse (renewable green energy) and development discourse, combined with compensation payments and mitigation measures, has stimulated the Food and Agriculture Organization to consider the case an example of best practice (UNFAO 2011).

However, the findings reveal gendered implications in terms of access to land and land-related common-pool resources (CPRs), as well as adverse effects on the livelihood strategies of women who have become marginalised through the formalisation of customary land rights. This formalisation has resulted in the privatisation of land and land-related CPRs in the hands of a few individuals, restricting access for other (cf. Toulmin 2008; Peters 2009; De Schutter 2011; Alden-Wily 2008). In line with the theoretical approaches of Political Ecology (Robbins 2012), Political Economy (Bernstein 2008) and New Institutionalism (Ensminger 2002; Haller 2013), this chapter argues that the loss of access to commons is not compensated by the company's compensation payments, job creation and development programmes. Consequently, it does not fulfil all of the demands of an ecologically and economically sustainable investment that would bring development. Rather, the loss of access to water has reduced livelihood options and undermined the resilience of different groups in this setting, especially women. Nonetheless, evidence demonstrates how women and other groups of affected people have proven capable of organising within a highly complex institutional setting. Through shopping around different pre-colonial, colonial and post-colonial institutions, as well as making alliances with various international civil society actors, they perform acts of resistance that influence and alter the intended outcomes of the bioenergy project and prevent further deterioration of their livelihoods. However, criminalisation of the resistance movement has had severe repercussions for the relationship between the company and the community, further restricting employment possibilities for community members.

The first part of the chapter comprises a brief introduction to the concepts of moral economy and political ecology, which will enable the reader to understand the institutional transformation and the (gendered) implications of different forms of access and resource control for local livelihoods in the northern region of Sierra Leone. Having presented the research questions and methodology, the discourses, narratives and alliances of the investor will be outlined. The impacts of the investment and the privatising effects of the formalisation of land rights on livelihoods will also be assessed in order to understand the formation of resistance.

Moral economy and political ecology

Pre-colonial socio-political institutions

In the year 2010, a Swiss-based bioenergy company signed a Memorandum of Understanding (MoU 2010) with the Sierra Leonean government for the lease of 57,000 hectares of land in the northern part of the country. In order to grasp the institutional transformation triggered by the investment project, we need to understand traditional (or pre-colonial) assumptions of moral economy and political ecology that inform local conceptualisations concerning access to wealth and land for the members of a community.

Rural livelihoods in the operational area were (and to a considerable extent remain) organised along the lines of age, gender and wealth. Indeed, elder people are generally more respected than their younger counterparts and tend to enjoy greater authority in decision-making processes within the community.[1] As regards gender, men are generally more actively involved in public and political decision-making processes than women, although it should be noted that gender and age are intersecting categories and the social position of a person is additionally influenced by his or her marital status. Wealth is a criterion that shapes acts of solidarity, not only within the (extended) family, but also in all interpersonal relations. Emically, wealth is understood as "the ability to accommodate people" (Jackson 2004: 146), hence means that provide comfort and security. Wealthy and successful people have the obligation to support their family and friends, and receive loyalty and respect in return. Scott (1985) theorised this informal institution, conceptualising it as "the moral economy of the peasant'. This model entails the moral obligation of solidarity on the one hand, and the right to subsistence on the other.[2] According to the resource economist Fafchamps (1992: 160), these patron-client relationships are "a formal way of organizing the compensation of wealthy individuals for their continued participation in the solidarity system". The wealthier (or more powerful) are obligated to help less wealthy (or less powerful) persons in times of hardship. Given that the wealthy have little to gain from such an arrangement, the less wealthy must reciprocate in certain other ways. Hence, they make small gifts, provide labour or pledge social or political loyalty to their patrons. The institution can be interpreted as mechanism of redistribution, in which everyone can claim his or her "rightful share" (Ferguson 2015). If local ideas regarding patron-client relationships are scaled up to the regional and national levels, they translate into a political system that is described as patrimonialism (Richards 1998). This has a range of policy implications because it affects rural welfare and behaviour with regards to food aid, prices, risk, technology and new institutions (Fafchamps 1992).

(Pre-)colonial land rights institutions

Prior to the implementation of the bioenergy project, local land rights patterns in the region were organised in the form of customary law. Family tenure and communal tenure determined rights and access to land through complex and interlinked pre-colonial and colonial customary institutions (Renner-Thomas 2010).

The upland, which is cultivated with different crops according to the shifting cultivation method of farming,[3] was formerly held under family tenure. Landowning families draw the legitimacy of property rights by claiming to be descendants of the founders (first-comers) of the community who cleared the bush and converted it into a field or a village. First-comers are deemed as having established a special relationship with the spirits of the land by offering sacrifices to guarantee its fertility and the well-being of the community. Subsequent immigrants, or strangers, were then allowed to use the land, build houses and bury their dead (Kuba and Lentz 2006 see also the comparison of six case studies in African floodplains in Haller 2010). Hence, the property rights of landowning families do not equate to exclusive ownership, but rather represent a combination of owner and stewardship with coordination functions as a means of redistributing access to the land and related CPRs in the form of common property. The first-comers are embedded in complex layers of secondary rights and claims for co-ownership as well as a rightful share of non-owning people (late-comers) such as women, migrants, pastoralists and individuals from other communities (cf. Haller 2007, 2010).

Under the indirect rule of the British Protectorate, the first-comers were transformed into landowners and the immigrants, strangers and women were labelled land-users (cf. Peters 2013). In spite of the colonial coining of the term, the pre-colonial institution of moral economy continued to determine the land and resource use. In this sense, landowning families have the obligation to grant access to non-landowning (late-comers) families that are secondary rights holders. A landowner's reputation is damaged if he refuses to give land and access to resources to a land-user (unless he wants to cultivate it himself) but he gains prestige by offering land to others.

Unlike the family-hold upland, the sacred bushes, common grazing lands, streams, *bolilands* (seasonally flooded lowland) and swampland around perennial water sources were under communal tenure and "belong to everyone, to all the landowners and all the land-users" (Personal communication of a landowner). Crops such as palm seeds from naturally grown palm trees were also a common good, even if they grew on land belonging to a landowner. Käser (2014) insightfully describes how the palm seeds constitute an important source of resilience for non-landowning poorer households. The same principle applies for deadwood that anyone can collect and use as firewood. Hence, even if a land-user was unable to attain a patch of land for farming, he or she could still benefit from the commons, which are "possibly the only capital asset of the poor" (Alden-Wily 2008: 46), and the right to a share of the local cultural landscape with all of its resources (cf. Haller 2013). In Worreh Yeama, the community where this study was conducted, the most important common-pool resource was a perennial water source that turned the surrounding land into a highly fertile swamp. All of the people of the community and the neighbouring villages had the right to obtain water for drinking from the source, and anyone who wanted to have a garden in the swampy area was allocated a plot. A *land-user* who occupied the role of the steward made sure to distribute the land evenly.

Gendered spaces in the rural landscape

In the customary land tenure system of this region, women are deprived of the right of land ownership due to lines of patrilineal inheritance. However, women enjoy access and user rights like other *land-users* and manage specific types of land and resources at specific times, from which they can derive personal benefit. Such gendered spaces and places in the rural landscape can constitute *in-between spaces*, such as the small garden plots adjacent to homes, and fallow land and CPRs where firewood, medicinal plants and wild food can be found. In order to identify these spaces, it is necessary to examine the products produced and controlled by women and men, respectively. This can vary seasonally, and access to land and associated resources may also vary according to the season, depending on the time of year that the crops grow. A patch of land may not benefit women when it is being used by men, but when it becomes fallow, women can harvest medicinal plants and wild foods (Rocheleau and Edmunds 1997).

In the region under study, gendered spaces are also connected to the traditional gendered division of labour in agriculture. Men generally perform more physically demanding tasks such as the burning, brushing and clearing of the land for cultivation. Women engage in tasks like planting, transplanting, weeding and post-harvest processing. While the harvest of the rice is primarily undertaken by men, vegetable planting is traditionally conducted by women. In the case of the commonly used swampland, men and women jointly plant rice in the swamp in the rainy season, but in the dry season, the swamp becomes a (non-exclusive) women's space due to the gendered division of labour. Women cultivate vegetables during the dry season and derive numerous advantages (see below).

Research questions and methodology

The research project "Ethnography of land deals" sought to examine the diverse ways in which land investment projects affect the livelihoods of heterogeneous groups of local people. The project comprised three different case studies in Kenya, India and Sierra Leone, which investigated local perceptions, consultation processes, impacts on land rights and livelihoods and the coping strategies of the people affected. Theoretically, it adopted the approaches of Political Ecology (Robbins 2012), Political Economy (Bernstein 2008) and New Institutionalism Theory (Ensminger 2002; Haller 2013). Beyond these common parameters, this research focused on gendered experiences and impacts and the emergence of local resistance. Based on literature concerning the potential exclusionary effects of the privatisation of customary land rights, which often accompanies development and investment projects (Toulmin 2008; Peters 2009; Deininger and Byerlee 2011; De Schutter 2011; Alden-Wily 2008), two hypotheses were developed:

1) Through misreading "customary" land rights institutions in the formalisation process, the company transforms customary landowners with pre-colonial obligations of sharing into exclusive landowners. This can exclude groups with mere secondary rights and institutionalises unequal gender patterns in land rights. This alteration to institutions and bargaining power has adverse effects on the distribution of the benefits of the land (Ensminger 2002), especially in terms of gender. With the formalisation of land rights, the investment project contributes to the situation of legal pluralism as formal institutions do not suspend but rather overlap (pre-) colonial institutions. The latter may be revitalised if it appears necessary to actors affected by the institutional change.
2) In a pluralistic institutional setting, women and land-users who are deprived of livelihoods shop around different institutions in order to cope with the transformed situation (Toulmin 2008; Haller 2010, 2013). This enables them to develop a collective action strategy to claim back land and prevent further deterioration of their livelihoods. The claims are legitimised by the use of ideologies based on discourses and narratives (for example, of being a member of the first-comers as "sons of the soil", or having access rights to the commons even as late-comers). However, investors and the state also use different ideologies to legitimise their investment.

In order to operationalise the research questions and test the hypotheses, three months of fieldwork were conducted in an affected community in the operational area of Addax Bioenergy Sierra Leone towards the end of 2013. A mixed-methods approach consisting of various qualitative interview techniques (Schlehe 2003), focus group discussions (Crang and Cook 2007) and participant observation (Hauser-Schäublin 2003) allowed for the gathering of data on emic perspectives of affected people. Data analysis was informed by the inductive grounded theory approach (Strübing 2008) and methods of coding and memoing (Emerson et al. 1995).

Discourses, narratives and alliances of the investor

The interplay of politics, economy and personal networks

Addax Bioenergy Sierra Leone (ABSL) is a Swiss-based company with a registered office in Geneva. In 2010, ABSL leased an area of 57,000 hectares for a 50-year period, with the possibility of a 25-year extension. The company cultivates the land with sugarcane and processes it into biofuel in an ethanol factory built in the operational area. The biofuel is taken to the port in Freetown and exported to Europe. The manufacturing of 90,000m^3 of ethanol creates a by-product called bagasse, which produces 32 MW of electricity per year. According to the project plan, 32 MW is to be fed into the national power grid and constitutes an important complement to the Bumbuna dam, which produces 50 MW of hydroelectric power per year.[4] The project justification promised that the approximately 25,000 people living in the project area will benefit from job opportunities and road infrastructure (AfDB n.d.; English & Sandström 2014). But why did the company choose to invest in Sierra Leone in general and in this specific area in particular?

When ABSL started to become interested in developing the large-scale biofuel project in 2007, the country had only recently escaped from a brutal, decade-long civil war: 70,000 people were killed, 2.6 million people were displaced, and the country's economy and infrastructure were completely devastated (Pham 2006). The agricultural sector was destroyed because people had fled their villages and farmers had abandoned their fields, afraid of being caught by rebels searching for food. When they returned after the war, they found their land overgrown, their tools broken and their seed stock looted (Bolten 2009). The country ranked absolute bottom of the United Nations Human Development Index[5] and the government was incapable of rebuilding the country. Consequently, Sierra Leone was flooded by international aid organisations, which provided relief to the suffering population. The government embraced a liberal peace agenda, promoted by international donors, fostering economic growth through the opening of markets (Millar 2015) and decentralisation politics (Fanthorpe 2005). Other pillars of the "sustainable peace agenda" include market liberalisation and additional measures that seek to enhance growth in gross domestic product (GDP). In 2007, the Sierra Leone Investment and Export Promotion Agency (SLIEPA) was founded with the aid of the World Bank, the International Trade Centre and the International Development of the United Kingdom (UNCTAD 2012). As advertised on SLIEPA's website, Sierra Leone is "Africa's new investment destination", providing "arable land in abundance". The government offers favourable conditions for investment by providing very cheap land (from US$5 per hectare per year), flexible labour regulations, free utilisation of water and resources (i.e. open access according to Hardin 1968), and very attractive tax exemptions (Oakland 2011; Bürgi 2015). This strategy has been successful, and several long-term, large-scale land leases were contracted between the Sierra Leonean government and investing companies, among them ABSL.

At around the same time, the European Union (EU) adopted the Renewable Energy Directive, obligating its member states to achieve a share of renewable energy sources of 20 per cent, including a 10 per cent target in the final consumption of transport energy by 2020 (predominantly but not exclusively from biofuel substitution) (European Commission 2009). The Renewable Energy Directive can be understood as a consequence of the biofuel discourse that emerged in the early 1980s and that depicts biofuels as a green, renewable alternative to gasoline. With the new directives and the absence of custom barriers, the EU created a stable market and a continuous demand for biofuels, a fact that substantially influenced the notion of ABSL. Sierra Leone has become considered an ideal location for the cultivation of sugarcane due to its biophysical factors, including high annual rainfall, hot temperatures, high insolation and availability of water for irrigation from the country's largest river during the dry season. Furthermore, its relatively short shipping distances to Europe (indeed, AOG built shipping facilities at the port in Freetown), the adequate Freetown-Makeni highway constructed by the EU after the civil war, favourable investment climate (which includes tax exemptions and very low prices for land) all represented important factors that saw Sierra Leone chosen as the host to a biofuel project (Lustenberger 2015). The necessary funding for the €258 million project (BusinessWire 2011) was largely obtained from various development finance institutions (DFIs),[6] which were brought on board through the amalgamation of the biofuel discourse (renewable green energy) and the development discourse in justifying the project (ebd.; cf. AfDB n.d.). ABSL complied with the different requirements of each DFI with regard to social and environmental standards,[7] "reflecting its ambition to become a benchmark in responsible investing" (English and Sandström 2014: 6). The FAO (2011) considers the bioenergy project an example of best practice.

Lustenberger (2015) has convincingly argued that in addition to the requirement of a marketable product, a suitable site of production and necessary funding, personal relations were of primary importance for the realisation of ABSL. These relations were established through the mother company, Addax and Oryx Group (AOG), a Geneva-based company that has specialised in oil and gas trading in Sub-Sahara Africa since 1987. AOG worked closely with the Sierra Leone National Petroleum Company (NP) and established strong ties to influential businessmen and high-ranking politicians in the country over the course of two decades. The interconnectedness of European businessmen, diplomats who are familiar with development politics and development finance institutions, and Sierra Leonean tradesmen and politicians born in the operational area and linked to local people through a patronage-clientele relationship, was crucial for the realisation of the large-scale agricultural investment project. The local political authorities ("sons of the soil") acted as brokers between the company and the government on the one hand and the company and the people in the affected communities on the other. People in Worreh Yeama recalled their local authorities coming to the community, introducing ABSL representatives as friends who were bringing development to the region. They promised jobs and compensation if the farmers gave their land to the company. Local people were tired of the hardship of war, poverty and arduous subsistence farming, and so welcomed the company and its offer. As one informant living in the affected community stated:

> It was very easy [for the company] to capture land because people were living abject poverty. When the investor says we will come tomorrow we would be saying: "No no don't come tomorrow, come today!" We were desperate to get this benefit.
>
> (Personal communication)

The impacts of the investment on land tenure and common-pool resources

The exclusionary privatising effects of "land reform"

Sierra Leonean Law (Land Act Cap 122) requires the rent for leased land to be divided among the landowners (50 per cent), Chiefdom Council (20 per cent), District Council (20 per cent) and government (10 per cent). On a voluntary basis, ABSL introduced an additional payment for the landowners via an Acknowledgement Agreement (AfDB, n.d.; English and Sandström 2014: 3). In order to guarantee fair payments and the correct calculation of the complex compensation mechanism, ABSL had to formalise the land so that everyone would know "which land belongs to whom". ABSL started mapping and titling the land and conducted what a member of the ABSL upper management proudly called "a land reform" (personal communication). The company started issuing titles to landowners as they asserted ownership through discourses of being the descendants of the first-comers and of "belonging to the land" (cf. Ribot and Peluso 2003; Berry 2008; Toulmin 2008). Consequently, the male household heads of the landowning families in every community were identified by the company and given land titles. Contrary to the discourse that had been attached to land titling mechanisms, which claimed that titling would protect the poor from dispossession, the formalisation process amounted to a mechanism of exclusion (cf. De Schutter 2011). It added an element of exclusivity to the benefit of the primary right holder whilst marginalising secondary right-holders such as women, migrants and male land-users (cf. Benjaminsen et al. 2008). In Worreh Yeama, women were unable to claim formal ownership via discourses of belonging and traditional ownership because their access to the land was indirect and dependent on a male relative. Hence, patterns of unequal rights to the land based on gender were reinforced through the titling process, and women were thus stripped of their customary user rights (cf. Lastarria-Cornhiel 1995; Peters 2010; Haller 2013).

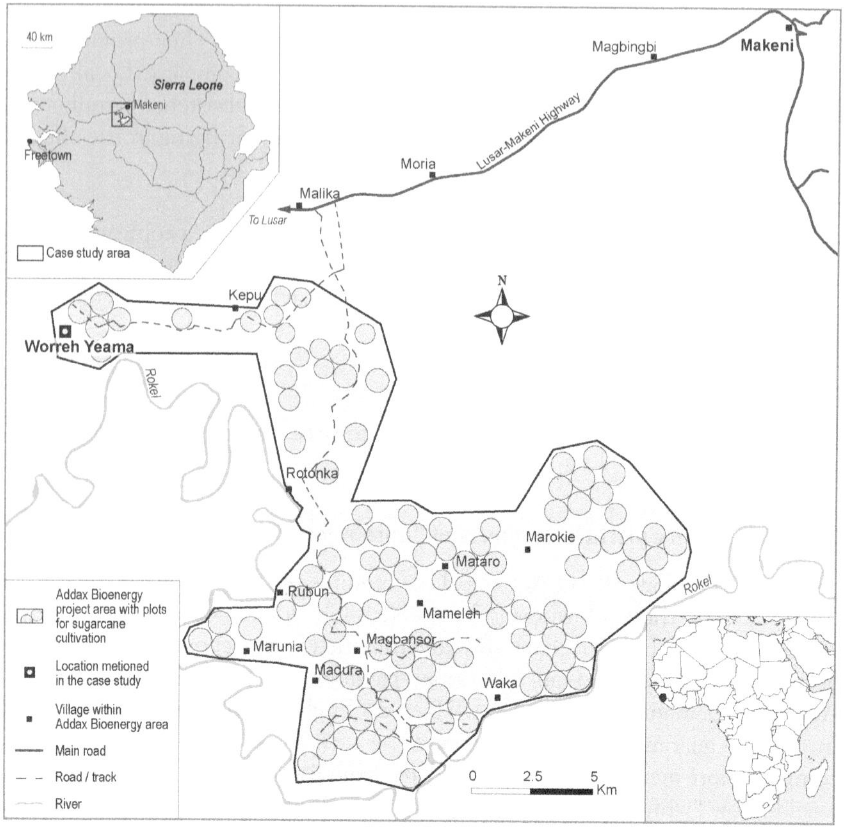

Figure 16.1 Map of case study.

Source: Map compilation and design by Ulla Gämperli Krauer (2018), CDE, University of Bern, Bern, Switzerland. Area of Addax Bioenergy satellite data sources: *ESA Copernicus Open Access Hub* (2018). European Space Agency/EO Ground Segment and Mission Operations Department. EO Common Services Section, Rome, Italy. Further geodata source: *OSM data by OSMaxx (2018)*. Geometalab, HSR University of Applied Sciences Rapperswil, Switzerland.

In the case of the ABSL project, the newly created institution of exclusive land ownership constituted the basis for the remainder of the project plan: only landowners were consulted and compensated for the land, and jobs were even "given to landowners preferentially" (personal communication by member of upper management). The women and other land-users were neither consulted nor compensated, even though the land used to be pivotal for their livelihoods too.

The Sierra Leone Network on the Right for Food (SiLNoRF) that from the beginning was monitoring developments in the ABSL case was aware of the exclusionary tendencies of the project. The NGO helped land-using people from different project-affected communities to create the Affected Land User Association (AFLUA). The purpose of the AFLUA was to provide land-users with information while fostering their capacity to discuss and organise as a group in order to articulate their interests vis-à-vis the landowners and the investor.

Gendered impacts of lost common-pool resources and access to wage labour

The main problem in the community of Worreh Yeama was the loss of the commonly owned perennial water source and surrounding swamp. The company's announcement to drain the swamp and use the land for a large sugarcane field sparked protests among landowners because the source was essential to many people's livelihoods, especially women. Indeed, the water source provided drinking water for the community and the surrounding villages and allowed women to plant vegetables during the dry season when agricultural products are particularly scarce due to the aridity. The vegetable harvest covered the subsistence needs of many households and contributed to the food security of the community. The selling of the surplus harvest at local markets generated cash income. Cash was bitterly needed in the context of low-income subsistence farming and enabled women to pay for their children's education and purchase household items for their chores. Women felt that this income had increased their bargaining power and their overall status in the community.

Many people recalled the meeting in which the community leaders informed the company of the importance of the water source and swamp, and explained why they wanted to retain the area for the community's use. The formal social manager reacted by promising mitigation measures in the form of two water wells in the village, as well as a dam where women would be able to pond water for vegetable cultivation during the dry season. Faced with these promises, the landowners agreed to lease the swampy area to ABSL. Having drained the swamp, the company built one of the water wells, but it failed to construct the dam. At the time of research, the company constantly denied the promise of a dam for vegetable plantation by claiming that dams did not constitute part of their mitigation measures. In an interview, a member of the upper management stated that he did not understand the complaints of the community because it had retained considerable land and a functioning well. His statement revealed his lack of understanding of the diverse benefits of water being a CPR. In his opinion, it was "only" water, and could easily be replaced with two wells in the village. However, he lacked awareness that the value of the land was due to the combination with water as a resource. If soil and water are fragmented, the benefit of the land is reduced. Certainly, Worreh Yeama retains considerable land for farming, but the issue arises because the components of the swamp, land and water are disconnected from one another (cf. Haller 2013; Mhlanga et al. 2014), and a well in the village does not compensate for the benefits of a swamp in the dry season.

During interviews, women consistently stated that the lost swampland and the non-implemented but promised mitigation measures, in combination with the creation of very few jobs, had severe consequences for the community.

Certainly, the creation of employment facilities was emphasised in the company's project justifications and in initial community meetings with the people affected. Distribution and access to wage labour, as well as the dismissal of workers, comprised constant and emotional topics of conversation in the village. Generally speaking, and regardless of whether people were young or old, land-user or landowner, male or female, all individuals wanted to work for ABSL, as the only option of attaining a steady income and fleeing the hardship of agricultural work. However, the number of jobs provided fell well below the expectations of the community members. During interviews and in everyday conversations, I developed the impression that people had anticipated jobs for everyone following news of the initial meetings with ABSL. In particular, landowners expected their sons and daughters would be employed, as the company had stated that it would show preference to them. Thus, the reality at the time of research was rather disenchanting, with only five men in the community employed by ABSL or one of its subcontractors. People were very frustrated with the job situation in general, but women who had previously been engaged in vegetable planting in the swamp were especially uncomfortable: "For us it's difficult to survive because at Pivot Five [swamp] we usually benefited a lot before. But we women didn't get employment facilities. So we don't have anywhere to plant and we don't have employment facilities". These women feel that they have lost a lot but won nothing from the project. They were very critical of the fact that not a single woman was employed at the time of research, "although we are not unable to plant sugarcane!" (personal communication).

Resistance to the company and renegotiation of the land lease contract

Approximately two years after the start of operations, ABSL staff visited Worreh Yeama and began to demarcate the commonly used *bolilands* for a second sugarcane field on village land. Following swampland, the bolilands are the most fertile type of land for rice production, the staple food in the region. The group of women passing by to collect firewood recalled the day in a focus group discussion:

> We met those pegs [piles] there and we knew Addax wanted to take the boliland, but we, the women, did not agree. So we came and told the men about these pegs. We told the men it is not possible for Addax to come and take this boliland, because they had [already] taken the Pivot Five land [swampland] where we used to benefit a lot. But some wanted the land to go.
>
> (Personal communication)

When the women returned to the village, a meeting was called and the demarcation of additional land was discussed within the community. One of the elder landowners wanted to give the boliland to the company because he would benefit from an increase in land lease payments. However, numerous women fiercely opposed this option and explained to the landowner that they did not want to lose more fertile land, especially given that they were unemployed. Many men from land-using families and members of the Affected Land User Association (AFLUA) supported the women and emphasised the requirement for fertile land in the absence of employment possibilities. Together, they managed to convince the landowner to reject the lease of the boliland so that they could keep it for their subsistence farming. The landowner conceded because he and another landowner were accused of having stolen the land lease payment in the past. He later explained in an interview that he did not want to run the risk of further damaging his reputation and therefore fulfilled the wish of the women, refusing to give the land to the company:

> People are accusing me of having eaten [stolen] a large amount of money, which is not true, so I faced this embarrassment at the police station, so really I am unhappy. Still, we are with the same poverty as before but people feel you collect a huge amount of money. This is why I agreed not to give this boliland to Addax, so I can rest in peace.
>
> (Personal communication)

The community meeting concluded with an agreement to refuse the boliland and the village headman ordered the removal of the piles. The women and some land-using men went to uproot the piles while the landowner remained in the village. The members of the AFLUA informed SiLNoRF about the resistance movement. SiLNoRF is collaborating with NAMATI,[8] a global legal empowerment network. Representatives of SiLNoRF and NAMATI came to the village to re-discuss the issue. "We did not have to pay Mr Sonkita [lawyer from NAMATI] and we agreed for him to fight for us," recalled a community member (personal communication). With the final renegotiation on 15 July 2013, the NAMATI lawyer represented the community in negotiations with the company. These concluded with the company cancelling the second sugarcane field on village land and returning the *boliland* to the community.

Although the company agreed to renegotiate the land lease agreement, several staff members and local authorities exposed the community to negative labelling, with people deemed "enemies of progress", "backward farmers", "opponents of the president and the welfare of the Sierra Leonean nation", "saboteurs of the company" or more prosaically "liars". Local authorities were offended that the community had not accepted "their project". Given that they were the ones entrusted with the recruitment of local people for jobs at the company, they simply stopped considered people in Worreh Yeama when hiring. Indeed, they believed that people in Worreh Yeama lacked loyalty and refused to "accommodate them", as Jackson (2004: 146) might say.

Discussing multiple institution shopping as a coping strategy

This article has described the consequences of the formalisation of highly complex customary land tenure and access systems through a foreign investing company. The heads of landowning families, who formerly had assumed the role of stewards granting land access and associated resources to land-users, became the sole landowners through the titling process. The landowners were not only given land titles and exclusive property rights, but were also the only individuals to be compensated for the land by the company. Due to the patrilineal inheritance line, women were unable to claim ownership of the land and were thus excluded from compensation, even though they had benefited considerably from the swamp and water source that is now used for sugarcane plantation. With the loss of the swampland, women became deprived of a vital means of ensuring their households' food security and of benefiting economically from selling their surplus harvest.

These far-reaching and gendered impacts, instigated through the institutional change triggered by the formalisation of the land rights, corroborate hypothesis one, presented in the first section.

However, the loss of the perennial water source and the swampland became the ultimate problem because the few job opportunities developed failed to compensate for the lost livelihood. In this situation, women realised that they would be forced back into subsistence farming. With the *boliland* incident, women sought strategies to avoid an undesirable outcome, and so made use of three different institutions:

First, they made reference to pre-colonial common property institutions, according to which not only first-comers but everybody is entitled to a rightful share. Indeed, they emphasised the importance of the *boliland* for already deteriorating livelihoods. Second, they made use of the formalised colonial institution of private property holders by approaching a landowner who was willing to give away some land. Certainly, it was apparent that his reputation would be damaged if he failed to behave in the sense of moral economy, in which wealthier or more powerful individuals are expected to provide for the less wealthy or marginalised. Given his already tenuous reputation due to accusations of stealing land lease payments, he suffered from a very weak bargaining position and thus was convinced to meet the needs of the land-users. Third, women and land-using men made use of the post-colonial institutions and international network of NGOs. Through the Affected Land User's Association (AFLUA), they were linked to the Sierra Leonean Network on the Right to Food (SiLNoRF) and NAMATI, a globally connected legal advocacy movement. NAMATI provided free services as well as the language, information and analytical capacities that the community lacked. The lawyer was able to embed the claims of the community within the paradigm of human rights, in particular the discourse on the right to food, which is additionally informed by the right to subsistence (pre-colonial institution). For McKeon (2013), this connection between local struggles and global policy spaces is crucial to the former's success.

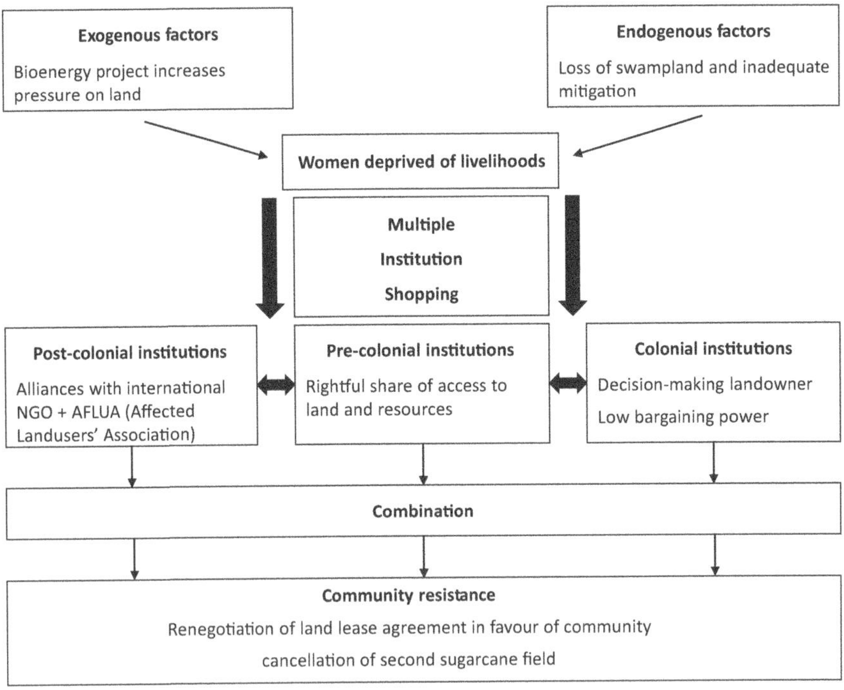

Figure 16.2 Model of multiple institution shopping.
Source: Drawn by the author.

This analysis has emphasised the robustness of this pre-colonial institution of a rightful share, which influenced both colonial (landowners) and post-colonial institutions (NGOs), as depicted by the horizontal arrows in the diagram. Due to the pluralistic institutional setting, women were forced to shop around different institutions in order to secure their livelihoods. I therefore suggest amending Toulmin's (2008) and Haller's (2013, 2016) notion of institution shopping to "multiple institution shopping". The fact that CPRs here were not only subjected to (pre-) colonial conceptualisations of land, access to land and the distribution of wealth, but also became the subject of "glocal" and overlapping bodies of rules through the investment project, appears to confirm the dynamic of "multiple institution shopping", as actors use discourses and institutions at different levels in order to secure their rights.

Notes

1 However, the position of a person is less fixed than this scheme might suggest. An elder could lose the right to be considered superior if he or she behaves immaturely. As Jackson (2004: 46–47) observed: "If he acted like a child, he *was* a child. Superiority

(...) derived not only from being born first or from being big and powerful; it also stemmed from the way one behaved".

2 Of course, these "acts of sharing" are not always voluntary and may instead signify social obligations from which people may attempt to escape. Stories of people hiding parts of their rice harvest or cash from family members in other households are not uncommon. Customarily, as soon as wealth becomes publicly known, the individual has the obligation to share. However, Fafchamps (1992: 147) does not consider this a contradiction, and instead holds that "mutual solidarity can be sustained by existence of lasting relationship between self-interested members". This is in line with the perspective of Ensminger (2002), who departs from the premise that the behaviour of people can reflect both self-interest (ensured through the reciprocal arrangement) and concern for the well-being of others (expressed through redistributional acts of solidarity).

3 The FAO defines shifting cultivation as "a system in which relatively short periods of continuous cultivation are followed by relatively long periods of fallow". Following the cultivation of a patch of land, it is left fallow without any form of active cultivation in order to allow the soil to regain its fertility. The cultivation and fallow periods are contingent on the type of crops grown and the fertility of the soil.

4 In the year 2012, less than 10 per cent of the Sierra Leonean population enjoyed access to electricity (UNCTAD in Sierra Leone, 2012).

5 The HDI is composed of the dimensions *health* (indicator = life expectancy at birth), *education* (indicator = average years of schooling and expected years of schooling) and *living standard* (indicator = gross national income per capita) (Human Development Index, www.hdr.undp.org). In 2014, Sierra Leone was ranked 183rd out of 187 countries (UNDP 2014: 162).

6 This includes The Swedish Development Finance Institution Swedfund, The Netherlands Development Finance Company (FMO), the African Development Bank (AfDB), The German Investment Corporation (DEG), The UK-based Emerging Africa Infrastructure Fund (EAIF), The Infrastructure Fund managed by Cordiant Capital (IDC) and the Belgian Development Bank (BIO) (English and Sandström 2014).

7 The International Labour Organization Conventions, the UN Declaration on Rights of the Indigenous Peoples, the Stockholm Convention on Persistent Organic Pollutants and the FAO International Code of Conduct on the Distribution and Use of Pesticides, among many others (cf. AfDB n.d.).

8 NAMATI Innovations in Legal Empowerment (www.namati.org).

References

African Development Bank Group (AfDB) 'Executive Summary of the Environmental, Social and Health Impact Assessment', www.afdb.org/fileadmin/uploads/afdb/Documents/Environmental-and-Social-Assessments/Addax%20Bioenergy%20-%20ESHIA%20summary%20-%20Final%20EN.pdf, accessed 9 October, 2018.

Alden-Wily, L. (2008) 'Custom and commonage in Africa rethinking the orthodoxies', *Land Use Policy*, vol 25, no 1, pp 43–52.

Benjaminsen, T., Holden, S., Lund, C., and Sjaastad E. (2008) 'Formalisation of land rights: some empirical evidence from Mali, Niger and South Africa', *Land Use Policy*, vol 26, pp 28–35.

Bernstein, H. (2008) 'Agrarian questions from transition to globalization', in Akram-Lodhi, A.H. and Kay, C., (eds), *Peasants and Globalization. Political Economy, Rural Transformation and the Agrarian Question*. London: Routledge, pp 239–261.

Berry, S. (2008) 'Privatization and the politics of belonging in West Africa', in Kuba, R. and Carola, L. (eds), *Land and the Politics of Belonging in West Africa*. Leiden: Brill, pp 241–265.

Bolten, C. (2009) 'The agricultural impasse: creating "normal" post-war development in northern Sierra Leone', *Journal of Political Ecology*, vol 16, no 1, pp 70–86.

Bürgi, E. (2015) 'Sustainable investment in land in the Global South: what would it require from a coherence perspective? The case of Sierra Leone', *Questions of International Law*, vol 21, pp 17–37.

BusinessWire (2011) Addax Bioenergy Signs Loan Agreement for €258 Million Renewable Energy Project in Sierra Leone. www.businesswire.com/news/home/20110617005346/en/Addax-Bioenergy-Signs-Loan-Agreement-%E2%82%AC258-Million.

Crang, M. and Cook, I. (2007) *Doing Ethnographies*. London: Sage.

Deininger, K. and Byerlee, D. (2011) *Rising Global Interest in Farmland. Can It Yield Sustainable and Equitable Benefits?* Washington, DC: World Bank.

De Schutter, O. (2011) 'How not to think of land-grabbing. Three critiques of large-scale investments in farmland', *The Journal of Peasant Studies*, vol 38, no 2, pp 249–279.

Emerson, R., Fretz, R., and Shaw, L. (1995) *Writing Ethnographic Fieldnotes*. Chicago, IL: University of Chicago Press.

Ensminger, J. (2002) *Making a Market. The Institutional Transformation of an African Society*. Cambridge: University of Cambridge Press.

European Commission (EC) (2009) Commission Decision of 30 June 2009 establishing a template for National Renewable Energy Action Plans under Directive 2009/28/EC of the European Parliament and of the Council, Brussels.

Fafchamps, M. (1992) 'Solidarity networks in preindustrial societies: rational peasants with a moral economy', *Economic Development and Cultural Change*, vol 41, no 1, pp 147–174.

Fanthorpe, R. (2005) 'On the limits of liberal peace. Chiefs and democratic decentralization in post-war Sierra Leone' *African Affairs*, vol 105, no 418, pp 27–49.

Ferguson, J. (2015) *Give a Man a Fish. Reflections on the New Politics of Distribution*. Durham, NC and London: Duke University Press.

Haller, T. (2007) Understanding Institutions and Their Links to Resource Management from the Perspective of New Institutionalism. NCCR North-South Dialogue, no. 2.

Haller, T. (2013) *The Contested Floodplain. Institutional Change of the Commons in the Kafue Flats, Zambia*. Lanham, MD: Lexington Books.

Haller, T. (2016) 'Managing the commons with floods: the role of institutions and power relations for water governance and food resilience in African Floodplains', in Ostegard, T. (ed.), *Water and Food – Africa in a Global Context* (pp 369–397). Uppsala: The Nordic African Institute. London: Tauris.

Haller, T. (ed.) (2010) *Disputing the Floodplains: Institutional Change and the Politics of Resource Management in African Floodplains*. Leiden and Boston, MA: Brill.

Hardin, G. (1968) 'The tragedy of the commons', *Science*, vol 162, pp 1243–1248.

Hauser-Schäublin, B. (2003) 'Teilnehmende Beobachtung', in Beer, B. (ed.), *Methoden und Techniken der Feldforschung*. Berlin: Reimer, pp 33–54.

Jackson, M. (2004) *In Sierra Leone*. Durham and London: Duke University Press.

Käser, F. (2014) 'Ethnography of Land Deals. A Village Perspective on the Addax Bioenergy Project', Master's Thesis, University of Bern.

Kuba R. and Lentz, C. (2006) *Land and the Politics of Belonging in West Africa*. Leiden: Brill.

Lastarria-Cornhiel, S. (1995) *Impact of Privatization on Gender and Property Rights in Africa*. Madison, WI: University of Wisconsin-Madison.

Lustenberger, S. (2015) 'Addax Bioenergy Sierra Leone. Analysis of the Implementation Process of a Large-Scale Land Acquisition Project from Perspective of Assemblage Theory', Master's Thesis, University of Bern.

McKeon, N. (2013) 'One does not sell the land upon which the people walk: land grabbing, transnational rural social movements and global governance', *Globalizations*, vol 10, no 1, pp 105–122.

Memorandum of Understanding (MoU) (2010) 'Government of the Republic of Sierra Leone (GoSL) Addax Bioenergy Sierra Leone Ltd (ABSL) and Addax &Oryx Holdings BV (AOH)', Makeni.

Mhlanga, L., Nyikahadzoi, K., Haller, T. (2014) *Fragmentation of Natural Resources Management: Experiences from Lake Kariba*. Vienna and Berlin: LIT Verlag.

Millar, G. (2015) 'Knowledge and control in the contemporary land rush: making local land legible and corporate power applicable in rural Sierra Leone', *Journal of Agrarian Change*, vol 16, no 2, pp 1–19.

NAMATI (n.d.) 'Innovations in Legal Empowerment, Offical Website', https://namati.org, accessed 19 October, 2018.

Netherlands Development Finance Company (FMO) (2011) 'DFIs announce financial close of pioneering Addax Bioenergy Project in Sierra Leone', www.fmo.nl/k/news/view/877/538/dfis-announce-financial-close-of-pioneering-addax-bioenergy-project-in-sierra- leone.html, accessed 9 October, 2018.

Oakland Institute (2011) *Understanding Land Investment Deals in Africa. Country Report Sierra Leone*. Oakland, CA: The Oakland Institute.

Peters, P. (2009) 'Challenges in land tenure and land reform in Africa: anthropological contributions', *World Development*, vol 37, no 8, pp 1317–1325.

Peters, P. (2010) 'Our daughters inherit our land, but our sons use their wives' fields: matrilineal matrilocal land tenure and the New Land Policy in Malawi', *Journal of Eastern African Studies*, vol 4, no 1, pp 179–199.

Peters, P. (2013) 'Land appropriation, surplus people and a battle over visions of Agrarian futures in Africa', *The Journal of Peasant Studies*, vol 40, no 3, pp 537–562.

Pham, P. J. (2006) *The Tragedy of Sierra Leone: History and Global Dimensions*. New York: Nova Science Publishers.

Renner-Thomas, A. (2010) *Land Tenure in Sierra Leone: The Law, Dualism and the Making of a Land Policy*. Milton Keynes: AuthorHouse.

Ribot, J. and Peluso, N. (2003) 'A Theory of Access', *Rural Sociology*, vol 68, no 2, pp 153–181.

Richards, P. (1998) *Fighting for the Rain Forest: War, Youth and Resources in Sierra Leone*. London: International African Institute.

Richards, P. (2005) 'To Fight or to Farm? Agrarian Dimensions of the Mano River Conflicts (Liberia and Sierra Leone)' *African Affairs*, vol 104, no 417, pp 571–590.

Robbins, P. (2012) *Political Ecology: A Critical Introduction*. Chichester: John Wiley & Sons Ltd.

Rocheleau, D. and Edmunds, D. (1997) 'Women, men and trees. Gender, power and property in forest and agrarian landscapes', *World Development*, vol 25, no 8, pp 1351–1371.

Sandström, J. and English, C. (2014) 'Implementing a Large Land Based Investment in Sierra Leone Land Grab or Real Development – an Investors Perspective', presentation at the 2014 World Bank Conference on Land and Poverty. Washington, DC, 24th–27th March, 2014.

Schlehe, J. (2003) 'Formen qualitativer ethnographischer interviews', in Beer, B. (ed.), *Methoden und Techniken der Feldforschung*. Berlin: Reimer, pp 71–93.

Scott, J. (1985) *Weapons of the Weak: Everyday Forms of Peasant Resistance*. New Haven, NC: Yale University Press.

Sierra Leone Investment and Export Promotion Agency (SLIEPA) 'Official Website', www.investsierraleone.biz, accessed 9 October, 2018.

Strübing, J. (2008) *Grounded Theory. Zur sozialtheoretischen und epistemologischen Fundierung des Verfahrens der empririsch begründeten Theoriebildung*. Wiesbaden: Verlag für Sozialwissenschaften.

Toulmin, C. (2008) 'Securing land and property rights in sub-Saharan Africa: the role of local institutions' *Land Use Policy*, vol 26, no 1, pp 10–19.

United Nations Conference on Trade and Development (UNCTAD) (2012) 'National Energy Profile of Sierra Leone', www.sl.undp.org/content/dam/sierraleone/docs/focusareadocs/undp_sle_energyprofile.pdf, accessed 9 October, 2018.

United Nations Food and Agricultural Organization (UNFAO) (2011) 'Good Socio-Economic Practices in Modern Bioenergy Production: Minimizing Risks and Increasing Opportunities for Food Security', www.fao.org/docrep/015/i2507e/i2507e00.pdf, accessed 9 October, 2018.

17 The open cut

Mining, transnational corporations and the commons

Thomas Niederberger, Madlen Kobi and Tobias Haller[1]

Introduction

Mining of metals and minerals relies on a process of institutional converting of common-pool resources (CPRs) to state property. This leads to the privatization of ore deposits, while the associated deterioration of the resource base and related costs are externalized onto local communities. Corporate and state actors collude in pushing forward this process, often against considerable local resistance, referring to an ideology of 'development', which has aptly been criticized as an 'anti-politics machine' (Ferguson 1994). In the newest wave of mining-related expropriation of commons, these mechanisms have become more sophisticated under the modality of corporate social responsibility (CSR), which includes compensation and micro-development plans designed to foster the consent of a critical number of local actors, inciting expectations and emphasizing short-term financial gains, often captured by local elites, while discounting the long-term damages involved. However, we also observe how local and indigenous people's responses to major mining projects, amplified by strategic 'glocal' alliance-building (with local, national and international actors), can re-open spaces for political deliberation about resource ownership regimes, thereby turning mining conflicts into actual 'politics machines' capable of inducing cracks in the neo-liberal discourse on national scales and beyond.

For this study we adopted an emerging New Institutional Political Ecology (NIPE) perspective in social anthropology that foregrounds power constellations in institutional settings (Haller 2017). This approach analyses the interaction between external variables that affect relative prices for resources (markets, environment, technology, colonial context) and internal variables defining the relevant institutions for bargaining (resource ownership and management, compensation and benefit sharing, consultation, etc.). Based on the specific local interaction of external and internal variables, a NIPE approach is useful to explain how powerful external actors are able to exclude local actors from the use of CPRs. This is mostly a result of the legal disconnect created by the fragmentation of cultural landscape ecosystems into separate realms governed under different property rights regimes (Haller 2017, 3–4).

The two cases presented in this chapter illustrate two very different ways in which power coalitions between states and corporations, local populations and international campaigns can influence local CPRs management. In Zambia, historical marginalization and dependence on mining incomes leaves local people with little leverage over the interests of mining corporations. In the Philippines, local people were able to mobilize an internationally recognized and legitimized indigenous identity, combined with legal recognition of collective ownership rights, at an early stage of mine development. This relatively favourable power setting is met by the company Xstrata, with its extensive CSR programme designed to manufacture consent through local leaders and institutions.

The presented findings have evolved from a collective learning process that started in 2010 and at its peak involved over 50 students, researchers, professionals and activists of various backgrounds (Niederberger et al. 2016). The increasing numbers of reports about conflicts around Swiss-owned mining projects reaching the public through NGOs, solidarity groups and news media provoked our interest.[2] Taking up the example of an earlier collective scientific investigation into fossil fuel extraction, transnational companies and indigenous peoples (Haller et al. 2007), our group started analysing a wide range of sources, combining ethnographic literature on the specific regions with scientific literature as well as grey papers, online sources from corporate websites, campaign blogs and first-hand fieldwork data. Reflecting the transnational character of the mining business, our own work extended globally too through personal exchanges with activists from affected regions and the participation in ongoing campaigns.[3]

The chapter is structured as follows: the first section briefly introduces the context and current debates around large-scale mining, focusing on four aspects: colonial continuities, technological innovation, legal pluralism and CSR, and socio-environmental impacts. Following NIPE, this will set the framework for understanding the kind of knowledge actors have and will be used 'strategically to select among the different institutional options they anticipate will yield the best results and which are consistent with their power frame' (Haller 2017, 4). Second, we present the two case studies: the Mopani Copper Mines in Zambia and the proposed Tampakan copper-gold mine in the Philippines, both owned by the Swiss company Glencore/Xstrata. In the conclusion, we compare the two cases with regard to current debates around CPRs. Further, as the proliferation of mining-related conflicts indicates a failure of the relevant institutional settings, we explore their potential for building fairer and more sustainable 'glocal' institutions.

Large-scale mining compared

Since the turn of the millennium, the academic literature on mining conflicts has grown significantly.[4] To understand the current impacts of large-scale mining investments on the commons, we have to consider the histories of colonial resource exploitation, the characteristics of the affected ecosystems and the local population, as well as the institutional context (resource ownership, property regimes and management, customary rules and regulations) prior to a given investment. Our data highlights the importance of specific project characteristics such as the type of extracted mineral, its size, extraction technology and waste management. These aspects interact with changing global market prices, government policies and legal frameworks. Our comparison shows that the following underlying conflict fields shape mining activities in most places, linked to what institutionalist approaches call external and historical variables.

First, the political and historical environment of mining is marked by colonial expropriation and exploitation. Local people often perceive mining projects in continuity with prior, traumatic experiences, and not merely as 'governance problems' that can easily be addressed on a policy level (Acuña 2015). Indeed, it is no coincidence that today, companies from the USA, Canada, the UK and Australia are leading the sector in terms of investments and headquarters of mining corporations as well as related (academic) knowledge production. In recent years, however, mining has also reflected geopolitical power shifts, with new players from China but also Brazil, Russia, India and South Africa rapidly gaining ground (all of them expanding from their domestic mining industries).[5]

Second, the newest cycle of extractivist expansion since the mid-2000s has been ignited by a series of technological innovations. Many of the new projects are open-pit (or open-cut) mines of an unprecedented scale, using highly mechanized processes to extract previously commercially unviable, low-quality ore grades. The quantity of extracted rock required to obtain a ton of refined metals, as well as the necessary inputs (water, energy, fuels, explosives, chemicals), has multiplied (e.g. Mudd 2007, for gold). Such mines can be extremely capital intensive, with investments going up to several billion USD over two or three decades. Moreover, they need a lot of support infrastructure for transport, electricity generation and so forth, which makes mining a major contributor to global greenhouse gas emissions – even without considering the climate impact of certain mining products such as coal and tar sands (for a summary, see Niederberger et al. 2016, 15–34).

Third, the legal-institutional environment has become more complex. On the one hand, in almost all the case study countries, the neo-liberal turn of the 1990s, with World Bank advisors promoting reforms specifically crafted to attract foreign investments, laid the ground for the following expansive phase (Bridge 2004). Often, these packages included tax breaks, subsidized rates for energy and water, and other specific incentives that turned out to be detrimental to the national economies and prevented them from taking their fair share of revenues when resource prices exploded after the mid-2000s. Further, over 3,000 bilateral and regional investment treaties were signed between states, giving rise to an obscure system of investor–state dispute settlement mechanisms that allow extractive companies to sue states for restricting their access to land and mineral concessions (Cotula 2013, 2015, 10). In parallel, multilateral agreements, voluntary standards and CSR policies evolved into a framework rich in declarations of intent, but poor in enforceability.[6] Meanwhile, indigenous rights movements advocate for free, prior and informed consent (FPIC) to be included in national legislation, especially in view of enabling local peoples' fair participation in decision-making about extractive projects (Doyle and Whitmore 2014).

Minerals are typically claimed as property of the state, to be leased out to private investors. Forms of local common property regimes for minerals sometimes exist for artisanal mining, but they are usually overruled once a private company acquires permits to mine the same area (e.g., for Guinea, see Bolay 2016). However, the institutional change from commons to state and private property in mining mostly concerns common land and, critically, water resources. Customary rights are frequently ignored, especially when dealing with historically marginalized people whose rights have never been formalized legally. Such a situation of legal pluralism facilitates the 'shopping' of favourable frameworks by all involved actors in order to legitimize their positions (see Toulmin 2009; Haller ed. 2010; Haller 2016). Ultimately, the enforcement of regulations remains mostly in the hands of national governments and courts, who all too often decide in favour of the economically more powerful (Gambon and Kobi 2016).

Fourth, large-scale mining creates extremely complex, interconnected, multilayered socio-environmental impacts on a wide range of CPRs (Franks et al. 2014; Niederberger et al. 2016). Mining activities affect the underground and surface hydrology, liberate toxic elements naturally contained in the rock such as arsenic, cause acid mine drainage and so on. The best available technology to prevent contamination is often prohibitively expensive. Even if companies are required to hand in detailed Environmental Impact Assessments, those reports are often criticized as incomplete. Kuipers et al. (2006) show in their comparison of the predicted and actual environmental impacts of hard rock mines in the USA that in most cases mines predicted low or no impacts, but in the majority, impacts on surface water quality, groundwater quality and acid drainage have occurred. Local residents observe changes to the natural surroundings, microclimate, water availability, wildlife and so forth on a daily basis. This can also amount to epistemological conflicts where the landscape has spiritual significance for locals, as in the case of the Australian Mirarr aborigines (Brom and Bucher 2016), or the 'Apu' mountains in Peru (De La Cadena 2015; Li 2015). Many mining conflicts evolve around the use and contamination of direct water sources or the headwaters of river systems (Kemp et al. 2010; Li 2015), followed by land use and resettlement conflicts (Hilson 2002).

A further dimension to the mentioned aspects is the 'politics of time' (a term we adopt from Kirsch 2014): a mine's impacts change dramatically over the lifespan of a project, and they are perceived very differently by different actors and groups involved over large geographical areas. Often, the early stages – exploration to planning and construction – come with the promise of development and employment, welcomed especially by local elites or male youth with professional training who are best prepared to benefit from jobs and subsidiary commerce. When the mine starts running, this is often followed by a discrepancy, as few jobs are really available to locals, while social impacts linked to the sudden influx of cash – such as alcoholism and prostitution – destabilize local communities. People who heavily rely on commons, for the likes of farming, fishing or animal husbandry, may find that subsistence in peril due to their loss, restricted access and/or contamination. Nevertheless, some actors might profit from short-term cash gains. Mining thus tends to amplify previous heterogeneity, making collective action more difficult. Even if mining impacts are perceived negatively by a majority, reactions depend on the bargaining power that can be mobilized against interest groups that have become economically dependent on mining and who defend the activity fervently (Niederberger et al. 2016, 489–507).

A recurrent pattern is that people further away from the mining site, such as fishermen and farmers living downstream, residents along drive-through roads or long-distance transhumance populations only perceive the negative impacts on their commons, without benefitting economically.[7] Other disadvantaged groups include women and poor populations. Women feel the negative impacts first and more dramatically, particularly where they depend on access to and quality of CPRs needed for household modes of production. At the same time, their relative position may be weakened when male wage labour is gaining importance (Jenkins 2014).

The impacts of mining activities on the different actors involved and the various power constellations between them have to be considered for the discussion of the following two case studies. Different actor groups develop their collective organizing, narratives and strategies based on their economic, social or cultural resources. We find heterogeneity among both corporate/state actors and local communities. Coalitions and loyalties are shifting and ambiguous.[8]

Case studies

We present two case studies that mark the extremities on a timeline of mining-related impacts. The investment in the first case, Mopani Copper Mines in Zambia, was undertaken shortly before the beginning of the commodities boom of the 2000s by the Switzerland-based multinational Glencore. The second concerns the Switzerland/UK-based company Xstrata, which invested in the Tampakan copper-gold project in Mindanao, Philippines. While both cases concern copper extraction, the technology differs: 'old school' underground mining in Zambia versus modern open-pit mining in Mindanao. The two cases exemplify two different histories of ore bodies that influence investment strategies, known in industry language as 'brownfield' and 'greenfield' strategy. These sites became linked through the takeover of Xstrata by Glencore in 2012, creating the largest integrated mining and commodity trading corporation based in the Swiss tax haven of Zug. Our analysis focuses on the impacts of large-scale mining on the local commons and on the institutional frameworks that regulate them; specific attention is given to changes in power relations between population groups, as well as between them and state and corporate actors. We show that undermining the commons physically as well as institutionally (through loss of local common property rights) in the first case leads to massively reduced bargaining power of local people, from which they cannot recover due to the lack of legal protection and allied local partners, together with the prevalence of a 'development' ideology that works as an 'anti-politics machine'. In the second case, the same process of loss triggers local alliances and mobilization, as a kind of 'politics machine'.

Mopani copper mines in Zambia

The case of Zambian copper mining is a dramatic example of the impacts of long-term dependency on the extraction of a single metal, based on the expropriation and marginalization of its original inhabitants. From 1890, the region came under the rule of the British South African Company, which built a railway to connect the Katanga mines in the Belgian Congo with the coast. The region that was to become today's Copperbelt Province was largely inhabited by the Lamba, an ethnic group belonging to the Bemba language family (Doke 1931). They practised extensive agriculture and some animal husbandry on common property land based on shifting cultivation technology. Access to land was regulated under a system of village headmen and group-level chiefs, overseen by a paramount chief. The authority of chiefs, clan membership and access to the common lands were based on matrilineal descent, and residency after marriage was matrilocal.

In 1928, the colonial authorities forced Lamba chief Mushili II to retreat with his 23,000 people into a native reserve, thereby losing 80 per cent of the communal territory. This prepared the ground for the emerging mining industry and urban settlements. The intention of this proletarianization was to create a pool of cheap labourers, hoping that Lamba people would accept working in the mines, rather than in small-scale agriculture (Siegel 1988). For the loss of their territory 'no compensation was ever paid, apart from some small celebration hosted by Chirupula Stephenson around or before 1930', as stated by Siegel, and the Lamba felt that 'their wealth was stolen from them by the mining companies, backed by the Northern Rhodesian government'.[9]

With the development of the copper industry from the mid-1920s onwards, the Copperbelt rapidly urbanized and thousands of workers immigrated from all over the country. The beginning of what James Ferguson (1999) described as 'expectations of modernity' goes along with the further marginalization of the Lamba, who continued to shy away from working in the mines, adopting other activities such as beer brewing and prostitution to earn cash instead. To add insult to injury, their evasiveness bolstered the stereotype of them being lazy and weak (Siegel 1989). The mining industry moved on to become the economic cornerstone of independent Zambia after 1964, with strong mine worker unions that had a key role in the country's political development. Under the autocratic rule of President Kenneth Kaunda, the mines were nationalized and copper revenue became the base for an ambitious welfare state of socialist inspiration. It only experienced crisis with the sharp drop in copper prices after 1975. As we shall see, and as the NIPE approach emphasizes, such price fluctuations significantly affect the bargaining power of involved actors.

The downturn of the economy in the 1980s created social unrest and prepared the ground for the swing to multi-party democracy and a change of government in 1991. Zambia became a test site for a neo-liberal World Bank/IMF policy mix. This included the privatization of the copper mines, which were in bad shape after years of underinvestment. The Zambian government was desperate to attract foreign investors as copper prices further dwindled to historic lows in the late 1990s. At the same time, external debt and inflation rose to exorbitant levels. This is when, in 2000, Glencore secured a mayor share of the Zambian mining industry – namely, the Mopani Copper Mines – at a ridiculously low price of USD 43 million.[10] Glencore was at the time still a privately-owned commodity trader following an opportunistic policy of buying up defunct mines and storage infrastructure on the cheap, the so-called 'brownfield strategy'. Glencore benefitted from undisclosed 'development agreements', which granted the company tax holidays and exemption from responsibility for environmental liabilities for long periods (Fraser and Lungu 2007). Four years after the deal was signed, copper prices started to soar from below USD 2,000/ton to almost USD 9,000/ton in 2011.

With the increase of world copper prices, mine workers pressured for salary increases through unions and wildcat strikes. In 2011, the left-wing Patriotic Front party under President Michael Sata won an election and promised to make multinationals pay their fair share (Uzar 2017a). Sata's demands addressed taxation and the rampant air and water pollution. Glencore, as owner of the largest mine in Zambia in terms of turnover and employees, played a key role in all these debates. Opposing the demands for wage increases, the company threatened to close the mine and fire most of the employees, playing with the economic dependency of both workers and the government. In 2014, half of Mopani's total of roughly 20,000 workers was subcontracted (and mostly not unionized), and salary increases from 2000 to 2016 just matched inflation rates (Uzar 2017a, 301) despite the rise in world copper prices. Mining companies paid merely 1 per cent of Zambia's total tax income from 2000 to 2008. After the new government had reformed tax law, company tax incomes increased to an average of 14.8 per cent of the total tax income from 2010 to 2014. In 2012, Glencore paid company taxes for the first time (Uzar, 2017b, 76–77). Local media, authorities and international NGOs denounced the fact that the Mopani Copper Mines smelter continued to emit large quantities of unfiltered sulphur dioxide. Further, they criticized a series of accidents, leaking tailing ponds and pipelines as well as the contamination of groundwater (Simpere 2010). The company's strategy was to deny any allegations, taking shelter under the development agreement from prior to 2000 that exempts it from liabilities.

Due to the lack of reliable, independent data, the allegations from different NGOs that corporations such as Glencore were reaping tremendous profits from Zambia by avoiding tax payments through complex offshore structures and transfer-pricing were never fully clarified. The populations of Mufulira and Kitwe (137,000 and 362,000 respectively, in the census from 2000) meanwhile continue to live right next to huge tailing ponds and slag heaps. Poverty, poor health and lack of education as well as environmental contamination continue to haunt the Copperbelt. Little is known about the situation of the urban poor or about gender-specific impacts, but prostitution and informal scrap mining seem to be important survival strategies (Mususa 2010). Many residents lack property titles and risk being expelled as 'squatters' whenever they happen to be in the way of construction projects. Decades of acid rain from sulphur dioxide emissions have made the soil unsuitable for agriculture and also affected the river systems and groundwater quality. Glencore, however, claims that the creation of employment and investments are a fair enough contribution to the wellbeing of Zambia. CSR efforts were limited to some health and retirement programmes, mainly benefitting the company's regular employees, rather than contract workers and the general population. Glencore only published its first, rudimentary sustainability report in 2010, when going public at the stock market, but detailed reporting about the situation in Mopani was not available (Schober et al. 2016).

Summing up, the Mopani case shows how investments by transnational corporations in an existing industry can capitalize on long-standing, unaddressed injustice, which in the case of Zambia goes back to the beginning of Western colonial expansion. Companies cannot be made responsible for the expropriation of the Lamba peoples' common lands at the beginning of the twentieth century. Nonetheless, they are co-responsible for the perpetuation of this injustice and have benefitted from easy access to land while the Lamba have been further marginalized. The investments expose residents to severe health risks through the contamination of natural resources. Moreover, the degradation of CPRs in the mining areas increases the pressure on fisheries and wildlife from migrants seeking an income in other regions, including the ecologically sensitive Kafue floodplains (see Haller 2013).

Urban populations are affected in their access to health, education, adequate housing, water and sanitation, pensions and other services, mainly because the tax-avoiding strategies of companies deprive the state and municipal authorities of the possibility to deliver these basic public goods (Sikka 2010). Similarly, corporations share a responsibility when it comes to their workers: the policy of subcontracting labour and frequent (threats of) mass layoffs to keep salaries low and avoid social security payments create precarious living conditions for thousands of families.

The Tampakan copper-gold mine project on Mindanao, Philippines

Tampakan district is situated in South Cotabato, on the island of Mindanao, Philippines. The area is hilly, covered with tropical forest of high biodiversity and wildlife protection value, and hosts the headwaters of six river catchments. It is also geologically unstable, with seismic activity and an active volcano (Mount Matutum). The indigenous Bla-an landowners practice shifting cultivation (the main crops are dry-grown rice, maize, millet, sweet potato) combined with foraging. The forest and the land for cultivation are collectively owned, with use rights inherited through clans, forming village units with a dispersed settlement pattern. The territory is understood in an integral manner, as being inherently connected to its people (closely resembling the 'Ancestral Domain' as defined in the Indigenous Peoples' Rights Act, see below). In this tropical climate, water is abundantly available and there are no local institutions known that regulate its use. The area directly affected by the proposed mining project covers nine administrative village units (known as *barangay*) with about 17,000 inhabitants. There is not much information on Bla-an political institutions prior to the investments, but it seems that they had relatively weak village headmen, whose role largely covered dispute settlements, without any supra-level chiefs. In the 1980s and 1990s, the government encouraged the formation of tribal councils as new institutions to formally represent indigenous communities, next to the elected local administrative bodies of the *barangay* (Wenk and Scherler 2016).

Mindanao has been subject to a long history of colonization, with several waves of Filipino settlers occupying indigenous lands up until the second half of the twentieth century. These settlements were actively encouraged by the central state. The population today is mixed, with settlers mostly practising market-oriented, intensive agriculture on privately owned land. The Bla-an had controlled much greater areas of land before the arrival of settlers, but then retreated more and more into the hilly areas. The region downstream from the proposed mining site is today a major rice production zone, with about 80,000 potentially affected Filipino farmers organized in farmers' associations.

The Philippines are representative of the two main trends in legal developments mentioned in the introduction: on the one hand, the Mining Act of 1995 promoted a neo-liberal agenda, heavily influenced by World Bank advisers and designed to attract foreign corporations through a package of tax exemptions and direct agreements with the central government. On the other, the Indigenous Peoples' Rights Act of 1997 belongs to the most progressive international legal protections, recognizing indigenous peoples' collective ownership and making it mandatory to obtain their free, prior and informed consent for extractive projects.

Among the most interesting features of this case study is that the Western Mining Corporation (WMC), a mid-level Australian company, actively supported the Bla-an tribal councils in acquiring collective land titles in the early stages of project development in the 1990s. The company provided funding to the local councils through its CSR programme, enabling headmen to build up clientelist relationships and increase their power. WMC was then able to obtain 'principal agreements' in the late 1990s from all but one of the affected communities (the resisting one was subsequently punished by exclusion from payments). When the Swiss/UK-based Xstrata entered the project in 2007,[11] it built upon this policy and added an extensive CSR programme covering a wide range of development projects (healthcare, education, scholarships) as well as direct cash payments to the tribal councils and *barangay*.

While the corporate co-optation of indigenous institutions contributed to the marginalization of a silent minority of opposing Bla-an, there was also open resistance from some Bla-an groups, including a faction that took up arms and – informed by a tradition of tribal warfare – launched sabotage attacks against construction camps. On top of this, the leftist New People's Army guerrilla group was active in the area. Although there is no proof of collaboration between the two, the state quickly responded by militarizing the area, in close co-operation with the mining company, which led to displacements and the extra-judicial killings of several alleged Bla-an insurgents.

Rather than by the directly affected Bla-an people, the Tampakan project was heavily challenged by the downstream settler population, which demanded to be included in compensation schemes and/or opposed it due to concerns about the potential impact on CPRs, in particular the contamination of irrigation water. Together with a strong anti-mining movement in Mindanao City and at the national level – motivated inter alia by the disastrous Marcopper tailings spill of 1996 – this led to a ban on open-pit mining in South Cotabato Province in 2010. The campaign against the Tampakan mine also received international support through NGOs and the Catholic Church.

The Tampakan mine, set to become one of the largest copper-gold mines in the world and the single largest foreign investment in the Philippines, remained in the planning stage, and has stalled after Glencore sold its stake in 2015 to the Philippine Alsons Group. The effects of the project on CPRs remain low – just exploration and some preliminary infrastructure works – but the potential impacts, once in operation, would be tremendous: apart from resettlements and loss of access to land and forests, the project's location poses very serious risks of disasters (catalysing earthquakes and volcanic activity) and water contamination. Regardless of this, the case illustrates a rather unexpected impact of a mining project on local common pool management institutions long before its actual construction: an apparent strengthening of 'traditional' leadership, albeit at the price of dependence on compensation payments. The case study authors termed this situation an 'actively negotiated dependence' (Wenk and Scherler 2016). Finally, the main impact of the project was an exacerbation of social conflict and the militarization of the region.

Undermining local commons institutions

The two case studies at first evince few similarities and seem illustrative rather in their differences. Two different investment strategies are at work – 'brownfield' and 'greenfield' – which go along with a limited CSR programme in the Zambian example and a very sophisticated one in the Philippine case. However, in both cases, local institutions for the governance of CPRs, conflict resolution and democratic deliberation are undermined by corporate interests, allied with certain sectors of the state.

In line with NIPE theory, the shifts in relative prices for metals over the period in question have had different impacts, affecting corporate strategies and local peoples' resistance. In Zambia, the autochthonous Lamba people appear totally dependent on mining-related incomes, having lost their commons a long time ago. There was only a short period when demands for better salaries, taxes and environmental regulations were successful due to a combination of high market prices, combative trade unionism and the election victory of a new political party. With the decline of copper prices, things have reverted to 'business as usual': meagre salaries and continuing contamination are reported, with companies threatening mass layoffs in light of attempts to improve taxation or enforce environmental regulations.

In the Philippines, the resistance from local groups in alliance with national and international NGOs delayed the development of the planned Tampakan mine, until the sudden drop in copper prices made retreat the best option for the owner, Glencore. Therefore, the Tampakan case could be considered an example of successful defence of the commons. Still, the abandoned mining project had considerable impacts: reinforcement of pre-existing local conflict lines and asymmetric power relations within indigenous communities, capturing of benefits by local elites and further accentuation of divisions between indigenous peoples and Filipino settlers. This led to an extremely volatile situation followed by the militarization of the area. The necessary resettlement of about 5,000 people (67 per cent of them being Bla-an, the rest Christian Filipino settlers) would certainly have exacerbated this further. With the reputation risk from international observers ready to point out the company's co-responsibility for any acts of violence, for Glencore the high investments under lower profit expectations no longer seemed worthwhile: a combination of internal and external factors had weakened its power position in the actor constellation.

The juxtaposition of the two cases thus shows how local groups that are able to mobilize resistance in alliance with NGOs at an early stage of mine development can succeed in defending and regaining the common ground in a context of decreasing relative prices for mining products. However, if local bargaining power is and remains low due to previous dependence on mining incomes, such resistance is less likely to make a difference. This aspect of the 'politics of time' (Kirsch 2014) confirms that social movements are more likely to be successful if they gain momentum before mining projects are operational. On a longer timeframe, one has to consider also changes in the institutional setting, as 'the state disconnects, legally, land from water, fisheries, and so on, and creates separate legal spheres, regulations, and organizations' (Haller 2017, 4). Such disconnects can then be exploited by more powerful actors, while they reduce the resilience of local groups who struggle to defend their CPRs over long cycles of resource price fluctuations. The challenge therefore lies in bridging these cycles, with their long- and short-term developments on different scales, in order to resist the fragmentation of commons into separate realms ready to be appropriated by powerful external actors.

Conclusions: 'glocal' responses

Mining played a crucial role in the establishment of neo-liberal policies during a phase of historically low metal prices in the 1990s. The subsequent push of the 'frontier of extraction' into ever remoter regions was just one expression of the expansion of corporate power which is linked to state and private property institutions, the ideology of Modernity and the discourse of development. Due to their tremendous impacts on large numbers of people over wide geographical areas, mining projects have become crucial sites for challenging contemporary capitalist accumulation, and the defence of the commons stands at the core of all this. Our combined NIPE approach involves external changes and heterogeneous bargaining power relations in examining the selection of institutional options (institution shopping) and the ideological production of legitimacy (discourses and counter-discourses). Thereby, NIPE allows the researcher to go beyond analysing specific projects and to engage with cross-cutting, underlying tendencies.

The invention of a discourse of CSR in this context can be interpreted as a reaction to criticism of large-scale mining (Kirsch 2014), while at the same time legitimizing its expansion. In our understanding, CSR forms an extension of the 'classic' narrative of modernization, and thus an essential part of the 'anti-politics machine' – in other words, the mystification of 'development' as a technocratic endeavour seemingly unrelated to 'politics', or the debate about how, where and which resources should be extracted. CSR provides an ideological justification for corporations to lobby for their interests, and selectively to take over state functions for the advancement of their interests (e.g. providing security, infrastructure, health and education), leading to the phenomenon of 'corporate state-building' (Niederberger et al. 2016, 460ff). This corporate–state nexus helps to co-opt local actors and institutions and to implement divide-and-rule tactics.

At the same time, CSR and the emergence of an entire business sector specializing in 'community relations' can also be interpreted as a tacit recognition that mining *does* cause severe impacts on cultural landscape ecosystems, CPRs and local livelihoods. CSR has created new rules of engagement and increased local peoples' leverage in some cases, most notably to claim compensations. However, it needs to be analysed in detail who is compensated and under what conditions. Often, such compensations not only have negative social impacts themselves, but act as part of the 'anti-politics machine' as well, by diverting attention from long-term environmental damages and irreparable loss of CPRs to the distribution of monetary benefits. Without negating the relevance of the latter, we insist on the necessity to keep in mind the larger issues at stake, for which a NIPE approach can provide a suitable framework.

An interesting aspect of the Tampakan case points to the limits of corporate state-building and CSR: these strategies tend to nurture demands by large population groups outside of the core interest area of a mining project, which can exceed a company's willingness to pay. The stagnation or decrease of metal prices and the shelving of countless mining projects over recent years cast doubt on the continuity of these policies, which were developed during a boom period when they served companies to gain competitive advantage to access new mineral deposits. For further research, we suggest being attentive to how companies and governments are preparing the ground for the next cycle of expansion. The rise of autocratic regimes, together with ever more powerful private security companies and the use of sophisticated technology for surveillance and repression of protests give plenty of reason for a sombre outlook. Extractive and mining transnationals hereby play a key role.

On a more positive note, we can also state that local actors sometimes succeed to contest seemingly all-powerful corporations by using strategic knowledge exchange, alliance-building and the formation of heterogeneous, 'glocal' coalitions. Indeed, where affected or potentially affected people are able to learn about mining impacts and resistance from other cases, and to mobilize this knowledge strategically, the anti-politics machine can be turned into a 'politics machine' capable of demystifying the neo-liberal spell and giving way to a re-politicization of resource use and ownership. Social learning allows the bridging of time cycles and geographic distances and enables local actors to select the best combination of 'glocal' connections and institutional options to pursue their interests. Further, it helps locals to better understand and counteract the strategies of powerful external actors.

Despite mining being a business dominated by transnational companies acting on global scales, the socio-environmental impacts remain local. As much as external actors take advantage of the fragmentation of the commons at the local level, it is those very commons – water, air, land and forests – that become the core battleground for resistance movements. Thus, the reconfiguration of legally disconnected 'natural resources' into culturally meaningful ecosystem landscapes managed as interrelated commons by local people, increasingly comes to the fore as a unifying theme in mining and related conflicts. This ideological ground allows for local specificity while at the same time exploring linkages elsewhere and embracing heterogeneity. And from here, we argue, comes the creative power that can drive the emergence of new institutional arrangements for the sustainable management of the 'glocal' commons.

Notes

1 The authors would like to express their thanks to the contributors of the detailed case studies about the Mopani Copper Mines in Zambia (Eva Schober and Amadea Tschannen) and of the Tampakan Mine in the Philippines (Irina Wenk and Loretta Scherler) which are used in this chapter to exemplify the change in CPRs.

2 In particular, a publication by Berne Declaration (2012) targeted the involvement of Switzerland-based companies in the commodities boom, an issue previously largely ignored by the Swiss public.

3 On-site ethnographic research was not possible in all cases due to lack of resources, but some chapters are based on field research by advanced scholars. Some students started MA or PhD research involving fieldwork on a mining-related topic after their literature-based research for our study.

4 For a literature overview, see e.g. Bebbington (2012); on the 'anthropology of mining': Ballard and Banks (2003); recent monographs include Golub (2014); Kirsch (2014); Welker (2014); De La Cadena (2015); Li (2015).

5 The role of Switzerland is thus somewhat special: based on a long tradition of commodity trading and banking, with its gold refineries holding a key position (Lindt 2016), only

recently did it become host to transnational mining companies' headquarters (Berne Declaration, 2012). Domestic mining was mostly abandoned by the 1950s.

6 See, e.g., Cochrane (2017); Dolan and Rajak (2016); Kirsch (2014); Welker (2014) for social anthropological approaches to CSR; Hilson (2012) for a review.

7 E.g., the case of the Areva uranium mines in Niger shows how the lifestyle of Tuareg nomads is affected by radioactive contamination and depletion of crucial water and salt sources on the migration routes (Oxby and Walentowitz 2016).

8 Some ethnographies examine the internal heterogeneity and tensions within transnational corporations (Welker 2014; Welker et al. 2011). Other authors analyse corporate strategies from the outside and focus on the relationship between corporations and their critics as being characteristic of contemporary capitalism (Benson and Kirsch 2010).

9 Email correspondence with Brian Siegel, 15 May, 2013.

10 Mopani Copper Mines consists of several underground mines and related industries in the towns of Kitwe and Mufulira. Glencore Plc holds 73.1 per cent of the mines, the Canadian First Quantum Minerals Ltd 16.9 per cent, and the Zambian state-owned Zambia Consolidated Copper Mines 10 per cent . See: www.zccm-ih.com.zm/copper-cobalt-gold/mopani-copper-mines/ (accessed on September 5, 2018).

11 In 1999, WMC sold the project to the Philippine-based Sagittarius Mines Inc., a subsidiary of the Australian Indophil Resources NL, which in 2007 sold a 62.5 per cent stake of the project to Xstrata.

References

Acuña, R. (2015) 'The Politics of Extractive Governance: Indigenous Peoples and Socio-Environmental Conflicts.' *Extractive Industries and Society* 2(1): 85–92.

Bebbington, A. (ed.) (2012) *Social Conflict, Economic Development and Extractive Industry: Evidence from South America*. New York: Routledge.

Benson, P. and Kirsch, S. (2010) 'Capitalism and the Politics of Resignation.' *Current Anthropology* 51(4): 459–486.

Berne Declaration (ed.) (2012) *Commodities: Switzerland's Most Dangerous Business*. Zürich: Berne Declaration.

Bolay, M. (2016) 'Artisanal Gold Miners Encountering Large-Scale Mining in Guinea: Expulsion, Tolerance and Interference', in Niederberger, T. et al. (eds), *The Open Cut: Mining, Transnational Corporations and Local Populations*. Zürich: LIT, 187–204.

Bridge, G. (2004) 'Mapping the Bonanza: Geographies of Mining Investment in an Era of Neoliberal Reform.' *Professional Geographer* 56(3): 406–421.

Brom, I. and Bucher, D. (2016) 'The Ranger and Jabiluka Uranium Mines in Northern Territory, Australia', in Niederberger, T. et al. (eds), *The Open Cut: Mining, Transnational Corporations and Local Populations*. Zürich: LIT, 301–328.

Cochrane, G. (2017) *Anthropology in the Mining Industry: Community Relations after Bougainville's Civil War*. Cham, Switzerland: Palgrave Macmillan.

Conde, M. (2017) 'Resistance to Mining: A Review.' *Ecological Economics* 132: 80–90.

Cotula, L. (2013) 'The New Enclosures? Polanyi, International Investment Law and the Global Land Rush.' *Third World Quarterly* 34(9): 1605–1629.

Cotula, L. (2015) *Land Rights and Investment Treaties: Exploring the Interface*. London: IIED.

De La Cadena, M. (2015) *Earth Beings: Ecologies of Practice across Andean Worlds*. Durham, NC and London: Duke University Press.

Dolan, C. and Rajak, D. (eds) (2016) *The Anthropology of Corporate Social Responsibility*. New York: Berghahn.

Doke, C.M. (1931) *The Lambas of Northern Rhodesia: A Study of Their Customs and Beliefs*. London: Harrap.

Doyle, C. and Whitmore, A. (2014) *Indigenous Peoples and the Extractive Sector: Towards a Rights-Respecting Engagement*. Baguio: Tebtebba, PIPLinks and Middlesex University.

Ferguson, J. (1994) *The Anti-Politics Machine: 'Development', Depoliticization and Bureaucratic Power in Lesotho*. Minneapolis, MN: University of Minnesota Press.

Ferguson, J. (1999) *Expectations of Modernity: Myths and Meanings of Urban Life on the Zambian Copperbelt*. Berkeley, CA: University of California Press.

Franks, D.M., Davis, R., Bebbington, A.J., Ali, S.H., Kemp, D., and Scurrah, M. (2014) 'Conflict Translates Environmental and Social Risk into Business Costs.' *Proceedings of the National Academy of Sciences* 111(21): 7576–7581.

Fraser, A. and Lungu, J. (2007) *For Whom the Windfalls? Winners & Losers in the Privatisation of Zambia's Copper Mines*. Lusaka: Civil Society Trade Network of Zambia.

Gambon, H. and Kobi, M. (2016) 'Legal Frameworks and Regulations for the Mining Industry', in Niederberger, T. et al. (eds), *The Open Cut: Mining, Transnational Corporations and Local Populations*. Zürich: LIT, 35–58.

Golub, A. (2014) *Leviathans at the Gold Mine: Creating Indigenous and Corporate Actors in Papua New Guinea*. Durham, NC and London: Duke University Press.

Haller, T., Blöchlinger, A., John,M., Marthaler, E., and Ziegler, S. (eds) (2007) *Fossil Fuels, Oil Companies, and Indigenous Peoples*. Berlin: LIT.

Haller, T. (2010) 'Institutional Change, Power and Conflicts in the Management of Common-Pool Resources in African Floodplain Ecosystems: An Introduction', in Haller, T. (ed.), *Disputing the Floodplains*. Leiden and Boston: Brill, 1–75.

Haller, T. (2013) *The Contested Floodplain. Institutional Change of the Commons in the Kafue Flats, Zambia*. Plymouth: Lexington Books.

Haller, T. (2016) 'Managing the Commons with Floods: The Role of Institutions and Power Relations for Water Governance and Food Resilience in African Floodplains', in Tvedt, T., Oestigaard, T., and Bakke, J. (eds), *Water and Food – Africa in a Global Context*. London: I.B. Tauris, 369–397.

Haller, T. (2017) 'Perceptions and Control of Assemblage in a "Glocal" World.' *Dialogues in Human Geography* 7(2): 207–211.

Hilson, G. (2002) 'An Overview of Land Use Conflicts in Mining Communities.' *Land Use Policy* 19(1): 65–73.

Hilson, G. (2012) 'Corporate Social Responsibility in the Extractive Industries: Experiences from Developing Countries.' *Resources Policy* 37(2): 131–137.

Jenkins, K. (2014) 'Women, Mining and Development: An Emerging Research Agenda.' *The Extractive Industries and Society* 1(2): 329–339.

Kemp, D., Bond, C.J., Franks, D.M. and Cote, C. (2010) 'Mining, Water and Human Rights: Making the Connection.' *Journal of Cleaner Production* 18(15): 1553–1562.

Kirsch, S. (2014) *Mining Capitalism*. Oakland, CA: University of California Press.

Kuipers, J. R., Maest, A.S., MacHardy, K.A., and Lawson, G. (2006) *Comparison of Predicted and Actual Water Quality at Hardrock Mines: The Reliability of Predictions in Environmental Impact Statements*. Butte: Kuipers & Associates and Buka Environmental.

Li, F. (2015) *Unearthing Conflict: Corporate Mining, Activism, and Expertise in Peru*. Durham, NC and London: Duke University Press.

Lindt, A. (2016) '"Speech is Silver, Silence is Golden": Gold Refining in Switzerland', in Niederberger, T. et al. (eds), *The Open Cut: Mining, Transnational Corporations and Local Populations*. Zürich: LIT, 91–102.

Mudd, G. (2007) 'Global Trends in Gold Mining: Towards Quantifying Environmental and Resource Sustainability.' *Resources Policy* 32(1–2): 42–56.

Mususa, P. (2010) 'Contesting Illegality. Women in the Informal Copper Business', in Fraser, A. and Larmer, M. (eds), *Zambia, Mining and Neoliberalism*. New York: Palgrave Macmillan, 185–204.

Niederberger, T., Haller, T., Gambon, H., Kobi, M. and Wenk, I. (eds) (2016) *The Open Cut: Mining, Transnational Corporations and Local Populations*. Zürich: LIT.

Oxby, C. and Walentowitz, S. (2016) 'Uranium Mining in Niger. Undermining Pastoralist Lifeworlds', in Niederberger, T. et al. (eds), *The Open Cut: Mining, Transnational Corporations and Local Populations*. Zürich: LIT, 155–186.

Schober, E., Tschannen, A., and Niederberger, T. (2016) 'The Mopani Copper Mines in the Copperbelt District, Zambia', in Niederberger, T. et al. (eds), *The Open Cut: Mining, Transnational Corporations and Local Populations*. Zürich: LIT, 127–154.

Siegel, B. (1988) 'Bomas, Missions and Mines: The Making of Centers on the Zambian Copperbelt.' *African Studies Review* 31(3): 61–84.

Siegel, B. (1989) 'The "Wild" and "Lazy" Lamba: Ethnic Stereotypes on the Central African Copperbelt', in Vail, L. (ed.), *The Creation of Tribalism in Southern Africa*. London: James Curry Ltd., 350–371.

Sikka, P. (2010) 'Smoke and Mirrors: Corporate Social Responsibility and Tax Avoidance.' *Accounting Forum* 34(3–4): 153–168.

Simpere, A. (2010) 'Project Mopani (Zambie). L'Europe au Coeur d'un Scandale Minier. Raport de Mission. Les Amis de la Terre France.' <www.amisdelaterre.org> accessed on 28 January, 2012.

Toulmin, C. (2009) 'Securing Land and Property Rights in Sub-Saharan Africa: The Role of Local Institutions.' *Land Use Policy* 26(1): 10–19.

Uzar, E. (2017a) 'Contested Labour and Political Leadership: Three Mineworkers' Unions after the Opposition Victory in Zambia.' *Review of African Political Economy* 44(152): 292–311.

Uzar, E. (2017b) 'Politische Mobilisierung gegen Rohstoffkonzerne in Sambia.' *Forschungsjournal Soziale Bewegungen* 30(1): 71–80.

Welker, M. (2014) *Enacting the Corporation*. Berkeley, CA: University of California Press.

Welker, M., Partridge, D.J., and Hardin, R. (2011) 'Corporate Lives: New Perspectives on the Social Life of the Corporate Form.' *Current Anthropology* 52(S3): S3–S16.

Wenk, I. and Scherler, L. (2016) 'The Tampakan Copper-Gold Mine Project in Mindanao, Philippines', in Niederberger, T. et al. (eds), *The Open Cut: Mining, Transnational Corporations and Local Populations*. Zürich: LIT, 363–405.

18 Are green energy investments levelled by the 'new commons'?

Compensations, CSR measures and gendered impacts of a solar energy project in Morocco

Sarah Ryser

Introduction

The quote below draws attention to the gendered perspective of the complexity of the set of problems which the largest solar energy project Noor Ouarzazate brings besides the production of green solar energy.

> I live in a 'Douar'[1] close to the Moroccan solar energy project 'Noor Ouarzazate'. In Amazigh, my Berber dialect, we say 'Douar' for a village. (…). We prefer the Amazigh expression and we do not call ourselves 'Berber'. Now, Noor Ouarzazate is the largest solar energy project in the world with a surface of more than 3000 ha on our land. The land was previously owned by my Amazigh[2] clan, which are the Aït Ougrour, our ethnic community. They say the project is on wasteland, but we used to collect fire wood there and graze on it. When it rains, a beautiful yellow plant grows there, and people transplant it to their gardens … Now, that the King wants the solar project; we won't say no. All that God and the King want, we say, welcome. It's a project and our goal is that our youth can work. But now, people say that they lost their land. And now it is fenced and you can't go on it.
>
> (Chama,[3] female living in a village close to the Solar Project Ouarzazate, Morocco, excerpt from an interview conducted in 2015)

This chapter outlines the processes that led to the development of 'Noor Ouarzazate' ('The light of Ouarzazate' in English), the largest solar project in the world built on the common land of an *Amazigh*-Berber group in Morocco. The case study shows how Large-Scale Land Acquisition (LSLA) impacts existing power constellations and gender relations governing access to land and land-related, common-pool resources (CPRs) (pasture, veldt products, animal fodder) and leads to a loss of these resources previously managed as common property. On paper, this loss is compensated by payments and state-driven development programs, as well as by contributions from the operating company's Corporate Social Responsibility Projects (CSR). It is argued that these projects provide a form of 'new commons', which promise to deliver on-going revenue and development. The chapter focuses on the land acquisition process that includes the institutional change from common to state and semi-private property. By considering the emic perceptions of access to and distribution of the 'new commons', the paper addresses the question of whether the loss of the old commons is really compensated by the 'new commons'. It further aims to identify the scope and scale for action in transformed gender relations that are now challenged by the solar investment, supported by the state of Morocco (especially the King), the World Bank, the EU and other international donors. The chapter aims to advance our understanding of ongoing large-scale green infrastructure investment processes and how historically-evolved and transformed common property institutions and gender roles are exposed to 'glocal' changes as well as to discourses of development and gender equality via forms of compensation and CSR schemes. Considering the historical changes and transformations of land rights during the pre-protectorate, protectorate and the post-protectorate period, I focus on women's options to negotiate depending on how their bargaining power changes after the arrival of the investment. This depends on the perception of options available to them and how they make use of these possibilities within the context of the emerging investment project. In this regard, the analysis goes beyond simply assessing the effects of such large-scale investment by also considering the dynamics of land use change and its implications on gender relations in the context of renewable energy production and emerging development projects. Therefore, I pay close attention to the way compensations and CSR act as an anti-politics machine (Ferguson 1994) at the local level, as it masks losses by pointing at inaccessible benefits, especially for more marginal groups and women. To assess this impact, the chapter argues that we need to know more about the women's political and economic positions and how these were affected by institutional changes before the investment took place (see Charrad 2001; Sadiqi and Ennaji 2011). Finally, I look at the national and local Moroccan context within which the Solar Complex Project Noor I, II, III and IV Ouarzazate operate, and how this affects changes in property rights from common property land ownership by the Aït Ougrour to state and semi-private ownership used for the solar project.

Theoretical perspective on gender and large-scale land acquisitions/ land grabbing

Preliminary evidence from within the growing LSLA literature indicates that women's access to land, including CPRs (i.e. fuelwood, water, wild foods) is threatened by (CSR) investments (see Schoneveld 2010; White and White 2012). There is, however, a notorious gender blindness in the LSLA literature as well as in literature generally (Doss et al. 2014). Women's experiences are mostly conflated with men's, or women are treated as a homogenous group (Doss et al. 2014) and/or add-on category without having an explicit contextualized gender focus. Research questions in relation to LSLA and land grabbing topics, including a concrete gender focus, are widely missing in property rights and development literature (see Behrmann et al. 2012; Doss et al. 2014). In order to fill this gap, I use a combination of new institutionalism in social anthropology (Ensminger 1992; Haller 2010, 2013), neo-Marxist inspired feminist political ecology (Meillassoux 1982; Narotzky 2016) and approaches focusing on 'modernity' (Appadurai 1996; Abu-Lughod 1998, 2013), as well as post-colonial transnational feminist theories (Chrisman and Williams 2015). This approach enables me to explain the drivers of institutional change and how they shape people's access to resources. As well, it reveals impacts on 'households'[4] mode of production/reproduction and gender relations of exploitation (Meillassoux 1982). New institutionalism combined with political ecology (Haller 2002, 2017) leans on an actor-oriented approach. It discusses how external factors lead to changes in the value of a resource context (changes in relative prices) which further shape internal bargaining power of actors as well as ideologies of gender relations, impacting the institutional choice and finally leading to unequal distributional effects. The combined approaches predict that the rise in the value of land via external effects or interests contributes to lowering the bargaining power for women.

Under the common property regime, women had collective ownership and access rights to CPRs. Furthermore, the rising value of land decoupled from CPRs and legitimized by discourses of male representation, property and control has led to property and rule changes in favour of more powerful actors. In the context of these newly organizing gender relations, the research reveals evidence of different perceptions of men and women. Evidence suggests increased attempts by state actors to enact gender policies aimed at increasing the bargaining power of women in Morocco since colonial times, but this has yet to be translated into practical policy outcomes (Charrad 2001; Agnaou 2004; Sadiqi and Ennaji 2011; Sadiqi 2014). Therefore, despite rising bargaining power for women as a category since colonial times, women still get excluded and at the same time are not given access to the new institutional options after such changes (Bergh 2017).

Feminist political ecology sheds light on the power dynamics and ideological processes that create legitimacy for 'gender sensitive' investments. While project designers indicate that they create development opportunities for women by creating opportunities for redistribution of access to 'new' resources, such as job opportunities or benefits from CSR schemes, they often come at the cost of CPRs for women. These new benefits are often labelled in general terms such as 'development' that LSLA investments provide to local areas, but the kind of development they provide needs to be analysed from a local perspective. Are these new opportunities a valuable substitute for the loss of access to CPRs with their continuous reproduction? Or do these, on the contrary, hide the loss of the old commons, which are not compensated for as promised? The interesting issue here is that in the discourse of investors and the state, the promise of betterment, not compensation, is key. In this betterment discourse women are used as a positive norm orientation but are not understood for their role in production and the reproduction of the means of labour for the local economy.

Meillassoux (1982) has indicated that, 'women, despite their crucial role in reproduction, never appear as vectors of the social organization. They are hidden behind men, behind fathers, brothers and/or husbands' (Meillassoux 1991: 75). This adds to the gender discourse aspect because, according to Meillassoux, the situation of women 'is not a natural given condition but one which results from changing historical circumstances, and always linked to the exploitation of women's reproductive functions' (Meillassoux 1991: 75).

Following the argument of Ferguson's anti-politics machine, it may be suggested that LSLA projects can hide the fact that they increase the ability of the state, local leaders, NGOs and the company, etc. to gain access and control over the rural economy, which could reduce local options and bargaining power, especially for women. In this regard, I draw on feminist discourse on modernity (Abu-Lughod 1998) to investigate whether such claims of women's empowerment and gender equality discourse by the Moroccan state, in partnership with the parastatal company, MASEN (Moroccan Agency for Sustainable Energy), are translated into practise at the local level. This analysis is linked to discourses of development (see Escobar 2011) in which 'women's empowerment', combined with a discourse promoting national and international versions of modernity and the notion of sustainable development are used to attain legitimation for green energy projects. Such legitimations of LSLA investments are then perceived not only to bring green energy and development in general but also to enhance a gender-sensitive development outcome. Insights from feminist critique of development (Prügl 2016) help us to shed light on questions about who has the power to define and to decide what is necessary for gendered development and how CSR projects should be framed to bring development to a rural *Amazigh* woman. It asks the question, who are the women who shall benefit and how should this lead to a gender-balanced, sustainable development? Therefore, the theoretical perspective does not only address the issue of green grabbing and its impacts on women but also considers the gendered power asymmetries created as a result of such commercial investments.

Methodology

The research was carried out in in the surrounding area of the largest solar project in Morocco, situated close to Ouarzazate (see Figure 18.1). Fieldwork comprised of more than 12 months of data collection in the field between 2015 and 2017. As a qualitative research project, I relied heavily on participant observation in combination with narrative interviews, informal conversations with experts, associations and NGO representatives and focus group discussions with both men and women (e.g. sex workers, local women, solar project workers, etc.). This was then augmented with in-depth biographies and secondary data from the literature review.

The research process first involved an exploratory phase at the beginning of July 2015 when I travelled to the study area to make initial contacts with local actors and tested the questionnaires. This was then followed by the second phase of intensive data collection a month later, after all, adjustments to the questionnaire were completed and updated. A prolonged period of stay in the local community with members of my extended family opened many doors during the research and granted access to the local community members.

The natural and infrastructural aspects of the case study area

The case study area is near Ouarzazate in a semi-arid and arid climate zone with little and erratic rainfall. It is enclosed in the North of Ouarzazate by the mountain chain of the High Atlas and to the south, by the Anti Atlas and the Saghro Mountains. In the past, the long stretches of mountains with lowland areas in between have resulted in the creation of a few rivers including the rivers Draâ and Dades, as well as the Oued Izerki. Today, the rivers Draâ and Dades serve as the main sources of water to the local inhabitants. Oued Izerki serves as an important source of water for agricultural irrigation, as well as meeting household needs for water in the area and its nearby villages. The area has not experienced many investments yet; the 'Mansour ed-Dahbi' Dam on the territory of different Berber groups such as the Aït Ouarzazate and Aït Bou Dellal is the only noteworthy infrastructural development besides the solar project investment. The dam became the major source of water for all-year-round commercial and subsistence irrigation farming. During dry seasons, this dam serves a critical role in the distribution of water to the Ouarzazate township and the whole of the Draâ Valley area. It is also the main water reservoir in the area during heavy rains (often in winter) and helps to prevent floods in the Ouarzazate and the Draâ Valley area. Still, water scarcity is a big challenge for the region and is becoming much worse. It remains unclear how the water needs of the solar project will add to the demand for water during the dry seasons.[5] Apart from the dam, and dating back to the time of the French protectorate, large tracks of land from the area had been separated out for the construction of an airport and the military base in 1978, both of which were previously part of the Aït Ouarzazate land (Tawrirt land is a subgroup of Aït Ouarzazate). Therefore, the current extensive extraction of 3,000 hectares of land for the solar project adds to land loss for local people (see Figure 18.2).

On first sight, the ecosystem consists of a large stretch of dry plateau containing little vegetation. However, the plateau hosts more than 20 locally known plants during the rainy and after-rainy seasons (for example *izri*/wormwood,[6] *astay*/ Haloxylon-Hannad Elegans[7]). Several dry river beds containing water in the rainy season cross the hilly areas but not the plateau. More fertile land is found close to these river beds based on higher humidity but also water sources which serve for irrigation and all year plantations (temperatures average 20°C with a maximum temperature range between 42–45°C in July/August and minimal temperatures in February of about –9° C). High sun radiation on the plateau makes it seemingly ideal for installing a large solar plant for concentrated solar power (CSP).

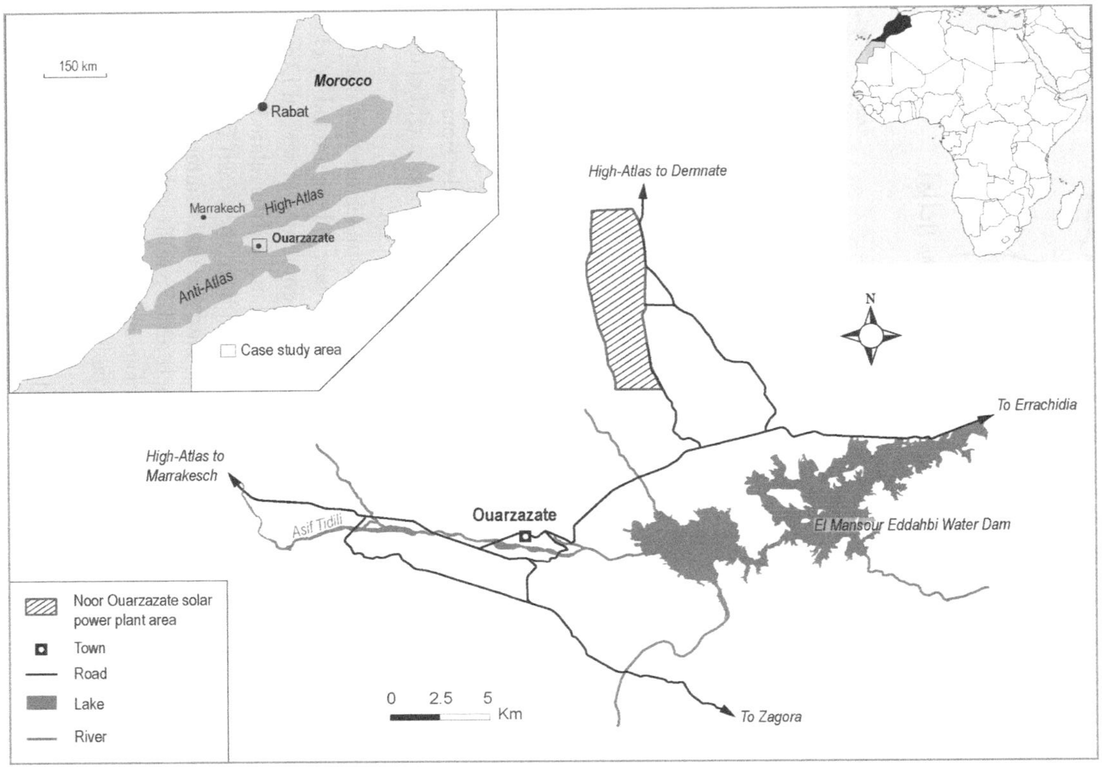

Figure 18.1 Location and map of the wider case study area.
Source: Map compilation and design by Ulla Gämperli Krauer (2018), CDE, University of Bern, Bern, Switzerland. Geodata source: OSM data by OSMaxx (2018). Geometalab, HSR University of Applied Sciences Rapperswil, Switzerland.

Figure 18.2 Segmentary Moroccan *Amazigh* Clan near Ouarzazate, between the High Atlas, the Anti Atlas and the Saghro Mountains.
Source: Drawn by the author, based on Hoffman (1958) and Bidwell (1973).

Local livelihoods and common property institutions

The land of the area in which the solar energy project is located belongs to the Aït Ougrour ethnic group, which in turn is part of a larger sub-group or 'confederation' called *Imghrane*, which is again part of the 'Berber people' who call themselves *Amazigh* (see Figure 18.2). However, as Moroccan scholars claim, *Amazigh* identities can be mixed because Morocco has never been totally 'homogenized', 'Arabized' or 'Islamized' (Sadiqi 2003: 17). Therefore, local groups might claim to be of several sources of descent, including Arab and *Amazigh*.

Households among the Aït Ougrour are organized in segmentary clan and lineage groups living in village settlements called *Douar*. These are led by elder men who establish a council. The Aït Ougrour are composed of several *Douars*. These village groups are communal co-owners of the land and the related common-pool resources (pasture, water, veldt products) and have their own independent decision-making processes on matters such as land use and communal land and resource governance. At the times of French and Spanish colonization, *Caïds* (plural of *Caïd*, which means leader or head of an authority office) were appointed in communes to maintain control. It was a type of indirect ruling office structured of men representing the state. In every district, the *Caïd/a* exercises a great deal of power, keeps the peace, manages people's daily affairs in terms of security, policing and gendarme armed forces, and represents the authority at the communal level. However, the *Na'ib* or *Na'iba* (female in very few cases) is the elected person to speak on behalf of the people.[8] For example, the communal land representative is the *Na'ib aradi aljoumoua*. The *Na'ib* or *Na'iba* is elected by his/her community as the person responsible for managing the collective land affairs and maintaining contact with the state's Directory of Rural Affaires (DAR). Once the land has been divided into private properties there is no need for such *Na'ib/Na'iba* anymore. M'Hassni et al. (2003) has cautioned against considering such selection processes as 'elections' because the selection is done by a local council comprised exclusively of older men and is therefore not gender and age representative. As such, the established structure prevented women and youth from articulating their will and has curtailed their bargaining power. Nevertheless, when it comes to gaining access to the commons, the younger generation and women were able to co-own and use the commons as members of a local commune.

The Aït Ougrour communities live from small-scale agriculture, small animal husbandry, the sale of handcrafts and jewellery, and tourism. In formal terms, these groups also include formally educated[9] people working in government administration. For the larger part of the population, however, the most important cash income stems from farming, cultivating dates, almonds, pumpkin, stone fruit, Lucerne,[10] carrots, tomatoes, henna,[11] melon, aubergine, animal rearing (e.g. goat, sheep, cattle) for meat, as well as milk production for subsistence and cash. Prior to the solar project investment, settled people (among others) depended on the area – now labelled as 'wasteland' by the state and the company – to collect and hunt CPRs such as firewood for cooking and heating, small animals, and animal feed under common property regimes of clan, lineage and village-groups. In fact, local herders could identify more than 20 plants growing in the area of the solar project. Parts of these plants are aromatic and said to give the goat meat a special taste, rendering the meat valuable for the local markets near Ouarzazate. Several plants were also used for traditional medicine before the solar project was constructed. In both businesses – traditional medicine and meat production – women were involved.

Herders and semi-nomadic groups with goats and sheep used the area on a reciprocal basis under *Amazigh* institutions of reciprocal communal access. Some herders from a neighbouring community had reciprocal use rights with the Aït Ougrour community. They used the now fenced 3,000 hectares as a temporal wet season pasture. Other herders from a neighbouring community had rights for reciprocal passage for transhumance. While sedentary farmers and people on passage used the land around the solar project for sustenance activities, such as hunting small animals and extraction of related resources including but not limited to grasses and herbs, the semi-nomadic people migrated to the area with their animals and stayed there for some time while caring for their animals. The area not only served as a grazing site for food for their animals, but it also provided shelter and close proximity to water sources. This allowed people to move between seasons, balancing their livelihood strategies to maintain sustenance. In addition, the area was also used as a trading centre where people used to herd their animals. They sold animals to people from the towns, who, in turn, purchased them or exchanged other consumable goods for livestock, which was later sold at the local market in Ouarzazate township. As such, the area did not only serve for transhumance or as passage but also as a forum for the exchange of goods and services and a strategic junction for local trade facilitation.

In summary, the data reveals that this area was not 'wasteland' or 'worthless'. Instead, the land served diverse and unique purposes and was regulated under the local common property institutions of the Aït Ougrour community. These findings underline the importance of the old commons. This changed in 2009 with the establishment of the solar energy project, which was based on external processes linked to the national government policies for rural development and energy sufficiency.

Creating national and local identities via investment projects

The Royal Family in Morocco, especially King Mohammed VI, can be considered as an 'arbitrator' among the different Moroccan populations. During protectorate times, the French had created tensions between the *Amazigh* and Arabs, which were reconciled by Hassan II, who chose an *Amazigh* from Khénifra for a wife. This clever move can be considered as one of the first steps toward national unity of the Moroccan people after French protectorate ended in 1956. Hassan II also constructed the dam in the region, which was contested until local people realized its utility for irrigation. Nevertheless, King Mohammed VI is locally admired for his social projects and thus the solar project under his banner was not criticized at first. The current King made efforts to: a) fight poverty, which meant bringing infrastructure and work into marginal regions; b) introduce a new family law (*Moudawana*), which was intended to change, at least on paper, women's inheritance rights; and c) normalize the political situation after King Hassan II, which contributed to local conflicts (Lugan 2011). The current King made the promise of gains from the energy policy, such as the creation of new jobs for the local population and at the same time, to make Morocco an energy independent country as well as generate surplus green energy for export to other African and European countries. With this plan, the Moroccan government endorsed a local infrastructure development policy called 'regionalization' allowing regional autonomy through devolution of power to regional provinces. This move makes the large-scale land acquisition a relevant issue in the politics of regionalization; it comes at a time when regional provinces have some autonomy over the land. However, despite this politics of regionalization it was still under the control of the King, who did (and still does) not focus exclusively on solar energy. In the King's discourse, the solar project is also linked to local development initiatives, which are labelled to represent regional development that combines 'traditional' and 'modern' approaches. One element of that discourse is the promotion of traditional handicrafts as a means of keeping local traditional knowledge and to create jobs in rural areas. Together with the solar energy project, these strategies constitute a holistic policy which the Moroccan national government, controlled by the King, uses to promote its view of rural development. Furthermore, national level development projects involving large-scale land investments in agriculture and infrastructure are on the top of the political agenda. Such projects include the largest ports in Africa (the 2009 Tanger Med and Tanger Med II) and the national railroad network. The railroad (TGV) network will link the port of Tanger Med with other important industrial centres and should guarantee fast transportation of passengers as well as goods and services. All these investments are legitimised by the argument that they are boosting the nation's green development and provide new job opportunities on all levels, last but not least, also for women. Furthermore, the infrastructure projects are also legitimized by the argument that they reduce the country's dependency on foreign supplies to meet its growing energy demands (Bahgat 2013: 291). The specific geographical location of Morocco is argued to be very suitable to produce solar, wind and water energy. In addition, Morocco is also the only African country with a power cable link to Europe (Bahgat 2013: 29). The project is financed by a public-private partnership consisting of the Moroccan State, the Clean Technology Fund, the World Bank, the African Development Bank Group, KFW (Kreditanstalt für Wiederaufbau), Entwicklungsbank Bundesministerium für wirtschaftliche Zusammenarbeit und Entwicklung, European Investment Bank, European Commission, AFD Agence Française de Développement, investors from Japan, other commercial banks and global private partners and players.[12] These investments are done from the King's perspective. On the national level, the Moroccan King Mohammed VI and the government demonstrate a strong intention to realise these ambitious renewable energy projects. Additionally, the plans to use green energy for water desalination in favour of irrigation systems and to serve inhabitants are also part of the Moroccan energy policies. The projects in energy production and water desalination are linked to agricultural development programs. In public discussions about the subject, the King and politicians underline the importance of these projects for all Moroccans. In their discourse, they aim to stimulate a sense of community among Moroccans.

The land deal and its implications: development or commons grabbing?

The implementation of green energy sufficiency policies often involves the loss of local land for said developments. Local people have to endure this loss, despite the national development discourse, of which they are also part. In order to establish the solar project, *Amazigh* land held in common property had to be expropriated by the state, in this case via the *Office National de l'Electricité et de l'Eau potable* (ONEE). The ONEE then transferred the land to the national company, 'Moroccan Agency for Sustainable Energy' (MASEN), which is financed by the Moroccan state and private partnership of mostly European Investment banks (including strong German participation). This transfer of land was done because, in Morocco, land in common property can only be rented but not sold to private individuals, especially not to foreign companies. However, in this case, the transfer of the common land was allowed as it was done via the state represented by ONEE, which acquired the land legally in order to hand it over to MASEN, which is also seen as a private-public partnership. According to MASEN, the acquired land from the Aït Ougrour community was mainly owned by the Izerki *Douar* and, to a lesser degree, by the Tidgheste *Douar*. However, it is also said that Izerki encompasses three villages (Tasselmante, Essour and Oum Romane). These three villages share one *Na'ib*. The two *Douars*, Izerki and Tidgheste, are geographically the closest to the project site. Furthermore, three more *Douars* – Tiflit Timatdit, Ighram Amallal and Tafarghoust – are affected as they also have a stake in this common land where the solar energy project was established. Five *Nuwwab* (plural of *Na'ib*), all elderly men of the closest local communities, were invited by the local authorities (probably the Governor, the President of the community of Ghessate and a local government representative *Caïd* of Ghessate) and had to sign a sales agreement of the land belonging to the five communities (see Figure 18.2 for locations and names). Interestingly, MASEN mentions in a report (MASEN Report 2011) that only three representatives signed the sales agreement (Izerki, Tidgheste and Tafarghoust). The report states that two representatives felt bad for not having enough information about the project and did not sign. Nevertheless, the government perceived this as a positive response based on the argument that a majority of the villages had signed, which is legally questionable since for a legal sale two signatures are missing.[13] For instance, among the three Nuwwab who signed, one of them confessed to having signed the document under duress because of an anonymous threat he received about the land deal (Informant, January 2016). This information became clear to me after the *Caïd* himself cautioned us against talking to the local people in the communities. He also informed the local people ahead of my visit to the communities to not welcome or talk to any foreign person about the project (Informant, Interview, February 2016). In addition, there was a disagreement on the price of the land. The transaction price was fixed by the *Direction des Affaires Rurales* (DAR) from the Ministry of the Interior first at 1 DH per m^2.[14] This was the price for about 2,500 hectares. Later, the DAR increased the price to 3DH per m^2, which was the price paid for the remaining 500 hectares. The initial price, as well as the increased price, are argued to be common for the region and the type of land. However, even 3 DH per m^2 seems to be low considering the investment planned and the fact that this land involved reciprocal arrangements with external people such as the Semi-Nomads from the Atlas Mountain and local communities. In addition, the labelling of the land as 'wasteland' does not consider the losses for meat production and options for mobility, which might be another reason why two Nuwwab refused to sign. They shared the opinion that the administration's price was too low. However, the ministry refused the demand to increase the price. As one relative of a village leader recalled:

> My uncle (the Na'ib of our village) told me, that the Ministry of the Interior contacted the Nuwwab. At least one Na'ib asked them to increase the price but they refused. One Na'ib told them it should be 10 DH per square meter, instead of only 1 DH.
>
> (Informant, male, Tasselmante, Morocco, excerpt from interview conducted in April, 2016)

Interestingly, for reasons that are unclear, in 2014 MASEN paid a three-times-higher price for the second part of the land. This raises questions about the process of fixing a low price in 2011. Regarding the official documents, MASEN justifies the value of the land in 2011 and 2014 based on the same legal paragraph (Masen 2014).

The non-transparent means of fixing the land price is not the only issue. Another problem stems from the fact that money was not paid to the local groups: MASEN transferred two instalments to a special account belonging to the Ministry of the Interior who hold the money on behalf of the Aït Ougrour community.[15] The DAR manages the account and decides with the Community Supervisory Council on how the money will be used. In this way – according to the discourse of the administration – local communities' expectations are better addressed. However, local people do not feel that they are asked how the money shall be used and thus do not feel involved in the decisions. While my fieldwork indicates that the DAR argues that it organizes meetings to review local needs, this did not happen often and the demands of the local people have not been taken up during these sessions according to the local views. DAR administrators seem to prioritize development projects which they believe will profit local people. However, locals do not feel included; rather, they think that development decisions are imposed upon them. There is no transparency regarding the use of the money and there are no direct cash payments; people are told to make official requests, to which there is either no positive response or not in due time.[16]

Three major hegemonic state discourses further justify the land deal, which is locally perceived as unfair. First, the implementation of the solar project is linked to the discourse of agricultural development. It states that community members, who have lost access to land and land-related CPRs, after the land deal, will become incorporated into modernized agricultural projects. Furthermore, the state argues that these projects will generate encompassing agricultural development, helping to link small-scale farmers to the national and global market, including the use of sustainable energy (Rignall 2012). In addition, state discourse refers to the gender-sensitive national agrarian policies related to sustainable green energy production. Second, the government tries to reduce open resistance through a national identity discourse. One example is the government's claim that so-called 'worthless land' (*terre inutile*) has become 'valuable land' (*terre utile*), in order to develop marginalized Berber-Land in rural areas. As such, officials combine the development discourse with ethnic identity discourse. They argue that the solar project is also in the interest of the local *Amazigh*. By converting the land into solar fields, people from these marginalized ethnic groups become – in their distinct way – citizens of the state who contribute to a common good. However, this does not mean that their distinctiveness is undermined. This became symbolically evident when the *Tamazight* Berber language was declared an official language in 2011, which meant it could now be taught in schools. Thus, the *Amazigh* are citizens but remain a distinct ethnic group. Third, the government and MASEN representatives legitimize the solar project with a gender-equality discourse by arguing that women in local communities are given the possibility to participate in the new initiatives supported by MASEN, especially in female cooperatives governed by the *Initiative Nationale pour le Développement Humain* (INDH) and 'Dar Taliba of Ghessate',[17] a student centre (*Banque Africaine de Développement* 2014 & INDH *note d'orientation* 2011–2015). In this centre, there will be an educational establishment for professional education (for example, in carpet weaving), literacy, nursery education and sanitation education, linked to the gender-beneficial outcome of the solar project.

CSR *as a green and gender specific anti-politics machine?*

The positively-labelled outcomes linked to the sale of the commons were not all that was presented to the local groups. In addition, MASEN included a total of 38 projects, which are financed by the company and which are labelled as part of the company's corporate social responsibility (CSR) program for the local communities. These are not said to be additional compensations but are set up as a sign of the company's voluntary responsibility. These projects include the provision of mobile health units, such as a mobile hospital stationed for two days each year in the principal village Ghessate. They also include school buses, boys and girls' dormitory, stables for sheep and goats, welding courses, sponsorship of a local marathon, vacations for children and allocation of funds to NGOs that are implementing rural agricultural development projects (both national and international NGOs). The projects are aimed at different groups and should bring benefit to women and men at the local level as well as youth and elderly people (see Table 18.1).

Despite the state and the company's claim of a double benefit for local people ('wasteland' purchased for development including CSR projects), the institutional change from the commons to state and private ownership of land did not bring the expected benefits. The payments for the land, and especially the CSR development projects, hide the fact that women now lose access to the commons. At the same time, they are being excluded from the direct decision-making processes. CSR policies are not crafted in cooperation with the women; rather, they are developed in a top-down manner for them. In addition, benefits are not directly accessible to all women and especially not to marginal groups such as poorer villagers and transhumant pastoralists. Therefore, the state and company's image of the process as being fully participatory is not reflective of the actual practices on the ground and acts as an anti-politics machine (Ferguson 1994). This is particularly the case given that the state officials exercise full control over the project on all levels. The parastatal company also hides behind the CSR projects, using it as an example to legitimise their development position as gender sensitive and 'green'.

Local perceptions on the implications of 'new commons'

On paper, locals could benefit from projects financed through funds acquired from the land sale and through MASEN/Acwa Power's[18] CSR projects. Several social and local development activities have been conducted or are ongoing,

Table 18.1 Table showing the planned benefits of CSR measures for different target groups

Publication for MASEN and Acwa Power, the operator of the Solar Project	*For children*	*For the whole community*	*Gendered development programmes (mostly for young people)*	*Support and activities in regional/local organisations and associations*
Sponsoring the Ultra-Marathon 'International d'Ouarzazate'	School buses	Distribution of warm clothes	Welding courses (addressed to men)	Support for the Ouarzazate orphanage (les yeux de l'Aicha) 100m^2 of photo-voltaic panels and insulation
Sponsoring the Moroccan Solar Festival 'Ouarzazate'	Scholarships for Bachelor students from low income families (25,000 Dirham per year per student for five years)	Mobile hospital	Collaboration with the French NGO *Agrisud*. Different development projects in the agricultural sector in local villages Stock farming, pomiculture (mostly Almond trees), aviculture, transferring technical knowledge in agriculture and knowledge in business management	
Sponsoring the Municipal Club d'Ouarzazate	Distribution of warm clothes, bicycles and school equipment for children	Basic infrastructure like paved roads to villages		

(*continued*)

Table 18.1 Cont.

Publication for MASEN and Acwa Power, the operator of the Solar Project	*For children*	*For the whole community*	*Gendered development programmes (mostly for young people)*	*Support and activities in regional/local organisations and associations*
Sponsoring the IGRAR Festival for the Development of the Arts	Summer Camps	Opération 'Zero Mika'. No plastic bags (awareness-raising campaign regarding negative impacts of plastic bags in the environment)		
	Linguistic sustainability in the schools (to learn and make progress in foreign languages)			
	School renovation and investment in equipment for the local library			
	Sponsoring the IGRAR Festival (Association IGRAR pour le développent et l'Art)			

Source: Table drawn by the author.

namely: irrigation, agricultural water management, village water supplies, crop area protection, construction of village highways and feeder roads, and the construction of health and school infrastructure. However, what looks like a gender-sensitive 'new commons' on paper, from local perspectives it does not seem to be a commons giving continuous access. Different local stakeholders realize that they did not receive what they were promised and thus began to question the rhetoric of participation and to assert their rightful share. My social anthropological research data show that local actors – men and women from all the villages where I conducted fieldwork – did not seem to feel involved in the gains of the 'new commons'. On the contrary, what was stated on paper and in interviews with state representatives as well as with those from MASEN, and especially with DAR, was not visible on the ground. Therefore, local villagers addressed me openly and asked me to help them by acting as a go-between to contact the persons responsible for the implementation of the CSR projects from MASEN, and to transmit to them what local people needed or wished. Others asked if there were possibilities to take part in the decision-making process for what forms of projects should be initiated, in order that they could derive some benefits from these initiatives. What was implemented by the state and MASEN was not visible for them. The following citation from a young man shows that participation in local initiatives was in vain:

> A group of young people had the idea to ask for the financing for a project of our own. But they (the DAR), along with the Caïd (the local state authority), are the ones responsible for the money. We wanted to fund an association with the money but the Caïd refused to grant us permission. We have the right to be the ones responsible for the fund, but they refused.
>
> (Informant, male, living in a village close to the Solar Project Ouarzazate, Morocco, excerpt from interview conducted in March, 2017)

The statement indicates that locals perceive a combination of disinterest from the DAR and MASEN to get local involvement. It also discloses local power relations with elites, such as among village-level state authorities who seem to control the funds and did not want local, bottom-up organizing of local people. The discourse of having ideas and wanting to set up their own projects indicates the will to have local control over the funds and projects. It also shows a desire for responsibility to be held not by state administrators but by local actors; a desire which is refused by the more powerful. This led to the feeling of being unable to get involved and generate development ideas, which was the goal of the money from the sales to be managed by the state department in the first place. Another citation even goes further:

> It was said that an international association was given 3 million dirhams! It's absolute nonsense. We know that they're stealing the money. Now they say the money is no longer available.
>
> (Chama, female, Tasselmante, Morocco, excerpt from interview conducted in March, 2017)

Alongside these local discourses of being side-lined and of money being stolen by the state or NGOs (who are often managing MASENs CSR projects), the state's discourse of gender-sensitive development in the CSR is challenged by locals who are of the opinion that claims of the process as being participatory is not evident since development has ignored the views and wishes of local women and youth. The CSR projects seem to enforce hegemonic, male-biased stereotypes and gender ideologies, which influence the choice of CSR projects. Whereas MASEN's CSR projects are said to be for the benefit of the entire local population, which should include projects focussing on men, women, girls, boys and the whole community. The evidence suggests the contrary. The projects are structured in a manner that there is a clear distinction between what activities should be reserved for women, and men. Most projects, which focussed on training and skills development for future job acquisition were promoted primarily for men whilst projects such as collective handcrafts (e.g. carpet weaving, sheep rearing, etc.) were targeted mostly at women – with the expectation that the women remain at home or within the locality. Through these courses, MASEN argues that men might have job opportunities in the future in other Moroccan green energy projects. Professional education is geared toward less valuable agriculture and handcrafts to promote local products and is oriented towards women; traditional handcrafts are considered to be an important gender-specific income source, which are also socio-culturally accepted and perceived as women's activities. Furthermore, there is no local discussion between the project agent and local people to learn if rural women would like to be educated in this way or to know what form of education they want. As Chama, a young women from the village, summarizes in her statement, most projects for rural women are focused on local handcrafts, which do not offer them the possibility to work in the project in the future.

> It's easier for the urban women to get a job in the solar project and to benefit from specific technical training programs but for us here, we can only participate in the handcrafts and mat weaving training. If this is the modernity the project will bring, then it is not for us here.
>
> (Chama, female, Tasselmante, Morocco, excerpt from an interview conducted in March, 2017)

Since the project aims to be the engine for growth in the rural area, the rural women also wished to have been given the opportunity to become employed in the sector by learning welding and other related technical training, but such demands are often frowned upon by the management of these CSR projects. Other forms of CSR projects include the provision of a medicinal caravan and some school equipment with services such as summer camps for children from the Aït Ougrour community. Such services are not accessible to all villages and especially not all women have access and only very few profits from these rare opportunities. Another way community members can profit from MASEN is through employment in the project itself but these are only short-term jobs. Evidence indicates that labour demand is only available for temporary employment, which lasts for a few months in the construction industry. In the entire project, 400 people are employed from the whole area. It can be argued that before the project there were no job opportunities for local men and women and most of them had to search in other parts of Morocco for employment. However, these few jobs do not seem to compensate for the loss of access to the land and CPRs for the wider population. Especially for women with very few job opportunities, grazing and feeding goats to be sold in the city and thus earning a small but stable income – an option that is now limited – might still be a much better choice. Finally, the reciprocal arrangements for access to pasture with semi-nomads from the Atlas Mountain from autumn to spring are compromised by the investment. These semi-nomadic people are not part of the *Amazigh* Clan Aït Ougrour, but nevertheless, they had access to the land. Because of the ways in which local common property institutions have been undermined – for the so-called 'benefit' of all Moroccan citizens – the irony is that now, the *Amazigh* Clan Aït Ougrour will lose access to this land.

Discussion and conclusion

The land acquisition process in Morocco is sold as green infrastructure investment bringing benefits to local people via common land sales, of which the revenues are pooled in a fund that shall generate participatory development and jobs. In addition, the para-statal company initiates voluntary CSR projects. All these benefits can be considered as 'new commons' since they promise to offer continuous gains generated by the loss of the old commons via compensation. From a local perspective, however, the investment is part of a new form of green grabbing pursued by the Moroccan government and a private investor supported by European funding and technology. The major problem is that the old cannot be replaced by the 'new commons' because the promised 'continual gains' do not materialize.

Although compensation payment funds and CSR projects both promise development, they act as anti-politics machines (Ferguson 1994) and hide power asymmetries in the sale of the land and the decision to install the solar project. In addition, they also function as anti-politics machines in that they ultimately reduce the capacity of women, nomadic groups, and men in the community to use the commons for subsistence. This prevents and/or reduces a steady flow of income, leaving people more vulnerable to shocks. Compensation payment funds and CSR projects hide 'encountering development' (Escobar 2011) with hegemonic discourses of how green and gender-sensitive development should be (see Doss et al. 2014). Finally, they obscure the fact that land that is not at all 'wasteland', but rather, already useful, has been grabbed from local people.

In relation to a gender and social justice analysis, there are two final implications of this process to consider. 1) Women's options in their relationship with men are negatively influenced by the project as it limits options of livelihoods for women contrary to what is advocated in the CSR projects. The same is true for semi-nomads descending from other *Amazigh* Clans who are part of reciprocal arrangements that do not hold anymore because they can no longer get access to the territory. Similarly, CSR policies provide guidelines to create collective compensation mechanisms that we refer to as 'new commons' but these redistribution processes have resulted in a change in local people's relationship to their natural environment and have created winners and losers. Despite the initial argument of the Moroccan state and companies involved that the land is 'worthless' and is now being put to use for the benefit of all Moroccans and for the people in the affected area, realities on the ground seem to differ as: (a) Aït Ougrour women were not involved in the sales negotiations; (b) semi-Nomads descending from other (*Amazigh*) Clans who also depended on the land are entirely left out; (c) in practice, not all local actors get to participate in the CSR project because of cumbersome administrative procedures. Additionally, job opportunities also limit participation and yet no form of targeting is used to identify the most affected by the land loss and the most vulnerable; and (d) women and semi-Nomads from other ethnic communities both face difficulties accessing additional CSR projects because the management of such projects propagates existing (gender and other) stereotypes and nomadic cultural perceptions, which inhibit their full participation in some of the CSR projects. 2) At first glance, the investment seems to be beneficial because CSR broadens livelihood opportunities for the entire locality and different users to have access to the programmes, which may be seen as a benefit in the form of commons. However, not all people who used the commons previously are now able to get full access to the projects and gain benefit from them. In addition, there is a high degree of uncertainty regarding the benefits in the long-term as MASEN bases its activities on its voluntary commitment linked to the solar project. One cannot, therefore, consider such development resources as a desirable form of new CPRs that can replace the losses. Notwithstanding all the environmental and social benefits being derived from the solar project in general, one cannot neglect the lack of involvement of women in the decision-making processes and that this has resulted in the neglect of women's access to veldt products as well as loss of land for transhumance and pastoralism. Women's concrete participation in the decision making would have helped to identify the contentious issues confronting women related to their livelihood. Regulations and conditions of gaining access to the new forms of commons require specialised training which most local actors do not have or have not yet gained through specific formation, as mentioned in different CSR projects. As a result, outsiders (for example, international NGOs or non-members of the local ethnic groups and state representative such as the *Caïds*) turn to manage the project at the moment. The following citation summarizes the impact of the change from old to 'new commons':

> This project is good for everyone, but we, the local population, want to receive something back because Noor is on our land and we are the

ones who are taking the risk if there are some illnesses,[19] and we have our conditions.

(Addi, male, living in a village close to the Solar Project Ouarzazate, Morocco, excerpt from interview conducted April to July, 2016)

The citation shows clearly that my interlocutor, a young man from the Tasselmante village, still considers the land in use for the solar project as the common property of his *Amazigh* clan. Additionally, as solar energy is produced for many users who do not have a relation to this land, he expresses the sense of taking risks for the majority. This connects with local perceptions about the project as not giving 'something back' in the sense of development for the local people. His comment that they 'have conditions' highlights that they want something in return for giving up their land. Addi's reflection also suggests what many of the interviewees revealed – namely, that they do not have (enough) control of the process over what is labelled fair compensation and development for all in the form of CSR projects.

The case study has shown the miscellaneous, the uncertainities, and the dynamic multifaceted context of the Moroccan solar energy project Noor Ouarzazate. It indeed reflects the reality and includes different sentiments and opinions from all kind of actors with their own view of the situation in the research area. However, the effects are not simply 'negative' or 'positive', they are rather very dynamic complex context. The citation, at the beginning of the article, illustrates, the contrariness stays for the complexity of the set of problems which the solar project Noor Ouarzazate brings besides the production of green solar energy and this is also evident in Chama's sentence which reflects versatility. She mentions how proud locals are, that the project is on their land and how important several infrastructural investments are, her relation to King Mohammed VI and his national investment policies but she also fore cited the loss of her land, the feeling in looking at the enclosure. Furthermore, she mentions the interpretation or classification of her land as 'wasteland', which is not adequate to her perception, because she knows for whom the land was certainly not wasteland and how it was used before the investment.

Overall, the research demonstrates that we need to know more about how green infrastructure might create a tragedy for the former commoners in the name of 'glocal' development, which does not compensate for the losses on several levels. Such policies in the name of rural development and women's empowerment are masked by transferring the commons into private property for a parastatal company at a very low price. At the same time, the company and the state fail to provide a 'new commons' based on the adequate and fair distribution of gains that merit the potential for 'new commons'. And last but not least, the emic perceptions of modernity and development should be seriously embraced to enhance real gender-sensitive development outcomes.

Notes

Research for this paper was financed by the Swiss National Science Foundation, Grant: 10001A_152773.

1 « Le douar est officiellement défini comme « un ensemble de foyers réunis par les liens réels ou fictifs de parenté, correspondant à une cellule territoriale comportant ou non des modes d'exploitation communautaires et dirigée autant que possible par un muqaddam. » Circulaire du Ministère de l'intérieur, datée du 27 décembre 1964 et relative à un projet de publication d'un recueil de circonscriptions administratives [Rfass 1996] ».

'The Douar can officially be defined as an 'assemblage of households linked by real or fictitious kinship relations, and correspond to a territorial unit, which may include community modes of (farming) work, and is managed as far as possible by a Muqaddam (local representative of the Ministry of the Interior)' (Ministry of the Interior, 1964).

2 Since 1980, *Imazighen* (sing. *Amazigh*) has come to refer to all Berbers from North Africa. Regional subgroups are distinguished with specific names (Hart 1996). *Imazighen*, in its original meaning, also refers to the Berber of central and southeast-central Morocco where I conducted my field work. Other Berber subgroups in Morocco are *Ishilhayen* in southwestern Morocco and *Irifiyen*, the Rifians in north-eastern Morocco (Hart 1996).

3 All names and genders of research participants were changed/chosen by the author. By using a fictitious identity to refer to the different actors, the anonymity of the interlocutors are ensured.

4 Our definition of a household: people eating from the same pot.

5 According to MASEN's 2011 impact study in the actual case of wet cooling for Noor I, and dry cooling for Noor II and III, water consumption will be around 3.7 million m^3 annually. This is around 1 per cent of the water storage of the dam for a normal year and between 5 and 6 per cent for a dry year. The simulation used for 'normal and dry years' is from Armin Kuhn, Institute for Agricultural Policy, University of Bonn. This does not mean that the loss of water is neglected because it is possible that one dry year can be followed by several dry years and can influence the dam's water storage with smaller inflow (Kuhn et al. 2010).

6 The common name for *astemisia herba alba* is wormwood. It is used for grazing (sheep, goats, cattle) and has many medical and therapeutic uses and purposes as it contains important foodstuffs and compounds which are beneficial to human health. For example, it contains iron, potassium, calcium, phosphorus, manganese, magnesium, lead and nickel. The most important uses and benefits of wormwood to human health include: antiparasitic, hair nourishment and the treatment of colic. It also treats indigestion, works as an appetite suppressant, and relieves bloating. Women and girls who suffer from menstrual pain and disruption are advised to use wormwood as it works to relax tense muscles and relieve suffering. Finally, it is used as tea.

7 Haloxylon is used both for animal feed and medicinal purposes for colds, fever, fatigue, wounds, aching bones, weak hair, high blood sugar and as an anti-fungal.

8 Any person chosen by villagers to be in charge of any talks or communications with another party – be it with investors or in the case of dividing the land to individuals with rights.

9 This is not meant as a value judgement; it is used to distinguish formally educated from non-formally educated people.

10 Lucerne/*alfalfa medicago sativa* is a flowering plant used for animal feed. It is also a cover crop to manage soil fertility, soil erosion and is important for biodiversity. Lucerne is also used in traditional Moroccan dishes.

11 Henna is not only used for beauty culture but also in traditional medicine.
12 Interview with Amina Benkhadra, Minister of Energy, Mines, Water and Environment and www.marcopolis.net accessed 20 February, 2015, www.morocco.com accessed 23 February, 2016.
13 Although all five didn't sign the document, the project acquisition plans mention that it was legally sanctioned (Masen 2011).
14 10 Dirhams = 0.92 Euro or 1.07 Dollar, www.postfinance.ch, accessed 4 September, 2018.
15 'In accordance with Article 6 of the Royal Dahir of 19 March 1951, by a commission of expertise comprised of the following members: (i) the pasha or caïd as president, (ii) a local representative of the Ministry of Finance, (iii) local representatives of the Ministries of Agriculture and Commerce and the High Commission for Water and Forests, and (iv) a local representative of the Ministry of Public Works. The community tutelage council will decide on the use of funds from the sale of land to the benefit of the community of Ait Oukrour Toundout' (Moroccan Agency for Solar Energy (MASEN) 2014: Plan d'Acquisition de Terrain).
16 Local community members in Tasselmante and Tiflit. Interviews conducted between September, 2015 and May, 2017.
17 In Arabic *Dar Taliba* means the 'House of Female Students.' *Dar Talib*, 'House of Male Students', also exist in Morocco.
18 Acwa Power is a developer, investor and operator of power generation plants: in the case of Ouarzazate, photovoltaic (PV) solar plants and concentrated Solar Power (CSP). It has been selected for the construction of Noor I and Noor II and III together with the Spanish company Abengoa. In 2016, Acwa Power won the tender also for the last phase of Noor Ouarzazate with the construction of Noor IV. www.acwapower.com, accessed 25 July, 2018.
19 He means illnesses like cancer or unintended side effects from the solar project. He compared the solar project with a nuclear power station and the threat of radiation. He is also worried about the phreatic water quality, not knowing if certain chemical substances coming from the solar project will infiltrate the soil and impact humans, animals and plants.

References

Abu-Lughod, L. (1998) *Remaking Women: Feminism and Modernity in the Middle East*. Princeton, NJ: Princeton University Press.

Abu-Lughod, L. (2013) *Do Muslim Women Need Saving?* London: Harvard University Press.

Acwa Power Ouarzazate, www.acwapower.com, accessed 25 July, 2018.

African Development Bank Group (AfDB) 'Projet de complexe solaire d'Ouarzazate-Phase II. Maroc', www.afdb.org/fileadmin/uploads/afdb/Documents/Environmental-and-Social-Assessments/Maroc_-_Projet_centrale_solaire_d%E2%80%99Ouarzazate_II_-_R%C3%A9sum%C3%A9_EIES.pdf, accessed 15 September, 2018.

Agnaou, F. (2004) *Gender, Literacy, and Empowerment in Morocco*. New York: Routledge.

Appadurai, A. (1996) *Modernity at Large: Cultural Dimensions of Globalization*. Vol 1. Minneapolis, MN: University of Minnesota Press.

Bahgat, G. (2013) 'Morocco energy outlook: opportunities and challenges', *The Journal of North African Studies*, vol 18, no 2, pp 291–304.

Behrman, J., Meinzen-Dick, R., and Quisumbing, A. (2012) 'The gender implications of large-scale land deals', *The Journal of Peasant Studies*, vol 39, no 1, pp 49–79.
Bergh, S.I. (2017) *The Politics of Development in Morocco. Local Governance and Participation in North Africa*. London and New York: I.B. Tauris.
Bidwell, R. (1973) *Morocco Under Colonial Rule, French Administration of Tribal Areas 1912–1956*. London: Routledge.
Charrad, M. (2001) *States and Women's Rights: The Making of Postcolonial Tunisia, Algeria and Morocco*. Berkeley, CA: University of California Press.
Chrisman, L. and Williams, P. (2015) *Colonial Discourse and Post-Colonial Theory: A Reader*. London: Routledge.
Doss, Ch. (2011) *The Role of Women in Agriculture*. FAO 2013. Governing land for women and men: A technical guide to support the achievement of responsible gender-equitable governance of land tenure, ESA Working Paper no 11–02 Rome.
Doss, Ch., Meinzen-Dick, R., and Bomuhangi, A. (2014) 'Who owns the land? Perspectives from rural Ugandans and implications for large-scale land acquisitions', *Feminist Economics*, vol 20, no 1, pp 76–100, DOI: 10.1080/13545701.2013.855320.
Ensminger, J. (1992) *Making a Market. The Institutional Transformation of an African Society*. Cambridge: University of Cambridge Press.
Escobar, A. (2011) *Encountering Development: The Making and Unmaking of the Third World*. Princeton, NJ: Princeton University Press.
Ferguson, J. (1994) 'The anti-politics machine: "development" and bureaucratic power in Lesotho', *The Ecologist*, 24, pp 176–181.
Haller, T. (2002) *The Understanding of Institutions and their Link to Resource Management from a New Institutionalism Perspective*. NCCR North South, IP 6 Institutional Change and Livelihood Strategies, Working Paper No 1.
Haller, T. ed. (2010) *Disputing the Floodplains*. Leiden and Boston: The Netherlands, Brill.
Haller, T., Gerber, J-D., Gmür, D., Lanz, K., and Ryser, S. (2013) 'Large scale land acquisitions and gender in Africa', *SNF Research proposal*. Bern, Switzerland: University of Bern.
Haller, T., Acciaioli, G., and Rist, S. (2016) 'Constitutionality: conditions for crafting local ownership of institution-building processes', *Society and Natural Resources*, vol 29, no 1, pp 68–87.
Haller, T. (2017) 'Perceptions and control of assemblage in a "glocal" world', *Dialogues in Human Geography*, vol 7, no 2, pp 207–211.
INDH (2011) *note d'orientation phase II 2011–2015*: Mise en oeuvre de l'Initiative Nationale pour le Développement Humain, Royaume du Maroc, www.indh.gov.ma, accessed 1 May, 2016.
Kouz, K., Cherkaoui Dekkaki, H., and Cherel, S. et al. (2011) *Étude d'Impact Environnementale et Sociale Cadre du Projet de Complexe Solaire d'Ouarzazate*. Rabat: Moroccan Agency for Solar Energy. Phénixa.
Kuhn, A., Heidecke, C., Roth, A., Goldbach, H., Burkhardt, J., Linstädter, A., Kemmerling, B., Gaiser, T., Speth, P., Christoph, M., and Diekkrüger, B. (2010) 'Importance of resource management for livelihood security under Climate Change in Southern Morocco' in Speth P., Christoph, M., and Diekkrüger, B. (eds), *Impacts of Global Change on the Hydrological Cycle in West and Northwest Africa*. Heidelberg: Springer, pp 566–591.
Lugan, B. (2011) *L'histoire du Maroc. Des origines à nos jours*. Paris: Ellipses.
Meillassoux, C. (1982) *Femmes, grenier et capitaux*. Paris: F. Maspéro.
Meillassoux, C. (1991) *Maidens, Meal and Money*. Cambridge: Cambridge University Press.

M'Hassni, M., Feljy, M., and Khalali, H. (2003) 'Le système foncier au Maroc'. Une sécurité et un facteur de développement durable, Au milieu urbain et rural. Paper presented at the 2nd FIG Regional Conference, Morocco, Marrakech.

Moroccan Agency for Solar Energy (MASEN) (2011) *étude d'impact*. Rabat: Phénixa, www.masen.org.ma, accessed 20 July, 2018.

Moroccan Agency for Solar Energy (MASEN) (2011) *Plan d'Acquisition de Terrain*. Rabat: Phénixa, www.masen.org.ma, accessed 20 July, 2018.

Moroccan Agency for Solar Energy (MASEN) (2014) *Plan d'Acquisition de Terrain*. Rabat: Phénixa, www.masen.org.ma, accessed 20 July, 2018.

Morocco Renewable Energy Megaprojects, www.marcopolis.net/morocco-renewableenergy-megaprojects.htm, accessed 20 February, 2015.

Narotzky, S. (2016) 'Where Have All the Peasants Gone?', *Annual Review of Anthropology*, no 45.

Prügl, E. (2016) 'Neoliberalism with a feminist face: crafting a new hegemony at the World Bank', *Feminist Economics*, pp 1–24.

Revue internationale de l'économie sociale, www.recma.org, accessed 29 March, 2016.

Rignall, K. (2012) *Land Deal Politics Initiative LDPI. Theorizing Sovereignty in Empty Land: The Land Tenure Implications of Concentrated Solar Power in pre-Saharan Morocco*. Ithaca, NY: Organized by the Land Deals Politics Initiative (LDPI) and hosted by the Department of Development Sociology at Cornell University.

Rignall, K. (2012) *Land, Rights, and the practice of making a living in Pre-Saharan Morocco*. Lexington, KY: University of Kentucky.

Ryser, S. (2011) 'Kulturelle Konzeption und Symbolik des Klimas in Marokko', Master's Thesis. Bern, Switzerland: University of Bern.

Sadiqi, F. (2003) *Women, Gender, and Language in Morocco*. Leiden: Brill.

Sadiqi, F. (2006) *The Impact of Islamization on Moroccan Feminisms*. Chicago, IL: University of Chicago Press.

Sadiqi, F. and Ennaji, M. (2011) *Women in the middle east and North Africa. Agents of change*. London and New York: Routledge.

Sadiqi, F. (2014) *Moroccan Feminist Discourses*. New York: Palgrave and Macmillan.

Schoneveld et al. (2011) 'Land-based investments for rural development? A grounded analysis of the local impacts of biofuel feedstock plantations in Ghana', *Ecology and Society*, vol 16, no 4.

Währungsrechner, www.postfinance.ch, accessed 4 September, 2018.

White, J. and White, B. (2012) 'Gendered experiences of dispossession: oil palm expansion in a Dayak Hibun community in West Kalimantan', *The Journal of Peasant Studies*, no 39, pp 995–1016.

Wind-Power as a Renewable Energy Resource in Morocco, www.morocco.com, accessed 23 March, 2016.

19 Global changes in local governance of the commons

The case of the African Parks Foundation engagement in Nech Sar National Park, Ethiopia

Girma Kelboro and Till Stellmacher

Introduction

Management, conservation and governance of biodiversity and landscapes often follow global values and standards. The relations between commons such as forests, wildlife, water and local communities is often redefined in that context. International conservation NGOs are among the carriers of these global values and standards, particularly when they are applied to protected areas in the Global South. In this regard, the work of the African Parks Foundation (APF)[1] in Nech Sar National Park (Nech Sar NP) is an interesting example (see Figure 19.1).

Nech Sar NP was established in South Ethiopia in 1974 for wildlife conservation. The park was first managed by the Ethiopian Wildlife Conservation Organization, which was later developed into the Ethiopian Wildlife Conservation Authority. In 2005, the management responsibility for Nech Sar NP was handed over to the APF. The park is located in the region Southern Nations, Nationalities and People's Regional State (SNNPRS) and shares a border with Oromia Regional State.

The APF took over management responsibility for Nech Sar NP in 2005 for a period of 25 years – based on an agreement signed between the Federal Ministry of Agriculture, Office of the President of the SNNPRS and APF on February 17, 2004. The Ethiopian state considered the involvement of APF in managing protected areas in the country as an opportunity to bring in international expertise, experience and capital to make protected area management and governance more effective and integrated. The expectation from the APF's engagement with Nech Sar NP was to establish a show case model in the country.

Global–local connections in conservation

International organizations have become increasingly active in negotiating, implementing and monitoring environmental management and governance affairs – an area previously often reserved to state agencies (Raustiala 1997: 719; Betsill and Corell 2001; Tamiotti and Finger 2001). This makes them part of a global 'machinery' which introduces new conservation models based on values and standards set elsewhere. According to Zimmerer (2007), organizations such as the Worldwide Wildlife Fund, The Nature Conservancy and the International Union for the Conservation of Nature (IUCN) are influential in setting conservation agendas throughout the world. IUCN, for example, sets standards for protected area categories that are used worldwide (Dudley 2008). IUCN guidelines, however, require that protected areas should be managed by working closely with local and indigenous communities. This means that protected area governance should be an outcome of local consultation and discussion processes. Performance of countries in conservation are evaluated based on the IUCN standards and protected area categories. In Ethiopia, national parks belong to IUCN protected area category II from which direct use of resources is prohibited, but visitors are allowed. The same term 'national park' is used for Dipperu National Park in Australia, which belongs to IUCN category Ia (strict nature reserve) in which all kinds of human activities are prohibited.

The direct and indirect influence of international organizations on the conservation of protected areas in the Global South has been increasing in the last decades. State authorities align their conservation policies in this direction also in order to be eligible for financial support through bilateral and multilateral donor agreements as they are chronically underfunded. Most protected areas in the Global South face severe problems – poaching, illegal logging and livestock grazing, just to name a few. Opening new sources of financial and technical support is hence essential and urgent.

The establishment of protected areas in Ethiopia gained momentum after country's participation in the 12th session of the UNESCO General Conference in 1962 in Paris. Ethiopia supported the resolutions on natural resource conservation for economic development and provision of support to developing countries by international organizations. In 1963, UNESCO experts came to Ethiopia and proposed the establishment of several protected areas throughout the country. Along with other national parks, Nech Sar NP was proposed in 1966 on 514 km^2 between Lake Abaya and Lake Chamo in southern Ethiopia. The proposal was accepted by the government in 1974 to establish Nech Sar as a national park. The park consists of grassland plains, acacia savannah woodlands, rivers, riverine forests, groundwater forests and parts of the lakes. The grassland plains in the centre of the park are important feeding grounds for wildlife including Grant's zebra (*Equus quagga*) and Swayne's hartebeest (*Alcellaphus buselaphus* swaynei). Guji Oromo ethnic group live inside the park and use primarily the grassland plains for cattle grazing; Kore ethnic group reside on the south-eastern borders outside the park boundaries but carry out traditional irrigation-based small-scale farming along River Sermelle inside the park. Administratively, Nech Sar NP is situated in SNNPRS, but borders Oromia Region.

Ever since the 1960s, the number of protected areas of different categories has increased in Ethiopia, now covering about 14 per cent of the total surface area of the country (Kelboro and Stellmacher 2015). On the one hand, it is recommended that the number and size of protected areas in Ethiopia should be increased to help conserving the countries' ecological richness and natural resources (Vreugdenhil et al. 2012). On the other hand, the existing protected areas in the country badly suffer from overutilization and degradation. The expansion of farmland, settlements and harvesting of different kinds of products within the protected areas are absolutely common (Stellmacher and Nolten 2010; Fetene et al. 2016; WoldeYohannes et al. 2018).

International conservation NGOs serve as vehicles to bring global values and standards to the local level. In Ethiopia, the engagement of international NGOs gained momentum after the 1973/1974 drought (Berhanu 2002). The 1984/1985 drought also added to the influx of NGOs. Since then the number and diversity of international NGOs continued to increase. Although there was a strict control over activities of international NGOs during the communist era between 1974 and 1991, they were accepted or even welcomed as a means of bringing in expertise, technical support and funding, particularly for post-famine recovery and rehabilitation projects. Despite some restrictions such as the 'NGO law' in 2009, ever since, international NGOs play vital gap-filling roles in capacity building, relief, and conservation activities in the country.

Management, conservation and governance of protected areas in Ethiopia inherit a long path dependency of central and authoritarian state control (Assefa and Gebre-Egziabher 2007). Since 1991, decentralization and privatization policies allow to 'hand over the stick' founded on the justification that state management and conservation efforts over decades ended in failures. First, there has been compelling evidence of massive degradation and destruction of natural resources – in protected areas and outside – which increasingly negatively affected the economic development of the country (Senbeta and Tefera 2001–2002; Taddese 2001; Teketay 2001; USAID 2008; Gessesse 2010). Second, it became more than evident that the existing management and conservation approaches were utterly ineffective, calling for alternatives (Jagger et al. 2003). Third, over the decades there had been a massive shortage of financial, technical and human capacities (Hillman 1993; ÖBF 2009). The engagement of organizations such as APF in Ethiopia can, therefore, be seen as a part of the solution to such challenges.

Methodology

For this study, we collected empirical data through semi-structured interviews with 120 pastoralists and farmers living within and around Nech Sar NP (see Figure 19.1); in-depth interviews with 50 (actual and former) park staff and experts in regional and national authorities; participant observation of daily practices

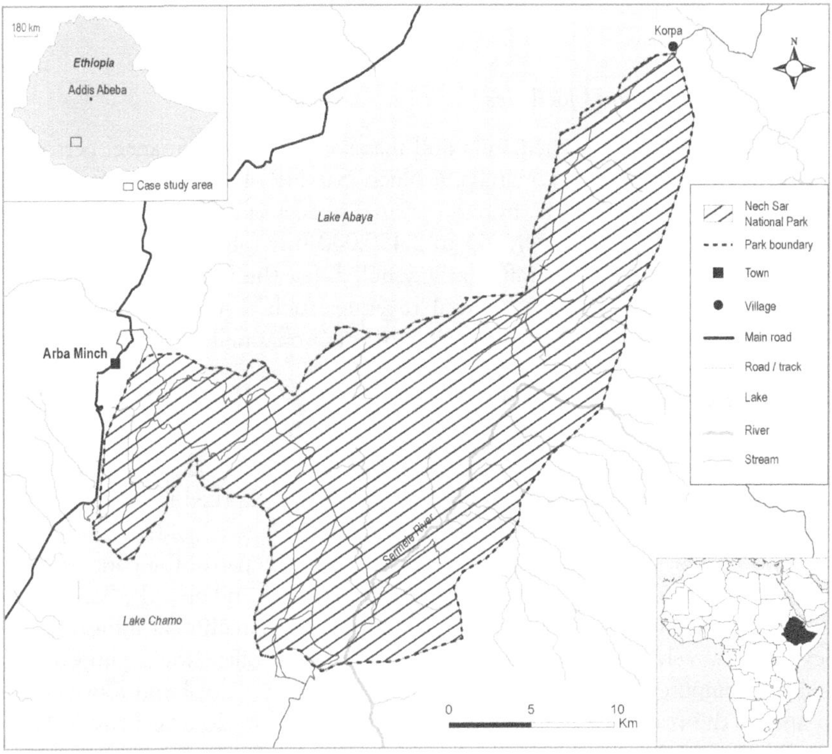

Figure 19.1 Location of Nech Sar National Park.
Source: Map compilation and design by Ulla Gämperli Krauer (2018), CDE, University of Bern, Bern, Switzerland. Geodata source: OSM data by OSMaxx (2018). Geometalab, HSR University of Applied Sciences Rapperswil, Switzerland.

in the park; and review of archival sources. Additional data was collected from reports and letters produced by park and conservation authorities.

The African Parks Foundation engagement in Nech Sar National Park

After taking over the management responsibility for Nech Sar NP in 2005, the first steps for the APF were defining short-term and long-term objectives and plans for the park management as well as establishing a research unit for the Nech Sar NP. Some of the plans included reorganizing park management structures; rehabilitation of degraded habitats; public awareness raising and establishment of close working partnerships with stakeholders; promotion of tourism; as well as anti-poaching activities. In the following we look in detail at the activities and achievements during the APF engagement in Nech Sar NP.

Finance, personnel and facilities

The APF engagement immediately and massively increased finance, personnel and technical facilities for managing Nech Sar NP. The budget of the park increased by nearly 15 times, from an annual budget of 250,000 Ethiopian Birr (about 30,000 USD) in 2004/2005 to 3,450,000 Birr (about 430,000 USD) in 2005/2006. The number of staff nearly doubled. For the first time, the park was authorized to utilize its own internal revenues such as tourist park entrance fees. This was in stark contrast to the tradition of channelling revenues fully to centralized state treasuries.

Our interviews show that the increase in budget and personnel helped to enhance the commitment of the park employees. The morale of the park staff seemed very high during the APF period. Wildlife rangers stated their satisfaction with the system, and a park ranger's team leader who worked for Nech Sar NP before, during and after APF described the time of APF as a 'golden opportunity' that was missed. They were equipped and supported to patrol the park 24 hours a day with additional payments for extra working hours and periodic bonuses.

APF engagement also helped to improve the park's technical infrastructure. Several new vehicles were bought. Practically, every office was equipped with desktop computers, which was previously unthinkable. Patrol and tourist roads to and in the park were built or repaired. A main obstacle to get inside Nech Sar NP (for both park staff and tourists), especially during rainy seasons, was the weakness of the bridge over the Kulfo River. Under the APF engagement, the old bridge was restored. A park rangers' team leader stated that "crossing the Kulfo River was a problem both for tourist cars and our own vehicles before APF constructed a bridge. Construction of the bridge to cross the river is one of the unforgettable achievements of APF's works."[2]

Stakeholder partnerships

APF categorized the park's stakeholders as either primary or secondary. The primary stakeholders include local communities who live in or around the park as well as state organizations whose operational areas bound the park. Accordingly, primary stakeholders among the local communities were:

- Kore people (about 1,100 households) in Abulo and Alfacho *kebele*[3] who were resettled outside the park.
- Arba Minch town residents (about 20,000 households) who depend on fuel and construction wood, animal fodder, and fish from the park; seasonal migrants[4] who travel from Gamo highlands[5] to Arba Minch town and rely primarily on wood and grass they collect from the park and sell in the town as a source of their livelihoods.

- People in five rural *kebeles* in Amaro Special *wereda*[6] that borders Nech Sar NP and Kore people who have economic ties with Arba Minch town connected via a footpath that crosses the park.
- Guji people who live in Nech Sar NP and depend on the park for cattle grazing and farming.
- People settled on the shore of Lake Chamo who use the park for agriculture and fishing.

The primary stakeholders of the state organizations were:

- State bodies in Gamo Gofa Zone, Arba Minch Zuria *wereda*, Arba Minch town, Gelana *wereda*, and Amaro Special *wereda*.
- Departments of Agriculture and Rural Development of the respective *weredas*.
- Office of Small-scale Industry and Micro-business Enterprises.

The secondary stakeholders were potential partners of the Nech Sar NP management. They included private organizations using the parks' resources (such as fishery cooperatives; the Ethiopian Fishery Corporation; boat operators on Lake Chamo; tour guides and travel agents), and other organizations concerned with natural resource conservation (such as GIZ; Forum for Environment, Arba Minch; and SNV-Ethiopia, South Portfolio).

APF organized stakeholder workshops and discussion forums in which both primary and secondary stakeholders participated and shared their views on the future and management options of the park. A series of meetings was held in which Guji and Kore representatives, local and regional state delegates of Oromia and SNNPRS, officers of APF and Nech Sar NP authorities took part. The aim was to develop conservation strategies based on a public-private partnership business model and, thereby, reduce the local resource use pressure on the park, mainly those perceived to be the cause for loss of vegetation cover and wild animals.

The first large-scale stakeholder workshop was held on March 25, 2007, at Arba Minch University. Representatives of Guji and Kore communities; *wereda*, zone and regional officers from Oromia and SNNPRS; as well as delegates from higher education institutes and NGOs took part in this workshop. The first author of this article also participated in this workshop as a delegate of Wondo Genet College of Forestry and Natural Resources, Hawassa University. One recommendation of the stakeholders' workshop was that detailed discussions should be held with Guji and Kore communities whose livelihoods depend primarily on cattle raising and farming in the park respectively. The outcome of the discussions was to be communicated to the other stakeholders. Consequently, APF held separate discussions with Gujii and Kore. However, the discussions further developed to the level where APF worked in more detail with Guji people and Oromia Regional state authorities while discussions with Kore people and SNNPRS Region state authorities were stopped at a certain stage. The outcomes of the discussions were not communicated to the other stakeholders the contrary to the agreement reached in the workshop on March 25, 2007. Four different meetings were held with Guji people and Oromia Region state authorities in 2007 in which different boundary and zoning concepts for Nech Sar NP were presented and discussed.

The African Parks Foundation zoning concept

The idea of national park zoning started with the pioneering proposal of a national park management zoning concept at the World Conference on national parks at Seattle (Beltrán 1962). However, this concept was not implemented in many countries including Ethiopia. National parks are considered in Ethiopia as non-negotiable conservation areas and not as common-pool resources to be managed in consideration of local knowledge, institutions and livelihoods. However, the engagement of APF allowed for a negotiation of a zoning concept under which the park's area would be divided into a core area and a Community Conservation Area (CCA). The core area would include the groundwater forest, the lake areas and the grassland plains. Only tourism, research and educational activities would be allowed in this part. The CCA would serve as a buffer zone in which local people could exercise traditional resource use, management and conservation practices with technical and material support from the park's authorities. Two such CCAs were proposed: a North East CCA to be managed by Guji people, and a South East CCA to be managed by Kore people.

The negotiation of a zoning concept was, however, highly conflictual and complex. APF proposed to define the grassland plains in the middle of the park as the main part of the core area, meaning that it had to be free from consumptive uses. This part of the park is, however, used by Guji people for cattle grazing. From interviews with Guji and Kore people as well as with experts in regional state authorities of both SNNPRS and Oromia, we realized that defining the grassland plains as a core area was considered highly problematic. On the one hand, both local user groups (Guji and Kore) and state authorities perceive that the park boundary ends at the core area, and Guji people are allowed to live inside and to use the grassland plains for grazing. This view contradicts the zoning concept. In addition, for Kore people it was irritating to see negotiations largely taking place between the park authorities and Guji people.

Separation of people and park: the road to failure

The seemingly participatory process in Nech Sar NP, particularly with regard to the zoning concept, tried to follow the world-wide common sense of local participation. However, it did not function as intended. The zoning would have fundamentally changed the use and management patterns of local people in the park, likely with massive negative impacts on their livelihoods, and would have ultimately led to the resettlement of local people, particularly Guji. The (forceful) resettlement of people for conservation reasons (eco-resettlement) can be a severe violation of human rights. For example, based on empirical evidences from 12 case studies from six countries, Cernea and Schmidt-Soltau (2006) conclude that eco-resettlement aggravated impoverishment of local people. On the other hand, there is an increasing sense of urgency to increase effectiveness of protected areas in order to save the last remaining natural and biological resources of the world from degradation and loss.

In the case of Nech Sar NP, resettlement plans were driven by the vision of minimizing human impacts on the park's core habitats. On the one hand, the resettlement of people outside the park was recommended by Freeman (2006: 36) as the park is relatively small. On the other hand, resettlement was interpreted as a risk to peoples' livelihoods and a human rights violation.[7] The complexity and sensitivity of eco-resettlement in the form of forceful eviction from national parks has a long tenure in Ethiopia. For example, in 1978 thousands of people were forcefully evicted from the Simien Mountains NP. Eight villages were destroyed. However, six years later many of the people returned back to the park (Hurni and Ludi 2000). Between 1995 and 1998, a project called 'National Parks Rehabilitation in Southern Ethiopia' planned resettlement of people from Mago NP and Omo NP. However, it also did not succeed (Biodiversity Indicators Development National Task Force 2010).

In Nech Sar NP, the first and most prominent case of forced resettlement is the action taken by the military regime (known as the *Derg*) in the 1980s in which militiamen and park rangers devastated all settlements in the park and burned down the crops. However, as the *Derg* lost power in 1991, Guji and Kore people returned back to the park.

Another resettlement action came in 2004. This time, the park was able to move about 1,000 Kore households to an area located 15km south of Nech Sar. In our interviews, the resettled Kore people were unhappy with the life in their new location. They particularly suffered from a lack of access to drinking water and road connections to towns. They further explain their dissatisfaction with the resettlement move by the fact that Guji people are 'allowed' to continue living in the park. Guji resisted the resettlement plan. This prompted forceful actions from the park authorities including burning of hundreds of Guji residential houses after which the people moved away from their residences and stayed in peripheral areas surrounding the park (Tsegaye et al. 2017). Nevertheless, they returned back to the park again to see the reaction, and they found out that they could negotiate to continue living in the park.

The prior experience of forced resettlement and back-and-forth movement proved to both Guji and Kore that the APF conservation zoning and eviction plan were a repetition of history. The plans also received some international media coverage. For example, the homepage of the Dana Declaration[8] wrote:

> If African Parks is to be accepted internationally as a responsible conservation agency, it must ensure that the principles and policies articulated by the relevant international conventions and widely understood best practices are adhered to in the projects it manages. The responsibilities incumbent on

> African Parks in this respect are neither diminished nor abrogated by the fact that it may be the government with which it collaborates that actually carries out the evictions or lesser forms of loss of access that violate the rights of local communities … In particular, there should be a clause in the agreement which states that there will be no evictions, and no denial of access to agricultural and grazing land lying within the park boundaries.[9]

The negotiations for the zoning concept were influenced by deeply rooted historical and contemporary belonging of Guji and Kore people to the area and strong socio-economic and cultural attachments to the land they use and live on. Burial sites, religious places and trees of traditional worship exemplify the strong bonds to the land (Kelboro 2013).

African Parks Foundation's decision to end the Nech Sar NP journey

After lengthy negotiations with Guji people representatives, a consensus was reached in September 2007. The consensus alluded to a revised zoning map, which consists of a newly proposed core area. The revised zoning plans were presented to delegates of both Oromia and SNNPRS region. APF made the approval a precondition to continue their Nech Sar NP endeavour. The SNNPRS delegates, however, felt not to be sufficiently involved in the negotiation processes and finally did not accept the revised zoning concept. APF argued that a delegate conservation expert was working with them and updated the SNNPRS authorities accordingly.

Our interviews showed evidence that the lacking direct involvement of the SNNPRS authorities in the zoning and resettlement negotiation processes endangered the whole endeavour. Apparently, the implications of negotiating on the management and conservation of a protected area which connects two different regional states and hosts two different ethnic groups – each one attached to one of these regions – was underestimated. At the end of all negotiations, the plans were ultimately rejected by the SNNPRS. Immediately thereafter, APF decided to end its activities not only in Nech Sar NP but also in Ethiopia entirely. Instead of 25 years, as signed in the agreement, the APF journey ended after three years. Freeman (2006: 35) predictively stated, 'The main issue that is stopping APF from fully implementing its conservation work within the park is the fact that the Guji people, and their thousands of cattle, are still living in the park'.

A former APF executive officer summarized the APF perspective as follows:

> APF waited patiently for several years for the state authorities to fulfil the resettlement requirement put on the agreement. After waiting for long in vain, APF asked to agree on the negotiation it is going to make with the Guji. The APF got 'a go ahead sign' from the Vice Prime Minister of the

> time. That led to the negotiation processes with the Guji that ended up with a newly proposed boundary. Every step of the negotiation process was also clear to the SNNPRS authorities through the project head as well.[10]

The position of the SNNPRS authorities was revealed in a letter written by the SNNPRS Office of the President on 27 November, 2007.[11] The letter stated the following:

- APF acted beyond the limits of responsibilities given in the tri-lateral agreement signed by the Federal government, the SNNPRS, and APF. In particular, Article 8(3) clearly states that relocation of the park inhabitants is fully the responsibility of the government. Accordingly, the regional government had relocated communities of 1359 households outside the park area by providing the necessary farmland and other support.
- APF unnecessarily intervened into an on-going Guji resettlement process that also requires negotiations between two regions. The SNNPRS had been working with Oromia to relocate the Guji community.
- The boundary negotiation agreement signed between APF and Guji community complicated the government's efforts to resettle the people outside the park boundaries. Recently, an *ad hoc* supervisory committee and a technical committee had been set up with representation from both sides (Oromia and SNNPRS) to work jointly on the Guji community relocation process. However, the reports of the committee specified that the Guji community who are living inside the park had strongly opposed the recent government relocation plan and process since they have already agreed on a new boundary in advance with APF.

SNNPRS representatives argued that negotiating park boundaries, which are partly also regional administrative boundaries, should be dealt with in accordance with Article 48 of the Federal Constitution, saying that border disputes between two regions can only be resolved when both respective regions agree on the case. When they do not, a binding decision can only be made by the House of Federation.

The decision of APF to end its engagement in Nech Sar NP remained a puzzle to international conservation organizations. For example, a letter from the IUCN to the Prime Minister Meles Zenawi on 5 February, 2008, asked the Ethiopian Federal and regional authorities to come to a constructive 'dialogue, consultation, and collaboration with local communities in view of setting up an equitable and effective governance system for the Nech Sar NP'. In the letter, IUCN emphasized that,

> in most countries of the world, no protected area could or would be established or managed by forcefully displacing the local residents. This principle is

> internationally recognized in a number of legal instruments, including the Convention on Biological Diversity, of which Ethiopia is a signatory.

IUCN referred, in particular, to the Program of Work on Protected Areas approved at the CBD COP 7 in Kuala Lumpur in 2004, namely Section 2.2.5 which states a commitment of all its parties to 'ensure that any resettlement of indigenous communities as a consequence of the establishment or management of protected areas will only take place with their prior informed consent that may be given according to national legislation and applicable international obligations'.

On the one hand, the concern of IUCN and other international conservation bodies apparently added pressure on the negotiators. On the other hand, in the year-long negotiations, the main stakeholders were not able to come to a consensus.

Causes for the failure of the African Parks Foundation engagement

First and foremost, causes for the failure were not in the financial or technical realm. The whole argument around the very early ending of the APF engagement can be summarized as a divide between international conservationists, federal and two bordering regional state authorities and different local user groups, with apparently different interests and understandings. Technical fixes to govern protected areas offer seemingly clearer models, when compared to the local and regional realities with dynamic and complex socio-political human dimensions. The difficulty of finding a delicate balance between conservation of protected areas on the one hand and satisfying political national and regional interests as well as local livelihood needs of different groups on the other remains a key challenge. Alternative strategies need to be worked out in collaboration with all relevant stakeholders to reach consensus. Stakeholder interests should be clarified well before negotiations. However, this can only happen when the parties reach a common understanding and build trust with each other.

Protected area governance has always been very difficult in Ethiopia. The role of the state authorities in Ethiopia is a difficult one. People, especially in the rural areas, tend to perceive the state mainly as an 'alien authority' that punishes them, collects taxes, and controls land and other resources (Vaughan and Tronvoll 2003). By the terms of the constitution, all land practically belongs to the state. Therefore, when it comes to the protected areas, the different state authorities take the lead with their individual interests. Path dependency plays a significant role here. Historical experiences such as the forceful expansion of Menelik II to the south, the exploitative land-lord tenant system, and the forced resettlement and villagization measures of the *Derg* regime have negatively affected the relationship between state and people in Ethiopia (Stellmacher 2007a). More recent but no less radical changes in land use since the 1990s – planned and enforced by central state bodies – also contributed to this (Stellmacher 2015). The land use trajectory and its consequences for the governance of protected areas are similar to the findings of other studies in the country (Stellmacher 2007b). The conflicts in Nech Sar NP mirror all these developments.

However, the lack of cooperation and communication between Oromia region and the SNNPRS played another important role in jeopardizing the work of APF in Nech Sar NP. Developing a zoning concept for the park became a hot regional political issue. Negotiators from both regions had to be very careful with their decisions in order not to avoid future land claims from the other region. Under this situation, the question of zoning and resettlement had become the major obstacles and main stumbling blocks to the APF engagement.

Nech Sar NP after African Parks Foundation

The withdrawal of APF in 2008 put Nech Sar NP temporarily in no one's hand. In this transition period, natural resource overuse sharply increased. Local people considered the power vacuum as an opportunity to collect wood, grass and other resources from the park and use its grazing areas.

After the chaos in 2008, the park's management was taken over by the SNNPRS Bureau of Culture and Tourism. Due to the parks' location between two regional states, the management responsibility was transferred in 2009 to the federal Ethiopian Wildlife Conservation Authority. However, the tension between Oromia and SNNPRS continued further and are still not solved. Conflicts between park authorities and people, and the degradation and loss of biodiversity and natural resources in the park due to the overuse by local people are still continuing – until today (Fetene et al. 2014; Fetene et al. 2016; WoldeYohannes et al. 2018).

Conclusion

Management, conservation and governance of biodiversity and landscapes in Ethiopia, and governance of protected areas in particular, are associated with global conservation values, concepts and approaches. For decades many protected areas were rather 'paper parks', chronically underfinanced, and often badly managed, with devastating ecological and socio-economic consequences. Upon this backdrop responsible state authorities increasingly orient their policy and decision making on global criteria and concepts – also in order to attract funding and technical support. International NGOs can act as a vehicle to implement global conservation concepts on the local level. However, implementation of global standards and values is dictated by local realities and can be impeded by national and regional state interests.

Focus is often given to financial means and support. However, negotiating, and actually changing, protected area governance patterns needs well-developed and organized capacities and expertise. The meaning of protected area degradation and the role of people living in and around protected areas and (over)using their resources should be understood in the specific contexts in question. General intellectual views on population-degradation links provide foundations for arguments in policy-making, which in turn influence practices that can lead to (more) conflicts or effective cooperation between stakeholders.

Our example shows how a mosaic of different interests, concepts and perceptions among state authorities, different local user groups and an international conservation organization made effective governance of a national park in Ethiopia impossible. Different situations need different approaches, especially in politically and historically heavily loaded environments. Whenever there is resistance in the negotiation processes, it may be necessary to go back and analyse the underlying reasons and the historical relations between the parties. Generally, understanding the context, flexibility and adaptive approaches are helpful in dealing with diversities.

Acknowledgements

This study was funded by the German Academic Exchange Service (DAAD) for which we are grateful. We also thank Wondo Genet College of Forestry and Natural Resources, Hawassa University, for the support we received for our field research. We are thankful to the local communities particularly Guji and Kore for being our partners to collect data and host us during the field research. The cooperation of Oromia and SNNPRS authorities to be interviewed and to provide us with secondary sources of data is highly appreciated.

Notes

1 APF was founded as the African Parks Management and Finance Company, then became an NGO called African Parks Foundation, and was later renamed African Parks Network. We use the term African Parks Foundation (APF) as the organization as known in Ethiopia at the time of this empirical study. Based in Johannesburg, APF takes long-term management responsibilities for protected areas in Africa with the ambition of achieving both conservation and development objectives. In Europe APF is represented by the Stichting African Parks Foundation (SAPF), with its main office in Amsterdam.

2 Interviews with F043 (the Nech Sar NP ranger team leader) on 23 October, 2010, Arba Minch.

3 The smallest unit in the government administrative system in Ethiopia.

4 These are mostly young women and men from farming families who live in the surrounding highlands. They travel to Arba Mich town in search of alternative sources of income at the end of farming seasons.

5 A chain of mountainous areas (on the western side of Nech Sar NP) named after the ethnic group that lives there.

6 Comprises of a collection of *kebeles* (see note 3, above) in the Ethiopian government administrative system.

7 www.conservationrefugees.org/lawsguidelines.html accessed 14 February, 2013.

8 Dana declaration is the outcome of an international meeting of social and natural scientists and NGOs that took place in Wadi Dana Nature Reserve in Jordan in early April 2002. http://danadeclaration.org/ accessed 9 January, 2018.

9 www.danadeclaration.org/pdf/omotakeover.pdf accessed 1 August, 2011.
10 Interviews with F045 (a former APF executive officer) on 3 March, 2011, Addis Ababa.
11 Written by the SNNPRS Office of the President on 27 November, 2007.

References

APF (2005) 'African Parks Annual Report 2005', www.africanparks-conservation.com, accessed 29 October, 2018.

Assefa, T. and Gebre-Egziabher, T. (eds) (2007) *Decentralization in Ethiopia*. Addis Ababa: Forum for Social Studies.

Beltrán, E. (1962) 'Use and conservation: two conflicting principles', in Adams, A.B. (ed.), *Proceedings of the First World Conference on National Parks held at Seattle, Washington* (June 30–July 7, 1962). Switzerland: IUCN, Gland, pp 35–43.

Berhanu, K. (2002) 'The role of NGOs in promoting democratic values: the Ethiopian experience' in Zewde, B. and Pausewang, S. (eds), *Ethiopia: The Challenge of Democracy from Below*. Uppsala and Addis Ababa: Nordiska Afrikainstitutet and Forum for Social Studies, pp 120–129.

Betsill, M.M. and Corell, E. (2001) 'NGO influence in international environmental negotiations: a framework for analysis', *Global Environmental Politics*, vol 1, no 4, pp 65–85.

Biodiversity Indicators Development National Task Force (2010) *Ethiopia: Overview of Selected Biodiversity Indicators*. Addis Ababa: Biodiversity Indicators Development National Task Force.

Cernea, M.M. and Schmidt-Soltau, K. (2006) 'Poverty risks and national parks: policy issues in conservation and resettlement', *World Development*, vol 34, no 10, pp 1808–1830.

Dudley, N. (2008) *Guidelines for Applying Protected Area Management Categories*. IUCN, Gland, pp x and 86. with Stolton, S., Shadie, P., and Dudley, N. (2013) *IUCN WCPA Best Practice Guidance on Recognising Protected Areas and Assigning Management Categories and Governance Types*. Best Practice Protected Area Guidelines Series No. 21, IUCN, Gland.

Dudley, N., Belokurov, A., Higgins-Zogib, L., Hockings, M., Stolton, S., and Burgess, N. (2007) *Tracking Progress in Managing Protected Areas around the World: An Analysis of Two Applications of the Management Effectiveness Tracking Tool Developed by WWF and the World Bank*. Gland: WWF International.

Fetene, A., Hilker, T., Yeshitela, K., Prasse, R., Cohen, W., and Yang, Z. (2016) 'Detecting trends in landuse and landcover change of Nech Sar National Park, Ethiopia', *Environmental Management*, vol 57, pp 137–147.

Fetene, A., Yeshitela, K., Prasse, R., and Hilker, T. (2014) 'Study of changes in habitat type distribution and habitat structure of Nech Sar National Park, Ethiopia', *Ecologia*, vol 4, pp 1–15.

Freeman, D. (2006) *Natural Resource Management in and around Nech Sar National Park: A Situation Analysis. Final Report*. Arba Minch: Forum for Environment.

Gessesse, B. (2010) 'Review of land degradation and land management in Ethiopia up to 2008/09', in Edwards, S. (ed.), *Ethiopian Environmental Review*, No. 1. Addis Ababa: Forum for Environment, pp 187–214.

Hillman, J.C. (1993) *Ethiopia: Compendium of Wildlife Conservation Information, Vol. 2 Information on Wildlife Conservation Areas*. Addis Ababa: Ethiopian Wildlife Conservation Organization.

Hurni, H. and Ludi, E. (2000) *Reconciling conservation with sustainable development. A participatory study inside and around the Simen Mountains National Park, Ethiopia.* Produced with the assistance of an interdisciplinary group of contributors. Centre for Development and Environment, Bern, p 476.

Jagger, P., Pender, J., and Gebremedhin, B. (2003) *Woodlot Devolution in Northern Ethiopia: Opportunities for Empowerment, Smallholder Income Diversification, and Sustainable Land Management.* Discussion Paper No. 107. Washington, DC: EPTD.

Kelboro, G. (2013) *Unraveling the Parks and People Dichotomy: Local Interests and Conflicts in Nech Sar National Park, Ethiopia.* Weikersheim: Margraph Publishers.

Kelboro, G. and Stellmacher, T. (2012) *Contesting the National Park theorem? Governance and land use in Nech Sar National Park, Ethiopia.* ZEF Working Paper 104, Bonn.

Kelboro, G., Stellmacher, T., and Hoffmann, V. (2013) '"Conservationists" and the "local people" in biodiversity conservation: the case of Nech Sar National Park, Ethiopia', *Ethiopian Journal of the Social Sciences and Humanities*, vol 9, no 1, pp 29–55.

Kelboro, G. and Stellmacher, T. (2015) 'Protected areas as contested spaces: Nech Sar National Park, Ethiopia, between "local people", the state, and NGO engagement', *Environmental Development*, vol 16, pp 63–75.

ÖBF (2009) 'Assessment of the Value of the Protected Area System of Ethiopia', www.hoarec.org/index.php/resources/library/42?task=download&format=raw, accessed 29 October, 2018.

Raustiala, K. (1997) 'States, NGOs and international environmental institutions', *International Studies Quarterly*, vol 41, pp 719–740.

Senbeta, F. and Tefera, F. (2001–2002) 'Environmental crisis in the Abiyatta-Shalla Lakes National Park', *Walia*, vol 22, pp 28–36.

Stellmacher, T. (2007a) 'The historical development of local forest governance in Ethiopia – from imperial times to the military regime of the *Derg*', *Afrika Spectrum*, vol 42, no 3 pp 519–530.

Stellmacher, T. (2007b) *Governing the Ethiopian Coffee Forests: A Local Level Institutional Analysis in Kaffa and Bale Mountains.* Aachen: Shaker Publishing.

Stellmacher, T. and Nolten, R. (2010) 'Forest resource use and local decision making in the Bale Mountains Coffee Forests, Ethiopia', in Eguavoen, I. and Laube, W. (eds), *Negotiating Local Governance. Natural Resources Management at the Interface of Communities and the State.* Berlin: Lit Publishing.

Stellmacher, T. (2015) *Socio-Economic Change In Rural Ethiopia. Understanding Local Dynamics in Environmental Planning and Natural Resource Management.* Frankfurt: Peter Lang.

Taddese, G. (2001) 'Land degradation: a challenge to Ethiopia', *Environmental Management*, vol 27, no 6, pp 815–824.

Tamiotti, L. and Finger, M. (2001) 'Environmental organizations: changing roles and functions in global politics', *Global Environmental Politics*, vol 1, no 1, pp 56–76.

Teketay, D. (2001) 'Deforestation, wood famine, and environmental degradation in Ethiopia's highland ecosystems: urgent need for action', *Northeast African Studies*, vol 8, no 1, pp 53–76.

Tsegaye, G., Dondeyne, S., Lemenih, M., Marye, A., Nyssen, J., Deckers, J.A., and Maertens, M. (2017) '"Facing conservation" or "conservation with a human face"? People–park interactions in southern Ethiopia', *Journal of Eastern African Studies*, vol 11, no 2, pp 290–309.

USAID (2008) 'Ethiopia Biodiversity and Tropical Forests 118/119 Assessment', www.encapafrica.org/documents/biofor/Ethiopia2008.pdf, accessed 29 October, 2018.

Vaughan, S. and Tronvoll, K. (2003) *The Culture of Power in Contemporary Ethiopian Political Life*. Stockholm: Sida Studies.

Vreugdenhil, D., Vreugdenhil, A.D., Tilahun, T., Shimelis, A., and Tefera, Z. (2012) *Gap Analysis of the Protected Areas System of Ethiopia*, with technical contributions from Nagelkerke, L., Gedeon, K., Spawls, S., Yalden, D., Berhanu, L., and Siege, L. of the World Institute for Conservation and Environment.

WoldeYohannes, A., Cotter, M., Kelboro, G., and Dessalegn, W. (2018) 'Land use and land cover changes and their effects on the landscape of Abaya-Chamo Basin, southern Ethiopia', *Land*, vol 7, no 2.

Zimmerer, K.S. (2007) 'Agriculture, livelihoods, and globalization: the analysis of new trajectories (and avoidance of just-so stories) of human-environment change and conservation', *Agriculture and Human Values*, vol 24, pp 9–16.

20 Discourse and entanglement in a transnational conservation arena

Deciphering the ideologies and narratives behind conservation discourse in the 'glocal' commons in Kenya

Samuel Weissman

Introduction

There is a rift in how conservation is thought about and consequently practiced when we follow the discourse from global to local concepts. Rather than simply following a linear devolution of the hegemonic discourse from the global to the local spaces of conservation, as in an Agrawalian environmentality approach (Agrawal 2005), the conundrum of deciphering the actual complexities of discoursive interlinkages needs another way of being dealt with. This chapter proposes to do this in a form comparable to describing entangled particles, to borrow the term from quantum physics.[1] By this concept, the chapter here will propose that the elements of discourses[2] are entangled similarly to the way that pairs or groups of particles are thought of as entangled in that the state of each particle cannot be described apart from the system as a whole (Bub 2017). Applied to conservation discourse, this would mean that the elements that govern the discourse today are not only found in narratives on the state-of-the-art or best-practice defined by the techniques and technologies deemed to be the newest and therefore modern by the powers that be. We need to consider larger correlations considering the various actors with divers bargaining power positions, working according to different sets of rules, organising along different ideologies and acting within and across different economic, political and social spaces. These might be seen as separate and conflicting at times, but are interwoven nonetheless through historical and social ties, which transcend local or national boundaries. However, what can be discerned between conservation interest groups from global to local arenas of interaction are the numerous levels of influence on discourse. These seem to be based on constructions of traditional and modern institutions. In regard to discourse, Foucault (2010) suggests that '[t]radition enables us to isolate the new against a background of permanence [...]', whereby in this case the 'new' seems to gain popularity in an arena of global proportions regarding nature and wildlife as a common property facsimile in the shared interest of saving the planet, while the actual commons in the local context, which were in place long before an ideology of conservation existed, are being pushed to the background, immobilised and all but forgotten, save some romantic, marketable re-inventions of what is thought to be tradition.

Re-examining landscapes through political ecology

Since colonialization through western powers in many parts of the world, land- and resource ownership and -use have shifted greatly. Especially the forms of common-pool resource management systems, generally known as commons, which have been appropriated by outside agents and privatised or nationalised (see Ostrom 1990, 2005; Haller 2010; Ensminger 1992). Similarly, in recent years, the appropriation of land and resources has come to include new valuations on certain natural resources, found in terms such as 'nature', 'wildlife', 'biodiversity'. and has extended the concept of land grabbing to one of 'green grabbing' (see Ojeda 2011; Benjaminsen and Bryceson 2012; Fairhead et al. 2012; German et al. 2017). Therewith, land and resources include newly constructed valuations and are not just privatised, but commodified, as Fairhead et al. (2012) frame it, whereas former commons, such as pastures, forests, water and wildlife become economically viable and valuable goods to be sustainably managed and marketed (see also Brockington et al. 2008). This 'modern', e.g. neo-liberal discourse on sustainable resource use rises against a backdrop of 'primitive' degradation narratives, which often legitimize the actions undertaken by conservation efforts (Benjaminsen and Bryceson 2012). In recent years, therefore, the conservation community have opted towards more politically accepted models of community conservation, evolving the above narrative to include the education of these "degraders of land" in sustainable management strategies (see Brockington 2002; Galvin and Haller 2008).

As political ecologists such as Galvin, Haller, Paulson, Robbins and others argue, this approach to community-based conservation, albeit discoursively preferred by international interest groups, has not yet received its final stamp of approval. The critique falls mainly on the restrictions and interventions by the conservation agenda in the form of institutional change undermining the lifestyles of locals and their economic practices, and at the same time altering previous cultural landscapes the conservation agenda wanted to protect in the first place (Haller et al. 2013). To uncover how such disadvantages are brought about by the conservation agenda, this chapter deals with discourse in the Foucauldian (1971) sense, which is influential in shaping legitimacy and political action through narratives and power relations within and across interest groups, which in this case span a global arena in a pluralistic institutional setting (see Ensminger 1992; Haller 2007). By this definition, the predominant powers and interests in conservation discourse would refer to those whose narratives portray the benefits to be gained through 'modern' techniques of conservation said to be an institutionally adaptable model capable of evolving to meet the views, interests and goals of a transnational, complex community, managing a former local commons via the means of a global one (see Fletcher 2010). Therefore, referring to Foucault's quote in the beginning, this narrative of the functioning 'modern' method of conservation can only work in opposition to an unmoving, conservative, even 'primitive' system, often referred to as tradition. This chapter proposes to look beyond arguments juxtaposing modern and traditional, sustainable and unsustainable and other binary simplifications that legitimize discoursive best-practice models of conservation and consider constructions of reality through the use of narratives and ideologies (Haller 2010, 2013). It is suggested, therefore, to adapt research and practice methods, which consider the various narratives and ideologies that lead to understanding about where the constructions of nature and culture come from and how this enables legitimating conservation action within the power spectrum created by discourse.

As a guide of reference for this argument, the paper will first portray the case study undertaken in Kenya, then highlight the situation so as to problematize the complexity of an interaction arena between the different interest groups, before breaking down the narratives and ideologies, which not only lead to understanding the dominant discourse on conservation, but also a counter-discourse, which illustrates how normative measures for conservation are not internalised by all actors. This discontinuity of conformity shall be finally discussed in order to demonstrate the need for differentiated methods to analyse complex interaction arenas.

The case study area and research methods

The case study was undertaken during three months from November 2015 to February 2016 and focused on a protected area on the borders to the three counties Meru, Laikipia and Isiolo in Kenya, East Africa. The protected area, in this case the Lewa Wildlife Conservancy (LWC), founded in 1995, was chosen for several reasons. First, it is referred to as a model for community conservation as it has inspired other protected areas to follow its lead.[3] Second, the conservancy co-founded an organisation in 2004 known as the Northern Rangelands Trust (NRT), which has since grown to support more than 30 community conservancies throughout the north-west rangelands and other places.[4] And third, and possibly most importantly, these conservation organisations rely heavily on a transnational donor network for financial support (Weissman 2017). Hereby, the research focused on the Zoo of Zürich as one the main contributors, as it not only represents an interesting example of an influential donor from a small European country, but also shows the relevance for transnational networks connected to the research, which originated from a Swiss university (2017). Both the LWC as well as NRT conservancies are founded on land which was arguably mostly governed as commons by pastoralist societies in pre-colonial times, then partly annexed as private land by colonial settlers and, lastly, redistributed to state assets, left to former colonial land owners as in the case of the LWC or declared public land under the land-reforms after independence (see Leifer 1977; Hornsby 2012; McAuslan 2013). Best cases allowed for communities to form group-ranches in some counties, but even after the land law reforms of 2010 few considerations in regard to community land were addressed (McAuslan 2013).

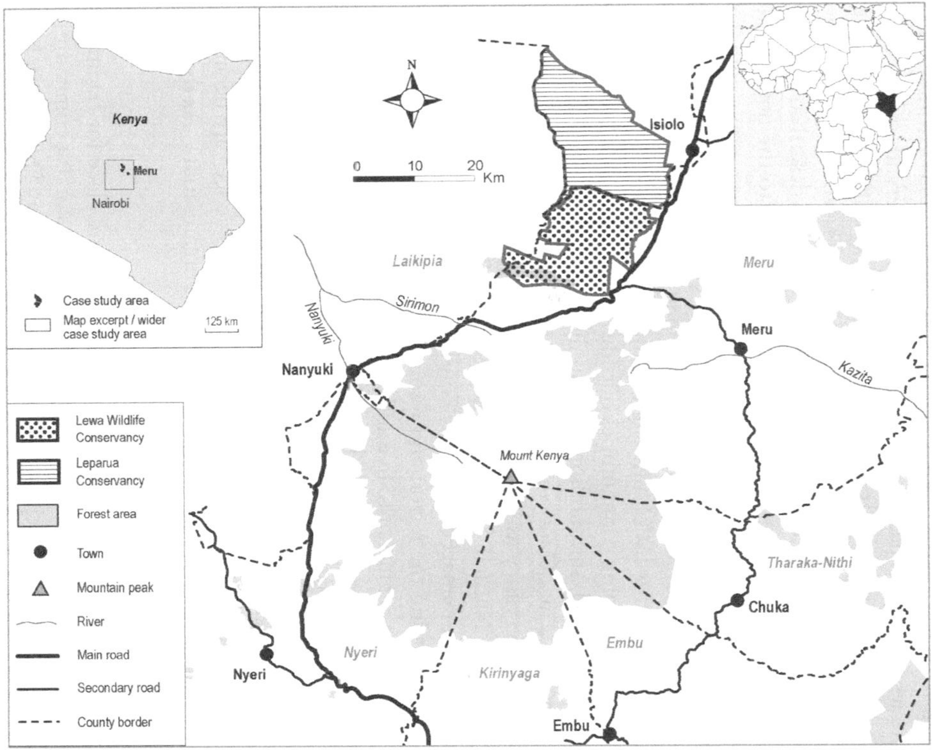

Figure 20.1 Map of the wider case study area north of Mount Kenya.
Source: Map compilation and design by Ulla Gämperli Krauer (2018), CDE, University of Bern, Bern, Switzerland. Geodata source: OSM data by OSMaxx (2018). Geometalab, HSR University of Applied Sciences Rapperswil, Switzerland.

In order to ensure a thorough understanding of how the conservancies work, the research was conducted to a part inside the LWC, the founding conservancy of NRT, and to another in an adjacent conservancy called Leparua conservancy founded in 2011 to the north of LWC, part of NRT. Both conservancies are listed as community conservation by IUCN,[5] but have quite different approaches with regard to how community conservation can be categorised.

Through employing the methods of participant observation, various interview techniques, discourse analysis and historical analysis, the data gained during my fieldwork in the area between two conservancies mentioned above, provided for substantial results regarding the complexities of social interaction in this particular context.

Between the frontiers of conservation discourses

By choosing to not treat the observed groups as homogenous entities, subtle inconsistencies in the expected behaviour towards conservation that exist between individuals ascribed or said to belong to a specific organisation or community could be observed. Therefore, this chapter will focus on these inconsistencies between the interest groups LWC, NRT, a so described Ndorobo Massai community by NRT inside the Leparua conservancy to the north of the LWC, and the Zoo of Zürich. In order to describe, how these groups interact, the perspective will set out first from the LWC, the conservancy which operates as a private conservancy but is listed as a community conservancy.[6] By this form of governance, the LWC, although managing a privatised conservation area, invests in developing communities outside the enclosed area. It also forms the focal point for all surrounding communities and conservancies, having established itself as a model for conservation in the region.[7] Brockington (2002) would list the conservancy as 'Conservation with Development' as opposed to 'Community-Based Natural Resource Management' (CBNRM). The latter is an institution where management of the conservancy falls to the community or communities governing the region and resources designated within a protected area, meaning people are meant to live and pursue sustainable economic practices inside the protected area (2002: 8) but without being the owners of the area, although it once belonged to them (see Mbeyale and Songorwa 2008). This form of community conservation is more akin to the concept of NRT. Their aim is to have a combination of sustainable and traditional land-use practices, while controlling which aspects of so-called traditional practices are deemed as beneficial by internal expert knowledge within NRT's organisation.[8] As the CEO of NRT explained in 2016:

> The world is a changing place, so there's a very strong place for traditional systems and traditional knowledge, but a lot of what pastoralist communities in the north here are faced with are completely new. They don't have traditional systems for dealing with that.
>
> (CEO-NRT, LWC-Headquarters)[9]

This perspective framing tradition as being incapable of adapting to a changing world, and at the same time, on the one hand using a superficial discourse in respect of traditional knowledge (i.e. discourse of being differentiated responsible actors), legitimises, on the other hand, intervention by agents with the ability and knowledge judged as capable of doing so (see Haller 2007; Galvin and Haller 2008). This perspective on tradition is coupled with the narrative of unsustainable land and resource-use practices, whereby NRT experts see the inability of the mostly pastoralist societies in the region in their supposed overstocking of herds for reasons of prestige and wealth as well as in overpopulation.[10] This narrative, at the same time, ignores the huge loss of land during the colonial and post-colonial history. In this context, as arguably in many other cases in African countries (see Haller 2013), before colonial governance, the ancestors of the local pastoralists treated the entire region around Lewa, between the fertile highlands from Laikipia to Mt. Kenya as an area held in common property before European settlers upset the system in place by not recognising any form of ownership of land by locals.[11] Ever since, disownment and disregard have continued and are no longer discussed, as under today's Kenyan law, the conservancies have right to land or land-use, while non-conservation agents struggle with open access situations (Weissman 2017). The past remains buried in the past.

In the LWC, the same narratives are used, and, by their perspective, they deem these issues of overstocking, overpopulation as well as undereducation and lack of infrastructure as driving factors responsible for land and resource overuse.[12] While NRT aims to enable communities to manage their own conservation institutions, the LWC excludes communities from the direct physical interaction with the protected space and what it entails. The protected area is surrounded by different types of fencing, has several controlled entry points, and uses most of its staff for security purposes, such as rangers and scouts. Additionally, the LWC employs an Anti-Poaching Unit (APU), uses a helicopter with night-vision technology, trained tracking dogs and works closely with NRT's Rangers and Kenya's Wildlife Service to coordinate efforts together with other conservancies in an area spanning roughly 45,000km^2.[13] In order to keep the LWC running, the organisation relies on an income of around five to six million USD yearly, whereof almost two-thirds are used for security purposes (Weissman 2017).

As mentioned above, this income is generated mostly through donors and bursaries, while tourism is outsourced to private lodges within the conservancy itself. Although the LWC is continuously working on creating new avenues of revenue, international stakeholders can claim large portions of influence through their contribution and support. In this case, the Zoo of Zürich, for instance, has invested mostly in security for especially endangered wildlife, such as rhinos and elephants, by equipping the security forces with communications gear, providing the helicopter with night vision technology, or investing in a securely fenced corridor for elephants that migrate to the Mt. Kenya National Park. The Zoo has recently begun building an enclosure in Zürich, named 'Savanne Lewa', in order to increase and secure their ongoing support long-term (Weissman 2017).

This situation highlights the nature of security that comes with this form of conservation, as discussed by Brockington et al. (2008) and others, described to be more akin to fortress conservation. And although in response to the critique on fortress conservation, the participatory approaches employed now by conservationists such as the LWC or NRT still remind us of a fortress-like system, it is closer to Foucault's (1977) concept of panopticism, whereby the power to transform the deviant actors in society are based on methods of self-regulation brought about by a moral standard set through various surroundings that aim to discipline not through punishment but constant watchfulness and judgement (see Fletcher 2010). The actors in and around the LWC are for that matter bound through a net of ever-present security measures as well as development initiatives aimed at transforming the public's mentality and morality to one favouring conservation. The clinics, schools and other infrastructure are important tools in this endeavour and lead to an important point made by Fletcher (2010) in regard to neo-liberal environmentality, whereby incentives to act in accordance with conservation ideology are no longer aimed at only ethical but also cost-benefit methodology. The LWC does this by aiming to have the communities as the first line of defence in the fight against poaching at a low cost. By suggesting a landscape where conservation is beneficial to all people and providing the incentives to economically profit from it under certain circumstances, the LWC creates a discourse of friendship and team-work as the core of their business, acting not unlike a neo-liberal government agent and what Ostrom (1990) referred to as a 'nested enterprise'.

For the purpose of demonstrating how this friendship-discourse takes hold, in the following, an example from an actual event, which took place inside the LWC in 2016 will be described in order to deconstruct the pluralistic legal and institutional settings involved. It will also demonstrate to some extent the discontinuance of normative transformation, ethically or otherwise, as suggested by an environmentality approach.

Encountering conservation discourse

In the following excerpt from a field journal entry in January 2016 a situation is described through participant observation, where a herder meets two rangers on a public road running through the southern part of the LWC. While the conservancy has private land rights, the road is public and herders passing through the conservancy can graze their livestock on marked stretches, each 30 meters on either side of the road. This example highlights frequent interactions between the rangers patrolling the road and herders passing through. The rangers patrol the road daily in order to mitigate potential conflicts that can occur and to ensure that there is no trespassing, as in the excerpt below:

> One day in early January I joined two rangers who were on patrol duty. We left from the Headquarters and crossed the airfield on our way to a hill on Samba Ridge. From there, after having a look out for animals and spotting

> a rhino to be called in via radio, we continued to T.M. gate on the south-eastern end of the conservancy [...] The rangers' job was to patrol along the roads and make sure people only graze their cattle within 30 meters on each side of the road. Mostly, they said, their presence was enough, but that some people still made trouble. As they explained to me, it was often young people who challenged them, the older people were easier to handle. As we continue from T.M. Gate to Rugusu Gate north of our location, the rangers tell me that their relationship with the people around the conservancy is friendly, that it needs to be. Because when they talk to the herders and they have a good relationship with them, they can ask them about news on poachers and the herders will in turn tell them if they know of any or heard of anyone. When I ask how they maintain friendly relations, the rangers answered by saying that people generally benefit from Lewa and can be rewarded for information. There are always chances of getting employment, they can be rewarded with money, and their children could receive sponsorship at school. Along the way we meet a man who is grazing past the 30-meter marker. I had to ask the rangers myself, where the marker was, as I didn't see it. They showed me a small wooden stump sticking out of the ground saying people who graze here regularly know exactly where the border is. This man, however, had his own reasons for crossing the marker. He was dressed in torn shorts and a t-shirt, in his hand a long stick with which he directed his small herd of a dozen goats along the way. He was explaining to the rangers that animals coming from LWC had damaged his property, invaded his shamba[14] and destroyed some plants and fences. Now he felt he should be compensated and decided to graze his animals a little further in where the grass was better. The rangers who were wearing military-style khaki gear and one holding a semi-automatic weapon, the other a shotgun, told him that he could not do that. What he could do was appeal to the community committee, which could make a proposal for LWC to be considered. Normally, however, they tell me, the KWS is responsible for compensation as the wildlife is the property of the state of Kenya, but that the KWS wasn't very reliable in this regard. We met a few more people along the way who weren't grazing inside the markers and they all were warned. If the rangers catch serial offenders, mostly they already know them, they are fined 5,000 KES [roughly 50 USD or CHF].
>
> (Field notes by the author, January, 2016)

The herder, who had damage done to his property by a wild animal, decides to graze beyond the 30-metre marker on the side of the road. The action is illegal by law and Lewa reserves the right in this case to punish trespassers. Mostly, herders are warned by the rangers, but if occurrences are too frequent with certain herders, they are then fined as noted above. In this case the conservancy is empowered by the judicial institution of Kenyan law to protect their property. The herder may explain, that he only trespassed, because an animal which came from the LWC damaged his property and now he seeks to compensate by grazing his animals on better pastures, which will make them healthier, fetching a better price and thereby giving him the necessary means to repair his garden. In this case, however, the conservancy disagrees with the herder's logic. As they are only responsible for the animals inside the conservancy, they cannot be held liable for what happens outside. Therefore, the herder must seek his compensation not with them, but with the state, the so proclaimed owner of the animal that damaged his property.

However, the herder knows that in order to receive compensation he must prove that his garden was destroyed by a wild animal. Even the rangers admit that compensation often falls through, mostly due to the rigorous procedure which requires claimants to produce evidence of damage and have it assessed by an official representative of the KWS in the county.[15] Another alternative suggested by the rangers was that the claimant could appeal to his community committee. These committees are also part of the LWC's community development efforts to establishing a stronger relationship with surrounding groups of people living around the conservancy,[16] with the goal of creating incentives for the landholders outside to conserve natural resources as intended by the LWC (see Emerton 2001). Both Emerton (2001) and Brockington (2002), who discuss community development as part of the benefits gained through conservation, argue however, that the economic gains to be had can rival the costs which amount through damage to land and property as well as restrictions of protected land and resource use. Brockington (2002: 105) hereby states that the benefits emerging from conservation should at least match or exceed the costs it incurs, if the conservancy aims at creating good relationships with its neighbours. As according to the rangers this is described above as mostly being the case, as herders know how they can profit from working with them and therefore mostly follow the LWC's rules.

Therefore, the KWS, the LWC as well as the Community Committees following the status quo, whereby conservation is seen as the more beneficial economic practice, can establish a more powerful bargaining position as opposed to the herder, who has neither legal nor institutional backing. Additionally, the herder is not only in a low bargaining power position but rendered powerless in the above situation by the fact that the LWC is not the owner of the wildlife, but simply it's caretaker. If we imagine, for arguments sake, the conservancy were a day-care facility, the parent of the child – in this case the government owning the wildlife it seeks to protect – would give their child to a day-care facility in order to pursue their job, putting responsibility over their child into the hands of the facilities caretakers. Therefore, the day-care facility would be liable, should a child damage the property of someone outside the facilities boundaries. Analogously, if the child in the conservation scenario, e.g. the animal crossed this boundary and damaged property outside, we could expect the caretaking facility to compensate the affected party, as it has incompletely managed to secure its infrastructure. This, however, is not the case in Kenya, since the legal apparatus of the state assumes responsibly to compensate, the LWC's narrative is to accept their bargaining position to be void in the matter. The power, however, within the hands of the conservancy, is the ability to invest in securing the boundaries and reducing the number of incidents. This is achieved in several ways, which all rely on financial support to achieve heightened security measures. The conservancy seeks investment in areas, such as better equipment for rangers, state-of-the-art monitoring technology, secured fences surrounding the entire conservancy, as well as development programs in the surrounding communities. This investment, then again, comes from donors, including the Zoo of Zürich, who's main goal is to secure its interest: the protection of endangered species as one of its selling image and funding goals. If this, in turn, means that also investing in development programs for the LWC's neighbours guarantees a higher security outcome, the investment can be seen as beneficial to all interest groups and the circle closes (Weissman 2017).

This narrative by the rangers, who explain the system of reward whereby the conservancy, its donors and the herder are beneficiaries, also creates a 'positive feedback loop' from the perspective of the LWC (see Haller 2013). The more investment coming from donors, the better are the security measures and the development initiatives implemented by the conservancy. Therefore, the properties on the outside are safer and the state does not have to compensate any damage by wildlife. Additionally, the members of the communities are more content and see the benefit of having schools, roads, clinics and other infrastructure donated through development initiatives and therefore these members will relay information to the conservancy regarding potential threats about outside agents looking to harm them or the wildlife they take care of. Therefore these risk factors, such as poaching, can also be reduced by additionally having a highly trained anti-poaching unit. And the donors, such as zoos and conservation NGOs, who seek the protection of wildlife and eco-systems, can be rest assured that their investments are worthwhile *and* for a good cause (2017).

This positive and self-reinforcing feedback loop is the basis for the LWC's and NRT's narrative, which, within conservation discourse, creates a strong incentive for an industry to invest further in these models of community conservation. From the perspective of the conservation elite (e.g. the LWC, NRT and donors) this feedback-loop theoretically demonstrates where the critique on neo-liberal conservation comes into play, by considering the actual circumstances as gathered by the empirical material in the field (see Fletcher 2010). As the example above shows, the observable reality is less ideal than the narrative of community conservation suggests.

The counter-discourse

The herder in the scenario above uses a counter narrative to legitimate his actions. By grazing outside the marked boundary of the public road, the herder demonstrates that he does not seek to apply for compensation by the state or by the LWC by means of a formal request. He rather sees the conservancy as the immediate party responsible for the damage done to his garden, as the animal came from there, through the fence, and since the LWC did not try to respond to the incident, in his mind, this legitimates his action of allowing his herd to graze inside the conservancy on a quid pro quo logic. However, his appearance in everyday clothes and only carrying a stick to herd his livestock without any intention of taking a more confrontational stance against the heavily armed rangers along with only the argument he uses, indicates that his bargaining power position carries little weight. The counter narrative herein lies with the ideology represented in the discourse used often by pastoralists, which will be described in the following.

In the participant observation done to the north of the LWC in the Leparua Conservancy, a clearer picture emerged regarding the perspective that the Ma-speaking agro-pastoralists living there have. The emic view there, is that the people are Massai and originally moved throughout the region since long before colonial times. By accounts of elders, their ancestors used to be hunter-gatherers that came to be Massai through alliances and trade and started a pastoralist lifestyle speaking Ma-language and using Massai institutions. In their account, as hunter-gatherers as well as pastoralists, they used to move between the foothills of Mt. Kenya and up into regions that are today Samburu county. As they say, back then they knew where the boarders between different groups of pastoralists were. After the British came, however, their land was used by them as well; they established their own cattle farms but stayed where they settled and didn't move on. The LWC originally was such a ranch run by a British settler, an ancestor to the later founder of the LWC. Their own region was then declared a quarantine zone by the colonial government to control cattle movement from the north of Kenya to Nairobi in the south. The elders say they were not allowed to move out of, or into, the quarantine area. They also never learned that the quarantine ended. By the time they found out, a government department was moving into the region, laying claim to it. By this time, they were no longer recognised by the government as being indigenous to the region and denied any rights to land or resources (Weissman 2017: 46–49).

Even outside of their area, narratives by the LWC and NRT experts tell a story where the Massai didn't live there until 1971, when they suddenly appeared. By this view, the Massai are: First, not indigenous to the region, and, second, not true Massai. They are referred to as Ndorobo Massai, which is a derogatory term originating in Ma-language and later used by the British settlers to describe hunters thought to be a lesser form than true Massai, which came with their cattle given to them by god from the heavens (ibid: 52–57).

However, the Massai in Leparua did not accept the claims by the government department that was trying to grab the land and they took their claim to court. According to their history, they have always had to fight for their land but have yet to be successful in receiving official recognition. It was declared public land in 1995, which for the Massai was only a partial victory, because although the government department claiming the land had to leave, the land was now, as they say, 'no man's land', because the court ruling did not define a specific group that could own it. In 2011, the people in Leparua and adjacent valleys accepted offers by NRT to become a community conservancy, a status which at least partially secured some identity with and claim to the land, although it does not give legal rights to specific land titles. Therefore, the situation is one of open access leading to conflicts over grazing rights and other resource use with neighbouring groups of pastoralists (ibid: 46–49).

Through their narratives, the conservation discourse seems to be integrated into their lifestyle, as they say their way of dealing with resource allocation and usage has in the past been regulated through their age-set system (e.g. institution) in a sustainable manner. In this semi-arid to arid region the further north it goes, movement and redistribution were key to successful resource management. By their accounts, the age-set system works on generational authority over decision-making while the eldest had the highest authority, but each age-set would decide on their own issues as long as they were manageable within their group. Their elders remember that it was the game and hunting parks the British established, which mostly led to the degradation of wildlife and that restriction of their movement and disruption of their institutions following the colonial government led to overstocking and consequently overgrazing (ibid: 54–55).

Through this discourse, the Massai tell the story of a lifestyle, which has been beneficial to them and their environment long before todays labels of sustainability and conservation appeared. However, in their view, what the early conservationists that came to the region did and continue doing today does not seem new and was rather a curious experiment, which did not seem to work out well in the beginning. They disrupted the system in place for a long time before it got better, as the Massai recall. From their perspective, the LWC made many mistakes, especially because they closed-off their land and held wildlife separately from livestock and people. When they later reintroduced them to each other, this caused a lot of human (including livestock)–wildlife conflict, as they recount (ibid: 65–66).

Today, the Massai are mainly interested in securing their land rights and if the status as a conservancy can help them, they will continue to partially follow the institution set forth by NRT. Although they do not agree with all terms set by this institution, they see the political benefit, should they gain more bargaining power through their recognition as a conservancy and the ideologies thereof (See Haller 2010, 2016; Weissman 2017: 103).

This institutional arrangement, which the Massai and other groups agree to with NRT creates hybrid institutions between different agents, as is also described by Bersaglio and Cleaver (2018) who describe the re-shaping of institutional arrangements between these different actors as a form of green-grabbing and see its formation in what Cleaver (2012) terms 'bricolage: the conscious and unconscious reshaping of institutional arrangements for managing communal lands and natural resources by bricoleurs, who make use of whatever resources and tools happen to be at their disposal' (Bersaglio and Cleaver 2018). While I agree with this interpretation between institutions for the sake of conservation as a form of green grabbing and the complexity of this can be described by 'bricolage', I prefer to use Haller's (2013, 2016) idea of institution shopping, whereby making ends meet is justified through the use of ideology and discourse.

Disentangling conservation discourse

The two above perspectives; on the one hand by the conservation enterprises LWC and NRT and on the other hand by the local agro-pastoralists, each of which form different versions of conservation discourse, show a juxtaposition between ontologies of a 'pure nature' on the one hand and a 'cultural landscape' managed in common property on the other. The contrast between these two positions, which have existed alongside each other for at least 47 years,[17] is reflected in other perspectives as well.

By recording the narratives from various stakeholders connected to the conservation efforts in the region, the case-study helped in discerning, which stakeholders held to which discourse, revealing commonalities in worldviews and ideologies. However, it also showed a discontinuation from the standard narrative prevalent in the global north to a contradictory degree of understanding of conservation the closer to the local context of the conservancy in question (Weissman 2017: 105). For example:

- Where the major investors such as conservation organisations and zoos are concerned, conservation, from this perspective, should aim to halt the loss of bio-diversity and ensure a sustainable future for wildlife and, in effect, for humankind. The ideology herein proposes that there is a form of 'pure' nature being mishandled by humans and is therefore threatened but can be protected through their efforts and support.
- Where local agro-pastoralists are concerned, their life style should be understood as a part of conservation, which they understand they have practiced long before labels such as sustainable resource management existed in their world. In fact, these cultural landscapes and land related resources such as pasture were held in common property and managed by related common institutions based on membership and rules of mobility and local interactions. What they ask for is rather, that they are met on a level playing field where conservation can enable rights to land and resources. The ideology herein follows the description of the landscape as a cultural one, where the continuation of human and non-human entities is interwoven and hardly separable when considering the history of common property ownership and land-use (see Weissman 2017).

To display all the various positions in the context of the case-study, Table 20.1 gives an overview from the global to local-level.

While the Zoo represents an influential stakeholder within the global arena of conservationists, both the organisations LWC and NRT can be understood as a local governance factor in questions of land management, economic influence and policy shaping and are therefore situated within the governance level of discourse, together with the State of Kenya, which acts on a national level. The only actor remaining in this case is the local level, which in the case-study is represented by a Massai community within the Leparua Conservancy. All viewpoints, however, are highly variable as they depend on the affected party

Table 20.1 Overview of discourses used by the four main stakeholders of the LEWA-Conservancy (Weissman 2017: 104)

Stakeholders from global to local arena	*Main ideologies*	*Main discourses*	*Main narratives*
Global level: Int. Donors (Zoo of Zürich, CH)	'Pure nature' in peril	Biodiversity loss is to be stopped on a global scale by 'best practice' conservation.	Wildlife and endangered species need to be protected because of harmful practices by humans (such as poaching).
Governance level: Kenya State	Natural resources in peril	Natural resources are to be secured and controlled long term, for the benefit of the nation.	Conflicts over natural resources and between humans and wildlife are harmful to the economy and need to be mitigated.
LWC (Conservation with Development)	'Pure nature' in peril	Conservation preserves and furthers biodiversity through working with people outside the last islands of wilderness.	Population pressure, poverty and lack of education need to be addressed to enable modern solutions to conservation.
NRT (Community-Based Natural Resource Management)	Hybrid landscape between 'pure nature' and 'traditional' landscapes	Conservation can preserve biodiversity and secure the economy by combining professionalism with 'traditional' systems.	Traditional systems are fixed and need help to adapt to optimal resource and land management in a rapidly changing world.
Local level: Leparua Community Conservancy (Maa-speaking agropastoralists)	Cultural landscape managed in common property	Conservation has always been part of lifestyle in a landscape where people and wildlife have always lived together.	People have always practiced sustainable resource use but need a level playing field and rights to property to restore functioning institutions.

within each group, where each party associates its identity in regard to a particular organisation or community and its position regarding the form of conservation institution. These inconsistencies within the interaction arena will be discussed along with viewpoints from:

- The Zoo of Zürich, situated within the global level in the table above, and representing the hegemonic discourse in the global north together with international donors such as the Nature Conservancy, Tusk Trust, several zoological organisations and various others.[18]
- The NRT, situated between the LWC and the agro-pastoralists in Leparua, on the 'lowest' rung of the governance level in Table 20.1, as it represents a hybridised approach to participatory conservation.

The LWC and NRT play an important role by embodying what is thought to be a 'modern' approach to conservation. They represent the key instruments guiding best-practice conservation in the region. The LWC started out as a ranch owned and run by a British settler in the colonial period and then in 1995 part of the private property was turned into a rhino sanctuary by the descendants of the family who owned the ranch, which ultimately led to the entire property being transformed into a conservancy (see Paice 1995). Today the land has been placed in a holding company so as to secure the conservation land-use form in the long term.[19] By their form of conservation, what is inside the conservancy shall serve only the practice of 'pure' conservation, meaning that no land-use should be practiced for commercial activities, such as farming, as one executive explained.[20] Therefore, the conservancy imposes strict rules through their institution, founding a strong nature-culture divide, whereby the ideology that 'pure nature' is essential for a functioning eco-system and thereby guaranteeing the safety of highly endangered species, sets a precedent throughout the region (Weissman 2017: 74). Van den Akker (2016: 128) shows how certain endangered species, such as the black rhino, can generate enormous value for conservationists (see also Emerton 2001) and thus change relative prices for services provided in an area (see Ensminger 1992; Haller 2013). As Keller (2015) points out, inscribing such values onto nature or wildlife, as supporters such as zoological associations do, creates incentives to secure these values, which are threatened by humans. This further enhances the ontological separation of non-human and human, e.g. nature and culture, creating thereby a further imperative to establish and continue building on institutions, which aim to separate one from the other (see Descola 2013). While the Zoo of Zürich has mostly taken this stance, they have also indirectly supported the LWC in their community development efforts, shown in the positive feedback loop mentioned earlier, as the communities are part of the LWC's security measures as well. However, while the development measures seem to benefit the communities according to the zoo's and the LWC's view, some among those living around the conservancy rather see no other alternative than to follow the conservancies rules and guidelines, as they are faced with an anti-politics machine (Ferguson 1990). Hereby, what is hidden, is the fact that the former common property institutions were not only disabled during colonisation – the land and resources grabbed and reallocated and then transformed into state property after independence creating open access situations loaded with conflict potential and actual conflicts over mobility, rights to grazing and other land and resource-uses – but the newly, re-invented institutions have also widened to embrace a stakeholder group incorporating all mankind, seemingly entitled to partake in what is a global, common ownership of wildlife and nature in general. This enables the conservationists to act as middlemen in the local context, managing nature and wildlife on the behest of the world, while the previous owners and managers of the land are disabled or rendered invisible unless they become part of the conservation agenda.

This transformation aimed for by the conservation elite finds legitimacy through the dominant discourse in the global north and hopes to find traction locally through the methods of 'neo-liberal environmentality', as described by Fletcher (2010). However, as certain cases of counter-discourse show, the transformation is not internalised as expected but forms of resistance (see Scott 1985) suggest that there is an anti-politics machine in play (Ferguson 1990) and that green grabbing occurs not via linear conceptions of power enacted from above and reacted to from below (Bersaglio and Cleaver 2018), but by complex interaction patterns creating seemingly fluid power networks changing along practices of institution-shopping (Haller 2013, 2016). Therefore, as Fletcher (2010: 180) suggests, alternative forms of conservation can only emerge 'by clearly defining the parameters of various stakeholders' particular approaches to conservation, the beliefs and assumptions upon which they are based, and the differences among competing positions'.

Discussion

The Lewa Wildlife Conservancy has gained traction since its establishment by attracting investment from a global community through the means of endangered wildlife, which form the backbone for discoursive value ascription to 'natural resources' in need of protection. Additionally, since the conservancy is faced with danger of poaching and conflicts over land-use, it also engages with surrounding groups of people living around the conservancy by means of community development. Thereby, they promise to enhance the lives of all acting in accordance with the rules set forth by conservation institutions, which follows a hegemonic discourse saturated with an ideology of a romanticised 'pure' nature established and distilled through various investors in the global north. Since conservation, therefore, follows the ideological concept of sustainable resource use and development, their bargaining power position within national and international politics is favoured over those following other economic practices, seen as harmful to the environment, such as 'traditional' pastoralism or farming. NRT, for instance, among other programs, creates, however, further incentives for pastoralists to manage cattle rearing in a sustainable fashion by creating a market for fair exchange conditions.[21] This market, reserved for those communities within an established community conservancy under NRT, comes with incentives to reward those pastoralists who choose to reduce large herds in favour of smaller herds of a particular kind and showing good conservancy performance. Thereby, it is implied that the 'traditional' institutions are unable to adapt to today's challenges of overpopulation of livestock and loss of rangelands, by NRT's and LWC's narrative. Seemingly, this would explain the logic behind adapting to the conservation strategy and result in a straight forward movement where all land and resource-users see the benefit of conservation (Weissman 2017).

However, as argued in this article, the movement towards conservation is not as straightforward as it is suggested by the discourse representing the predominant interests and powers in this case. While the discourse makes a strong case for a reality which favours the 'new' over the 'traditional' way of land and resource management, and while it may shape a reality where 'conservation for the people' or 'by the people' attracts investment from the global north, this strategy within a neo-liberal environmentality approach would, as Fletcher (2010) points out, also not allow for true participation. Other strategies, however, exist on the local level, also influencing the discourse. People among the Massai for instance, have been confronted with land grabbing in the past and now realise that there is a hidden green grabbing process going on behind this conservation agenda in Kenya. For comparison purposes: When British occupation began in Kenya, around 8,000km^2 were annexed for European settlers in the highlands under an ideology of agricultural economic growth beneficial to the colony (See Morgan 1963; Duder 1993). Today, alone the LWC and NRT have grown to incorporate more than five times that amount at around 45,000km^2.[22] This fast-growing industry is putting a lot of pressure on previous lifestyles in Kenya. In the views of the conservancies this is necessary, as their goal is to protect nature from harm and the Kenyans' old lifestyles, such as pastoralism and farming are seen as harmful if practiced as before. Convincing communities to use their land for conservation instead of pastoralism, for example, is supposed to alleviate pressure on grazing land and help eco-systems in general. The promise by conservation supporters in return for this change in life-style, culture even, is the profit that flows from conservation. First, according to this narrative, the disburdened land can now generate enough food for the domesticated animals while also bringing back more wildlife, which is important for functioning eco-systems and, which in time should continue to improve. Second, well-managed areas gain support through umbrella organisations as in the case of NRT, who contribute by bringing donors and investors through various projects, encouraging development work. This means the conservancies attract projects which are seen as beneficial, such as school-buildings, clinics, electricity supply, and other infrastructure, as well as work-programs, employment, and so on. In this way, leading conservancies see their role as beneficial to the communities.

However, other actors within the same so-called 'communities' choose different strategies to benefit from this conservation agenda. As individuals and groups are faced with institutional pluralism by various organisations such as the LWC, NRT, several development agencies and programs as well as the state, they will opt to 'shop their institutions' accordingly (See Haller 2013, 2016). The herder from the example described earlier chose to follow a path independent of any formal institutions of the organisations mentioned above and was caught. In this instance he chose to explain his trespassing on private property by reasoning that if the LWC cannot keep the animals it takes care of in check and one of them damages his private property, a debt is owed. Others who act within groups may identify with a certain community or organisation, seeking to profit from either NRT's 'Livestock to Market' system or find employment as a ranger in a conservancy. Some might see benefits in building positive relationships with the conservancy development programs in order to get employment in a clinic, school or other facility. Some might seek benefits for their children, gaining a sponsorship for them through the international associations working with the conservancies (see Weissman 2017). The first reaction to conservation is, however, not the benefit of protecting endangered species per se – that is to say, having a healthy planet for the future of humanity – but making the most out of the current system dominating the economic land-use strategy. The main problem with the dominant conservation approach in this case is, arguably, that it is not having the intended outcome of an environmentality of any kind (see Fletcher 2010).

Most locals among the agro-pastoralists, and possibly most among all people neighbouring the LWC, are aware that the state creates a narrative which leads to believing that wildlife is a quasi-commons, which if protected benefits all and the risks that are endured are supposedly compensated for. This narrative is slightly adapted in the global north where the protection of wildlife is also portrayed as beneficial to all but must be supported for global reasons adjusting the local commons size in ideology to a global community. However, if all should be able to participate in the managing of these 'glocal' commons, the input by those most affected in the physical places where conservation is practiced seems marginal and mostly represented by expert knowledge, but rarely receives actual input from below. It seems paradoxical, to think that the Zoo of Zürich, which displays animals in a zoological garden out of context to their origin, in order to generate financing to protect these animals in their context of origin, without first considering the societies from those contexts, but the ones out of it.

This circumstance consolidates power located within the conservation elite and since they arguably continue to contribute to various forms of inequality (Fletcher 2010), the present argument in this paper would not only acknowledge this circumstance in the described case-study, but propose to use methods of identifying ideologies and narratives used to legitimize actions tied to various institutional arrangements in order to disentangle the discoursive web in a 'glocal' space. Conservation practice, however successful or not, is therefore not explained by contrasting binary opposites – such as conservationists vs. non-conservationists, modern vs. tradition, experts vs. laymen, enterprise vs. community, and so on – but rather through disentangling the complex interactions between fluidly moving actors by recording narratives and ideologies, and showing how these may be reasoned within one institution or another, dependent on the circumstance an individual or group is faced with in a pluralistic institutional system. What can be understood from examining this is that a global conservation commons charged with taking care of 'our' environment can only work if participants can engage in a discussion on conceptions, ideas and institutions, and relinquish imagined conceptions of permanence and adaptability being tied to discriminating identities.

Outlook

The counter-conservation discourse which were to build upon green grabbing arguments has to become a more discussed topic and become integrated into general conservation discourse on all levels. Most importantly, in the most distant places, such as a small country in Europe for example, where the discourse follows an old fashioned, romanticised ideology of a nature untouched by humans. This is not only important because most financial contribution to conservation comes from economically powerful nations, but also because the knowledge bias towards this type of conservation is established in the academic sphere of these nations.

It is necessary to conduct more research on how conservation is practised, based on what models and what methods. Have these improved since the early national parks in a sense that considers heterogenous interaction arenas, without discriminating against certain institutions or people? And are there possibly other sustainable land-use methods out there which are better suited for certain contexts, such as old or new commons systems? Treating wildlife in various localities as global commons – in the sense that we are all responsible for them or what they contain, protect, or nourish, and can help to alleviate pressure by investing in projects initiated by zoos, the WWF, the Nature Conservancy and others – should be connected to an obligation, which ensures responsibility is taken for the changes brought about in the spaces where conservation asks the most of people, who have been confronted with upheaval since the first colonial settlers started imposing 'new' and 'better' ways of land-use strategies without considering what made those, who came before them, so successful for such a long period of time. For, they are still here today, not as they were surely, but not giving in either.

Notes

1 As by the latest entry in 'The Stanford Encyclopaedia of Philosophy' (Bub 2017).
2 According to Foucault (1971), discourse is what produces and re-produces reality through language and is structured along particular sets of rules dependant on the predominant powers and interests in a society.
3 Annual Report 2014. Lewa Wildlife Conservancy (www. Lewa.org).
4 NRT 2018 (http://www.nrt-kenya.org/regions/), accessed 10 July, 2018.
5 International Union for Conservation of Nature, www.iucn.org/theme/protected-areas/our-work/world-database-protected-areas, accessed 12 July, 2018.
6 World Database on Protected Areas, https://protectedplanet.net/555555490 accessed 3 July 2018.
7 Annual Report 2014. Lewa Wildlife Conservancy (www. Lewa.org).
8 www.nrt-kenya.org accessed 3 July, 2018.
9 Interview with executive, NRT, January, 2016.
10 Interviews with executives, LWC and NRT, January, 2016.
11 Sometimes settlers didn't even recognise the presence of humans, deeming land as empty, waste-land or other. Most often, the fact that pastoralists or hunter-gatherers would move around as semi-nomadic people, legitimised the explanations by colonial settlers that the land was not owned by anyone (see Weissman 2017).
12 Interview with executive, LWC, January, 2016.
13 www.lewa.org/wildlife-conservation/security/, accessed 3 July, 2018.
14 A Shamba is a Swahili word for a garden or any form of cultivated land. Field notes by the author, November, 2015.
15 www.laikipia.org/wp-content/uploads/2015/09/KWS_-WILDLIFE-COMPENSATION-REGULATIONS-May.pdf accessed 5 July, 2018.
16 Interview with the author, January, 2016.
17 Only if understood that the official recognition of the Massai settling in Leparua took place in 1971. Otherwise, this time extends to the settling of the first British farmer in Lewa in the 1920s (see Paice 1995).

18 See Annual Reports 2011 through 2015 and Impact Report 2016. Lewa Wildlife Conservancy, www.lewa.org, accessed 3 July, 2018.
19 See Project 'Lewa Milele', www.lewa.org accessed 10 July, 2018.
20 Interview with executive, LWC, January, 2016.
21 NRT State of Conservancies Report 2014. Northern Rangelands Trust.
22 www.lewa.org, accessed 18 July, 2018. www.nrt-kenya.org, accessed 18 July, 2018.

References

Agrawal, A. (2005) *Environmentality: Technologies of Government and the Making of Subjects*. Durham, NC and London: Duke University Press.

Benjaminsen, T.A. and Bryceson, I. (2012) 'Conservation, green/blue grabbing and accumulation by dispossession in Tanzania', *The Journal of Peasant Studies*, vol 39, no 2, pp 335–355.

Bersaglio, B. and Cleaver, F. (2018) 'Green grab by bricolage – the institutional workings of community conservancies in Kenya', *Conservation and Society*, vol 16, no 4, pp 467–480.

Brockington, D. (2002) *Fortress Conservation: The Preservation of the Mkoazi Game Reserve, Tanzania*. Bloomington, IN: Indiana University Press.

Brockington, D., Duffy, R., and Igoe, J. (2008) *Nature Unbound: Conservation, Capitalism and the Future of Protected Areas*. London: Earthscan.

Bub, J. (2017) 'Quantum entanglement and information', in Edward N. Z. (ed.), *The Stanford Encyclopedia of Philosophy* (Spring 2017 Edition), https://plato.stanford.edu/archives/spr2017/entries/qt-entangle/, accessed 10 October, 2018.

Cleaver, F. (2012) *Development Through Bricolage: Rethinking Institutions for Natural Resource Management*. London: Routledge.

Descola, P. (2013) *Jenseits von Natur und Kultur*. Berlin: Suhrkamp. (Original: Par-delà nature et culture. 2005. Paris).

Duder, C.J. (1993) '"Men of the officer class": the participants in the 1919 soldier settlement scheme in Kenya', *African Affairs*, vol 92, pp 69–87.

Emerton, L. (2001) 'The nature of benefits and the benefits of nature: why wildlife conservation has not economically benefitted communities in Africa', in Hulme, D. and Marshall, M. (ed.), *African Wildlife and Livelihoods: The Promise and Performance of Community Conservation*. Oxford and Portsmouth: James Curry and Heinemann, pp 208–226.

Ensminger, J. (1992) *Making a Market. The Institutional Transformation of an African Society*. Cambridge: Cambridge University Press.

Fairhead, J., Leach, M., and Scoones, I. (2012) 'Green grabbing: a new appropriation of nature?', *The Journal of Peasant Studies*, vol 39, no 2 pp 237–261. DOI: 10.1080/03066150.2012.671770.

Ferguson, J. (1990) *The Anti-Politics Machine. "Development", Depoliticization and Bureaucratic Power in Lesotho*. Cambridge: Cambridge University Press.

Fletcher, R. (2010) 'Neoliberal environmentality towards a poststructuralist political ecology of the conservation debate', *Conservation and Society*, vol 8, no 3, pp 171–181.

Foucault, M. (1977) *Überwachen und Strafen. Die Geburt des Gefängnisses*. 1. Auflage. Suhrkamp, Frankfurt a. M. (Original: Surveiller et punir, naissance de la prison 1975. Gallimard, Paris).

Foucault, M. (2010) *The Archaeology of Knowledge and the Discourse on Language*. New York: Vintage Books (Original: L'Archéologie du Savoir 1969. Gallimard, Paris / L'ordre du Discours 1971. Gallimard, Paris).

Galvin, M. and Haller, T. (ed.) (2008) *People, Protected Areas and Global Change: Participatory Conservation in Latin America, Africa, Asia and Europe*. Perspectives of the Swiss National Centre of Competence in Research (NCCR) North-South, University of Bern, vol. 3. Bern: Geographica Bernensia.

German, L., Unks, A.R., and King, E. (2017) 'Green appropriation through shifting contours of authority and property on a pastoralist commons', *The Journal of Peasant Studies*, vol 44, no 3, pp 631–657.

Haller, T. (2007) *Understanding Institutions and Their Links to Resource Management from the Perspective of New Institutionalism*. Bern, Switzerland: NCCR North-South, Dialogue 2.

Haller, T. (2010) 'Institutional change, power and conflicts in the management of common-pool resources in African floodplain ecosystems: an introduction?', in Haller, T. (ed), *Disputing the Floodplains: Institutional Change and the Politics of Resource Management in African Wetlands*. African Social Studies Series. Leiden: Brill.

Haller, T., Fokou, G., Mbeyale, G., and Meroka, P. (2013) 'How fit turns into misfit and back: institutional transformations of pastoral commons in African floodplains', *Ecology and Society*, vol 18, no 1, p 34, DOI: http://dx.doi.org/10.5751/ES-05510-180134.

Haller, T., Acciaioli, G., and Rist, S. (2016) 'Constitutionality: conditions for crafting local ownership of institution-building processes', *Society & Natural Resources*, vol 29, no 1, pp 68–87, DOI: http://dx.doi.org/10.1080/08941920.2015.1041661.

Hornsby, C. (2012) *Kenya: A History since Independence*. London: I.B. Tauris and Co. Ltd.

Keller, E. (2015) *Beyond the Lens Of Conservation: Malagasy and Swiss Imaginations of One Another*. New York: Berghahn Books.

Leifer, W. (1977) *Kenia: Geographie, Vorgeschichte, Geschichte, Gesellschaft, Kultur, Erziehung, Gesundheitswesen, Wirtschaft, Entwicklung*. Tübingen und Basel: Horst Erdmann Verlag.

Mbeyale, G.E. and Songorwa, A.N. (2008) 'Conservation for whose benefit? Challenges and opportunities for management of Mkomazi Game Reserve, Tanzania', in Galvin, M. and Haller, T. (eds), *People, Protected Areas and Global Change: Participatory Conservation in Latin America, Africa, Asia and Europe* (pp 221–251). Perspectives of the Swiss National Centre of Competence in Research (NCCR) North-South, University of Bern, Vol. 3. Bern: Geographica Bernensia.

McAuslan, P. (2013) *Land Law Reform in Eastern Africa: Traditional or Transformative? A critical review of 50 years of land law reform in Eastern Africa 1961–2011*. London: Routledge.

Morgan, W.T.W. (1963) 'The "white highlands" of Kenya', *The Geographical Journal*, vol 129, no 2, pp 140–155, www.jstor.org/stable/1792632, accessed 22 September, 2017.

Ojeda, D. (2011) *Whose Paradise? Conservation, tourism and land grabbing in Tayrona Natural Park, Colombia*. Conference Paper: LDPI International Conference on Global Land Grabbing, University of Sussex.

Ostrom, E. (1990) *Governing the Commons. The Evolution of Institutions for Collective Action*. Cambridge: Cambridge University Press.

Ostrom, E. (2005) *Understanding Institutional Diversity*. Princeton, NJ: Princeton University Press.

Paice, E. (1995) *Where Warriors Met: The Story of Lewa Downs, Kenya*. London: Tasker Publications.

Scott, J.C. (1985) *Weapons of the Weak: Everyday Forms of Peasant Resistance*. New Haven, CT: Yale University Press.

Van den Akker, M. (2016) *Monument of nature? An ethnography of the world heritage of Mt. Kenya*. Faculty of Social and Behavioural Sciences, University of Leiden, http://hdl.handle.net/1887/39726, accessed 22 September, 2018.

Weissman, S.P. (2017) *The New Pastoralists: From Cows to Rhinos. An Ethnography of Ideological and Institutional Change through Conservation in Kenya*. Bern: Institute of Social Anthropology, University of Bern.

Wildlife Conservation and Management Act 2013 (2013) *An Act of Parliament to provide for the protection, conservation, sustainable use and management of wildlife in Kenya and for connected purposes*. Act No. 47, Nairobi, Kenya, www.kenyalaw.org/lex/actview.xql?actid=No.%2047%20of%202013, accessed 10 September, 2018.

21 Rain forest anomy

National parks, REDD+ implementation and the run to the forest in Jambi, Indonesia

Heinzpeter Znoj, Rahel Jud and Yudi Bachrioktora

Jambi – an Indonesian frontier of the 21st century

Until the 1990s, the provincial capital of Jambi was a quiet small town, a backwater in the Indonesian economy. In the last two decades it has become a sprawling boom-town where the money made from logging, mining and, above all, oil palm plantations is spent on education, consumer goods and a wide range of enterprises. Along with its capital city, the entire province has become a busy place due to the tripling of its population over the past 40 years. In part, this increase stems from settlers arriving from neighbouring provinces in search of land and employment opportunities. Some of these migrants have become workers on oil palm plantations, yet most participate in the booming economy as small independent producers of oil palms, or of other lucrative cash crops such as rubber, coffee and cocoa. Laying hold of a piece of land is their first step in making a living in Jambi; and this is achieved mostly by clearing forest themselves or by occupying land cleared by logging companies. The combined effect of large-scale logging, the establishment of oil palm plantations and the influx of pioneer farmers has led to a reduction of the forestland in the province from 72 per cent (around 1980) to roughly 30 per cent today.

In this chapter, we look at the ways in which access for settlers and local peasants to the forest is mediated: (1) through the state in form of private property, as in the transmigration programme of the 1980s; (2) through local traditional regulations of the forest as commons; and (3) through new forms of patronage, which turn all types of forestland into an open-access resource. These competing regimes of access often lead to violent disputes between indigenous populations, settlers, the local government, as well as logging, plantation and conservation companies. In the following three case studies we show how these conflicts have played out in three different regions in the province of Jambi, and how in all these cases, commons are being replaced either directly, by private property, or indirectly, via an open-access phase. These changing forms of access fundamentally alter the social organisation and cultural identities of rural societies – not only of Jambi's forest nomads but also of local peasants.

The three authors of this chapter have conducted long-term fieldwork in three different areas close to national parks, REDD+ and oil palm concessions.

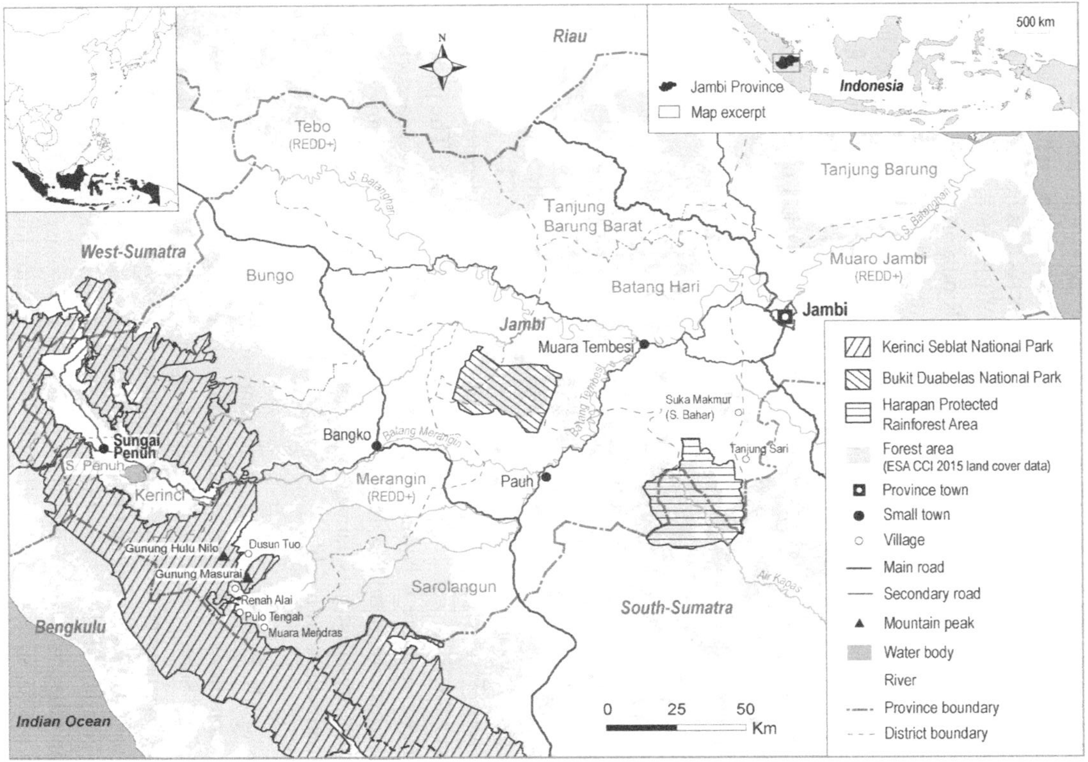

Figure 21.1 Map of the case study areas.

Source: Map compilation and design by Ulla Gämperli Krauer (2018), CDE, University of Bern, Bern, Switzerland. Data source forest cover: ESA CCI land cover data (2015). European Space Agency (ESA), ECSAT, Oxfordshire OX11 0FD, United Kingdom. Further geodata source: OSM data by OSMaxx (2018). Geometalab, HSR University of Applied Sciences Rapperswil, Switzerland. Further geoinformation provided by the authors of this article.

Heinzpeter Znoj, who has carried out fieldwork in Sungai Tenang in the highlands of Jambi since 1989, shows in the first case study that privatisation of land commons is a long-term process that began with the emergence of new village elites in the wake of the Village Law of 1979 and accelerated with the reforms of regional autonomy in the early 2000s, which created predatory provincial and district elites interested in concluding deals with logging and plantation companies. In the second case study, which is based on fieldwork carried out in 2015, Rahel Jud investigates how the establishment of the Bukit Duabelas National Park and its subsequent partial transformation into oil palm plantations has affected the former semi-nomads, the Orang Rimba. In contrast to agriculturalists, the Orang Rimbas' traditional subsistence strategy depends on access to large tracts of virgin forest which yield food, game, fish and forest products such as honey, rattan and various resins, all of which can be offered for sale. With the forests' enclosure, degradation and eventual disappearance the Orang Rimba are forced to become farmers and, thus, to adopt a way of life they have traditionally despised. Yudi Bachrioktora, who conducted fieldwork in Jambi between 2016 and 2018, in the third case study analyses how a group of immigrant farmers successfully first occupied land on a former logging concession as members of the Indonesian Peasant Union and later, following the concession's conversion into a REDD+ reforestation project, registered themselves as indigenous Batin Sembilan in order to gain access to the use of rights granted within the forest reserve to this group by the company.

Case 1: The demise of commons regimes for land and forest in Sungai Tenang (1977–2016)

> Oh, there were lots of buffalo, formerly. Almost everybody used to have buffaloes. Now it has become difficult to keep livestock. The population has grown. This is why the pastures have been turned into gardens. Years ago, over there, … wah … buffaloes roamed all over the place. The pasture was fenced in where it bordered the rice fields. Now you can't make such a pasture anymore, people make their gardens everywhere, and for any single cinnamon tree that gets damaged by your buffalo, they will be after you … ah … you will be in trouble.[1]

With these words, Hamzah, a 60-year-old man, in 1996 described the demise of the village commons in Muara Mendras, Sungai Tenang that had taken place 20 years earlier. Such commons, situated on upland plains, had been a defining feature of this remote part of the Central Sumatran highlands. In some villages they were used for temporary gardens and rain-fed rice fields followed by short periods of grassland-fallows. In others, such as Muara Mendras, irrigated rice fields in the adjacent alluvial plains replaced the temporary rain-fed fields on the upland plains, so that the latter could be turned into jointly managed and communally fenced-in pastures for individually owned buffaloes and cattle. These animals were an ideal source of cash income before roads were constructed to the lowlands. They could be herded to the nearest market town on week-long treks. In the 1970s a road was built that reduced the hiking distance to the market place to three, and later to two days. This made the transport and sale of cash crops such as cinnamon and coffee increasingly profitable. When an epidemic of rinderpest devastated the herds around 1970, the upland plain was again used for temporary gardens, and, with the market places coming closer and cinnamon prices going up sharply, some people started to plant cinnamon trees in the ageing gardens instead of allowing them to return to pasture. This violated the rules of the commons, as planted trees are acknowledged as the planter's personal property, and many trees on a piece of land also convert the land in practice into individually owned property. In spite of protests, this example was followed by others, and when, after a flood had ruined many rice fields close to the river, dozens of families at once occupied large plots on the upland plains in 1977 to grow rice for subsistence and subsequently planted cinnamon trees on these fields, the commons system could no longer be enforced and was eventually given up and replaced by a system of individual, permanent ownership of delimited plots of land. Old men such as Hamzah remember this period as spelling the end of village unity.

One reason why cinnamon trees were planted on the upland plains rather than in nearby forest clearings was the declaration of the Kerinci-Seblat National Park (TNKS[2]) in 1975, which made the conversion of the forests surrounding Sungai Tenang illegal. The original boundary of the TNKS all but encircled Muara Mendras at a distance of 2km, thereby blocking the anticipated expansion of the village. Furthermore, the declaration of a new village law by the central government in 1979 strengthened the central government's grip on the villages through the election of government-approved village heads (Galizia 1996: 51), thereby also weakening traditional village authorities in charge of ruling the commons. But Hamzah was certainly also correct in pointing out strong demographic growth as a factor leading to the demise of the commons. In oral history the late 1970s are remembered as a period of severe conflict within the village which, ultimately, was over land. Some families were denied a share of the upland plains – on the grounds of being "late-comers" – and, as a consequence, some 20 per cent of the village population departed to settle on pioneer land within the forest, at a hiking distance of two days from Muara Mendras (Znoj 1995).

The TNKS was always disputed by the inhabitants of Sungai Tenang. Yet, fear of the ruthless government under Suharto for a long time prevented people from encroaching upon the national park too openly. Then, during the late 1980s and 1990s, when logging companies connected to, or even owned by, members of the governing elite acquired logging concessions within protected forests as well as within the TNKS, villagers began to feel entitled to shift the boundary of the TNKS ever farther from the villages. As the logging companies came closer to Sungai Tenang in the mid-1990s, and increasing numbers of migrants flocked to the highlands, villagers began to cut down the forest so as to lay claim to large plots of land as reserves for themselves, for their children and grandchildren. Such "signal clearings" largely became secondary forest after a few years as their owners lacked the labour force and capital to turn them immediately into plantations. In this way, within a few years the forest commons surrounding the villages, which until then had periodically been cleared in small patches to create temporary gardens, was divided up into individual claims that anticipated their permanent conversion into privately owned coffee or cinnamon plantations.

In 1998, after the end of Suharto's dictatorship, local leaders and migrants in some villages took advantage of the central government's diminished power to block huge tracts of forest within the TNKS. Particularly in Dusun Tuo and Pulo Tengah village heads "blocked" thousands of hectares of forest and soon sold such blocks to settlers from southern Sumatra who were well organised and endorsed by a retired army general and the Indonesian Peasant Union, and who were partly financed by a coffee trader from Bengkulu in south-western Sumatra. Since the early 2000s, they have squatted on approximately 100km^2 of former TNKS territory on the slopes of the volcanoes Gunung Masurai and Gunung Ulu Nilo within and adjacent to Sungai Tenang. The influx of settlers again increased after 2010 as a consequence of the rise of large-scale land acquisitions in other parts of Sumatra by globally active investors: land had become so scarce and expensive in many parts of Sumatra that enterprising farmers from affected areas started to look for farming opportunities elsewhere. In the few years since 2010, whole new villages have been established for the now over 3,000 families of settlers.

To this day a handful of old villages remain opposed to the sale of land to migrants – amongst them Muara Mendras and Renah Alai. In Renah Alai this opposition resulted in violent clashes with settlers that led to several killings. In 2016, a unit of special army troops[3] was called to pacify an escalated conflict which had erupted over the clearing of a last stretch of TNKS forest, about 1km in width along the main road that separated the plantations of Renah Alai from those of the migrants. When the troops declared themselves unable to stop the settlers from clearing the forest, the villagers of Renah Alai mobilised all available men to cut it down on their own and thereby lay claim to it themselves.

The inability of even the most feared state authorities to stop encroachment upon the TNKS is in line with earlier failures. In one instance, in 2014, the district head had succeeded with the help of police and military troops to evict several hundred migrant families from within the TNKS. Their huts were burned down and their coffee trees felled. A few weeks later, most settlers returned to rebuild their huts and replant their fields. Under duress from the powerful backers of the settlers, the district head renounced on repeating the eviction.

As the following two case studies also show, the establishment of new national parks and the granting of REDD+ concessions in the province of Jambi – however flawed their management may be in the end – has further increased the pressure on the forests. We argue that this represents a situation of open access in the remaining village forests and within the TNKS – even if this is an anomy that is largely restricted to the access to the forest. Once the forest has been cut down, the person who cleared it is acknowledged as the land's rightful owner by their neighbours, fellow villagers and, in practice, also by the state. Once the forest is converted into a permanent plantation, this ownership becomes indisputable.

Yet, even if villages cannot enforce commons rules pertaining to the forest and are forced to condone its conversion into private property, they are nevertheless able to exclude non-locals from free access: in the villages, free access to the forest is restricted to the original inhabitants of those villages. Migrants must buy "blocked" forest, or forest that has already been cleared, from villagers. The villages have thus transformed themselves from institutions that jointly manage the forest and its resources into alliances of "original" villagers, who grant each other the right to grab as much forest as they are able to while excluding outsiders. Sooner or later however, in most villages, owners of large pieces of cleared forest begin to sell land to outsiders when they are in urgent need of cash.

At this point, several drivers of the village-level privatisation of the forests must be mentioned: money politics, education expenses, the high frequency and high costs of the Hajj, and the purchase of government jobs and political office. The costs of the Hajj are often covered by selling the bark of a hectare of cinnamon trees, planted for that very purpose 15–20 years before. In recent years, some members of the village elites of Sungai Tenang have occupied prominent positions in the district government. They invested large sums into their election campaign and into buying votes – the sale of land claims in the forest was one way to raise the necessary cash quickly. In some villages, the number of people with a university degree and of government employees has risen sharply in the recent past. The investment into a son's and daughter's university education corresponds to the annual sale for five to six years of the produce of one or two hectares of cash crops; the bribes demanded for the acquisition of a position as a teacher or village nurse currently amount to that position's salary over the course of five years, or roughly US$5,000 to US$10,000. Upon the announcement of job openings these sums must be raised at short notice, and typically candidates' parents or other close relatives are obliged to sell several hectares of land. It follows that not only migrants, logging and plantation entrepreneurs and foreign corporate investors put pressure on the forests of the province, but also high-ranking bureaucrats capitalising on their position by selling offices, as well as their clients in the villages, who all wish to gain security, status and influence in the emerging capitalist society and, hence, sell the forest to raise the necessary money.

Case 2: The Bukit Duabelas region: from indigenous habitat to National Park (1980–2016)

> I think there is a big difference in the condition of the air. I felt a lot fresher when the trees still stood and didn't get cut by anyone, the air felt fresh. I feel a lot warmer now, that's how I know there must be something wrong with the trees and the jungle.[4]

This is how Rahman, a 53-year-old man, in 2015 expressed his grief about the on-going disappearance of the forest in the Bukit Duabelas region over the last decades. The vast deforestation in Bukit Duabelas started in the early 1980s, when approximately 25,000 families settled in this region in the wake of the national transmigration programme. Each family was provided with housing and 2.5 hectares of land for subsistence gardening. By allocating land certificates to the transmigrants, the Indonesian government had suddenly turned part of the forest land in the region, which was originally used commonly, into private property (Steinebach 2012: 36). As a result of this, competition for the use of the remaining forest areas arose between the local population and the new settlers.

The local population of Bukit Duabelas included communities of the semi-nomadic indigenous people called the Orang Rimba ("people of the jungle"). One group of Orang Rimba from the southern Bukit Duabelas region appealed to the provincial government to secure their habitat against encroachment by settlers. In 1984, the claim of the Orang Rimba was acknowledged and nearly 30,000 hectares of lowland rainforest were formally declared as an area for the Orang Rimba to live (*kawasan pengembaraan Suku Anak Dalam*). But already in the following years the status of the forest area assigned to the Orang Rimba changed several times. In 1993, the same area was converted into a protected area (*suaka margasatwa*) for endangered animal species such as the Sumatra Tiger. This conversion was confirmed in 1995 in the national conservation plan (*perencanaan konservasi nasional*). However, according to the Orang Rimba and the rest of the local population, changes in the categorisation of the Bukit Duabelas area had no practical consequences for the ways in which the forest was used, and such measures were of significance only at the bureaucratic level (Ibid: 184, 185). In this period the global interest in protecting the remaining rainforests also became a central concern in Indonesian politics. In 2000, the Indonesian government finally designated the Bukit Duabelas area, one of the last lowland rainforest areas in Indonesia, as a national park. Following the westernised concept of nature conservation, the conservation area (*suaka margasatwa*) was extended northwards into areas formerly marked for commercial use and declared a national park of approximately 60,500 hectares (Ibid: 183). In the same year, 34 further national parks were founded throughout Indonesia, with a total area of approximately 8.8 million hectares. All these national parks include the internationally valid criteria of the IUCN (International Union for Conservation of Nature[5]) in their regulations, which state that 75 per cent of a national park's territory is to remain unused by humans (Ibid: 38). Thus, with these new usage regulations, the Orang Rimba in the Bukit Duabelas area formally became more excluded from their previously assigned habitat than ever before.

In the same decades the oil palm industry arrived in the region, attracting several thousand migrant families. In this way the region surrounding the Bukit Duabelas National Park once again became the target of land appropriations. In 2005, the national park was surrounded by populated areas (amounting to about 50,000 hectares), rubber tree and oil palm plantations (over 90,000 hectares), and – in the north and east – it bordered upon forest areas (of about 27,000 hectares) assigned for commercial timber production. In 2006, the construction of an illegal logging road in the district Batanghari was reported; within a few months an area of over 6,000 hectares of protected natural forest was stripped for roadwork. Shortly after this, illegal loggers were at work in several places within the national park (Ibid.).

In its immediate vicinity the park is additionally surrounded by districts and areas in which REDD+ mechanisms have been implemented: the REDD+ districts of Tebo and Merangin, and the Harapan Rainforest REDD+ Project located in the Batang Hari district (see the next case study). In regard to REDD+ scholars have found that the establishment of protected areas through such policies often provokes a rush for the remaining unprotected land in the area (Fairhead et al. 2012: 237). This is also the case with the REDD+ projects close to the Bukit Duabelas National Park: the introduction of such projects have led to a further increase in the struggle for scarce land throughout the area, thereby leading to higher land prices as well as illegal and violent forms of land grabbing. Due to these processes in the region, the habitat of the Orang Rimba has diminished dramatically, leading to substantial changes in the livelihoods of this indigenous group. For the formerly exclusive habitat of the Orang Rimba (*kawasan pengembaraan Suku Anak Dalam*), coherent regulations on land usage no longer exist. Often, illegal land appropriations by the plantation industry or the local population are not followed by sanctions, especially because in most cases it was political elites from various administrative levels who granted formal and informal access to forest land. This state of anomy and the permanent battle for the remaining forest have resulted in frequent violent incidents. Forcible land acquisition by loggers are no longer a rarity, and cases have been reported in which Orang Rimba opposing such assaults on their land were murdered (Manurung 2008).

In view of their vanishing habitat, the indigenous community claimed the right to convert the protected forest into plantations themselves and received permission from the district government to cultivate their own oil palm plantations on the margins of the Bukit Duabelas National Park. For this the traditional collective-access rules of the indigenous group have been transformed by the provincial government into private land-use rights. This move follows a global trend whereby governments have shown a tendency to privatise land with unclear usage rights rather than assigning usage rights to a collective (see, for example, Dafinger and Pelican 2006: 128).

The conflicts over distribution of scarce forest land in the Bukit Duabelas region have not only changed the economic condition of the region, but also influenced social relations among the local population, and in particular amongst the Orang Rimba. Before the settling of transmigrant families, that is, prior to the large-scale clearing of forest areas in the Bukit Duabelas region, the subsistence strategy of the Orang Rimba had consisted almost entirely of hunting and gathering in small collectives. With the arrival of the new settlers the indigenous people and their way of life became severely restricted, and individual Orang Rimba households increasingly began to cultivate their own rubber trees and, later, oil palms. Initially, the local population planted rubber trees to safe-guard forest land from appropriation by newcomers because, according to customary law in the region, planted trees belong to the planters and, thus, the land upon which these trees stand cannot be sold. Trespassing on these patches of land is likewise prohibited by customary law, thereby it should present an obstacle that hinders the clearing of forested areas (Steinebach 2012: 161).

Today many of the Orang Rimba in the region own and maintain between one and four hectares of rubber trees and oil palms per person or household. Some also sell woodland in the forest border zones to the sedentary population as an additional source of income. Due to the increasing scarcity of land the prices for forest land are steadily rising; the sale of land, however, is controversial and has caused much resentment and conflict within the larger Orang Rimba community (Ibid: 164). The transformation of the collective rights of access into private land-use rights by the provincial government and, as a consequence of this, the opportunity for individual Orang Rimba to sell land to settlers, has led to an erosion of their solidarity and the worldview at its base.

In fact, for the Orang Rimba the rainforest is full of symbolic meaning that underpins their cultural and social order. The practical knowledge and skills, as well as the ideologies and rituals rooted in their life in the rainforest, are crucial reference criteria in their ascription as Orang Rimba, both by themselves and by outsiders. From the perspective of the Orang Rimba, the world is divided into "inside" and "outside" the rainforest. By differentiating these two worlds, ethnic differences between the Orang Rimba (people from the inside) and the non-Orang Rimba (people from the outside) are established and ethnic identities are produced (Ibid: 89). The selling of rainforest patches and, hence, the undermining of Orang Rimba worldview by some individual Orang Rimba directly contributes to the dissolution of them as a social group. More and more deprived of their social identity, the Orang Rimba nevertheless seek to retain control of the forest land despite the de facto control over it by external economic and political forces. This problem is exacerbated by the fact that some settlers register as Orang Rimba in order to gain access to land-use rights.

Due to the social and economic changes, the lack of usage rights to most of the remaining national park territory, and the influences from outside discussed above, many of the Orang Rimba have had to relocate to alternative habitats. While some continue their semi-nomadic lifestyle in the national park and the adjacent plantations, others now live in a neighbourhood of small huts on the edge of the national park recently built by the national Ministry of Social Affairs to persuade the Orang Rimba to settle. Those who have accumulated enough money from land sales live in houses they have built themselves. These developments have been brought about, inter alia, by the programmes of the Indonesian government, which seek to persuade nomadic groups to settle and convert to Islam.[6] The disappearance of the rainforest, and the economic and social transformations this brings to the region, as well as the government's programme of settling nomadic and semi-nomadic groups, has diminished the Orang Rimbas' ability to depend on other Orang Rimba. In addition to this, they are in the process of losing their autonomy as a group and, through the conflicts over land appropriation, they have already lost their unity. The fear exists that, sooner or later, these factors could result in the ethnocide of the Orang Rimba.

Fewer than 1,000[7] Orang Rimba cling to their original lifestyle today, and they classify sedentary Orang Rimba collectively as "external people", in other words: they no longer consider them to be part of the indigenous community (Ibid: 103–104). Rahman, too, realises that he has excluded himself from the indigenous community by joining Islam and building his own house. Nowadays he is a small-scale farmer and cultivates rubber trees and oil palms. His is a special position in the village: he mediates between the Orang Rimba and the rest of the local population, although he seems to harbour a sense of discomfort about his position and his decision to no longer live in the forest.

> I feel lost, Miss, I feel like I'm lost, but when I think about my children and my grandchildren I could do nothing but to let it happen. I clear the land step by step, I still feel sad about its condition now, but my economic dependency makes me powerless.[8]

The changes that Rahman has experienced in his life exemplify the intensity of the economic pressure on the Orang Rimbas' habitat and on the scarce forest resources in the area. Beyond this his experiences also illuminate how both deforestation and, in turn, the rigid protection of the ever-dwindling forest has dramatically transformed the social conditions and identity of the Orang Rimba. Through such far-reaching social changes, social differentiation amongst them is increasing and the binding force of traditional social norms has dissolved. Already, they are no longer an ethnic group with a social and cultural identity.

Case 3: The Harapan rainforest REDD+ project and its contestation by indigenous Batin Sembilan and settlers (1967–2017)

> Looking at the forest right now, for me it's a loss. The loss is not just that we run out of fruits and food, or wood for the roof and walls of the house. What we used to have [the forest], is finished.[9]

With these words, Seman, 48, a former *Temenggung* (or leader)[10] of the Batin Sembilan, an indigenous group of forest nomads in the eastern lowlands of the province of Jambi, explained the situation and circumstances of his forest, now well-known as part of the Ecosystem Restoration concession called the Harapan rainforest. The concession covers an area of 98,555 hectares in the two provinces of Jambi (46,385 hectares) and South Sumatra (52,170 hectares). Most of the area is secondary forest due to earlier logging. The Harapan rainforest is now considered "contested land" as various actors have been disputing its usage since 2010.[11] This section focuses on disputes between PT REKI,[12] the concession holder in the Harapan rainforest, the Indonesian Peasant Union SPI[13] and the indigenous people of the Batin Sembilan, who live in the area of Sungai Jerat within the province of Jambi (Hein 2013; Hein and Faust 2013; Beckert et al. 2014; Buergin 2016). In particular, it examines how migrant peasants and Batin Sembilan pursue their interests and legitimise their various claims over the forest. We show that in doing so they rely on traditional norms and land rights, spontaneous collective action as well as membership in nation-wide organisations such as the SPI or the NGO-sponsored Indigenous People Movement. This helps to illuminate the role and influence of local and migrant peasants as well as the indigenous community in shaping the land tenure system in the forest.[14]

The Harapan rainforest was once the habitat of the indigenous Batin Sembilan, who for generations hunted and farmed along the banks of the Batanghari river and its tributaries. They gathered non-timber forest products for sale, such as rattan, grapefruit, honey, resin gum and medicinal plants. Today, approximately 300 Batin Sembilan families live in this forest, dispersed in small semi-nomadic groups. Since 2002, 700 members of the SPI also claim a portion of the forest. Of the 46,385 hectares of Harapan rainforest in Jambi, the SPI claims an area of 3,689 hectares (or 8 per cent). Most of this area has already been cleared and is used for settlements and plantations of rubber trees and oil palms.[15] As the holder of the concession PT REKI is strictly opposed to the occupation by SPI members, even as it has over the years tolerated the traditional use of the forest by the Batin Sembilan. The company was formed in 2007 as a consortium of Bird Indonesia, BirdLife International and the Royal Society for the Protection of Birds (RSPB), chaired by the Prince of Wales. As stated in the website of BirdLife international, the Harapan Rainforest Initiative is funded by a wide range of donors, such as the German Ministry of Environment's International Climate Initiative, the Global Conservation Fund of Conservation International, the European Union, the Danish International Development Agency (DANIDA), BirdLife Partner NGOs in Germany, Switzerland, Belgium, Denmark, Singapore and the Netherlands, as well as Singapore Airlines and individual donors (Bird Life International). Singapore Airlines donated US$3 million in 2010, while, from 2009 to 2013, the German Ministry of Environment's International Climate Initiative supported the REKI Company with US$9.4 million (Buergin 2016: 284–285; Singapore Airlines Annual Report 2010/2011, 23; see also Lang 2012 (the REDD desk)). The main support for the company from 2013 to 2018 has been provided by DANIDA, with a total of US$42.7 million.[16]

Ecosystem restoration concessions have only recently been introduced as forest management institutions in Indonesia.[17] The policy was developed in the early 2000s by a small group of people representing the Ministry of Forestry, the Indonesian Bird Foundation and the Royal Society for the Protection of Birds. In 2004, the Ministry of Forestry finally issued ecosystem restoration concessions; PT REKI obtained the first concession over the South Sumatra part of the Harapan forest in 2007, and the part in Jambi in 2010. Such concessions emerged as part of the global action to reduce greenhouse gas emissions from the forestry sector due to deforestation and forest degradation commonly occurring in developing countries such as Indonesia. For the Harapan rainforest the Indonesian government issued REDD+ concessions (Indrarto et al. 2012: 34–35, 48; Mardiastuti 2012: 9–10;

Silalahi et al. 2015). However, due to the on-going conversion of forest into plantations within Harapan rainforest itself, the implementation of ecosystem restoration according to REDD+ standards is marred by great difficulties.

The struggle over the Harapan forest can be traced back to the time when it was designated as a "state forest area" in 1967. Based on Law No. 5 from 1967 on Forestry, the state has the authority to transfer ownership rights to the private sector in the form of Forest Concession Rights for a limited period of time. Through this regulation the state denied the Batin Sembilan's traditional rights over the forest which had been granted to them by the Sultanate of Jambi in the 19th century. According to the Laws of the Sultanate of Jambi, the *undang-undang Jambi*, the indigenous groups had the right to live in the forest as long as they paid tribute to the Sultan (Znoj 2001: 187–190), which the Batin Sembilan had done as long as there had been a Sultan who claimed it.

Ignoring the long-standing rights of the Batin Sembilan, in the 1980s logging concessions in the area now known as the Harapan rainforest were issued to PT Asialog and PT Inhutani (Beckert et al. 2014: 80; Kunz et al. 2016: 136–137). The 1967 forestry law generally prohibits local people from accessing their traditionally owned territory once a concession has been granted to a company. However, according to Seman, a member of the Batin Sembilan who lives in the concessions area, the logging companies allowed the Batin Sembilan community to utilise non-timber forest products in order to reduce the tenurial conflict.[18] This situation remained unchanged until the late 1990s, when the Batin Sembilan were relocated to the Sungai Bahar area as part of the government project to sedentarise indigenous peoples. Their new location lay between the PT Asialog concession and a regular transmigration area established by the government in the late 1980s – an area which was part of the nucleus-estate programme for oil palm plantations run by the state company Perseroan Terbatas Perkebunan Nasional (PTPN) VI. This programme was an integral part of the national agenda to boost Indonesia's economy through plantation commodities (Adnan 2008: 368). Most of the transmigrants were people of Javanese, Sundanese, Malay, Nias and Batak descent. Similarly to the transmigrants, the Batin Sembilan community was taught how to plant oil palms and rubber trees as a source of income. In 1999, PT Asialog's permit expired and a void in "ownership" of the forest arose. Batin Sembilan leaders took advantage of this situation by selling blocks of forest to transmigrants and others seeking to purchase land. Initially, these transactions, concluded through customary agreements, were mutually understood as an acknowledgement of the authority of the Batin Sembilan over the former concession area. Yet, these customary agreements were already no longer made according to Batin Sembilan traditions but instead based on the migrants' beliefs and traditions to which the Batin Sembilan leaders were forced to agree.[19] Their authority thus depended on the good-will of the migrant farmers. Some of these transmigrants became local SPI leaders in Sungai Jerat and attracted new settlers, to whom they sold plots of land which they had already cleared. In exchange for the land, they asked for a "compensation" per hectare of IDR 1–6 million (ca. US$80–500) (and sometimes more) to cover the cost of opening up the forest.

When most of the SPI's local leaders became members of the newly established Indigenous Groups organisation in 2014, this compensation for taking over land came to be called "grant money". Any individual can buy land by providing this grant money, starting from IDR 1 million per hectare, to the Batin Sembilan indigenous groups organisation.[20]

The expiry of PT Asialog's concession in 1999 coincided with the power vacuum following Suharto's reign in 1998. In the post-dictatorship period an Indonesia-wide movement of land occupation emerged, in an act by peasants to take back land from plantation and forestry companies which no longer held or used their concessions (Widjardjo and Perdana 2001). Between 2001 and 2006, small-scale occupation movements occurred within the area of the former PT Asialog forestry concession. After purchasing the land from Batin Sembilan leaders, the local SPI leaders cleared the forest and divided it into blocks for the purpose of housing and plantation. But it was after a national congress of the SPI in 2007 that the provincial leader of the SPI finally decided to "officially" occupy the concession area. SPI's strategy was that of the "3Ts": "*Tebang, Tanam, dan Tunggu*" ("Cut, Plant and Wait"). By applying this strategy, every SPI member would immediately plant crops after clearing the forest area and live permanently in the occupied location. This strategy seems similar to that of the latin american Movimiento Sin Tierra's (MST)'s "Occupy, Resist, Produce". However, differences pertain between the MST and SPI in how they recruit people. The majority of MST members consisted of landless families who organised and lived in encampments to build their militancy before the occupation began (Vergara-Camus 2009: 181–183; Dinerstein 2015: 181–184). Members of SPI in Jambi, however, come from various economic and social backgrounds and join for a wide range of reasons; their ranks consist of, for instance, agricultural entrepreneurs from the nearby transmigration settlement or those from farther away who wish to expand their oil palm and rubber plantations, as well as landless labourers working on such entrepreneurs' plantations to earn a living.[21] Unlike amongst MST members there is thus no class-based ideological bond between SPI members.

In 2012, after having been granted a forest restoration concession of the territory two years before, PT REKI attempted to forcefully evict the peasant and Batin Sembilan communities from the forest. The company accused them of being involved in deforestation and illegal logging. Meanwhile, peasant and indigenous communities stated that it was the PT REKI managers who had befriended the logging operators. It was rumoured that illegal logging had become no longer a sporadic activity based on necessity but instead an entrepreneurial activity backed by "*cukong*" (big bosses) in South Sumatra and "tolerated" by the forestry department as well as by PT REKI, in other words that their staff allowed logging to take place against the payment of bribes. Most of the logs were transferred upon rivers from Jambi to South Sumatra. The mutual accusations of being involved in illegal logging were clearly part of the struggle for legitimacy of land claims between the PT REKI, on the one hand, and SPI members and the Batin Sembilan, who strongly opposed the eviction plans, on the other. In October 2012, after having been accused as illegal loggers, SPI members burned down the PT REKI camp in Sungai Jerat, the site of their largest land claim. In response, PT REKI burned down a number of peasants' homes. During this period, the Batin Sembilan agreed to an alliance with the SPI to aid in their opposition against PT REKI, thereby profiting from their militancy and seeking their protection, since they knew that the SPI wield influence at the national level. Calm only returned after PT REKI withdrew from the area. Since the clash with SPI members in 2012, PT REKI has been campaigning for a resolution to the conflict (Silalahi and Erwin 2015). The company has promoted mediation with the Batin Sembilan communities by appointing independent facilitators or conflict resolution consultants as mediators. In some sections of the Harapan rainforest this strategy has worked well: four groups of Batin Sembilan in those regions signed partnership agreements with PT REKI in 2016 and are now acknowledged as their partners in managing a forest area of 1,435 hectares.

This strategy has also been pursued in Sungai Jerat village, among others, where PT REKI works closely with the NGOs Scale Up, Forest People Programme and AGRA.[22] AGRA aims to promote and fight for the rights of farmers, fishermen and minority ethnic groups in natural resource management in order to secure their prosperity. Scale Up is an organisation which actively assists many communities affected by oil palm and pulpwood plantations to assert their rights and to negotiate with project developers. Forest Peoples Programme has been cooperating with indigenous peoples and civil society groups in Indonesia on issues of community rights. These latter two organisations see themselves in the role of conflict resolution experts; AGRA, meanwhile, acts as the negotiator for the indigenous Batin Sembilan people in Sungai Jerat.

AGRA uses indigenous rights provisions in REDD+ to arrange for both Batin Sembilan and SPI to live in the concession area. For this purpose, it has set up a new indigenous organisation consisting of the Batin Sembilan community as well as the migrant peasants as a tool to claim the forest as customary territory, i.e. as a commons. Aki Asep, a former SPI base leader who now serves as a spokesperson for the indigenous organisation, revealed in an interview that it is not surprising to see a lot of people leaving SPI and joining the indigenous people organisation: "It is not a matter of making them indigenous people. It is a matter of providing protection for their lands".[23] Meanwhile, the Batin Sembilan claim that they merely defend their traditional rights over the forest. In the words of *Depati* Jupri, the current leader of the indigenous organisation:

> We ask for the rights to our customary territory. We do not want to stay in state[-owned] forest. We do not accept social forestry, and we do not know what social forestry is. They must consider us as the owners of this forest, not as squatters.[24]

The mobilisation of indigenous identity to claim land or forest is not a new phenomenon in Indonesia. Indigenous groups began to use this strategy towards the end of the Soeharto period. Although Indonesia is neither a signatory of the 1989 ILO Indigenous and Tribal Peoples Convention nor the 2007 UN declaration on the Rights of Indigenous Peoples, many firms and development agencies investing in Indonesia's plantation and mining sectors as well as in REDD+ projects originate from countries which have signed these conventions, and they take measures to apply them. It follows that the claiming of indigenous identity has rapidly become a political tool in the struggle over natural resources in Indonesia. As scholars have rightly pointed out, the formation of a new indigenous organisation that brings together Batin Sembilan and peasant communities is a political reaction enabling both groups to take part in global land deals (Borras and Franco 2013; Hall et al. 2015: 471–472). In other words, this is a strategy with which to maintain access to the forest for both peasants and indigenous Batin Sembilan and to allow both groups to profit from it. Moreover, the implementation of the Harapan rainforest as REDD+ projects has provided them with the opportunity to legitimise their customary claims over the forest. This notwithstanding, it remains noteworthy that by accepting this new indigenous organisation as their representative, and allowing immigrant farmers to become members, the Batin Sembilan have also definitively given up their formerly exclusive claim to the forest.

Conclusion

In the three case studies of this chapter we have identified an overall trend towards the privatisation of forest land and other land commons and their concomitant transformation into cash-crop plantations. The privatisation of former land commons managed by the community is a long-term process which increased in scale over time. The circumstances that accompanied the enclosure of the upland plains commons 40 years ago in a small village in the highlands of Sungai Tenang are strikingly similar to those currently affecting the Harapan rainforest and the forests of the Bukit Duabelas region. Both the upland plains and the forests were traditionally managed as a common-pool resource prior to their (ongoing) conversion into privately owned large- and small-scale plantations. The circumstances for this transformation are manifold: a boom in cash crops (earlier cinnamon, presently palm oil); the declaration of new forest reserves leading to land scarcity (earlier the TNKS, nowadays REDD+); the weakening of traditional local authorities and the regulations for which they stood (formerly through the Suharto regime's forest and village laws, at present through international agreements on carbon emissions and storage trading); a strong population growth (then local, now local combined with immigration); massive conflicts in rural society (formerly internal to villages, nowadays mostly between local and immigrant groups); and a heady scramble for the former common-pool resource, once the pioneers of such privatisation had been successful (in the past pasture and garden land in close proximity to the villages, in the present day forests at a greater distance from the villages including those formerly managed by forest nomads). Beyond this, in all three cases discussed in this contribution we have observed that local communities lose their rootedness in the joint management of land as communal lands and forests are being privatised. Such conversion has exhibited (and still exhibits) an open-access phase that is related to frequently violent conflicts over forest land between settlers, local farmers, indigenous people and corporate as well as government actors.

With the conversion to private ownership of former forest land, rural society undergoes marked structural and cultural changes. The indigenous inhabitants of the Harapan rainforest and the Bukit Duabelas region, as well the long-established peasant villages of Sungai Tenang, are forced to adapt to the presence of and the competition with immigrant groups from different ethnic, social structural and religious backgrounds, as well as, in some areas, to the nearby presence of powerful and well-connected private plantations and REDD+ enterprises. Furthermore, the influx of money through the sale of wood, land and cash crops results in a rapid stratification within both local and immigrant populations. Wage labour becomes a necessity for less well-off rural inhabitants. This capitalisation of agriculture among the rural population is a decisive factor in further deforestation as it allows local elites to expand their landholdings through the employment of paid labourers.

Moreover, the cases of the Bukit Duabelas national park and the Harapan forest restoration project document the re-establishment of a restricted commons regime in the protected forests, due to the fact that (under the provisions of the REDD+ and IUCN regulations, respectively) the indigenous Orang Rimba and Batin Sembilan are granted usufruct rights to non-timber forest products in sections of the protected area. In a certain sense, the REDD+ concessionaries and the national park services thus assume the traditional role of the Sultan of Jambi, who had formerly acknowledged the exploitation of the forest products as commons for the forest nomads. But this top-down installed commons regime is anything but robust. Not only are there no sanctions against those who privatise the commons, but also the boundary of the group holding the common usufruct rights has broken down, as a large number of migrant farmers within and in the surroundings of the concession area has assumed an indigenous identity to get access to the resource. In fact, neither the state's forestry laws, which grant exclusive private property rights to forest restoration concession holders, nor the internationally sanctioned commons rights granted to indigenous people within National Parks and REDD+ concession areas are strong enough institutions to fend off the general anomic struggle for land. Rather, an open access situation results that leads to the conversion of the remaining forest into plantations, whose private ownership is eventually acknowledged by the state as fait accompli.

Ultimately, we are led to the conclusion that, under the current conditions, the business opportunities offered by nearby logging contractors, palm oil mills and rubber and coffee traders, and the national and international corporate interests vested in them, are stronger determinants of the fate of the forests than the laws which oppose the complete commodification of land and forest in Jambi. Today, in 2019, we witness a state of anomy in the forests of Jambi that can be overcome as little by traditional laws as by those of the state – and which the newly-installed, globally organised and funded forest restoration policies are just as powerless to address.

Notes

1 Interview by Heinzpeter Znoj with Hamzah in Muara Mendras in 1996.
2 Taman Nasional Kerinci Seblat.
3 BRIMOB, an anti-riot unit.
4 Interview by Rahel Jud with Rahman in the Bukit Duabelas region in 2015.
5 The *International Union for Conservation of Nature and Natural Resources* (IUNC) is committed to the protection of nature and species and the sustainable use of resources.
6 The target of this Indonesia-wide programme is the sedentarisation and conversion to Islam of groups that are categorised as KAT, *kommunitas adat terpencil* (geographically isolated indigenous communities). Their way of life is seen as incompatible with the ruling concept of an advanced Indonesian nation (DEPSOS RI 2003: 9). Indonesian citizens are obliged to adhere to one of six religions acknowledged by the state.
7 In the years 2003–2005 the number of Orang Rimba living in the Bukit Duabelas National Park was estimated at 1200 (Steinebach 2012: 43).
8 Interview by Rahel Jud with Rahman in the Bukit Duabelas region in 2015.
9 Interview by Yudi Bachrioktora with Seman in Sungai Jerat in 2016.
10 The title Temenggung is used for a leader of subgroups within the Batin Sembilan community. The title of the head of the whole community is Depati (see Hidayat 2012: 71–72). When the Batin Sembilan and peasant communities formed a new indigenous organisation, they reinstated the Depati position in order to unify the whole community of Batin Sembilan in the area of Sungai Bahar.
11 There are five areas of Harapan rainforest in Jambi which are claimed by both local migrants and the indigenous community of Batin Sembilan, namely Kunangan Jaya I, Kunangan Jaya II, Tanjung Mandiri, Alam Sakti and Bukit Sinyal.
12 *PT Restorasi Ekosistem Indonesia* (Indonesia Ecosystem Restoration Company).
13 Serikat Petani Indonesia was originally named the Federation of Indonesian Peasant Unions (FSPI). This organisation was founded on July 8, 1998, in North Sumatra. The focus of their struggle is agrarian reform, peasant rights, food sovereignty and sustainable agriculture. SPI membership is individual. Presently, SPI is active in ten provinces, including North Sumatra, West Sumatra, South Sumatra, Jambi, Lampung, Banten, Yogyakarta, Central Java, East Java and West Nusa Tenggara. Most of Indonesia's activist and peasant organisations, including the SPI, were inspired by the land occupation movement of Movimento dos Trabalhadores Rurais Sem Terra (MST) in Brazil. The SPI is also a member of the international peasant movement *La Via Campesina* founded in 1993, and currently hosts the organisation's international headquarters.
14 For the role of local migrants and indigenous communities in shaping land tenure in the forest, see Galudra et al. 2010; Galudra et al. 2014. For the similar case in the Harapan rainforest, especially in Kunangan I and Kunangan II, see Steinebach 2017.
15 Interview by Yudi Bachrioktora with Sukiman in Jambi City in 2016.
16 Denmark in Indonesia; SitiNurbaya.com.
17 The idea of forest restoration is considered an innovation in the field of preserving natural resources; see Elliot 2013.
18 Interview by Yudi Bachrioktora with Seman in Sungai Jerat in 2016.
19 Interview by Yudi Bachrioktora with Opung and Depati Jupri in Sungai Jerat in 2016.
20 For detailed information on this process and how land grabbing was organised by local elites (Batin Sembilan and formal positions) and migrants, see Steinebach 2017.
21 On the case of local migrants, see Galudra 2014.

22 *Aliansi Gerakan Reforma Agraria* (Agrarian Reform Movement Alliance) is the most influential of these three NGOs. Established in 2004, it is a national organisation based in Jakarta, consisting of peasant and fishermen organisations as well as minority ethnic groups and indigenous peoples.
23 Interview by Yudi Bachrioktora with Aki Asep in Sungai Jerat in 2016.
24 Interview by Yudi Bachrioktora with Depati Jupri in Sungai Jerat in 2016.

References

Adnan, H., Tadjudin, D., Yuliani, E.L., Komarudin, H., Lopulalan, D., Siagian, Y.L., and Munggoro, D.W. (2008) *Belajar dari Bungo. Mengelola sumber daya alam di era desentralisasi*. Bogor: CIFOR.

Beckert, B., Dittrich, C., and Adiwibowo, S. (2014) 'Contested land: An analysis of multi-layered conflicts in Jambi province, Sumatra, Indonesia', *ASEAS – Austrian Journal of South-East Asian Studies*, vol 7, no 1, pp 75–92.

Bird Life International. 'Forests of Hope site – Harapan Rainforest, Indonesia', <www.birdlife.org/worldwide/projects/forests-hope-site-harapan-rainforest-indonesia>, accessed 27 August 2018.

Borras, S.M., Jr and Franco, J.C. (2013) 'Global land grabbing and political reactions "from below"', *Third World Quarterly*, vol 34, no 9, pp 1723–1747.

Buergin, R. (2016) 'Ecosystem Restoration Concessions in Indonesia: Conflicts and Discourses', *Critical Asian Studies*, vol 48, no 2, pp 278–301.

Dafinger, A. and Pelican, M. (2006) 'Sharing or Dividing the Land? Land Rights and Farmer-Herder Relations in Burkina Faso and Northwest Cameroon', *Canadian Journal of African Studies*, vol 40, no 1, pp 127–151.

Denmark in Indonesia. Danida Support to Harapan Rainforest, http://indonesien.um.dk/en/danida-en/environmental-support-program/harapan-rain-forest, accessed 27 August, 2018.

DEPSOS RI (Departement Sosial RI Direktorat Jenderal Pemberdayaan Sosial. Direktorat Pemberdayaan Komunitas Adat Terpencil) (2003) *Sejarah Perkembangan Komunitas*. Jakarta: Adat Terpencil.

Dinerstein, A.C. (2015) *The Politics of Autonomy in Latin America: The Art of Organising Hope*. Basingstoke: Palgrave Macmillan.

Elliot, S., Blakesley, D., and Hardwick, K. (2013) *Restoring Tropical Forest. A Practical Guide*. Kew: Royal Botanic Gardens.

Fairhead, J., Leach, M., and Scoones, I. (2012) 'Green Grabbing: a new appropriation of nature?', *Journal of Peasant Studies*, vol 39, no 2, pp 237–261.

Galizia, M. (1996) 'Village institutions after the law no 5/1979 on village administration. The case of Rejang-Lebong in South-Western Sumatra', *Archipel no 51*, pp 135–160.

Galudra, G., Sirait, M., Pasya, G., Fay, C., Suyanto, S., van Noordwijk, M., and U. Pradhan, U. (2010) *RaTa: A Rapid Land Tenure Assesment Manual for Identifying the Nature of Land Tenure Conflicts*. Bogor: World Agroforestry Centre.

Galudra, G., van Noordwijk, M., Agung, P., Suyanto, S., and Pradhan, U. (2014) 'Migrants, Land Markets and Carbon Emissions in Jambi, Indonesia: Land Tenure Change and the Prospect of Emission Reduction', *Mitig Adapt Strateg Glob Change*, vol 19, pp 715–731.

Hall, R., Edelman, M., Borras Jr., S.M., Scoones, I., White, B., and Wolford, W. (2015) 'Resistance, acquiescence or incorporation? An introduction to land grabbing and political reactions "from below"', *The Journal of Peasant Studies*, vol 42, no 3–4, pp 467–488.

Hein, J. (2013) 'Reducing Emissions from Deforestation and Forest Degradation (REDD+), transnational conservation and access to land in Jambi, Indonesia', *EFForTS Discussion Paper Series*, no 2, pp 1–28.

Hein, J. and Faust, H. (2013) 'Conservation, REDD+ and the struggle for land in Jambi, Indonesia', *Pacific Geographies*, vol 41, pp 21–25.

Hidayat, R. (2012) *Membangkitkan Batang Terendam. Sejarah Asal Usul, Kebudayaan dan Perjuangan Hak SAD Batin 9*. Yayasan Setara: Jambi.

Indrarto, G.B., Murharjanti, P., Khatarina, J., Pulungan, I., Ivalerina, F., Rahman, J., Prana, M.N., Resosudarmo, I.A.P., and Muharrom, E. (2012) *The Context of REDD+ in Indonesia: Drivers, Agents and Institutions*. Bogor: CIFOR.

Institute for Policy Analysis of Conflict (2014) *Indigenous Rights vs Agrarian Reform in Indonesia. A Case Study from Jambi*. Institute for Policy Analysis of Conflict.

Kunz, Y., Hein, J., Mardiana, R., and Faust, H. (2016) 'Mimicry of the legal: Translating de jure land formalization processes into de facto local action in Jambi province, Sumatra', *Austrian Journal of South-East Asian Studies*, vol 9, no 1, pp 127–146.

Lang, C. (2012) 'On-going land conflicts at Harapan Rainforest Project: as a key funder of Harapan, what is the German Government's response?', www.redd-monitor.org/2012/12/19/on-going-land-conflicts-at-harapan-rainforest-project-as-a-key-funder-of-harapan-what-is-the-german-governments-response/, accessed 27 August, 2018.

Manurung, B. (2008) www.insideindonesia.org/jungle-school. *Inside Indonesia*, 92, April–June.

Mardiastuti, A. (2012) *The Role of UN-REDD in the Development of REDD+ in Indonesia. Volume III: Highlight of REDD Projects in Indonesia*. Ministry of Forestry of Republic Indonesia.

Silalahi, M. and Erwin, D. (2015) 'Collaborative Conflict Management on Ecosystem Restoration Concession: Lessons Learnt from Harapan Rainforest Jambi-South Sumatra-Indonesia', *Forest Res*, vol 4, no 134, pp 1–9.

Silalahi, M., Walsh, A.T., Utomo, A.B., Barnard, J., and Woodfield, E. (2015) 'Hutan Harapan and Ecosystem Restoration Concessions as a new approach to Sustainable Forest Management in Indonesia', Paper presented in XIV World Forestry Congress, Durban, South Africa.

Singapore Airlines Annual Report 2010/2011, <www.singaporeair.com/saar5/pdf/Investor-Relations/Annual-Report/annualreport1011.pdf>, accessed 5 November, 2018.

SitiNurbaya.com. 'Indonesia-Denmark Perkuat Kerjasama Pengelolaan Lingkungan Hidup Berkelanjutan', www.sitinurbaya.com/kegiatan-menteri-lhk/529-indonesia-denmark-perkuat-kerjasama-pengelolaan-lingkungan-hidup-berkelanjutan, accessed 27 August, 2018.

Steinebach, S. (2012) *Der Regenwald ist unser Haus: Die Orang Rimba auf Sumatra zwischen Autonomie und Fremdbestimmung*. Göttingen: Universitätsverlag Göttingen.

Steinebach, S. (2017) 'Farmers and Pawns: the role of migrants in agrarian conflicts and rural resistance in Sumatera, Indonesia', *The Asia Pasific Journal of Anthropology*, vol 18, no 3, pp 228–245.

The REDD desk. 'Harapan Rainforest Project, Sumatra', <https://theredddesk.org/countries /initiatives/harapan-rainforest-project-sumatra>, accessed 27 August, 2018.

Vergara-Camus, L. (2009) 'The politics of MST. Autonomous rural communities, the state, and electoral politics', *Latin American Perspectives*, vol 36, no 4, pp 178–191.

Widjardjo, B. and Perdana, H. (2001) *Reklaiming Dan Kedaultan Rakyat*. Jakarta and Bekasi: YLBHI and RACA Institute.

World Bank (2011) 'Indonesia-promoting sustainable production forest management to secure globally important biodiversity project: indigenous peoples plan. Indonesia', <http://documents.worldbank.org/curated/en/104701468268218960/pdf/IPP5080P1124190s0Framework0P112419.pdf>, accessed 27 August, 2018.

Znoj, H. (1995) *Tausch und Geld in Zentralsumatra. Zur Kritik des Schuldbegriffs in der Wirtschaftsethnologie*. Berlin: Reimer.

Znoj, H. (2001) *Heterarchy and Domination in Highland Jambi. The Contest for Community in a Matrilinear Society*. Habilitation thesis, University of Bern.

Part IV

Commons, privatisation and international law

The right to water

Top left: Floodplain area in Zambia: The right to water of wetlands in drylands is important for the resilience of local people. Picture by Tobias Haller.

Top right: Irrigated sugarcane field for biofuel production. Picture by Franziska Marfurt.

Bottom left: Herders and semi-nomadic groups from the Atlas Mountains with goats and sheep using the area on a reciprocal basis under *Amazigh* institutions. Picture by Sarah Ryser.

Bottom right: Collective fishing in an African floodplain. Water links many other common-pool resources and rights issues. Picture by Tobias Haller.

22 A structured checklist to identify connections between land grabbing and water grabbing[1]

Insa Theesfeld

Introduction

In recent years there has been a substantial increase in the amount and frequency of land acquisitions in a number of countries by players from outside the country, operating at a large scale. Reliable figures suggest that as of 2017, as much as 3 per cent of the world's arable land and permanent crop land has been the subject of such deals (FAO, 2017: Land Matrix figures). The land is being secured for agricultural production – or in some cases, for environmental protection – and particularly affects both Africa and Southeast Asia (Cotula, 2012; Deininger, 2011; Margulis et al., 2013).

Yet when these acquisitions occur, what actually happens to the individual property rights of local people is not well understood (Teklemariam et al., 2017; Theesfeld, 2018). There is some debate about whether large-scale land acquisitions focus on land that has either plentiful rainfall or major irrigation potential. If that is the case, it would suggest that the land acquisition that is occurring is really water acquisition – and, depending on the situation, should be called water grabbing[2] (Breu et al., 2016; Dell'Angelo et al., 2018; Franco et al., 2013; Hertzog et al., 2017; Mehta et al., 2012; Skinner and Cotula, 2011; van Eeden et al., 2016; Woodhouse, 2012). Several scholars have pointed out, however, that land grabbing research needs to move into a phase that examines what is actually happening on the ground (Scoones et al., 2013; Teklemariam et al., 2017). Otherwise, as Oya (2013) points out, studies on land and water grabbing will continue to be weakened by implicit, untested assumptions and loose use of analytical concepts not necessarily applicable to the subject. Examination of what happens on the ground could very usefully replace national-scale estimates of possible water grabbing with more reliable data derived from individual instances.

What is needed now is a mode of analysis to explain how land grabbing in a particular situation can result in water grabbing. As a foundation for such analysis, what is proposed here is a checklist, based on property rights theory, that makes it possible to examine the connection between land and water. The key to this approach is to focus on the impact that land deals have on the local water sector in the country where the land is acquired. It seeks to provide a conceptual framework allowing researchers to analyse and compare cases of large-scale land acquisition (LSLA) in a systematic way.

The section below provides an introduction to the property rights approach used to create the checklist to aid in analysing LSLA cases for water grabbing. The next section proposes the checklist of eight analytical patterns to connect property rights shifts in land and water. Finally, two land acquisition case studies illustrate land use changes that have an impact on the water sector. One is from a hot-spot country of foreign direct investment – Ethiopia; and one is from a region where witnesses have reported great potential for land acquisition deals – Tajikistan. The penultimate section analyses those two cases using the checklist and discusses the results. Finally, the last section presents conclusions on the value of this approach for systematically predicting trends in how water grabbing may accompany land grabbing.

Property rights theory for an inter-sectoral view

Although, in principle, larger-scale farming and capital-intensive investments in the agricultural sector can provide social and economic development opportunities for poor countries (Baumgartner et al., 2015; Teklemariam et al., 2017), cases with negative impact on the local population or on the environment have attracted public attention and triggered national and international policy debates. In these cases, deals were often closed without consulting the local population and without adequate compensation to them, and involved conflicts with and without violence (Dell'Angelo et al., 2017). Although literature still does not provide a comprehensive definition of such land grabbing (Teklemariam et al., 2015), the one from Borras et al. (2012, p. 850) is very meaningful: "Land grabbing is essentially control grabbing: grabbing the power to control land and other resources such as water in order to derive benefit from such control". First, this definition highlights the link to other resources and second, it focuses on claiming exclusive and control rights to a resource (Friis and Nielsen, 2016). We can further define land grabbing by a change in ownership and tenure structure in the way new landownership authority is exercised, which can lead to negative effects.[3] The issue is not necessarily regarding the amount of land acquired, nor the total sum of the investment. Nonetheless, the size of the investment makes the consequences more obvious.

A careful distinction between LSLA with rather neutral or positive outcomes, compared to those with negative impacts for the local population and the environment, is not necessarily important when the focus is on exploring how to link the land and water sector. In that regard, Zetland and Möller-Gulland (2013) make a simple (maybe too simple) distinction between "grabbers" and "investors", where grabbers are interested in maximizing short-term production at the expense of long-term sustainability. This classification leads to the assumption that grabbers are more likely to deplete soils and abuse their preferential access to water and thus overuse water resources, whereas investors are regarded as saving water for the next season and invest in water conservation technology.

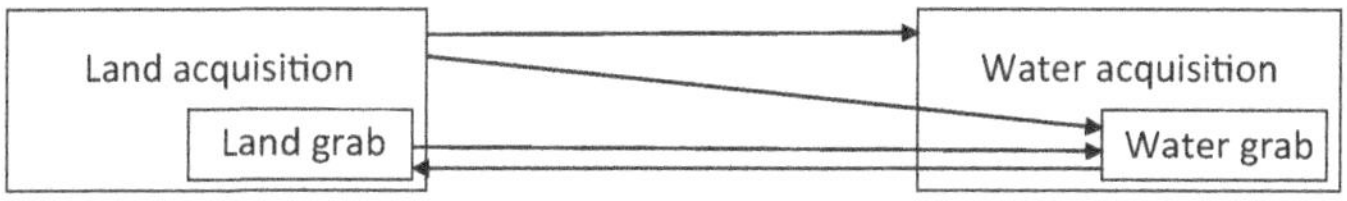

Figure 22.1 Land and water grabbing interactions.

But still, investors typically replace traditional farming with new processes, and have an impact on the intensification of the water consumption. So, whereas land grabbing is more likely to also lead to water grabbing, a rather fair large-scale land investment process can lead to both water acquisition and water grabbing, see Figure 22.1.

In principle, the relationship can also be thought of the other way around, starting with an investment in the water sector that grabs water and has a subsequent land grabbing effect. The most gigantic cases of such kind are the Great Man-made River in Libya that pumps fossil aquifer water to the Sahara Desert or the Three Gorges Dam in China. For the purpose of this contribution, the relationship studied proceeds from the investment in land.

This contribution seeks to present a checklist that helps to study the ways that water grabbing can accompany land grabbing. To study both, involves more than simply adding the water to the land grabbing debate. Even if we can estimate the global water flow impact of the foreign investment, we need a better understanding of the underlying processes (Breu et al., 2016; Scoones et al., 2013).

In order to be valuable from an investor's point of view, land needs to possess certain properties which taken together can turn land deals into lucrative business. One crucial characteristic is access to water resources (van Eeden et al., 2016) either rainwater but looking at LSLA and development of infrastructure most likely irrigation water (Dell'Angelo et al., 2018), which will be our focus here. Water access, in case rainwater is not sufficient, can determine crop selection. Without potential for irrigation, land investors would face high risks in drought periods. Thus, water plays a central, but not always explicit official role in formal land deals (Anseeuw et al., 2012, p. 37; Mehta et al., 2012; Smaller and Mann, 2009; Woodhouse, 2012). There are many different ways in which access or withdrawal rights to water can be obtained – and many ways those rights can be exercised. For instance, von Eeden et al. (2016), present in detail how land investors in Tanzania gain access rights, and dispossess other water users, through resettling, blocking access ways, closing infrastructure or not issuing permits. Depending on local conditions, investor approaches to water rights acquisition can range from careful advance planning with open discussion to unannounced misappropriation of water after the land deal is concluded.

To scrutinize the processes under investigation here, a disentangled property rights perspective is the most useful. Property rights do not necessarily imply full ownership but are composed of different bundles of rights that may be held by different claimants, such as the state, user groups, families, individuals (Meinzen-Dick, 2014) or are non-defined. Thus, property rights show a social relation among groups over land; a property rights analysis does not approach land essentially as a "thing" identified as a property. The new investor does of course not own the resource itself: with a sale or lease, it is only a bundle of rights to use a resource that has changed ownership (Alchian and Demsetz, 1973; Dell'Angelo et al., 2017; van Eeden et al., 2016).

In their efforts to secure water rights, foreign and domestic investors and local water users are often highly unequal actors in terms of bargaining power, knowledge, risk behaviour and economic endowments (Teklemariam et al., 2015). Both are looking for ways to protect their individual customary water claims or state-guaranteed rights to receive water in terms of access, withdrawal, management, exclusion and alienation (Ostrom and Schlager, 1992). The most important water rights are the access rights – the right to enter a defined physical area – and the withdrawal right – the right to obtain the resource unit (Ostrom and Schlager, 1992). By changing formal and de-facto land and water rights, actors' affiliation to a group may change as interests and power resources shift. Foreign or domestic investors as new users entering the arena are likely to undermine existing customary claims (Deininger, 2011; Ismar, 2013; Zoomers, 2010), including the existing water management system. Investors, taking water for granted or based upon agreements with governmental bodies, ignore the fact that most water is not likely to be freely available but that rights obtained to it must be reallocated from existing local users.

The following conceptual foundation concentrates on the local redistribution and structural effects land grabbing has on water grabbing.

Towards a checklist to connect land and water grabbing

The regimes of property rights in land can be very diverse and have impact on the situation of the water sector (Franco et al., 2013). Working with a checklist of patterns of interrelation between land and water will help to disentangle this analytical complexity.

The checklist contains eight patterns taken in the following from Theesfeld (2018): the ecological, the judicial, the resettlement, the use, the land use, the conjunctive use, the infrastructure and the governance one. Attention to the patterns allows a researcher to classify the impact of land grabbing on the various rights and claims in the agricultural water sector and facilitates the comparison of cases and thus makes reliable general predictions possible. These eight patterns should be investigated in numerous empirical case studies in order to test their applicability. They allow researchers to differentiate between direct and indirect impacts of land grabbing on the distribution of property rights on water and thus, help ensure that inter-sectoral effects are not overlooked in the further analysis on drivers of the investment or impacts on the rural population. This checklist may not yet be complete, but it is comprehensive enough to facilitate comparing cases and clustering similarities and differences.

Most land grabbing is reported in arid and semi-arid areas which require irrigation to achieve a profitable and reliable agricultural production (Cotula, 2011). Thus, while globally most agricultural production is based purely on rainwater, the framework put forth here assumes, in line with Dell'Angelo et al. (2018), Rulli et al. (2013) and Franco et al. (2013), that a large share of foreign direct agricultural investment in land shows a pattern where, after the investment, additional water – besides the annual rainfall – is required for efficient production. This is shown, e.g. for Tanzania and Sudan where blue water grabbing (water supplied by irrigation) is predicted (Rulli et al., 2013). This connection is framed under the *ecological pattern*.[4] Even in countries with at first sight favourable natural water conditions, however, the new cropping system or intensification of production can depend on additional water input. Investors are likely to turn to irrigation to help avoid vulnerability to the variations in water supply, which otherwise could lead to highly insecure rates of return of the investments.

Second, is the *judicial pattern*. Following the scheme for disentangling property rights by Ostrom and Schlager (1992), it is crucial to check whether access, withdrawal, management, exclusion or alienation rights on water are part of negotiations or explicitly mentioned in the contract for buying or leasing land. This dimension directly affects the resulting water acquisition, and it is the pattern most easily addressed by comprehensive land use policy measures. To study formal changes, it is important to understand the initial relation between officially assigned property rights to land and water even prior to experiencing an investment in land. In some countries, a farmer must pay a land tax to the government or rent land from another farmer to acquire rights to take water from canals or pump it from underground. In other countries the owner of surface land is the owner of the water under it. Such an unspecified property right would imply that the landowner can extract water without restriction (Theesfeld, 2010). In the Islamic school of thought it is rather common to clearly distinguish between land rights and water rights: The digger of a well – whether on his own land or on unoccupied land – automatically becomes the owner of the well water as soon as digging is completed (Caponera, 1992).

Further, as described above we find a pluralism of formal rights and customary claims that may even contradict one another but likewise influence each other (Meinzen-Dick, 2014; van Eeden et al., 2016). We often find de facto property rights functioning on the ground that could be made up of combinations of both formal secured rights and customary claims (Klümper et al., 2018). So even if the formal situation is laid out clearly it may be difficult for local users to defend their customary access and withdrawal claims (Franco et al., 2013).

Host countries may feel forced to make water deals along with any land deals, granting investors legal rights to water already governed by customary regulations (Skinner and Cotula, 2011; Woodhouse, 2012). Particularly in African countries, such legal appropriations ignore local needs (Ismar, 2013, p. 288), as is also shown for a case study in Ethiopia done by Baumgartner et al. (2015). In general, the rights allocated may differentiate between unrestricted access to canal and ground water and access conditioned upon issues such as dry season crops. Water payments might be fixed through volumetric billing or left to be negotiated per hectare depending on the type of irrigation used. Even when a land contract makes no specific reference to water, water may still form part of the deal, as governments may agree to invest in water-related infrastructure (Skinner and Cotula, 2011). Likewise, customary access claims to a river can simply become invalid, because the way to the river across land now owned by an investor can be denied, as shown by Kruchem (2013, p. 148) for a case in Zambia. Yet, the concrete formulation of agreements in the water sector is often subordinated and left to the local actors.

Looking at the water and land sectors at the same time during land deal negotiations is a big challenge due to the socio-ecological complexity of these systems. This is noted in the FAO Voluntary Guidelines on the Responsible Governance of Tenure of Land, Fisheries and Forests in the Context of National Food Security (FAO, 2012), which excluded the relation to the water sector in the fear of never being able to complete this Guideline (Franco et al., 2013). The more integrative Dakar Declaration Against Water and Land Grabbing (Watergrabbing.net, 2017) set up in 2014 by a number of civil society organizations in the framework of the Africa Social Forum and the further developed Declaration of the Global Convergence of Land and Water Struggles in Tunis 2015 (Tramel, 2016) has as such no place yet in the political and scholarly debate.

Third, land is often officially declared as abandoned, which results in the fact that the newly established formal recognition of property rights on land for the investor might lead to driving local farmers off that land (van Eeden et al., 2016). Attached to the land might have been access to water, either by natural access to a canal or river, or by pumping rights on that plot. The now-landless farmers have to look for new land (assuming they are not migrating to the towns) and that therefore increases not only the competition for land, but also the pressure on the water resources at neighbouring locations, which may lead to a change in the actual claims on water. This is called *resettlement pattern*. Instead of a positive spill-over effect to the agriculture sector, the big investment might also have the effect that more local small-scale farmers at neighbourhood locations leave agriculture. In short, the resettlement pattern highlights the effect of having more or less water users at adjacent plots.

Even if no resettlement or dispossession of land is involved, simply the fact that investors enter the arena leads to an increase of the number of farmers using irrigation. More important, the investor is not only one additional, but often a more intensive water user. This is often due to investors' more advanced techniques in water use. As a result, physical water availability is under more pressure and resource scarcity increases. When acquired land is irrigated, the associated appropriation of water can reduce the availability and the quality of irrigation water in the neighbouring and particularly downstream farmland areas. This *use pattern* stands for the strongest direct impact on water acquisition.

The most stand-alone analytical pattern of all other patterns is the *land use pattern*. It also describes a process which has a direct impact on water acquisition.

This means, no matter whether the new production is on humid fertile soils, or on dry soils that require irrigation, whether water rights have been mentioned in the land deal or not, or whether the land has been in fact abandoned or is occupied by other farmers, investors often favour more water-intensive crops (Ismar, 2013, p. 287) or monoculture forest plantations which extract large amount of water (Kruchem, 2013, p. 151). In a hydrologic model Rulli et al. (2013) assume that the grabbed water may range between a minimum value corresponding to crop water use in rain-fed agriculture and a maximum value corresponding to the case of irrigated agriculture in conditions that optimize crop yield.[5] Examples of a land use change which leads to massive impacts on the water sector are the establishment of flower farms in Ethiopia (Bues and Theesfeld, 2012), and the change from traditional cotton to rice production in Tajikistan, as shown in the next section.

The *conjunctive use pattern* notes that particular plant requirements can call for a higher share of surface canal water (diverted from rivers or rain-fed tanks) or groundwater. Certainly, pumped groundwater is cooler and cannot be used with sensitive crops. But, some crops require water at an exact point in time in the vegetation cycle and are very sensitive to a postponing of an irrigation run. This often requires individual groundwater pumping, which is more cost-intensive but more reliable compared to canal irrigation water which needs to be shared. The general shift in water sources has an indirect impact on the water available for other users and their withdrawal and access options. The surface-and-ground-water interaction makes it quite difficult to determine the actual effects of water re-allocations (Teklemariam et al., 2015) due to land deals.

One side-effect of a planned land use change accompanying the investment in land might be additional investments in infrastructure (irrigation, drainage, roads). This *infrastructure pattern* is of course a very crucial one and represents an indirect impact on water acquisition. With a change in the infrastructure and technology, almost every user faces a change in de facto water access and withdrawal rights, as shown in Laos with the destruction of the paddy rice irrigation system due to a land use change to banana plantation (Friis and Nielsen, 2016). Yet, from the investors' perspective it is their wish to secure not only withdrawal but particular access rights as shown for Mali (Nolte and Voget-Kleschin, 2014) or Tanzania (van Eeden et al., 2016). Depending on the individual location related to the infrastructure or canal outlet (Hertzog et al., 2017), the users might have at their disposal more or less water in various quantities and qualities, at different times, at different prices. Infrastructure investment comprises not only big dam, reservoir or canal system construction, but also irrigation techniques such as drip or spray irrigation, or even computer-driven water regulation to control humidity in greenhouses. It may also include development of boreholes for a farm (Teklemariam et al., 2017) or drainage systems, of which surrounding farmers can take advantage as well. Usually, also the management rights change with a reconstruction of the infrastructure. In line with that, the technical options to exclude users can become more effective.

Compared to rain-fed production systems, or the gravity-flooding of fields, the investment in canal systems and the upgrade in irrigation techniques allows the investor, primarily, to use irrigation water for increased harvesting seasons and crop rotation. Thus, the crop production and water acquisition becomes less season-dependent. This may change the actual withdrawal and access rights in the whole water and irrigation catchment area.

Finally, getting involved in irrigation infrastructure reconstruction and management leads in the long-run naturally to an involvement in water governance. Water grabbing is not only about the capturing of the water itself, but also about taking over the power to decide how water will be used (Franco et al., 2013). If the governance systems are inadequate, farmers can face what is called an institutional water scarcity, despite physically plentiful water availability (Theesfeld and Klümper, 2016). The previously informal or even formal agricultural water governance system may change. *Governance pattern* stands for this indirect impact and describes a medium or long-term change in formal water rights and in the customary water claims. A new governance pattern may include not only a new form of water user association, but also specific decisions and rule changes on how to set up irrigation turns, imposing new measures of water charging, or new workforce requests for maintenance. With the possible creation of a formal land market as a long-term consequence of opening up an area for foreign land investments often comes the creation of formal water rights (De Schutter, 2011), which in turn later serve as a basis for new governance arrangements. Alternatively, a change in the governance structures of an irrigation system might come first, and then require an infrastructure change.

Table 22.1 comprises a checklist to connect land and water grabbing locally. The eight patterns depict a change in certain bundles of property rights and thus imply a change in social relation among the involved actors, which can lead to a grabbing effect, see Figure 22.1.

Results and discussion based on case studies

Anseeuw et al. (2013) provide a selection of ways and procedures to decide what and how a land transaction might get included in any kind of database of large-scale land acquisition. The procedures are already so diverse that numbers for total land grab range from 67 million hectares reported in the *Transnational Land Deals for Agriculture in the Global South Report* to 230 million hectares reported in the *Land Rights and the Global Land Rush Report*. Adding to this diversity of procedures another layer – namely the one of the amount of water changing its potential user – makes a quantification hardly possible. Thus, the only way to go is either to calculate a hypothetical hydrological model as done by Dell'Angelo et al. (2018), Rulli et al. (2013) or Breu et al. (2016) for the global scale or to provide a conceptual basis to analyse the effects at a local scale. Yet, the tricky problem for analysts following the latter approach is that water grabbing is often subsequent to the change in land rights. This involves particular challenges for empirical field studies which usually do not cover a long enough time span.

Table 22.1 Checklist to connect land and water grabbing

Pattern	*Questions to the land deal that connect to the agricultural water sector*
1. Ecological pattern	Do the natural water conditions require irrigation?
2. Judicial pattern	Are water rights explicitly included in the land negotiation or contract?
3. Use pattern	Does competition for water and the total uptake increase?
4. Resettlement patterns	Does the competition of water users increase off-side the new investment farm?
5. Land use pattern	Does the change in cropping structure entail a change in irrigation?
6. Conjunctive use pattern	Is there a shift in the share of ground and surface water use?
7. Infrastructure pattern	Does the building of new infrastructure or its reconstruction change the actual irrigation habits of the users?
8. Governance pattern	Do informal or even formal water governance arrangements change?

Source: adapted from (Theesfeld, 2018)

The checklist proposed here to deal with this challenge is illustrated by two case studies given in (Theesfeld, 2018). One is from Ethiopia which has, in line with Africa as the most targeted continent, seen a significant rise in the number of LSLA (Baumgartner et al., 2015). The Land Matrix shows for Ethiopia alone a total of 66 land deals with contract size of more than 1 million hectares, whereof all of them have concluded the negotiations. Rahmato (2014, p. 226) in his development study for Ethiopia estimated that total land ceded to investors from mid-1990–2012 is rather in the order of 3 to 3.5 million hectares. He likewise confirms that not just land, but also water, has been acquired. But still, there is only limited evidence of the local impacts of the LSLAs (Baumgartner et al., 2015).

The other example is of Tajikistan, a rather unknown country for land acquisition research, but located in an area regarded as being very likely to experience land acquisition in the future, as land is available and a high yield gap has been estimated in combination with unclear property rights (Deininger, 2011, p. 231 ff.; Visser and Spoor, 2011). In Tajikistan, where arable land is scarce but water is available, especially Chinese, Iranian and Turkish investments in land are taking place, although much smaller in size compared to land deals elsewhere.

The empirical case study description along the checklist depicted above turned out to be very helpful in linking changes in property rights assignment between the land and the water sector. In the Ethiopian case study, foreign direct investment in land for horticulture production had an impact on small-scale irrigation schemes. Referring to the pattern classification, we will show that out of five patterns that helped to structure the case, the use pattern and the governance pattern are most crucial in describing the inter-sectoral links. The Tajikistan example of property rights change in water – as a result of the investment in land for rice production – will show that out of six patterns necessary to detect the sectoral linkages, the most crucial patterns for analysis are the infrastructure pattern and the land use pattern.

Flower farm investments in Ethiopia

Ethiopia is among the top ten target countries for agricultural foreign direct investment (Land Matrix, 2017; Rahmato, 2014). This is in part due to the water availability of the upstream Blue Nile location. In the central Ethiopian region of Oromia, investigated by the author (Bues and Theesfeld, 2012), since 2005, nine horticultural investment projects started to use water from an irrigation canal that was formerly managed under common property regime by local smallholders. The investors have been assigned a share of land from a former state farm and from local farmers. The average farm size is 20 hectares, which is a reasonable size for this high intensive production. The *ecological pattern* in this case is obvious as the investment in land for horticulture required and planned to use water from irrigation canals as rainfed production would not have been reliable and profitable enough.

Looking at the *use pattern* of the horticultural investment farms, the total irrigated area has not been increased, but investors use more water, leading to canal water scarcity in the remaining canal system (Bues and Theesfeld, 2012).

Water rights changed indirectly with the change in land rights. Water rights in the case study area are tied to land rights in that farmers may have land use rights with and without the right to water from the canal. For those investment farms which were established on former local farmers' land, land rights of the previous users changed and with this the access rights or customary claims to the canal may fade away. In the frame of this investigation, done in 2010, it was not possibly to determine whether that was explicitly written down in the negotiation contracts between the investors and the Ethiopian government. But the property rights connection, including the prevailing farmer's insecurity of land tenure as the rights of ownership are vested in the state, describes the initial *judicial pattern*. Likewise, farmers, who formerly used lands that have been transferred to various investors were dispossessed of their land and ended up with arable land elsewhere, but in fact limited water access rights to the canal. Rahmato (2014, p. 226) even talks about linked resettlement programs initiated by the Ethiopian government.

The change to a different land use, i.e. the flower and vegetable production, required not only more water but also a different quality of water. Instead of surface canal water, groundwater is substantially used, because pumping is more reliable than canal delivery. But groundwater is known for containing a high percentage of bicarbonates and has to be mixed with canal water before irrigating vegetables. The corresponding changes in shares between surface and groundwater sources, classified as *conjunctive use pattern*, have an impact on the de facto water rights of those farmers and others not engaged in the investment.

As shown by Bues and Theesfeld (2012), since the investors settled in the area, the institutional arrangements in the water sector have changed considerably. Investment farms employed their own guards to open and close the main regulatory gate at the dam, whereas local farmers still worked with the guards paid by their user group. The prevalence of these two different governance systems led to conflicts that finally led to the initiative of one of the horticultural farm managers to set up a new association. This association consisted of representatives from both the investors and local farmers in order to deal with the whole irrigation scheme. As a consequence of the new established association and the different bargaining power of the actors involved, the new watering arrangements comprise more restrictions for the small-scale farmers with more severe sanctions than those applying to the investment farms (Bues and Theesfeld, 2012). This is in contrast to the previous graduated sanctioning mechanism of the small farmers' user group. Besides the cut back on smallholders' water rights, the new institutional arrangement shows the outcome of distributional conflicts. The case shows that combined with an intensification in water use, particularly the *governance pattern* leads to a further reduction in local farmers' property rights to water and thus represents water grabbing in line with the definition given above by Dell'Angelo et al. (2018).

Rice production in Tajikistan

The public land grabbing debate almost neglects the Central Asia region. Nonetheless, in the vast grain production areas of Kazakhstan and the water-rich countries such as Tajikistan, foreign investors are more and more entering the formal and informal land markets (Hofman, 2013, p. 6; Visser and Spoor, 2011). Since the break-down of the Soviet Union, in Tajikistan all land is state property. Therefore, for Tajiks and for foreigners it is only possible to lease land. According to the national laws, foreigners can only lease state-owned land for up to 50 years, whereas local citizens (can) gain non-time-limited use rights. During the past five years, Tajikistan has experienced foreign land leases by primarily Chinese, Iranian and Turkish investors. Media reports available and an initial study by Hofman (2013) give insights to foreign land investments in the southern province of Tajikistan, Khatlon. With land leases between 45–550 hectares, those foreign land acquisition cases exceed the average private farm size of 30 hectares in an area of very limited arable land and increasing population density.

The following case study draws on own empirical data (Gerigk et al., 2013) and focuses on one part of the Amu Darya basin in the district Bokhtar where the agricultural structure is twofold. Here, large cotton farms and small household plots dominate the irrigation-intense agricultural sector. In 2009, a Chinese investor leased 120 hectares of land from a collective farm. This collective farm with a total of 340 hectares was reorganized in 2006 from a former kolkhoz, yet the kolkhoz manager remained in charge of the operation. In 2012, the Chinese cancelled the land leasing contract with the collective farm ahead of time after only three years in operation. In contrast to the formal regulation, this land had not been put under lease by the local Land Committee, but via direct negotiations with the collective farm manager, who still has strong ties to the local authorities and remains powerful in the village as a "Kolkhozi" elite (Boboyorov, 2012; Hofman, 2013). Although no population was dispossessed of land, this case should be considered a land grabbing case, where foreign investors got former kolkhoz land which could have been also distributed among the villagers and workers of the former kolkhoz farm, and thus social dissatisfaction has been noticed.

The following patterns facilitate the description how the Chinese investments in Bokhtar impact on water property rights. Due to the low precipitation (average annual rainfall 150–290 mm) and very dry summer months, irrigation by ground- or surface water is inevitable for all farms in this area – likewise for the Chinese rice and vegetable farm. Given that fact, without considering any interference in technical or organizational components of the irrigation system, the *ecological pattern* signifies already that an impact on the property rights on water is very likely. The former cotton land (70 per cent) from the collective farm was used to produce rice (80 per cent of the area) and vegetables (20 per cent of the area). According to estimations (Aldaya et al., 2010, p. 31; AquaStat, 2013) rice production in Tajikistan has a higher total water requirement (with an additional $300m^3$/hectare of total crop water use) than cotton.[6] Although the cultivated area did not increase in total, the total water uptake and thus competition for water withdrawal rights of all water users in the area increased. This is classified as the *use pattern*. This pattern comes along with the *land use pattern*, basically from cotton to rice, which in the Tajik case resulted not only in a higher water demand but also in a different organization of water flow and a necessity for improved quality of water in rice production. Thus, withdrawal rights of all adjacent farmers including those of the household plots are affected.

To achieve the required continuous water flow for rice production and efficient water use for vegetable production, the investors had to make changes in the irrigation infrastructure, being described under the notion of the *infrastructure pattern*. In Bokhtar, the Chinese farm has invested in off-farm infrastructure by rebuilding an upstream pump. This ensured that water from the river flowed again to the primary canal, and further via a secondary canal to the plots of the farm. Surrounding small farms and households benefited on the one hand in terms of water availability and access, but on the other hand, they were totally dependent to the irrigation schedule for rice set by the Chinese farm (Gerigk et al., 2013). Consequently, outside the rice season, water flow was not continuous anymore. This also means that in the water-scarce and highly variable time of the year, surplus water to downstream farmers was limited or even non-existent. With the off-farm investments, access, management and withdrawal rights were affected. The general bad national operation and maintenance of irrigation infrastructure in Tajikistan limited the Chinese incentive to invest further in water infrastructure. The overall water uptake demanded by the new rice fields could not be addressed by the Chinese investments alone. Ongoing water problems due to poor water infrastructure were one reason for the Chinese to leave the area. After the Chinese left, the manager of the collective farm took back the land and took over the Chinese irrigation system. The effective water rights for the collective farm thus were strengthened in terms of management and also access rights.

The *judicial pattern* in this case does not only encompass the joint (informal) agreement of land and water access. It also reveals the powerful position of the collective farm in terms of decision making even towards foreigners. Water rights were formally not included in the official land deal, instead they were directly negotiated between the collective farm manager and the Chinese investor.

Although there has not been any observed change in formal and informal governance structures, this case can be classified under the *governance pattern*, particularly this absence is important to be considered. The very limited options to change or reform the governance mechanisms could be seen as one reason why the Chinese left already after three years. Local commons initiatives, e.g. establishing Water User Groups or Farmer Associations is not wide spread in many parts of Tajikistan. In some instances, if Tajiks started to self-organize, the initiatives were shut down by local authorities or penalties had to be paid, framed as environmental taxes (Boboyorov, 2012). For the Chinese it was likewise not possible to change the local power structures and to intervene in any governance mechanisms to improve water access and availability neither for the watershed nor for their own farm.

This case shows, that after carefully disentangling the property rights relations along the lines of the linking patterns, particularly the land use change combined with the particular requirements for water use and the investments in the infrastructure made the local people quite dependent on the water schedule of the powerful investor. That in turn led to less secure local farmers' property rights to water and thus represents water grabbing.

Conclusion

Water grabbing tends to share the negative connotations associated with land grabbing. And indeed, the potential link between land and water grabbing, in the process of agricultural investment, can lead to long-term non-sustainable development from both a social and ecological view. This impact is not well-recognized. Farm managers and land investors might essentially be winners of an unannounced overhaul of property rights. They may follow up by seeking, through political processes new institutional arrangements in the water sector that favours them and do not allow self-organized commons management. The result may further disadvantage local actors who were already less powerful and could lead to social conflict.

Meanwhile, however, the link between the property rights of the land and water sectors also means that there can, in principle, be positive feedbacks for the water sector from the initial land acquisition. In fact, there are examples of land acquisition that can actually protect and manage water resources and supply (Ding, 2007). Investments in infrastructure are likely to happen and stay in the country even if the investor leaves. Knowledge and management capacities may increase due to training and more qualified jobs for local people.

More transparency in the process of how an irrigation schedule is elaborated, e.g. by a state-managed irrigation company, might develop at the request of a more powerful investor. Likewise, new governance forms, such as more participatory water user associations with all parties involved, could be, at least, an option. Empirical work with a focus on the interrelations between both sectors may bring to light evidence of these positive linkages, too.

The checklist proposed, with its eight analytical patterns, allows for a better understanding of the links between land acquisition and potential for water grabbing. Assuming analysis of a number of LSLA cases could be structured along these lines, various impacts on disentangled property rights and claims on water could be studied. With this approach, scholars can also come to reliable predictions about the impact of land grabbing on water grabbing. Those predictions would not implicate the impacts on economic activities, employment or income level among local people, which are the main socio-economic issues in the land grabbing debate. Using the patterns nonetheless facilitates an understanding of the social role of property rights and identifies important dimensions when linking two natural resource sectors. That is crucial for those who design policy to intervene on behalf of communities' access to resources. Those policy designers need to know how the process of marginalizing groups – excluding them from certain water and land resources – has happened.

Notes

1 The material in this chapter was previously published in Ecological Economics and is republished here with permission of that journal. Theesfeld, I. (2018). *From Land to Water Grabbing: A Property Rights Perspective on Linked Natural Resources*. *Ecological Economics* 154(2018): 62–70, doi.org/10.1016/j.ecolecon.2018.07.019.

2 Water grabbing is initially defined by Mehta et al. (2012) as the process in which powerful actors are able to take control of, or reallocate to their own benefits, water resources used by local communities [...] on which their livelihoods are based. Dell'Angelo et al. (2018) present an overview of current water grabbing definitions and propose as a common denominator among the different definitions that there is an aspect of injustice and power imbalance which is represented by the word "grabbing", leading to a disregard of local users and their customary rights.

3 In that respect the Tirana conference of the International Land Coalition in 2011 defined land grabbing as land acquisition that are in violation of human rights, without prior consent of the preexisting land users, and with no consideration of the social and environmental impacts (International Land Declaration, 2011).

4 As rainfall is strongly linked to other ecological conditions such as temperature range and storage characteristics of the soil, this pattern is identified here as the ecological pattern and not a humidity pattern.

5 But not every severe land use change does lead to a big change in the property rights on water. In Australia, where about 10 per cent of the total global land acquisition is happening, most of the land is used for livestock production. This land use change does – although large in size – only involve little amount of additional necessary water (FAO, 2017; Rulli et al., 2013).

6 The data are based on estimations and present the average for the country's cotton and rice total crop water use.

References

Alchian, A.A. and Demsetz, H. (1973) 'The property right paradigm', *The Journal of Economic History*, vol 33, pp 16–27.

Aldaya, M.M., Munoz, G., and Hoekstra, A.Y. (2010) *Water Footprint of Cotton, Wheat and Rice Production in Central Asia, Value of Water Research Report Series*. Delft: UNESCO-IHE Institute for Water Education.

Anseeuw, W., Lay, J., Messerli, P., Giger, M., and Taylor, M. (2013) 'Creating a public tool to assess and promote transparency in global land deals: the experience of the Land Matrix', *The Journal of Peasant Studies*, vol 40, pp 521–530.

Anseeuw, W., Wily, A.L., Cotula, L., and Taylor, M. (2012) *Land Rights and the Rush for Land: Findings of the Global Commercial Pressures on Land Research Project*. Rome: International Land Coalition.

AquaStat (2013) *Irrigation in Central Asia in Figures, Aquastat Survey – 2012*. Rome: FAO Land and Water Division.

Baumgartner, P., von Braun, J., Abebaw, D., and Müller, M. (2015) 'Impacts of large-scale land investments on income, prices, and employment: empirical analyses in Ethiopia', *World Development*, vol 72, pp 175–190.

Boboyorov, H. (2012) 'Personal networks of agricultural knowledge in cotton-growing communities of Southern Tajikistan', *Demokratizatsiya*, vol 20, pp 409–435.

Borras, S.M., Franco, J.C., Gómez, S., Kay, C., and Spoor, M. (2012) 'Land grabbing in Latin America and the Caribbean', *The Journal of Peasant Studies*, vol 39, pp 845–872.

Breu, T., Bader, C., Messerli, P., Heinimann, A., Rist, S. and Eckert, S. (2016) 'Large-scale land acquisition and its effects on the water balance in investor and host countries', *PLOS ONE*, vol 11, e0150901.

Bues, A. and Theesfeld, I. (2012) 'Water grabbing and the role of power: shifting water governance in the light of agricultural foreign direct investment', *Water Alternatives*, vol 5, pp 266–283.

Caponera, D.A. (1992) *Principles of Water Law and Administration: National and International*. Brookfield, Rotterdam: A.A. Balkema.

Cotula, L. (2011) *Land Deals in Africa: What is in the Contracts?* London: International Institute for Environment and Development (IIED).

Cotula, L. (2012) 'The international political economy of the global land rush: a critical appraisal of trends, scale, geography and drivers', *Journal of Peasant Studies*, vol 39, pp 649–680.

De Schutter, O. (2011) 'How not to think of land-grabbing: three critiques of large-scale investments in farmland', *The Journal of Peasant Studies*, vol 38, pp 249–279.

Deininger, K. (2011) 'Challenges posed by the new wave of farmland investment', *The Journal of Peasant Studies*, vol 38, pp 217–247.

Dell'Angelo, J., Rulli, M.C., and D'Odorico, P. (2018) 'The global water grabbing syndrome', *Ecological Economics*, vol 143, pp 276–285.

Dell'Angelo, J., D'Odorico, P., Rulli, M.C. and Marchand, P. (2017) 'The tragedy of the grabbed commons: coercion and dispossession in the global land rush', *World Development*, vol 92, pp 1–12.

Ding, C. (2007) 'Policy and praxis of land acquisition in China', *Land Use Policy*, vol 24, pp 1–13.

von Eeden, A., Mehta, L., and van Koppen, B. (2016) 'Whose water? Large-scale agricultural development and water grabbing in the Wami-Ruvu River Basin, Tanzania', *Water Alternatives*, vol 9, pp 608–626.

FAO (2012) *Voluntary Guidelines on the Responsible Governance of Tenure of Land, Fisheries and Forests in the Context of National Food Security*. Rome: Food and Agriculture Organization of the United Nations (FAO).

FAO (2017) http://faostat3.fao.org/download/E/EL/E, accessed 31 October, 2018.

Franco, J., Mehta, L., and Veldwisch, G.J. (2013) 'The global politics of water grabbing', *Third World Quarterly*, vol 34, pp 1651–1675.

Friis, C. and Nielsen, J.Ø. (2016) 'Small-scale land acquisitions, large-scale implications: exploring the case of Chinese banana investments in Northern Laos', *Land Use Policy*, vol 57, pp 117–129.

Gerigk, F., Herzfeld, T., and Theesfeld, I. (2013) 'Farmers' perceptions of land and water property rights: Discrepancies between de jure and de facto rights in Tajikistan', Paper presented at the conference on Natural resource use in Central Asia: Institutional challenges and the contribution of capacity building, Giessen.

Hertzog, T., Poussin, J.C., Tangara, B., and Jamin, J.Y. (2017) 'Participatory foresight to address long-term issues in a large irrigation scheme. An example in Office du Niger, Mali', *Land Use Policy*, vol 64, pp 13–28.

Hofman, I. (2013) *Understanding forms of contention in the post-Soviet setting: rural responses to Chinese land investments in Tajikistan, LDPI Working Paper*. Land Deal Politics Initiative (LDPI).

International Land Declaration (2011) *Tirana Declaration. "Securing land access for the poor in times of intensified natural resource competition"*. Tirana: International Land Coalition Global Assembly 2011.

Ismar, J. (2013) 'How to govern the global rush for land and water?', in Allan, T., Keulertz, M., Sojamo, S., and Warner, J. (eds), *Handbook of Land and Water Grabs in Africa*. London and New York: Routledge, pp 286–298.

Klümper, F., Theesfeld, I., and Herzfeld, T. (2018) 'Discrepancies between paper and practice in policy implementation: Tajikistan's property rights and customary claims to land and water', *Land Use Policy*, vol 75, pp 327–339.

Kruchem, T. (2013) *Land and Water*. Frankfurt am Main: Brandes & Apsel Verlag, 185 p.

Land Matrix (2017) 'Land matrix – get the detail by target areas', http://landmatrix.org/en/get-the-detail/, accessed 31 October, 2018.

Margulis, M.E., McKoen, N., and Borras, S.M., Jr. (2013) 'Land grabbing and global governance: critical perspectives', *Globalizations*, vol 10, pp 1–23.

Mehta, L., Veldwisch, G.J., and Franco, J. (2012) 'Introduction to the special issue: water grabbing? Focus on the (re)appropriation of finite water resources', *Water Alternatives*, vol 5, pp 193–207.

Meinzen-Dick, R. (2014) 'Property rights and sustainable irrigation: a developing country perspective', *Agricultural Water Management*, vol 145, pp 23–31.

Nolte, K. and Voget-Kleschin, L. (2014) 'Consultation in large-scale land acquisitions: an evaluation of three cases in Mali', *World Development*, vol 64, pp 654–668.

Ostrom, E. and Schlager, E. (1992) 'Property-rights regimes and natural resources: a conceptual analysis', *Land Economics*, vol 68, pp 249–262.

Oya, C. (2013) 'Methodological reflections on "land grab" databases and the "land grab" literature "rush"', *The Journal of Peasant Studies*, vol 40, pp 503–520.

Rahmato, D. (2014) 'Large-scale land investments revisited', in Rahmato, D., Ayenew, M., Kefale, A. and Habermann, B. (eds), *Reflections on Development in Ethiopia: New Trends, Sustainability and Challenges*. Addis Adaba: Forum for Social Studies and Friedrich Ebert Stiftung, pp 219–243.

Rulli, M.C., Saviori, A., and D'Odorico, P. (2013) 'Global land and water grabbing', *PNAS*, vol 110, pp 892–897.

Scoones, I., Hall, R., Borras, S.M., White, B., and Wolford, W. (2013) 'The politics of evidence: methodologies for understanding the global land rush', *The Journal of Peasant Studies*, vol 40, pp 469–483.

Skinner, J. and Cotula, L. (2011) *Are Land Deals Driving "Water Grabs"?* London: IIED Briefing.

Smaller, C. and Mann, H. (2009) *A Thirst for Distant Lands: Foreign Investment in Agricultural Land and Water, Foreign Investment for Sustainable Development Program*. Winnipeg: International Institute for Sustainable Development (IISD).

Teklemariam, D., Azadi, H., Nyssen, J., Haile, M., and Witlox, F. (2015) 'Transnational land deals: towards an inclusive land governance framework', *Land Use Policy*, vol 42, pp 781–789.

Teklemariam, D., Nyssen, J., Azadi, H., Haile, M., Lanckriet, S., Taheri, F., and Witlox, F. (2017) 'Commercial land deals and the interactions between investors and local people: evidence from western Ethiopia', *Land Use Policy*, vol 63, pp 312–323.

Theesfeld, I. (2010) 'Institutional challenges for national groundwater governance: policies and issues' *Ground Water*, vol 48, pp 131–142.

Theesfeld, I. (2018) 'From land to water grabbing: a property rights perspective on linked natural resources', *Ecological Economics*, vol 154, pp 62–70.

Theesfeld, I. and Klümper, F. (2016) 'Interplay between structural change in Central Asian agriculture and institutional scarcity of land and water: evidence from Tajikistan', *The German Journal of Economic Geography*, vol 60, pp 81–96.

Tramel, S. (2016) 'The road through Paris: climate change, carbon, and the political dynamics of convergence', *Globalizations*, vol 13, pp 960–969.

Visser, O. and Spoor, M. (2011) 'Land grabbing in post-Soviet Eurasia: the world's largest agricultural land reserves at stake', *Journal of Peasant Studies*, vol 38, pp 299–323.

Watergrabbing.net (2017) www.watergrabbing.net, accessed 31 October, 2018.

Woodhouse, P. (2012) 'Foreign agricultural land acquisition and the visibility of water resource impacts in Sub-Saharan Africa', *Water Alternatives*, vol 5, pp 208–222.

Zetland, D. and Möller-Gulland, J. (2013) 'The political economy of land and water grabs', in Allan, T., Keulertz, M., Sojamo, S., and Warner, J. (eds), *Handbook of Land and Water Grabs in Africa* (257–272). London and New York: Routledge.

Zoomers, A. (2010) 'Globalisation and the foreignisation of space: seven processes driving the current global land grab', *Journal of Peasant Studies*, vol 37, pp 429–447.

23 International investment agreements and mega-regionals

Promoting or undermining the right to water?

Rodrigo Polanco Lazo and Azernoosh Bazrafkan

Introduction

In recent years, there has been a definite movement headed for the recognition of a human right to water, beyond the mere cognizance of a human need, with emphasis in mechanisms to implement it as an effective right. But the capacity of States to provide water supply and sanitation depends largely on their ability to establish, operate and maintain water distribution services and related infrastructure (Boisson de Chazournes, 2015, pp. 80, 149). For that purpose, it is vital to secure sources of investment in this sector.

The last three decades have experienced an increase in public–private partnerships (PPPs) between States (and State agencies) and investors to pursue development projects, as one of the mechanisms used to address the lack of financing for utilities projects, particularly in the water and sanitation sector (United Nations Committee on Economic, Social and Cultural Rights (CESCR), 2017, para. 25). The PPPs in water supply and sewage services are characterized by the use of concession contracts, and the privatization of water services.

The liberalization of drinking water and sewage sectors pose questions concerning the relationship between States and private operators in charge of the distribution of this essential resource (Boisson de Chazournes, 2015, p. 78), especially with regard to safeguarding equal access to drinking water and wastewater services to citizens and residents including the vulnerable and marginalized groups (Winkler, 2012, pp. 108–111). At the same time, private investors involved in these projects have a protection regime in the form of international investment agreements (IIAs) or investor-State contracts (Boisson de Chazournes, 2015, p. 96).

This chapter examines how investment law interacts with the human right to safe drinking water and sanitation in IIAs, comprising both bilateral investment treaties (BITs) as well as preferential trade agreements (PTAs) with investment provisions. This analysis includes recently negotiated mega-regional agreements such as the Comprehensive Trade and Economic Agreement between Canada and the European Union (CETA) and the Comprehensive and Progressive Agreement for Trans-Pacific Partnership (CPTPP).[1]

The first section studies in what way, if any, investment protection standards as included in IIAs affect the right to water. The second part examines whether foreign investor's legitimate expectations are affected by a host state's measure to regulate the supply and maintenance of water on public policy objectives, including by implementing specific human rights obligations. The final section considers whether the human right to water can be used as a defence in investor-State dispute settlement (ISDS) by the respondent state. Special attention will be paid to the recent award by an arbitral tribunal under the rules of the International Centre for the Settlement of Disputes (ICSID), namely the *Urbaser v Argentina*.[2]

Relationship between international investment law and the right to water

While some legal research conceptualizes the relationship between international investment law and human rights as opposing fields of law with colliding policy interest as well as contradictory rules and regulations (Dupuy, Francioni and Petersmann, 2009), other argues the opposite, recognizing a human rights dimension of international investment law (Radi, 2013). In a fragmented international legal order, international commitments to protect foreign investment can potentially interfere with the duty of the States to fulfil its obligations under human rights treaties. Moreover, it is claimed that international investment law may trigger a 'regulatory chill' for national legislation (Tienhaara, 2011), particularly with respect to human rights law (Klein, 2012).

Investment and human rights law share some features, notably, the fact that they are asymmetrical by nature, as both almost exclusively grant individuals rights and protection from State interference while referring virtually all treaty obligations to the State, either the host State in the case of investment treaties or the own state, in the case of human rights treaties.[3]

Nevertheless, investment and human rights law also have important differences. Few human rights treaties provide for direct access to international tribunals (like the European and American human rights regional systems) and when they do, they usually require as a precondition the exhaustion of local remedies in the host state domestic courts.[4] In contrast, in the majority of the existing investment treaties, the investor often has immediate access to investor-State arbitration outside national courts of the receiving state, and without being required to exhaust local remedies (Van Aaken, 2013, p. 291), something that can be qualified as an 'anomaly' in the context of international law (Reiner and Schreuer, 2009, p. 37). The mega-regional treaties do not diverge on this issue. Neither the CETA nor the CPTPP require the investor to exhaust local remedies first.

While international investment law is one of the most prolific fields of international law, with a total number of IIAs of around 3,000 agreements after the first BIT between Germany and Pakistan was concluded in 1959 (UNCTAD, 2018a), the recognition of the human right to water has been a slow and progressive development. The starting point are Articles 11 and 12 of the International Covenant on Economic, Social and Cultural Rights (ICESCR), which includes a number of rights emanating from, and indispensable for, physical and mental health, as well as the realization of the right to an adequate standard of living 'including adequate food, clothing and housing'.[5]

Based on the abovementioned ICESCR provisions, as well as referencing the Convention on the Elimination of All Forms of Discrimination Against Women (CEDAW)[6] and the Convention on the Rights of the Child,[7] the UN Committee on Economic, Social and Cultural Rights adopted in November 2002, the General Comment No. 15 on the right to water, which states that: 'The human right to water is indispensable for leading a life in human dignity. It is a prerequisite for the realization of other human rights'. Comment No. 15 also declared that the human right to water encompasses the right of access to water 'in the amount and quality sufficient to meet vital human needs' such as 'drinking, the production of food, and sanitation' (United Nations, Office for the High Commissioner for Human Rights, 2002).[8]

The United Nations General Assembly only explicitly recognized the human right to water and sanitation on 28 July, 2010, through Resolution 64/292, and acknowledged that clean drinking water and sanitation are essential to the realization of all human rights. The Resolution calls upon states and international organizations 'to provide financial resources, capacity-building and technology transfer, through international assistance and cooperation, in particular to developing countries, in order to scale up efforts to provide safe, clean, accessible and affordable drinking water and sanitation for all'.[9]

Beyond human rights instruments, the recognition of the right to water is also found in conventions relating to the law of international watercourses. The 1999 Protocol on Water and Health to the 1992 Convention on the Protection and Use of Transboundary Watercourses and International Lakes; the 2002 Charter of Water of the Senegal River; the 2008 Water Charter of the River Niger Basin; and the 2012 Charter of Water of Lake Chad, also contain this right (Boisson de Chazournes, 2015, p. 151).

With the progressive recognition of the human right to water, the role of governments in managing water services and the scope of their regulatory autonomy has changed. The role of the state is now twofold: first, to meet its obligations to protect and promote human rights within its territory, including the right to water, water facilities and services affordable for all on a non-discriminatory basis (Meshel, 2015, p. 285); and second, to attract foreign investments to achieve its social and economic development needs, including the necessary infrastructure to establish, operate and maintain water distribution services.(Chaisse, 2017, p. 8)

For this reason, some scholars have cautioned that IIAs need to strike a proper balance between the human rights to water and the financing needs to implement it. This balancing between public and private interests, rights and obligations can be achieved through key provisions such as indirect expropriation, fair and equitable treatment and general exception clauses (Chaisse, 2017, p. 10). However, in practice there has been a limited relationship between the human right to water and international investment law.

In general, human rights issues have been relatively slow to emerge in international investment law. Indeed, no explicit reference to human rights is found in most Model BITs,[10] with the notable exception of the Norwegian Model BIT draft of 2007,[11] which expressly included the protection of human rights as part of provisions (Jacob, 2010, p. 10). In the mapping of 1959 available IIAs, UNCTAD has found only 188 that make reference of 'social investment aspects', in their preamble, a notion that not only includes human rights, but also labour, health, and poverty reduction (UNCTAD, 2018b).[12] Some recent exceptions to this silent treatment are eight Cooperation and Facilitation Investment Agreements (CFIAs), concluded by Brazil in 2015 with Angola, Mozambique, Malawi, Chile, Colombia and Mexico, and with Suriname, Guyana and Ethiopia in 2018; the Brazil-Chile FTA (2018); the Brazil-Peru Economic and Trade Expansion Agreement (2016); the Pacific Alliance Protocol (2014); and the Colombia-Costa Rica FTA (2013), which have expressly included the protection of human rights as part of the provisions on corporate social responsibility (CSR).[13] IIAs in general, devote almost all their provisions to defining the rights and protections afforded to foreign investors, with little or none provisions dedicated to human rights, or to the enforcement of CSR obligations – in the few treaties that include them, or provisions dealing with conflicting rights between foreign investors and locals (Abate Yimer, 2013, pp. 330–331).

Foreign investor's participation in providing these essential services has triggered a number of investment claims based on IIAs in the water sector that have led to final awards, mainly related to services concession disputes and significant tariff increases (Choukroune, 2017, pp. 196–197). However, reference to human rights in ISDS awards has been occasional, generally addressed in broader terms and limited to provide guidance to the arbitrator in order to interpret substantive protections established in favour of foreign investors (Peterson, 2009, p. 4). In the following section we will briefly examine those cases and analyse the relationship (or lack thereof) between investment protection and human right to water.

Early Investment Disputes and the Right to Water

In several instances, arbitral tribunals have decided investment disputes relating to water services for alleged breaches of IIAs by the host states. These disputes illustrate how international investment law can interfere with the management of water resources, especially through concession contracts. Most of the cases discussed below concern Argentina's economic crisis in 2000, which triggered a number of investor-state arbitrations, some particularly related to the supply of water and sewage services.

Compañía de Aguas del Aconquija and Vivendi v Argentina *(1996)*

In 1995, a French company and its Argentine affiliate entered into a concession contract in the Province of Tucuman (Argentina) to provide water and sewage services. In December 1996, both companies initiated investor-state arbitration under the Argentina-France BIT (1991). According to the claimants, shortly after the takeover of the project and continuing to the end of its concession performance, the provincial government breached the provisions of the BIT by repeated political and regulatory harassment, including imposing a unilateral tariff reduction and encouraging customers not to pay their bills in order to pressure the investor to renegotiate the tariffs of the concession.[14]

Although an initial award, dated 21 November, 2000, dismissed all the claims filed by the claimants against Argentina, that decision was annulled on 3 July, 2002. A second award, dated 20 August, 2007, awarded compensation in favour of the claimants, declaring that Argentina acted in breach of both fair and equitable treatment and full protection and security of the claimants' investments.

In the annulment application of the award, Argentina argued that the tribunal had disregarded fundamental issues related to the dispute between the parties, including that the dispute between the parties related to the right to water as an essential human right.[15] The annulment committee, however, failed to address the issue, and in 2010 rejected Argentina's annulment application.[16]

Zhinvali Development Project (2000)

This case was initiated in January 2000 by an Irish company (Zhinvali Development Ltd), based on the 1996 Georgian Investment Law which provides a general offer of ICSID arbitration. The claimant brought the dispute based on pre-investment expenditures incurred regarding the proposed rehabilitation of a hydroelectric power plant and its tailrace tunnel located near Tbilisi. After three years of negotiation between Zhinvali and the Georgian Government, the claimant was ultimately excluded from the project. Zhinvali initiated an ICSID arbitration to reclaim lost profits on the abandoned project, as well as expenses incurred during the negotiations (like costs of feasibility studies, consultancy, travel expenses and legal fees) (Lapiashvili, 2014).

In January 2003, the arbitral tribunal decided on a split decision with the dissenting vote of Zhinvali' s appointed arbitrator, that development costs did not qualify as an investment (neither under the 1996 Georgia Investment Law nor Article 25(1) of the ICSID Convention).[17] The award was neither published nor challenged (Boisson de Chazournes, 2015, p. 220).

The tribunal declined jurisdiction to decide the case, concluding that there was no agreement between parties to consider development costs as investments and that under the applicable law to the dispute (Georgian law), development costs could not be regarded as such, due to a lack of territorial presence. The tribunal also rejected claims of reimbursement of development cost based on Georgia's Civil Code, and that the draft concession and the content different documents shared with Georgia created an intellectual value or right within the meaning of Georgia's Investment Law (Ben Hamida, 2005, pp. 67–70).

Azurix v Argentina *(2001)*

In 1996, the Province of Buenos Aires (Argentina) initiated the privatization for the distribution of drinking water and sewage services. Azurix – a company headquartered in the US – won the tender for a 30-year concession through its Argentinean subsidiary (Azurix Buenos Aires – 'ABA'). A new authority was established to oversee and regulate the Concession (Organismo Regulador de Aguas Bonaerense – 'ORAB').

While ABA made a 'canon payment' for its tender bid (of 438.5 million Argentine pesos), the Province of Buenos Aires, in turn agreed to repair pre-existing problems with the filtering plants before Azurix took over the concession in July 1999. During 1999–2001, ORAB interfered with ABA's regime to charge customers for the water services, precluding the company from increasing its tariffs in excess of those billed prior to the granting of the concession, which the company consider as a denial that the canon was recoverable through tariffs. After levels of algae increased in the water in April 2000, ABA claimed the Province's failure to complete its infrastructure repair works was the main cause of the algae outbreak. In turn, the provincial politicians blamed the foreign investor and encouraged consumers not to pay their water bills. In 2001, Azurix terminated the concession contract with the Province of Buenos Aires, claiming that political concerns were always privileged over the financial integrity of the concession, and sought investor-State arbitration, under the Argentina-US BIT (1991).[18]

Argentina argued that commitments under human rights treaties to protect its citizens' right to water conflicted with the obligations under the BIT, and that these human rights obligations prevailed over the private interests of service providers. Azurix responded that the provisions in the concession contract protected the user's rights. In 2006, the arbitral tribunal concluded that Argentina had breached its fair and equitable treatment duty to the foreign investor noting that a measure, even if legitimate and serving a public purpose, could still give rise to a compensation claim.[19] The award pointed out that after the termination of the concession, ABA was replaced with a domestic service provider ('Aguas Bonaerenses'), which was allowed to raise tariffs (Boisson de Chazournes, 2015, p. 99).

Aguas del Tunari v Bolivia *(2002)*

In 1999, the Bolivian government granted a 40-year agreement for the exclusive provision of water services of the third largest city in the country (Cochabamba) to a private company ('Aguas del Tunari'), a subsidiary of Bechtel Enterprises of California. Some weeks later, the company raised the water rates by an average of over 50% and the local community started massive public protests, claiming inability to pay the invoices. In April 2000, in the midst of violence and riots that led to the declaration of a 'state of emergency', the Government deployed soldiers and police. In this so-called 'Water War', after more than 100 people were injured and a 17-year-old was killed, Aguas del Tunari executives left the country in April 2000 and abandoned the concession that was then rescinded by the Government (Polanco Lazo, 2014, p. 12).

Bechtel and its co-investor, the Spanish company Abengoa, filed a claim under ICSID Arbitration Rules, in February 2002, under Bolivia-Netherlands BIT, after 'migrating' their matrix from the Bahamas to Luxembourg, whose shares were in turn owned by a new company established in the Netherlands. In July 2002, an ICSID tribunal was constituted, which in October 2005 affirmed its jurisdiction over the dispute. Facing strong domestic and international pressure, Bechtel reached an agreement with Bolivia in January 19, 2006, accepting a token compensation for damages of two Bolivians – equivalent to 25 cents (de Gramont, 2006).

Given the significant weight of public interests in this dispute directly related to the human right to water, some NGOs requested permission to participate as amici curiae. The petitioners also demanded public disclosure of the parties' submissions, and the opening of the hearings – including a visit of the Tribunal to Bolivia to conduct public hearings. Although in January 2003 the ICSID Tribunal dismissed all these petitions, subsequent tribunals as well as the 2006 amendments to the ICSIS Arbitration Rules allowed amicus briefs in investor-state arbitrations (Boisson de Chazournes, 2015, p. 222).

Suez, Sociedad General de Aguas de Barcelona and Vivendi v Argentina *(2003)*

In April 1993, an Argentine company ('Aguas Argentinas S.A.' – AASA), established by a consortium of Argentinean and foreign investors (from France and Spain) entered into a 30-year concession contract with the Argentine government for water distribution and waste water treatment in Buenos Aires. By the year 2000, Argentina began to experience important economic difficulties that would lead to a financial crisis with serious consequences for the country, its people, and investors. As Argentina's financial crisis deepened, disagreement arose between the investor and the government over the freezing of water prices charged to consumers. In early 2001, after invoking emergency measures, the Argentine government refused to revise the water tariffs charged by the claimants. In 2003, the claimants filed a request for investor-state arbitration alleging that Argentina's action violated Argentina's obligations under both Argentina-France BIT (1991) and Argentina-Spain BIT (1991).[20]

The arbitral tribunal recognized that the protection and promotion of foreign investment was not the only purpose of the relevant investment treaty. The contracting parties pursue the broader goals of heightened economic cooperation also with a view toward achieving increased economic prosperity or development.[21] The tribunal recognized that the provision of water and sewage services certainly was vital to the health and well-being of the inhabitants and was therefore an essential interest of the Argentine state. However, it was not convinced that the only way that Argentina could satisfy that essential interest was by adopting measures that would subsequently violate the BITs rights of the claimant.[22] The tribunal considered Argentina's obligation to safeguard the human right to water as a necessity defence. However, it rejected such defence on the ground that the infringing measures were not the only way to safeguard the right to water. The tribunal instead found that Argentina was subject to both international obligations, i.e. human rights and treaty obligations, and must respect both of them equally. Under these circumstances of the cases, the award declared that Argentina's human rights obligations and its investment treaty obligations were not inconsistent, contradictory, or mutually exclusive.[23]

SAUR v Argentina *(2003)*

SAURI, a French limited liability company specialized in the production, treatment, distribution and sanitation of water, invested through a wholly-owned subsidiary (Société d'Aménagement Urbain et Rural, S.A. – SAUR), in the privatization of Obras Sanitarias Mendoza (OSM), in the Province of Mendoza (Argentina). OSM was a provincially owned water company holder of a concession for the supply of drinking water, sanitation and sewage services in the Mendoza Province. SAURI took part in a call for tenders as a member of an international consortium, where it was appointed technical operator. After winning the tender, SAURI incorporated a company under Argentine law, wholly controlled and named Aguas de Mendoza S.A., (AdM) which acquired the participation in OSM in June 1998.[24]

As mentioned, in the wake of the 2002 financial crisis, the Argentinean government enacted emergency measures to freeze all water prices charged to consumers. SAUR and the federal government reached an agreement to implement service tariff increases in the years following the crisis, because the peso's sustained devaluation made revenues insufficient to cover the costs. However, citing deficiencies by the French investor and its local concessionaire, like breaches of drinking water quality, in the provision of sewage services and of basic levels of consumer services, the provincial authorities declined to implement these increases and eventually nationalized the entity (Boisson de Chazournes, 2015, p. 222). In 2003, SAUR proceeded with a request for investor-State arbitration under the Argentina–France BIT.

In this case, the arbitral tribunal explicitly acknowledged that human rights in general, and the right to water in particular, are one of several sources that should be taken into consideration to settle the dispute, as these rights are embedded in the Argentine legal system and also formed part of the general principles of international law. The arbitral tribunal recognized that access to clean water is, from the standpoint of the state, a public service, and from the perspective of the citizen, a fundamental right. Therefore, the law can and should reserve for the public authority legitimate function of planning, supervision, police and punishment for the protection of the public interest in this matter.[25] However, the arbitral tribunal concluded that the fundamental right to water and the right of an investor to protection of its investment operate on different planes. Thus, the protection of the right to water must be combined with respect for the rights and guarantees granted to the foreign investor in virtue of the investment treaty.[26]

Bayview v Mexico *(2005)*

This case was brought up by several Texas Irrigation Districts, individual water users, and corporate investors (including a water supply company), which held water rights under a 1944 US-Mexico water treaty regulating the joint management and allocation by the US and Mexico of the waters of the lower Rio Grande River and six of its tributaries. This water was ultimately destined for business operations across the border in the US.[27]

After Mexican authorities diverted irrigation waters from this source for its domestic farmers over the course of a decade, in 2005 the claimants initiated investor-State arbitration against Mexico, under the North American Free Trade Agreement (NAFTA) Chapter 11. Due to Mexico's failure to supply water to that region in Texas, they claimed compensation, alleging have suffered irrigation water shortages since the mid-1990s, with a $1 billion in lost business and 30,000 lost jobs.[28]

The arbitral tribunal denied jurisdiction over the claim, holding that the Claimants were not 'foreign investors', but rather, they were domestic investors in Texas. 'The economic dependence of an enterprise upon supplies of goods – in this case, water – from another State is not sufficient to make the dependent enterprise an "investor" in that other State'.[29] For that reason, it concluded 'a foreign holder of a right to take host-State waters does not own or possess rights in a particular volume of water as it descends through those streams' (Boisson de Chazournes, 2015, p. 97).

Biwater v Tanzania *(2005)*

Another investor-state arbitration in connection with the human right to water was brought by Biwater Gauff (Tanzania), a company incorporated under the laws of England and Wales, against Tanzania. The dispute arose out of contractual disputes between Dar es Salaam Water and Sewerage Authority (DAWASA) and the local subsidiary of Biwater Gauff (City Water), to implement a comprehensive program of repairs and upgrades to, and the expansion of, the Dar es Salaam Water and Sewerage Infrastructure. Under a lease contract, City Water agreed to provide water and sewerage services on behalf of DAWASA since August 2003.[30]

The case was brought under the Tanzania – United Kingdom BIT (1994), based on a series of events that took place in 2005 (i.e. deportation of the investor's senior management, the termination of the lease contract, the seizure of the company's assets and takeover of its business). According to the claimant, these events constituted breaches by Tanzania of its obligations to grant fair and equitable treatment, not to take unreasonable and discriminatory measures, to grant full protection and security to investors and to guarantee the unrestricted transfer of funds.[31]

In *Biwater*, although Tanzania did not invoke a human right to water defence directly, it argued that the investor had created a real threat to public health and welfare and that considering the importance of the issue at hand, it acted well within the Republic's margin of appreciation under international law.[32] The tribunal addressed Tanzania's human rights defence as such but found Tanzania's occupation of the investor's facilities and the usurpation of management control was unreasonable and arbitrary, unjustified by any public purpose.[33] It also found that there was no necessity or impending public purpose to justify its deportation of the investor's executive staff.[34]

Impregilo v Argentina *(2007)*

In 1999, Aguas del Gran Buenos Aires ('AGBA'), an Argentine company in which the Italian company Impregilo had a dominating interest, entered into a concession contract with the Province of Buenos Aires to operate water and sewage services. The company took possession of the concession in January 2000. In 2001, AGBA had difficulties collecting payment from its customers in the midst of a severe economic crisis that affected Argentina that year. The Argentinian government undertook a series of measures facing that crisis, including a freezing of tariffs and a new regulatory framework for water services, suspending the right to cut off water services to customers that had unpaid bills. Renegotiation of the contract took four years but failed and the provincial authorities terminated the contract in 2006. The company initiated investor-state arbitration under the Argentina-Italy BIT (1990) in 2007.[35]

Like in *Azurix*, Argentina argued that the regulatory powers of the state were particularly important in order to guarantee its inhabitants the human right to water. It submitted that the commitments of the Argentine Republic regarding investments do not prevail over the obligations found in human rights treaties. In the case at hand, treaties providing for the human right to water must be especially taken into account in construing obligations arising from the BIT in accordance with the rules on protection of human rights.[36]

The arbitral tribunal declared that Argentina did not violate the Argentina–Italy BIT by expropriating or nationalizing AGBA's investment but did violate the fair and equitable standard recognized in the same treaty. In determining whether Argentina had breached its IIA obligations, the tribunal did not address the human rights arguments presented by Argentina.[37]

AbitibiBowater v Canada *(2010)*

This dispute was initiated in 2010 under NAFTA Chapter 11, by a US-registered firm (AbitibiBowater). The company claimed that after a legislation passed by Newfoundland and Labrador expropriating its water and timber rights (pulp mill) and hydroelectric assets in the province, Canada, did not provide fair value market compensation, also falling below the fair and equitable treatment and full protection and security standards.[38]

AbitibiBowater and provincial authorities differed as to whether the claimant had the right to sell its local water use permits, or whether such permits must revert directly to the local government. Canada settled the case early for CAN

130 million in December 2010.[39] Canada's decision to settle this case raised concerns in Canadian civil society, by groups who see market access rights intruding on a sovereign resource, as Canadian constitutional law affirms the 'public trust' doctrine for water resources, forbidding foreign investors to sell water use permits that authorities have granted them (Boisson de Chazournes, 2015, p. 98).

The disputes referred in this section illustrate that the private sector participation in water services provision is closely linked to the affordability of water services. They also show that a mere price increase does not give rise to a violation of the human right to water. This would be only if water services become unaffordable for parts of the population. It is of course the duty of the state to safeguard its citizen's right to water and to take necessary measures when this right threatens to be violated. Therefore, it is important for states to strengthen their regulatory framework governing water services and to clearly define public interest considerations in their concession agreements (Winkler, 2012, p. 139).

The abovementioned disputes indicate the lack of a holistic approach in regulating water services and access to water. In this regard, the recent award in *Urbaser v Argentina* can be viewed as ground-breaking since it – unlike previous tribunals – explicitly tackled the issue of the human right to water vis-à-vis investment protection, while deciding a counterclaim by Argentina.

The latest investment dispute on the right to water: Urbaser et al. v Argentina

In early 2000, the Province of Greater Buenos Aires granted a concession for water and sewage services in the province of Buenos Aires to Aguas del Gran Buenos Aires, S.A. (AGBA), a company established by foreign shareholders, including the Spanish corporations Urbaser and Consorcio de Aguas Bilbao Bizkaia.

But the concession soon faced several problems in the context of the financial crisis that hit Argentina in 2001, and that country ultimately terminated Urbaser' s concession in July 2006. In July 2007, Urbaser sought investor-state arbitration proceedings for the violation of the Argentina-Spain BIT, claiming that the emergency measures adopted by Argentina had caused Urbaser financial loss and insolvency. Argentina in turn, filed a counterclaim alleging that the concessionaire's failure to provide the necessary level of investment in the concession led to violations of the human right to water (Attanasio and Sainati, 2017, pp. 744–745).

Regarding the claim brought by Urbaser, the arbitral tribunal held that the fair and equitable treatment requirement in this BIT protects the investor's legitimate expectations. The arbitral tribunal held that the fair and equitable treatment requirement in the concerned investment treaty protects the investor's legitimate expectations, which are informed by the host state's constitutional and international legal obligations, as well as the social and economic environment at the time the investment was made.[40] Thus, any protected expectation must be legitimate in light of the regulatory and legal context of the investment, including human rights commitments.[41] The tribunal concluded that human rights obligations need not to be pretext to justify the state's actions but instead are a general and stable part of the investment environment, at least where the investor knows of the state's obligations. In this case, the tribunal found that Urbaser was aware of Argentina's obligations regarding the right to water. In that context, the protection of the human right to water constitutes part of the framework within which the claimant should frame its expectations.[42]

The arbitral tribunal attributed the concession's problems to sources other than Argentina[43] and concluded that there existed a situation of necessity as sufficient support for Argentina's emergency measures.[44] In the end, while the tribunal rejected most of the investor's claims,[45] it found Argentina's lack of transparency in renegotiating the concession agreement violated the Argentina–Spain BIT. However, this breach produced no damages,[46] and therefore no compensation was awarded (Attanasio and Sainati, 2017, p. 745).

Regarding Argentina's counterclaim, the tribunal first determined that it had jurisdiction over it, something that is not usual on investor-State arbitration.[47] The arbitral tribunal then continued and observed that international law accepts corporate social responsibility as a standard for companies operating in international commerce. In the arbitral tribunal's view, this standard also includes commitments to comply with human rights, rejecting the idea advanced by the claimant that 'guaranteeing the human right to water is a duty that may be born solely by the State, and never borne also by private companies'.[48] Thus, it can no longer be admitted that companies operating internationally are immune from becoming subjects of international law.[49] The tribunal referred to the Universal Declaration of Human Rights (UDHR) and the ICESCR and acknowledged the establishment of the right to water in both international instruments,[50] further declaring that the investment treaty concerned 'has to be construed in harmony with other rules of international law of which it forms part, including those relating to human rights'.[51]

However, in the end, the tribunal rejected Argentina's counterclaims on the grounds that they were directed at claimant's supposed failure to perform expansion work for its water concession.[52] Yet, the award found that none of the provisions of the Argentina–Spain BIT 'has the effect of extending or transferring to the concessionaire an obligation to perform services complying with the residents' human right to access to water and sewage services'.[53]

Urbaser v Argentina breaks with the perspective that international investment law and human rights are two separate areas of international law. In this case, the tribunal made clear that human rights could play a role in an investment dispute as part of the host State's counterclaims or defences against the investor (Farrugia, 2015, p. 21).

The novelty of the *Urbaser* award is that for the first time in international investment law, human rights considerations are integrated into the analysis of investment protections. The decision also departs from the previous investment disputes for its recognition that investors may have human rights obligations under international law. Consequently, the balancing exercise reflects the rights and obligations of the foreign investor and the host state rather than viewing those rights as conflicting or competing interests.

Conclusion: balancing investor's rights and the human right to water

As pointed out by Calamita, investment disputes regarding water supply can be divided into three categories (Calamita, 2017, p. 58). The first category encompasses cases involving the inability of governments and contracts to address external shocks. The economic crisis experienced by Argentina exposed the fragility of the concession contracts entered into by a number of Argentina provinces with foreign concessionaires, both commercially and politically. Particularly in *Impregilo*, *Urbaser* and *Suez* disputes, the economic crisis and Argentina's enactment of emergency measures placed significant pressure on the revenues of the investment. Argentina's liability was triggered because – in the view of the tribunals – renegotiations of the concession contracts, which protected both the provinces' interests, and those of the concessionaires were possible (Calamita, 2017, pp. 46–47).

The second category of cases involves 'regulatory opportunism' which can be found in the *Compañía de Aguas del Aconquija and Vivendi*, *Azurix* and *SAUR*, follow a change in policy and politics regarding water privatization. In *Vivendi*, the arbitral tribunal found that it is only possible to conclude that the local government, improperly and without justification, mounted an illegitimate campaign against the concession, and the foreign concessionaire from the moment it took office, aimed at either reversing the privatization or forcing the concessionaire to renegotiate and lower tariffs.[54] This was executed through the agency of the regulator by imposing excessive charges for water quality control breaches and through the dismissal of regulator's members.[55] In *Azurix*, the tribunal noted a repeated pattern of the provincial regulator giving in to political pressure from the governor in its decision-making and the governor threatening to remove regulator's members.[56] In *SAUR*, the tribunal found the pattern as revealing when the management of the regulatory agency seemed to be taking its political direction from the governor, particularly the timing and delays of its decisions.[57]

The third category involves 'contractual misadministration'. In *Biwater Gauff*, Tanzania had made the effort of establishing a detailed concession framework, which included mechanisms allowing termination for poor performance. Unfortunately, the Tanzanian authorities were unwilling and/or unable to restrict their decision-making to that legal framework with respect to the termination of the investor's concession. The Tanzanian government instead chose to take a number of politically opportunistic actions such as the occupation and seizure of City Water's facilities and the forcible deportation of management staff.[58] Further, it does not appear from the dispute that this failure was a result of a lack of capacity by the Tanzanian authorities. Consequently, if Tanzania officials had followed the normal contractual course of termination, there would have been no basis for an investment treaty claim.

A possible fourth category of investment claims – not yet presented – could be brought up because the investment has been directly or indirectly frustrated by other authorized uses of water at a domestic level (Abate Yimer, 2013, p. 333).

After reviewing the early and recent investment disputes relating to the human right to water, some conclusions can be drawn. Any process of privatization of water services needs to consider two objectives: ensure that non-economic concerns (like providing access to water for the entire population) remain in control of the State, and create a level playing field for the private sector to participate in a liberalized market in competitive terms. Private participation in water services does not imply a withdrawal of public power. It rather changes the intervention in the water services market through regulation (Chaisse, 2017, pp. 4–5).

The main purpose of water concession contracts is precisely to mitigate the tension between protecting the foreign investments of water suppliers and the obligation to adopt the regulations needed to protect the host State's own citizens. The international investment disputes in water services often relate to a sudden increase in the water prices or the quality of the water supply. In this regard, the tribunals often emphasized on the 'legitimate expectations' of the investor at the time the concession agreement was signed and which have become the basis for the investment in the first place.[59] However, they have paid less attention to the legitimate expectation of the host State that the investor would respect its public service obligations (Chaisse, 2017, p. 6). Naturally, any regulatory measure the host state takes in this regard has to be proportionate to the public interest it seeks to protect.[60] Conversely, non-State water service providers must consider the local context where their activities are carried out such as the institutional capacities of the government, as well as the actual and potential impact of their activities in the country and its inhabitants (Albuquerque, 2010, para. 27).

But, how to strike a proper balance? We can detect both *ex ante* (treaty making) and *ex post* balancing mechanisms (adjudication).

The first *ex ante* option to achieve a more balanced approach between the investor and the host State is to include more explicitly the rights of the parties to regulate with respect to the protection of health, safety and the environment, and the promotion of consumer protection and labour rights. This happens in the preamble of some investment treaties making explicit their objective to facilitate sustainable development (Meshel, 2015, p. 301).

Another *ex ante* option is to define more specifically what type of measures constitute a (regulatory) expropriation and what type of conduct could result in a possible breach of the fair and equitable treatment standard. For example, Article 9.6 (4) of the CPTPP states that the mere fact that a State party takes or fails to take action that may be inconsistent with an investor's expectations does not constitute a breach of the fair and equitable treatment provision. Similarly, Article 9.6 (5) CPTPP states that the mere fact that a subsidy or grant has not been issued, renewed or maintained, or has been modified or reduced, by a party, does not constitute a breach of fair and equitable treatment standard, even if there is loss or damage to the covered investment as a result.

A third *ex ante* option is to consider exception clauses in IIAs. One example is to include non-precluded measures clauses that are self-judging. This ensures deference to the State's subjective determination of what 'it considers necessary' when acting to protect its own essential security interests.[61] Nonetheless, the host state still has to demonstrate that it has acted in good faith. These carve-outs or reservations are often modelled after the WTO's General Agreement on Tariffs and Trade (GATT) Article XX that allows the host State to adopt measures relating to the protection of 'human, animal or plant life' or for the 'conservation of natural resources'. Other cross-sectoral general exceptions exist and include sustainable development provisions to preserve key policy priorities such as the right to water and sanitation, public order, environmental matters and public health (Farrugia, 2015, pp. 18–19). For instance, Article 8.9 (1) of the CETA is directly inspired from the GATT Article XX.

A fourth *ex ante* option is recommended by United Nations Conference on Trade and Development (UNCTAD) to engage in international investment policy-making such as including a reference to the UN Guidelines on Business and Human Rights as a way of balancing the parties' commitments and promote responsible investment (UNCTAD, 2015, p. 58). The United Nations Committee on Economic, Social and Cultural Rights also follows this approach (United Nations Committee on Economic, Social and Cultural Rights, 2017, para. 13).

With respect to *ex post* balancing mechanisms, to this date, most tribunals have applied a four-step proportionality analysis (the so-called 'new constitutionalism') to balance presumably conflicting rights, as suggested by Sweet and Mathews (2008). The first step in this analysis is a legitimacy test under which the tribunal needs to determine whether the government (i.e. the local authorities) were constitutionally authorized to take the disputed measure. The second step is the suitability test whereby the tribunal assesses whether the State's policy objectives are rationally related to the measure taken. The third step is the necessity test to examine whether the government has applied the least restrictive means. The last step is the actual balancing whereby the tribunal weighs the benefits of the measure against the costs incurred by the investor (Sweet and Mathews, 2008, pp. 75–76). However, this test might not be useful, if what is questioned is not the legality of a regulatory measure, but its unsatisfactory legal consequences to the disputing parties.

In several investor-State arbitrations related to water services, host States have raised the human right to water as a justification for state actions. However, until *Urbaser*, arbitral tribunals seem to have refrained from explicitly recognizing this right or discussing in any meaningful way its impact on States' investment protection obligations. Rather, when it comes to host States' defences, they seem to treat these two areas of international law as entirely separate (Meshel, 2015, p. 278).

An alternative *ex post* option to be considered in the absence of *ex ante* treaty-making, is that in order to balance investor's legitimate expectations against host State's legitimate and proper public policy objectives, the arbitral tribunal takes into consideration to what extent the investor's conduct has contributed, or exacerbated, a deterioration of the human rights and security situation in the host State. This may then affect the evaluation of that investor's right to damages in an investment tribunal or even the investor's right to avail itself of substantive investment protection at all (Farrugia, 2015, p. 20).[62] In this regard, the CPTPP has taken a progressive step, which allows State's counterclaims on certain issues.[63]

When it comes to the commodification of drinking water and sewage services, investment treaties and investor-state dispute settlement may not adequately provide protection for human rights and host state's public considerations (Fry, 2007, pp. 77, 112). Investment treaties traditionally provide foreign investors with rights and guarantees without any particular responsibilities in return which can be invoked in water related disputes (notably protection against expropriation and the fair and equitable treatment standard). If one recognizes the public nature of fresh water resources and its vital character for life, such disputes evidently involve a third-party interest that expands foreign investment protection arbitration beyond its general focus on investors (Boisson de Chazournes, 2015, pp. 81, 108). Some recent international investment agreements, including mega-regional agreements place more emphasis on human rights as well as on the host States' sustainable economic development objectives. The investor-State arbitration case law, it seems, is also changing in that direction, although it is maybe too soon to dimension the real impact of the *Urbaser* decision in future cases. A further recommendation would be to examine trade and investment treaties on a regular basis on their positive and negative impact on human rights and development through impact assessments (United Nations Committee on Economic, Social and Cultural Rights, 2017, para. 13).

Human rights and investment protection should not be seen as opposing fields. States should be able to respect both in a holistic and complementary way. Host States should not promise investment protection without respect for human rights, including the human right to water, and conversely the protection of this right and other human rights, should not be done disregarding commitments made by the State with respect to investments. Urbaser might be a first step in that direction.

Notes

1 The agreement was renamed after the withdrawal of the United States of the Trans-Pacific Partnership Agreement (TPP) in January 2017 (Polanco Lazo and Gómez Fiedler, 2017).

2 *Urbaser SA et al v Argentina*, ICSID Case No. ARB/07/26, Award, 8 December 2016.

3 On investment treaties, see Gazzini, 2012, p. 107; and on human rights treaties see Parlett, 2011, pp. 278–339.

4 European Convention of Human Rights, Arts. 34 and 35; American Convention on Human Rights, Art. 46; and Rules of Procedure of The African Commission on Human and Peoples' Rights, Rule 93.

5 International Covenant on Economic, Social and Cultural Rights. Adopted and opened for signature, ratification and accession by General Assembly resolution 2200A (XXI) of 16 December, 1966, 1976.

6 Convention on the Elimination of All Forms of Discrimination against Women Adopted and opened for signature, ratification and accession by General Assembly resolution 34/180 of 18 December, 1979, Art. 14.2.

7 Convention on the Rights of the Child, adopted and opened for signature, ratification and accession by General Assembly Resolution 44/25 of 20 November, 1989, Art. 24.2.

8 United Nations, Office for the High Commissioner for Human Rights, 'General Comment No. 15 (2002) The Right to Water (Arts. 11 and 12 of the International Covenant on Economic, Social and Cultural Rights)' (2002).

9 United Nations General Assembly, Resolution adopted on 28 July, 2010 (A/64/L.63/Rev.1 and Add.1) 64/292, the human right to water and sanitation.

10 For this chapter, the Models BITs of the following States were reviewed: Germany (2008), France (2006), China (2003), India (2003), the United Kingdom (2005) and the United States (2012).

11 In 2009, the Norwegian government decided to abandon its Model Draft BIT, following public input that was largely critical (Vis-Dumbar, 2009).

12 CETA reaffirms the importance of democracy and fundamental rights as laid down in the Universal Declaration of Human Rights in its preamble.

13 Other IIAs that include the protection of human rights within their CSR provisions are the Canadian BITs with Benin (2013), Cameroon (2014), Nigeria (2014), Serbia (2014), Senegal (2014), Mali (2014), Côte d'Ivoire (2014), Burkina Faso (2015), and in the investment chapters of the Canadian Free Trade Agreements (FTAs) with Peru (2008), Colombia (2008), Panama (2010), Honduras (2013) and South Korea (2014).

14 *Compañía de Aguas del Aconquija SA and Vivendi Universal* v *Argentina*, ICSID Case No. ARB/97/3, Award, 20 August 2007, para 5.2.4.

15 *Compañía de Aguas del Aconquija SA and Vivendi Universal* v *Argentina*, ICSID Case No. ARB/97/3, Decision on the Argentine Republic's Request for Annulement of the Award renderd on 20 August 2007, 30 July, 2010, paras. 57, 243.

16 Ibid paras. 248, 265.

17 *Zhinvali Development Ltd. v Republic of Georgia* (ICSID Case No. ARB/00/1), Award, 24 January, 2003.

18 *Azurix Corp v Argentina*, ICSID Case No. ARB/01/12, Award, 23 June, 2006, paras 38, 43, and 83.

19 Ibid para. 310.

20 *Suez and Vivendi v Argentina*, ICSID Case No. ARB/03/19, Decision on Liability, 30 July, 2010, paras. 34, 41 254.

21 Ibid para. 218.

22 Ibid para. 260.

23 Ibid para. 262.

24 *SAUR International SA v Argentina*, ICSID Case No. ARB/04/4, Decision on Jurisdiction and Liability, 6 June, 2012, paras. 47, 50, 51, 55, 330, and 413.

25 Ibid paras. 330, 413.

26 Ibid para. 331.

27 *Bayview Irrigation District and others v United Mexican States*, Ontario Superior Court of Justice, Reasons for Judgment (Application for Set-Aside), 5 May, 2008, para. 4.

28 Ibid para. 32.

29 *Bayview Irrigation District and others v United Mexican States* (ICSID Case No. ARB(AF)/05/1), Award, 19 June, 2007, para. 104.
30 *Biwater Gauff v Tanzania*, ICSID Case No. ARB/05/22, Award, 24 July, 2008, paras, 3,4, 5, 6, and 9.
31 Ibid paras. 15, 16 and 17.
32 Ibid para. 436.
33 Ibid para. 503.
34 Ibid para. 515.
35 *Impregilo S.pA v Argentina*, ICSID Case No. ARB/07/17, Award, 17 May, 2011, paras 1, 20, 21, 208 and 223.
36 Ibid paras 228, 230.
37 Ibid paras. 283, 330–331.
38 *AbitibiBowater Inc. v Government of Canada* (ICSID Case No. UNCT/10/1), Notice of Arbitration, 25 February, 2010, para. 99.
39 *AbitibiBowater Inc. v Government of Canada* (ICSID Case No. UNCT/10/1), Consent Award, 15 December, 2010.
40 *Urbaser S.A. et al. v Argentina*, Award, paras. 620–623.
41 Ibid para. 624.
42 Ibid paras. 720–723
43 Ibid paras. 638, 672–682.
44 Ibid para. 718.
45 Ibid paras. 1002, 1106, 1234.
46 Ibid paras. 845–847, 1234.
47 *Urbaser S.A. et al. v Argentina*, Award, paras. 1143–1155.
48 Ibid para. 1193.
49 Ibid para. 1195.
50 Ibid paras. 1196–1197.
51 Ibid para. 1200.
52 Ibid paras. 1211–1212.
53 Ibid para. 1207.
54 *Compañía de Aguas del Aconquija SA and Vivendi Universal v Argentina*, Award, para. 7.4.19.
55 Ibid paras. 7.4.22 and 7.4.26.
56 *Azurix Corp. v Argentina*, Award, paras. 92, 125, 137.
57 *SAUR International SA v Argentina*, Decision on Jurisdiction and Liability, para. 505.
58 *Biwater Gauff v Tanzania*, Award, para. 418.
59 *International Thunderbird Gaming Corporation v Mexico*, UNCITRAL, Award, 26 January, 2006, paras. 147–149, 160.
60 *Tecmed v Mexico*, ICSID Case No. ARB (AF)/00/2, Award, 29 May, 2003, para. 48.
61 Article 29.2 (b) of CPTPP. In this regard, Article 28.6 of CETA that defines what those essential security interests entail is clearer than CPTPP.
62 See also, *Urbaser S.A. et al. v Argentina*, Award, paras. 1194–1195.
63 Article 9.19 (2) of CPTPP.

References

Abate Yimer, B. (2013) 'Competing water use in large-scale commercial farms: Ethiopia', in Boisson de Chazournes, L., Leb, C., and Tignino, M. (eds), *International Law and*

Freshwater: The Multiple Challenges. Int edition. Cheltenham, UK; Northampton, MA: Edward Elgar Pub, pp. 319–337.

Albuquerque, C. de (2010) *Report of the Independent Expert on the Issue of Human Rights Obligations related to Access to Safe Drinking Water and Sanitation*. A/HRC/15/31. United Nations, p. 21.

Attanasio, D. and Sainati, T. (2017) 'Urbaser S.A. and Consorcio de Aguas Bilbao Bizkaia, Bilbao Biskaia Ur Partzuergoa v. The Argentine Republic ICSID', *American Journal of International Law*, vol 111, no 3, pp. 744–750. doi: 10.1017/ajil.2017.56.

Ben Hamida, W. (2005) 'The Mihaly v. Sri Lanka case: some thoughts relating to the status of pre-investment expenditures', in Weiler, T. (ed.), *International Investment Law and Arbitration: Leading Cases from the ICSID, NAFTA, Bilateral Treaties and Customary International Law*. London: Cameron May, pp. 47–76.

Boisson de Chazournes, L. (2015) *Fresh Water in International Law*. Reprint edition. Oxford: Oxford University Press.

Calamita, N.J. (2017) 'Are investments in water different? Sectoral economics, investment treaty architecture, and the role of governance', in Chaisse, J. (ed.), *The Regulation of the Global Water Services Market*. Cambridge: Cambridge University Press, pp. 27–69. doi: 10.1017/9781316678442.003.

Chaisse, J. (2017) 'Introduction', in Chaisse, J. (ed.), *The Regulation of the Global Water Services Market*. Cambridge: Cambridge University Press, pp. 1–24. doi: 10.1017/9781316678442.002.

Choukroune, L. (2017) 'Water and sanitation services in international trade and investment law: for a holistic human rights-based approach', in Chaisse, J. (ed.), *The Regulation of the Global Water Services Market*. Cambridge: Cambridge University Press, pp. 196–219. doi: 10.1017/9781316678442.009.

Dupuy, P.-M., Francioni, F., and Petersmann, E.-U. (eds) (2009) *Human Rights in International Investment Law and Arbitration*. Oxford: Oxford University Press (International economic law series).

Farrugia, B. (2015) 'The human right to water: defences to investment treaty violations', *Arbitration International*, vol 31, no 2, pp. 261–282.

Fry, J.D. (2007) 'International human rights law in investment arbitration: Evidence of international law's unity', *Duke Journal of Comparative & International Law*, 18, p. 77.

Gazzini, T. (2012) 'Bilateral investment treaties', in Gazzini, T. and de Brabandere, E. (eds), *International Investment Law. The Sources of Rights and Obligations*. Leiden: Martinus Nijhoff Publishers, pp. 99–132.

de Gramont, A. (2006) 'After the water war: the battle for jurisdiction in Aguas Del Tunari, S.A. v. Republic of Bolivia', *Transnational Dispute Management (TDM)*, vol, 3, no 5. Available at: www.transnational-dispute-management.com/article.asp?key=850 (accessed: 18 September, 2018).

Jacob, M. (2010) *International investment agreements and human rights*. INEF Research Paper Series Human Rights, Corporate Responsibility and Sustainable Development 03/2010. Institute for Development and Peace (INEF). Available at: http://humanrights-business.org/files/international_investment_agreements_and_human_rights.pdf (accessed: 10 November, 2016).

Klein, N. (2012) 'Human rights and international investment law: investment protection as human right', *Goettingen Journal of International Law*, 4, pp. 199–215.

Lapiashvili, N. (2014) 'Zhinvali Development Ltd. v. Republic of Georgia (ICSID Case No. ARB/00/1)', *Natia's Blog on Arbitration and EU Law*, 19 November. Available at: http://natialaw.blogspot.com/2014/11/zhinvali-development-ltd-v-republic-of.html (accessed: 18 September, 2018).

Meshel, T. (2015) 'Human Rights in investor-state arbitration: the human right to water and beyond', *Journal of International Dispute Settlement*, vol 6, no 2, pp. 277–307.

Parlett, K. (2011) *The Individual in the International Legal System: Continuity and Change in International Law*. Cambridge: Cambridge University Press.

Peterson, L.E. (2009) 'Selected developments in IIA arbitration and human rights', *IIA Monitor* (International Investment Agreements), 2. Available at: http://unctad.org/en/Docs/webdiaeia20097_en.pdf (accessed: 10 November, 2016).

Polanco Lazo, R. (2014) 'Is there a life for Latin American countries after denouncing the ICSID convention?', *Transnational Dispute Management (TDM)*, 11(1). Available at: www.transnational-dispute-management.com/article.asp?key=2037 (accessed: 17 March, 2014).

Polanco Lazo, R. and Gómez Fiedler, S. (2017) 'A requiem for the trans-pacific partnership: something new, something old and something borrowed?', *Melbourne Journal of International Law*, vol 18, no 2, pp. 1–50.

Radi, Y. (2013) *The 'Human Nature' of International Investment Law*. Grotius Centre Working Paper 2013/006-IEL. Netherlands: Leiden University. Available at: https://papers.ssrn.com/abstract=2278857 (accessed: 8 November, 2016).

Reiner, C. and Schreuer, C. (2009) 'Human rights and international arbitration', in Dupuy, P.-M., Francioni, F., and Petersmann, E.-U. (eds), *Human Rights in International Investment Law and Arbitration*. Oxford: Oxford University Press (International economic law series), pp. 82–96.

Sweet, A.S. and Mathews, J. (2008) 'Proportionality balancing and global constitutionalism', *Columbia Journal of Transnational Law*, vol 47, no 1, pp. 72–164.

Tienhaara, K. (2011) 'Regulatory chill and the threat of arbitration: a view from political science', in Brown, C. and Miles, K. (eds), *Evolution in Investment Treaty Law and Arbitration*. Cambridge: Cambridge University Press, pp. 606–628.

UNCTAD (2015) *Investment Policy Framework for Sustainable Development*. UNCTAD/DIAE/PCB/2015/5. Available at: http://unctad.org/en/PublicationsLibrary/diaepcb2015d5_en.pdf (accessed: 28 January, 2018).

UNCTAD (2018a) *International Investment Agreements Navigator, Investment Policy Hub*. Available at: http://investmentpolicyhub.unctad.org/IIA/IiasByCountry#iiaInnerMenu (accessed: 5 May, 2018).

UNCTAD (2018b) *Mapping of IIA Content, Investment Policy Hub*. Available at: http://investmentpolicyhub.unctad.org/IIA/mappedContent#iiaInnerMenu (accessed: 6 June, 2018).

United Nations Committee on Economic, Social and Cultural Rights (2017) *General comment No. 24 on State obligations under the International Covenant on Economic, Social and Cultural Rights in the context of business activities (E/C.12/GC/24)*. Available at: http://docstore.ohchr.org/SelfServices/FilesHandler.ashx?enc=4slQ6QSmlBEDzFEovLCuW1a0Szab0oXTdImnsJZZVQcIMOuuG4TpS9jwIhCJcXiuZ1yrkMD%2FSj8YF%2BSXo4mYx7Y%2F3L3zvM2zSUbw6ujlnCawQrJx3hlK8Odka6DUwG3Y (accessed: 20 September, 2018).

United Nations, Office for the High Commissioner for Human Rights (2002) 'General Comment No. 15 (2002) The right to water (arts. 11 and 12 of the International Covenant on Economic, Social and Cultural Rights)', *Agenda*, 11, p. 29.

Van Aaken, A. (2013) 'The interaction of remedies between national judicial systems and ICSID: an optimization problem', in Calamita, N. J., Earnest, D., and Burgstaller, M.

(eds), *The Future of ICSID and the Place of Investment Treaties in International Law*. London: British Institute of International and Comparative Law: Current Issues in Investment Treaty Law, pp. 291–324.

Vis-Dumbar, D. (2009) 'Norway shelves its draft model bilateral investment treaty', *Investment Treaty News*, 8 June. Available at: www.iisd.org/itn/2009/06/08/norway-shelves-its-proposed-model-bilateral-investment-treaty/ (accessed: 10 November, 2018).

Winkler, I.T. (2012) *The Human Right to Water: Significance, Legal Status and Implications for Water Allocation*. Oxford: Hart Publishing.

24 The human right to water in India

In search of an alternative commons-based approach in the context of climate change

Birsha Ohdedar

Introduction

Water is central to climate change. The impacts of climate change primarily centre upon too much water (e.g. floods, sea level rise), too little (e.g. droughts), or a shift in its composition (e.g. ocean acidification). The centrality of water to climate change and its importance to life mean that there are significant concerns about how climate change will impact the human right to water ("HRW") (Singh, 2016; UN High Commissioner for Human Rights, 2010). However, the underlying relationship between water and climate is also mediated by political, social and economic processes (Taylor, 2015; Boelens et al., 2016). Thus, the hydro-climatic injustices and rights breaches around water are intertwined with questions of gender, class, and caste, as well as the allocation of water for economic production, as much as they are caused by "environmental" or "climatic" processes (Linton, 2012; Taylor, 2015).

The dominant interpretation of the HRW both in legal and political discourse has been centred upon a fixed relationship between the rights-bearing subject and a particular quantity and quality of water (Linton, 2012). For example, in India, for many decades the dominant policy approach for implementing the HRW centred upon providing access to a fixed quantity of water for drinking, cooking, and domestic use (Cullet, 2017). As will be analysed, when one considers the way hydro-climatic change materialises this approach comes across as narrow and fails to respond to the way injustices and rights issues occur across the hydro-commons. Accordingly, it will be argued that the HRW needs to be reformulated to mean not just a consumption or entitlement right, but a right to transform the hydro-social conditions out of which water is currently accessed. Commons-based principles are integral to this remapping of the HRW, recognising the inter-connected social and ecological dimensions of water.

The second part of this chapter looks at the opportunities for reframing the HRW in this way in India. India is an interesting case study for several reasons. First, hydro-climatic issues in India have been particularly acute in recent years. Floods, droughts, heat waves and other climatic conditions have affected the country severely. India faces major issues around water pollution, scarcity, and access to basic water (Water Aid, 2016; Biswas, Tortajada, and Saklan, 2017). Second, India was one of the first countries to recognise the HRW, more than 20 years ago, having been read into the Constitutional right to life by the Supreme Court of India (*Subhash Kumar v State of Bihar & Ors*, 1991). The judiciary has also incorporated commons-based principles such as the common heritage of mankind and the public trust doctrine into Indian jurisprudence and linked them to the HRW. Finally, in recent years there has been an effort to update the national framework of water law and policy through the release of a Model Groundwater (Sustainable Management) Act, 2016, and Draft National Water Framework Bill, 2016. The second part of the chapter thus analyses the development of the HRW and common-based principles by the judiciary and legislature and the prospects of reframing the HRW.

Connecting the dots: water, climate, society and the commons

The impacts of climate change, such as extreme rainfall events, droughts, and sea level rise, will have significant implications for the global water cycle. The risks of climate change creating water scarcity is perceived as an important threat to the HRW (UN High Commissioner for Human Rights, 2010; Singh, 2016). However, it is not simply a physical water scarcity issue. Basic water use (for drinking domestic and subsistence livelihood) only takes a small fraction of physical water sources. Hence, while climate processes may undermine physical water availability in particularly regions, it does not follow that it will also undermine the HRW (Darrow, 2017, p. 175). As the United Nations Development Programme (UNDP) (2006, p. v) has stated, the roots of the "water crisis" can be traced to "poverty, inequality and unequal power relationships, as well as flawed water management policies that exacerbate scarcity".

The materialisation of climate processes and their impacts on access to drinking, domestic and livelihood water are dependent on several interacting processes. These include the climate and hydrological processes, and how social actors and institutions shape the flows of water (Linton, 2010; Linton and Budds, 2014). An example of how the impacts of droughts and floods on the HRW are tied to social processes such as power relationships, land ownership, and technology is provided by Taylor (2015). He observes a drought in 2012 in the Deccan Plateau in India. First, surface water becomes scarce. This scarcity is linked to climatic factors (the lack of rain) but also to the over consumption of water (by certain sectors) and water pollution which have dwindled water availability. Increasingly thirsty communities begin to extract more and more groundwater to meet their water needs. This undermines the production of fodder, which requires groundwater and is an important product for small farmers who hold cattle as their main livelihood. Distressed farmers are forced to sell cattle to richer merchants and landowners, producing relational dependency and increasing vulnerability. Moreover, rich merchants and landowners also start to control the extraction and distribution of groundwater. Drilling deep borewells is expensive and those who have access to capital can purchase the technology and gain significant power through not only depleting sources of their neighbours but also selling water extracted at a significant cost.

Taylor (2015) observes that the lines between what is hydrological, climatic or social in these situations become blurred. Moreover, the misallocation of water furthers inequitable relations around water and can be viewed as failure of adequately recognising water as a commons resource that is essential for all life. As will be analysed in the second half of this chapter, groundwater law in India gives individual landowners a licence to drill without concern for the wider social, ecological and commons-based dimensions of their groundwater use. Relational power between landowners, subsistence farmers, as well as the availability and access to credit and technology all play important roles in the production of hydro-climatic injustices that see the human rights of some realised over the rights of others. The intertwined hydrological, climatic and social processes described above thus underly the importance of a commons perspective towards water.

Developing a law for the commons

Commons-based principles put forward a governance system that recognises that water is a flow resource with social and ecological dimensions. It recognises that water has no owners and must be managed collectively. For example, Barlow (2011, p. 24) states that the water commons recognises that water:

> belongs to the earth, other species, and future generations as well as our own. Because it is a flow resource necessary for life and the health of the ecosystem, and because there is no substitute for it, water must be regarded as a public good to be preserved as such for all time in both law and practice.

There is no single definition of the commons. Different types of commons regimes exhibit different levels of individualism and communality, as well as exclusion (see Schmidt and Mitchell, 2014). For example, a communal pond shared by a specific group of individuals or households, akin to a common property resource, can be from the very start exclusive. On the other hand, a wider idea of water or atmospheric commons is more akin to prohibiting the appropriation of the resource. Furthermore, the role of the state in the commons can vary, and in some situations the state (or a group of states) could play an important role as a trustee of an area or resource, especially when we consider larger "global commons" (Bosselmann, 2015). The broader point here is the distinction between systems of water governance that are based on private property and individualistic rights to appropriate water and a system that recognises water as a resource that cannot be "owned" and is managed for the benefit of all (human and non-human natures).

Law and legal principles have an important role in the process of "communing". The principles of common heritage of humankind, as well as the public trust doctrine have often been used to challenge sovereignty-based or individual property-rights notions of water. The principle of common heritage emphasises that some resources or spaces are so important that they should not be appropriated. Rather the population manages such resources and shares in any rewards from exploiting them. While its exact scope can vary, certain elements are key characteristics when applied to common areas. These are that the resource cannot be appropriated, that it will be used for peaceful purposes, that its use and access will fall under a common management system and any benefits derived shall be equitably shared and reserved for future generations (Bosselmann, 2015, p. 76).

Similarly, the public trust doctrine reflects a commons-based understanding of water. The public trust doctrine holds that a group, a state or the international community holds a resource in trust for the public's use and enjoyment. Like a private trust, a trustee is identified (often the government) with a corresponding fiduciary duty. While the public trust doctrine is important in furthering a commons-based approach away from property rights, at the same time it has been highlighted that the notion of the public trust has no direct links with social concerns of human rights (Cullet, 2009, p. 44). Thus, how the public trust doctrine operates and how it can further the wider concerns of a HRW requires further clarification, for example through integrating human rights principles with the public trust doctrine.

Apart from changing the legal status of water, recognising the rights of people to participate in the governance of a common resource is integral to well-functioning commons. Participation can also be an important element to the HRW and, depending on the form it takes, a way to respond to the concerns about the individualism of rights. This aspect is explored later in this chapter. Participation and relatedly subsidiarity (the devolution of power to local democratic forms of government) are also integral if water is held by the state in public trust, to avoid giving unbridled powers to the State such that it would prevent democratic control over water.

Often, those who argue for a commons-based approach for water have also largely been supportive of the claims for a HRW (Barlow, 2011; Weston and Bollier, 2014). Weston and Bollier (2014) argue that a well-managed commons system should also, to the maximum extent possible, name human rights and nature rights an explicit and integral part of its governing system. However, it does not automatically flow that a commons regime and human rights are integrated. For instance, the dominant narratives on human rights have been criticised for being individualistic and having an underlying tension with the collectivists aspirations of the commons (Bakker, 2007). In the next section the tensions between the water commons and the HRW will be explored. However, a commons approach can be integrated with the human right to water, if the HRW itself is remapped or reimagined from its current form.

Human right to water – moving beyond a narrow approach

The HRW has received significant legal and political recognition over the last two decades. Under international law, the most significant development has been the adoption of General Comment 15 by the UN Committee on Economic, Social and Cultural Rights ("CESCR") recognising a right to water under the International Covenant on Economic Social and Cultural Rights ("ICESCR"). The CESCR interpreted a right to "sufficient, safe, acceptable, physically accessible and affordable water for personal and domestic uses" (UN Economic and Social Council, 2002).

In India, the judiciary has widened the scope of an existing constitutional right, the Right to Life that is guaranteed under Article 21 of the constitution, to include the HRW (*Subhash Kumar v State of Bihar* 1991). Over the last 20 years, the judiciary has confirmed the status of the right numerous times. However, with some exceptions, the Court has provided very little definition as to its scope and content. Furthermore, the legislature has largely ignored this development, failing to explicitly integrate the HRW into the National Water Policy 2012 or the National Rural Drinking Water Plan, the primary policy framework for implementing drinking water in rural areas.

Nevertheless, while the Government itself has not consistently and explicitly recognised the HRW in its policies, this does not mean it has not been active in efforts to implement the right (Cullet, 2017). For many years, drinking water policies (and thus the implementation on the HRW) centred upon providing access to a specific quantity of water for drinking, cooking, and domestic use. This reflects the broader paradigm at an international level, where the HRW has been interpreted in a narrow technocratic way. Even on discussions of expanding the HRW to recognise livelihood water uses, the quantification of allocation has been the focus (Woodhouse and Langford, 2009). A plausible explanation for this is that implementation of the HRW has been significantly influenced by development agencies, NGOs, and inter-governmental organisations that focus on indexes, indicators, and statistics in relation to developmental goals. However, a solely statistical approach to water access as a national policy to implement a constitutional right is inadequate, particularly as the right overlaps with several other rights such as the right to a healthy environment and the right to equality.

The inadequacy of this narrow approach can be observed in the seminal *Mazibuko* case from South Africa (*Mazibuko v City of Johannesburg*, 2009). This case centred upon the correct quantity of water the state had to provide to meet the requirements of the constitutionally guaranteed HRW. The Constitutional Court held that pre-paid water meters and supplying a minimum amount of water to residents was sufficient to meet the constitutional HRW. However, the Court deferred the question of "how much" water needed to be provided, stating that it was a matter for the government to decide. There was a significant focus by the plaintiffs on a minimum quantity of water and the responsibilities of the Government around providing it. Takacs (2016) contends that a major oversight by both lawyers and judges in the case was overlooking the ecological components of the HRW, particularly the responsibilities of the state under the commons-based public trust doctrine that is a principle of the National Water Act 1998. On the one hand, there may have been fears for the plaintiffs that reminding the Court of these requirements might reallocate water from the poor for ecological purposes, at least in the short term. On the other hand, Takacs argues that this presumption (of ecology versus providing for the poor) was false as only through managing water responsibly, according to public trust doctrine that is enshrined under South African law, could the minimum core quantity even be provided to all citizens. A holistic interpretation of the public trust would have forced the Courts to not only look at how water was shared between the poor, for water conservation, but also how it was wasted through leakages and the overconsumption of water by the rich. Thus, the public trust doctrine would require the Court to consider the HRW with reference to sustainability and stewardship of water and equity of water use. Such an interpretation moves us from the HRW as a "consumptive right" to recognising the ecological and social dimensions of water. The HRW would be closer to being recognised as a right to transform the hydro-social conditions out of which water is currently accessed (Linton, 2012; Bond, 2013; Goff and Crow, 2014). In the context of relational dependency to access water, inequitable water use, and climate vulnerability the right to water cannot be simply a right to a particular quantity of water. Rather the HRW needs to recognise water as a shared resource and crucially a right to participate and deliberate in the governance of the water commons.

This alternative framework for the HRW recognises water as first a flow resource constantly moves, from precipitation, condensation, infiltration, evaporation, and surface and subsurface flows. Second, that climatic, hydraulic, social, ecological and cultural processes also interact and transform some of the underlying assumptions around these flows. As water flows through the earth, it touches different domains and changes in use, management, and socio-political organisation. It is extracted, captured, polluted, manipulated, such that we cannot view it outside from these social, cultural and ecological worlds in what a number of scholars refer to as the "hydro-social" cycle (Swyngedouw, 2009; Linton and Budds, 2014). As the example provided earlier by Taylor (2015) illustrates, water access is mediated through this hydro-social cycle. In this way, the HRW can be used to frame a commons-based approach to water laws, moving towards a better recognition of the intertwined nature of these ecological, social and cultural dimensions.

Critical perspectives on the human right to water

While the HRW has been widely recognised and the above has argued for a reimagined HRW, there are significant doubts about the efficacy of human rights. Critical perspectives reflect on the value of human rights in general, as well as their compatibility with the collective aspirations of a water commons (Bakker, 2007). Human rights have been criticised for being individualistic and dependent on the autonomy of the individual as its subject (Bakker, 2007). Rights have been articulated in individualistic terms, designed to prohibit collective and interpersonal infringements. Linton (2012) writes that the HRW has been framed to mean a fixed relation between an individual human body and a quantity of water. This individualisation can be seen in how debate and litigation often centre around what quantity of water is adequate. "Rights talk" has also been accused of limiting the opportunities for real change to power, politics, and justice that give rise to hydro-climatic injustices (Douzinas and Gearty, 2012; Kennedy, 2012). In other words, it is argued that by articulating hydro-climatic concerns through rights, the processes that give rise to those injustices, are not adequately challenged.

To date, there has been a lively debate about the compatibility of the HRW with privatisation of water services (Sultana and Loftus, 2012). Movements against privatisation and commercialisation of water supply around the world have often asserted the HRW to defend water commons. However, Bakker (2007) has questioned whether framing water as a human right is the best way to recognise the ecological and social values of water. She argued that it is sloppy to assert that HRW was the opposite of privatisation. To oppose privatisation, she asserts, is to go to the heart of property rights. Individual human rights did not foreclose private sector involvement in the water sector. Rather, the HRW as "anthropocentric and individualistic" is fully compatible with the privatisation. Organisations, such as the World Bank that were influential in the drive for privatisation, recognised the HRW while also advocating privatisation and commercialisation of water (Salman and McInerney-Lankford, 2004).

Notwithstanding the shortcomings of human rights and the interpretations of the HRW to date, there is still merit in engaging with the right towards reframing it to a commons-based framework for several reasons. First, while the HRW may be compatible with privatisation under its current interpretation, this does not mean that it is the best strategy for the realisation of the HRW for all. In other words, a commons-based approach is also compatible with the HRW and may also be preferable. Academics and activists are thus constantly engaged in reframing the HRW (Sultana and Loftus, 2012; Harris, Rodina, and Morinville, 2015). Second, a common criticism is that "rights-talk" can narrow the radical possibilities of change, but this view is not universal. O'Connell (2018) has argued that while there is merit in the critique of rights narrowing radical change, when one looks closely at the practices of social movements and human agency in rights based activism, it does not necessarily prove a narrowing of claims. For example, he finds that the "Right2Water" struggle in Ireland mobilised both a human rights struggle and a broader campaign against austerity, neo-liberalism and structural causes of injustice. Similarly, Bakker (2012, pp. 37–38), in a post-script to her 2007 article that questioned the HRW, writes that defending and extending the HRW remains necessary and a crucially useful tactic in activism. Like O'Connell, she believes that rights can also be recaptured and interpreted more broadly. Third, while there are conceptual criticisms of human rights from critical legal studies scholars, there is still an enduring appeal to human rights. While critiquing the western human rights paradigm, Rajagopal argues that unlocking the transformative or counter-hegemonic potential of the human rights discourse requires to focus on the voices and perspectives of the historically marginalised, including through expanding the scope of socio-economic rights and solidarity rights (Rajagopal, 2005, 2007). Clark (2017) notes that the despite unfavourable result from the HRW point of view in the *Mazibuko* judgment in South Africa, the social movement does not regret its decision to frame the issue as a rights issue and litigate.

Human right to water and the commons in India

Building on the discussions above, this final section considers how the HRW in India has begun to link many aspects of the commons through both judicial decisions and legislative and policy reforms. Furthermore, this section considers the prospects for a reformulated HRW. The reforms to date need to be viewed in their socio-political context to ascertain the prospects towards reimaging the HRW. Accordingly, this section first introduces the legal framework in India and then analyses how commons-based perspectives have been introduced in the water law framework. Looking ahead, the discussion will emphasise the legislative direction being pushed by the Union government.

Legal framework of water in India

India lacks an umbrella water framework law to regulate water in all its dimensions. Rather, the laws and policies of water have developed in a piecemeal fashion. The Constitution of India delineates responsibilities and powers between the Union government, states, and local bodies (rural *panchayats* and municipalities). Under the Constitution, the primary competence for regulating water is given to each state. This reflects a need to regulate across a variety of conditions and needs. The Constitution also recognises a number of fundamental rights, and while it does not explicitly recognise the HRW, the judiciary has repeatedly confirmed its existence through reading it in under Article 21, the right to life (*Subhash Kumar v State of Bihar & Ors*, 1991; *MC Mehta v Kamal Nath*, 1997). The role of the judiciary has been integral to the development of human rights, and environmental rights, with rights-based public interest litigation since the 1980s leading to a substantial rights-based jurisprudence.

Ownership and control of water

The legal status of water is central to whether a common-based governance system is possible or not. A property right over water is a controversial subject because of how important water is to the survival of human societies. The history of water in Indian society suggests that private ownership of flowing water was generally prohibited even though various forms of private appropriation, such as through tanks, were accepted (Cullet and Gupta, 2009, p. 157). However, the need for some form of control and regulation within this system was also present (Cullet and Gupta, 2009, p. 162). Since the nineteenth century, access to water has been largely controlled by English common law principles. Rights over water have been linked to control over land. Water law has also developed along different rules for surface water and groundwater, partly due to a lack of knowledge around groundwater. While flowing surface water could not be owned, the development of riparian rights gave individual landowners the right to appropriate water flowing through a river for their own private use. Such riparian rights are indirectly codified in the Indian Easements Act, 1882, and have been further developed through litigation (Cullet and Koonan, 2017, p. 62).

A different set of rules developed for groundwater. Originally, this was because the link between surface water and groundwater was not known when the rules were developed. The prohibition of ownership of surface water did not apply and groundwater came to be seen as a chattel of the land, as it developed in both case law and the Indian Easements Act, 1882 (Cullet and Koonan, 2017, p. 62). Hence, the legal position in India today remains that landowners have an uncontrolled right to extract groundwater from the land. No legal action can be taken against a landowner who has overexploited groundwater and depleted neighbouring wells. This framework of land-based groundwater rights creates large scale inequalities, ecological depletion, and rights issues for landless and poor (Koonan, 2016). Groundwater law, in particular, remains conceptually disconnected from the global water cycle, and the existing legal framework for groundwater has been described as "insensitive to, and unprepared to deal with, climate-change related challenges and implications" (Cullet, Bhullar, and Koonan, 2017, p. 649).

Invoking the commons through the judiciary

There have been a number of water-related decisions in India, in particular since the early 1990s as public interest litigation, as well as water and environment related conflicts have risen. This sub-section considers the decisions relating to the HRW and commons-based legal principles, including the socio-political context of those decisions.

The judiciary in India have been active in recognising the HRW, as well as several commons-based principles. In 1997, the Supreme Court held that the public trust applied to all "running waters" (i.e. surface waters) and that the Government as the trustee is required to "protect the resources for the enjoyment of general public rather than to permit their use for private ownership or commercial purpose" (MC *Mehta v Kamal Nath*, 1997, pa ra. 25). Three years later, in *MI Builders Private Ltd v Radhey Shyam Sahu* (1999) the Supreme Court commented that the "public trust doctrine in our country, it would appear, has grown from Article 21 of the Constitution". Accordingly, in a short span of time, the Court not only confirmed the application of the public trust doctrine in India but directly linked it to the HRW that is recognised under Article 21. The Court has further clarified that there is a distinction between the Government's duties to act "for public benefit" and the "special, more demanding obligation which it may have as a trustee of certain public resources" (*Intellectuals Forum, Tirupathi v State of A.P. and Ors.*, 2006). Thus, the Court was asserting firmly that there was a different set of responsibilities for the state in its role as a trustee, being a duty to protect and manage such resources for the benefit of all.

The Courts have on occasions also recognised a common heritage of water. In both MC *Mehta v Kamal Nath* (1997) and *Intellectuals Forum* (2006, para. 76) the Supreme Court linked public trust to the principle of common heritage. The Court stated that:

> it is an affirmation of the duty of the state to protect the people's common heritage of streams, lakes, marshlands and tidelands, surrendering that right of protection only in rare cases when the abandonment of that right is consistent with the purposes of the trust.

Nevertheless, while this phrase introduced the idea of common heritage and linked this to public trust and water, it has never resulted in a more comprehensive approach towards recognition of water as common heritage of humankind. Groundwater (the primary source for people in India) was not included in the wording of the Court.

On the face of it, these cases suggest a shift towards a commons-based HRW. Takacs (2008) has suggested that the Indian judiciary has put the public trust doctrine "in service" of constitutionally guaranteed environmental rights, bolstering the demands on the government to advance the HRW on a common-based platform. However, there are reasons to be sceptical as there are still several major gaps and uncertainties. First, there is still uncertainty about the application of the public trust doctrine to groundwater. The Plachimada cases, which centred upon the Coca-Cola's over exploitation of a village aquifer initially extended the public trust doctrine to groundwater (*Perumatty Grama Panchayat v State of Kerala*, 2003). However, on appeal, the division bench of the High Court overturned the decision and rejected the scope of the public trust doctrine (*Hindustan Coca-Cola Beverages (P) Ltd v Perumatty Grama Panchayat*, 2005). Confusingly, in between these two cases, the Supreme Court mentioned in passing that the public trust doctrine extends to groundwater (*State of West Bengal v Kesoram Industries*, 2004). But, as the *Kesoram* case was not primarily about water, the question of the application of the public trust doctrine to groundwater remains uncertain. Second, national and state legislation have failed to recognise the change in legal status of water or asserted a different status. For the most part the legislature has ignored the Court's judgments. For example, the Jammu and Kashmir Water Resources (Regulation and Management) Act 2010 asserts that the government owns all the water in the state (section 3). Finally, the judiciary itself has also been inconsistent it its conceptual understanding of the public trust, for example stating that public trust doctrine "does not exactly prohibit the alienation of the property held as a public trust" and thus leaving scope of the appropriation of resources held under public trust (*Intellectuals Forum, Tirupathi v State of A.P. and Ors.*, 2006; see also: *Susetha v State of Tamil Nadu*, 2006).

The actions of the judiciary must be viewed in the socio-political context of public interest litigation ("PIL") in India today. The Court has been influential over the last 30 years in expanding Article 21, the right to life, to incorporate a range of human rights such as those relating to water, environment, food, health, and livelihoods. It has also introduced several commons-based principles and linked them to Article 21 as described above. However, critics have argued that the market-based neo-liberal ideologies and aspirations of the state have enveloped the Courts, including their interpretations of socio-economic rights. Whereas in the 1980s and the early 1990s, the rise of PIL was seen as a ray of hope for the poor, as well as for the broader landscape of Indian law and justice (Baxi, 1985), by the early 2000s the Court was said to be betraying a "lack of sensitivity towards the right of the poor and disadvantaged" (Bhushan, 2004). The Court became increasingly individualistic in its reasoning. For instance, when there were petitions to stop slum demolitions that would render poor, mostly migrant residents homeless, the Court shot back that "nobody forced you to come to Delhi. Is there a right to live in Delhi only? Stay where you can. If encroachment on public land are to be allowed, there will be anarchy" (Mahapatra, 2006).

Such an ideology has also affected the judiciary's interpretations of the HRW, in the *Narmada* case the right to water was affirmed but only for the beneficiaries of the dam (urban dwellers), not those who were affected by the construction of the dam and reallocation of water (mostly rural tribal populations) (*Narmada Bachao Andolan v Union of India and Others*, 2000). Similarly, the High Court of Delhi held that the pollution of the river Yamuna was caused by not just industrial and medical waste but also the "encroachments" of slums where sewage flows into the river (*Wazirpur Bartan Nirmata Sangh v Union of India*, 2006a, para. 9). The court chose not to focus on the rights issues of the people living in the slums, including their rights to a healthy environment, water, and sanitation. Rather, it stated that the slum dwellers had no right to their houses and that the residents of Delhi had a right to "clean potable water from the river Yamuna and health and friendly environment from its bed and embankment" as a constitutional right (*Wazirpur Bartan Nirmata Sangh v Union of India*, 2006b; see also Cullet, 2014). Thus, the right to water was protected for certain people over others. More recently, in 2014 an interim order confirmed that the HRW introduces the idea that the right can apply differentially depending on the legal status of one's dwelling, thus limiting the rights of slum dwellers (*Pani Haq Samiti v Brihan Mumbai Municipal Corporation*, 2014).

The Court has also been viewed as adopting a particular definition of the environment that has centred on protecting aesthetics and the "leisure and lifestyle" of the urban middle classes (coined as "bourgeoise environmentalism" by Baviskar (2002)). Accordingly, the "environment" has been used as a reason to demolish slums and render people homeless, based on the "unhygienic" conditions of not having drainage infrastructure that was never provided by the state. Moreover, Menon (2014) points out that in addition to privileging the "environment" over people, the Court has also adopted a formula of "development over environment". So, when it comes to large development projects such as the Narmada or Tehri dam, decisions are made without concern for the environment or local populations through utilitarian and economic rationality. Accordingly, the state is justified in its activities based on a logic of a "justifiable sacrifice" by affected communities (Rajagopal, 2005).

Opportunities to further a commons-based HRW through the Court in the future must be seen in this context. The Court, as the previous paragraphs indicate, have increasingly taken an approach to the HRW which does not recognise water beyond its utility as an economic good. Furthermore, ecological dimensions are only emphasised in a limited capacity. Recently, the Supreme Court had the opportunity to provide more content to the HRW, as well as confirm application of the public trust doctrine over groundwater in the appeal of the Plachimada case against Coca-Cola mentioned earlier. However, the Court avoided a substantive statement as it said Coca-Cola was no longer operating in the region ("Coca-Cola not to go back to Plachimada", 2017). The pronouncements of the Court over the years, through invoking the HRW and several common-based principles have been significant in giving legal authority to rights and principles. However, avenues of litigation through the Court can only be one tool in a box of strategies. Moreover, one that may provide significant pitfalls for activists.

Invoking the commons through the legislature

The development of water law in India has also been fragmented. First, many rules, statutes, and policies have been developed on a sectoral basis. Much of water legislation and policy has been driven by a need to harness the "productive use" of water (such as irrigation or industrial uses) and have not paid enough attention to socio-ecological and rights-based considerations (Cullet and Koonan, 2017, pp. 5–7). The underlying assumption has been a private right to use water without considering the impacts across a wider hydro-social cycle. Second, several critical gaps exist in the assumptions underpinning the rules around water. For example, there are different rules for ground and surface water, without sufficient understanding of how they are connected. Third, several different sources and instruments make up the water law framework in India. Apart from the statutes and judicial orders, administrative directions and policies have also played an important role in the development of water law in India. Particularly, in the context of the HRW, its implementation has largely been through administrative directions by the executive on drinking water.

As mentioned earlier, the legislature at both national and state levels have largely ignored the judicial developments on human rights and commons-based principles. Only recently, in the context of a widely acknowledged water crisis across the country, the legislature has been forced to take further action. Two proposed reforms are worth noting:

1. The Draft National Water Framework Bill 2016 ("Water Framework Bill") aims to provide a comprehensive approach to water regulation providing a set of common principles and assumptions.
2. The Model Groundwater (Sustainable Management) Act, 2016 ("Model GW Act"), which updates a 2012 version, offers a fresh approach to

comprehensive groundwater regulation in tune with its importance to communities and ecosystems (Cullet, 2012).

Constitutional division of responsibility between states and the centre over water means that the Water Framework Bill can only translate into actual national level law if two or more states pass a resolution recognising the Water Framework Bill. In that case, the Union government can adopt the law, and it would be applicable for the states that passed such resolutions. Otherwise, it aims to be a "model bill" for states to pass their own legislation based on the Water Framework Bill 2016 and adapted to local conditions. Similarly, Model GW Act is a model legislation that individual states can use frame their own legislation. Accordingly, individual states now have an important role in how these legislative reforms will move ahead.

Importantly, the legislative reforms seek to restate water as a human right, a public trust and common heritage resource. First, water is recognised as a fundamental right for all (Draft National Water Framework Bill, 2016, sec. 3). While the state may delegate the provision of water services to private agencies, it cannot delegate or evade from its human rights duties. Furthermore, under the fundamental HRW, the privatisation of water itself is prohibited (Draft National Water Framework Bill, 2016, sec. 3). Second, water is recognised as "a common heritage of the people of India" and held in public trust, not amenable to ownership by anyone (Draft National Water Framework Bill, 2016, sec. 4(1)). The state is entrusted to protect, preserve and conserve water, and passes this on to future generations (Section 4(4)). Accordingly, these sections assert the way water should be managed, that is through commons-based principles of stewardship, intergenerational justice, and sustainable development. Third, public trust and common heritage recognition also extends to groundwater, as the Water Framework Bill 2016 recognises the unity of groundwater and surface water (Water Framework Bill, 2016, preamble). Furthermore, the Model GW Act confirms the same and the prohibition of ownership of groundwater (Model Groundwater (Sustainable Management) Act, 2016, sec. 9). In the same section, a link is made towards ensuring that groundwater exploitation does not deprive neighbouring users' fundamental rights (Model GW Act, 2016, sec 9(4)). Thus, the Model GW Act 2016 also links the HRW directly with common heritage and public trust.

The recognition of water in public trust and as a common heritage resources is a significant step. Nevertheless, an earlier version of the Water Framework Bill, drafted in 2011 by a sub-committee of the Planning Commission of India, went further to recognise water as a "common natural heritage of humanity" rather than just of the "people of India". It went on to recognise water as "a bounty of nature to be shared with all other forms of life, with fellow humans of one's own and other groups, villages, States and countries, and with future generations" (Draft National Water Framework Act, 2011, sec. 3). The difference would be to recognise water as a true flow resource, in its global (and inter-generational) form. Such a provision conceptually links water from a national to the global hydro-commons, having implications for climate change and transboundary water issues. While the political nature of the reform procedure likely saw this wording change, from a commons-based perspective this represents a lost opportunity to truly connect water to the global water cycle.

Subsidiarity, participation and the HRW

Finally, as mentioned in the first part of this chapter, participation is important to the integrity of a commons-based governance system. Moreover, it is a key element to a HRW that is based on the right to transform the hydro-social conditions of access to water. Participation in water governance in India has taken two divergent paths. The first is under the 73rd and 74th Amendments to the Constitution, adopted in 1992, where local bodies of governance in rural and urban areas have been given significant power in the water context, including over drinking water supply and domestic water uses. These reforms envision a much more participatory water governance framework, through decisions around water being made through local democracy. Nevertheless, the implementation of such decentralisation has been uneven across the country. This can often be because states themselves have not incorporated such reforms into their own framework.

The second path developed is based on viewing water as an economic good to foster the management and efficiency of use. This has largely arisen from developments in international policy and pushed by developmental banks. Here, participation is separate from one's constitutional right, or democratic local body, but rather based upon one's ability to pay. For example, the *Swajaldhara* drinking water scheme is premised on the principle that the local community is able to participate, but only individuals who are able to pay their share of the capital costs can do so (Cullet, 2007). This form of participation is antithetical to the universality of the HRW, or the inclusiveness of commons-based governance systems.

To date, the participation in water governance in India has grown in these parallel paths and remains incomplete in making the relevant links to the public trust, commons-based governance, and the HRW. As discussed above, neither the judiciary nor the legislature have provided enough content or broadened the scope of the HRW. But, the recent law reforms attempt to bring some coherence to these issues as both the Water Framework Bill 2016 and the Model GW Bill 2016 link the principle of subsidiarity with the public trust. This means that a "trustee" should be the lowest possible democratically elected body that can regulate an entire waterbody or aquifer. The Model GW Bill (sec. 6) for example, integrates subsidiarity and decentralisation as a basic principle. Conservation and sustainability innovations, such as groundwater security plans, to protect groundwater protection zones are to be prepared by the lowest possible administrative level that encompasses the whole aquifer (Section 13, Model GW Bill 2016). These reforms try to reconcile the divergent trends in participation and foster a form of democratic participation that is based upon the Constitution.

For our purposes, it also furthers the goal of water being governed in commons (based upon public trust doctrine and common heritage) and grounded upon the HRW.

It is important to remember that reforms in legislation and policy are not a panacea. First, while subsidiarity means decisions are made at the lowest level, water as a flow resource is both local and global, as is hydro-climatic change. While local level decision making is seen as appropriate because access to basic water is often mediated at the local level, there is sufficient scope to integrate participative decision making across all levels of government, as well as improve the co-ordination between different local governments that are in the same watershed or river-basin. A second serious concern is that local democracy and decentralisation may not necessarily mean equal participation of all people on the ground. "Elite capture", where participatory processes are co-opted by powerful groups (including the elites of political parties), remain an important critique to decentralised and local commons based governance (Kothari, 2001; Chakrabarti, 2016). Moreover, concerns of gender and caste are particularly important in the Indian context. However, both these concerns do not negate the potential benefits of a participatory HRW. Rather, they emphasise the need for a greater emphasis of the principles of the commons, human rights, deliberation, as well as strengthening local democracy.

Conclusion

There is a long way to go for the HRW to be transformed into a commons-based human right. This chapter has argued that such a right is necessary to reconfigure socio-natural relations around water, climate and society. In the Indian context, the strides of the judiciary in the last twenty years to recognise and introduce concepts of HRW, public trust and common heritage provide the basis for a commons-based approach. However, the fragmented nature of water law and policy in India has meant this has never translated in a coherent way. Moreover, the judiciary's approach in recent times means activism towards a broader HRW through the Courts is problematic. Recent legislative reforms provide hope that further coherence can be brought to Indian water law. Importantly, such reforms have the fundamental elements to gradually transform water as a common resource, based on human rights and inclusive participation. Ultimately, much of this now rests with how such legislation is taken up by individual state governments. An opportunity remains for civil society and activists to use these legal tools and principles to demand for a broader commons-based human right.

References

Bakker, K. (2007) 'The "commons" versus the anti-privatization and the human right to water in the global south', *Antipode*, vol 39, no 3, pp 430–455.

Bakker, K. (2012) ‘Commons versus commodities. Debating the human right to water’, in Loftus, A. and Sultana, F. (eds), *The Right to Water. Politics, Governance and Social Struggles*. Oxford: Earthscan, pp 19–44.

Barlow, M. (2011) ‘Our Great Lakes Commons: A People’s Plan to Protect the Great Lakes Forever. Council of Canadians’, www.onthecommons.org/sites/default/files/GreatLakes-Final-Mar2011(2).pdf, accessed 16 August, 2018.

Baviskar, A. (2002) ‘The politics of the city. Seminar, (516)’, www.india-seminar.com/2002/516/516%20amita%20baviskar.html, accessed 16 August, 2018.

Baxi, U. (1985) ‘Taking suffering seriously: social action litigation in the Supreme Court of India’ *Third World Legal Studies*, vol 4, no 6, pp 107–132.

Bhushan, P. (2004) ‘Supreme Court and PIL’, *Economic & Political Weekly*, vol 39, no 18, pp 1770–1774.

Biswas, A.K., Tortajada, C., and Saklan, U. (2017) ‘India is facing its worst water crisis in generations’, https://qz.com/931878/india-is-facing-its-worst-water-crisis-in-generations/, accessed 16 August, 2018.

Boelens, R., Hoogesteger, J., Swyngedouw, E., Vos, J., and Wester, P. (2016) ‘Hydrosocial territories: a political ecology perspective’, *Water International*, vol 41, no 1, pp 1–14.

Bond, P. (2013) ‘The “right to the city” limits to rights talk and the need for rights to the commons’ *Theomai*, no 27–28, pp 42–62.

Bosselmann, K. (2015) *Earth Governance: Trusteeship of the Global Commons*. Cheltenham: Edward Elgar Publishing.

Chakrabarti, B. (2016) *Participation at the Crossroads: Decentralisation and Water Politics in West Bengal*. New Delhi: Oriental Blackswan.

Clark, C. (2017) ‘Of what use is a deradicalized human right to water?’ *Human Rights Law Review*, vol 17, no 2, pp 231–260.

Coca-Cola not to go back to Plachimada (2017, July 14). *The Hindu*. www.thehindu.com/news/national/kerala/not-going-back-to-plachimada-coca-cola/article19276028.ece, accessed 16 August, 2018.

Cullet, P. (2007) ‘Water Law in India: Overview of Existing Frameworks and Proposed Reforms’. IELRC Working Paper 2007-01. Retrieved from www.ielrc.org/content/w0701.pdf, accessed 16 August, 2018.

Cullet, P. (2009) *Water law, poverty, and development: water sector reforms in India*. New York: Oxford University Press.

Cullet, P. (2012) ‘The groundwater model bill: rethinking regulation for the primary source of water’, *Economic & Political Weekly*, vol 47, no 45, pp 40.

Cullet, P. (2014) ‘A meandering jurisprudence of the court’, in Suresh, M. and Narrain, S. (eds), *The Shifting Scales of Justice: The Supreme Court in Neo-liberal India*. New Delhi: Orient BlackSwan, pp 139–161.

Cullet, P. (2017) ‘The right to water in rural India and drinking water policy reforms’, in Langford, M. and Russell, A.F.S. (eds), *The Human Right to Water*. Cambridge: Cambridge University Press, pp 677–694.

Cullet, P., Bhullar, L., and Koonan, S. (2017) ‘Regulating the interactions between climate change and groundwater: lessons from India’, *Water International*, vol 42, no 6, pp 646–662.

Cullet, P. and Gupta, J. (2009) ‘India: evolution of water law and policy’, in Dellapenna, J. and Gupta, J. (eds), *The Evolution of the Law and Politics of Water*. Dordrecht: Springer, pp 157–173.

Cullet, P. and Koonan, S. (2017) *Water Law in India: An Introduction to Legal Instruments (2nd Edition)*. New Delhi: Oxford University Press.

Darrow, M. (2017) 'Climate change and the right to water', in Langford, M. & Russell, A.F.S. (eds), *The Human Right to Water: Theory, Practice and Prospects*. Cambridge and New York: Cambridge University Press, pp 144–173.

Douzinas, C. and Gearty, C. (2012) 'Introduction', in Gearty, C. and Douzinas, C. (eds) *The Cambridge Companion to Human Rights Law*. Cambridge: Cambridge University Press, pp 1–14.

Draft National Water Framework Act (2011) www.ielrc.org/content/e1148.pdf, accessed 16 August, 2018.

Draft National Water Framework Bill (2016) www.ielrc.org/content/e1604.pdf, accessed 16 August, 2018.

Goff, M. and Crow, B. (2014) 'What is water equity? The unfortunate consequences of a global focus on "drinking water"', *Water International*, vol 39, no 2, pp 159–171.

Harris, L.M., Rodina, L., and Morinville, C. (2015) 'Revisiting the human right to water from an environmental justice lens', *Politics, Groups, and Identities*, vol 4, pp 1–6.

Hindustan Coca-Cola (P) Ltd v Perumatty Grama Panchayat, 2 KLT 554 (High Court of Kerala 2005).

Intellectuals Forum, Tirupathi v State of A.P. and Ors., 3 SCC 549 (Supreme Court of India 2006).

Kennedy, D. (2012) 'The international human rights regime: still part of the problem?', in Dickinson, R., Katselli, E., Murray, C., and Pedersen, O.W. (eds.), *Examining Critical Perspectives on Human Rights*. Cambridge: Cambridge University Press, pp 19–34.

Koonan, S. (2016) 'Revamping the groundwater regime in India: towards ensuring equity and sustainability', *Socio-Legal Review*, vol 12, no 2, pp 45–73.

Kothari, U. (2001) 'The case for participation as tyranny', in Cook, B. and Kothari, U. (eds), *Participation: The New Tyranny*. London: Zed Books, pp 1–15.

Linton, J. (2010) *What is Water? The History of a Modern Abstraction (Nature/History/Society)*. Vancouver: University of British Columbia Press.

Linton, J. (2012) 'The human right to what? Water, rights, humans and the relation of things', in Loftus, A. and Sultana, F. (eds.), *The Right to Water: Politics, Governance and Social Struggles*. Oxford: Earthscan, pp 45–60.

Linton, J. and Budds, J. (2014) 'The hydrosocial cycle: defining and mobilizing a relational-dialectical approach to water', *Geoforum*, vol 57, pp 170–180.

Mahapatra, D. (2006, May 10) 'Encroachers have no right over public land', *Times of India*. https://timesofindia.indiatimes.com/city/delhi/Encroachers-have-no-right-over-public-land/articleshow/1523279.cms, 16 August, 2018.

Mazibuko v City of Johannesburg, JDR 1030 (CC 2009).

MC Mehta v Kamal Nath, 1 SCC 388 (Supreme Court of India 1997).

MC Mehta vs Kamal Nath & Ors, 1 SCC 388 (1997).

Menon, N. (2014) 'Environment and the will to rule: Supreme Court and Public Interest Litigation in the 1990s', in Suresh, M. and Narrain, S. (eds.), *The Shifting Scales of Justice: The Supreme Court in Neo-liberal India*. New Delhi: Orient BlackSwan, pp 59–78.

MI Builders Private Ltd v Radhey Shyam Sahu, AIR 2468 (Supreme Court of India 1999).

Model Groundwater (Sustainable Management) Act (2016). www.ielrc.org/content/e1605.pdf, 16 August, 2018.

Narmada Bachao Andolan v Union of India and Others, 10 SCC 664 (Supreme Court of India 2000).

O'Connell, P. (2018) 'Human rights: contesting the displacement thesis', *Northern Ireland Legal Qaurterly*, vol 69, no 1, pp 19–35.

Pani Haq Samiti v Brihan Mumbai Municipal Corporation, Public Interest Litigation No. 10 of 2012 (High Court of Judicature, Bombay 2014).

Perumatty Grama Panchayat v State of Kerala, 1 KLT 731 (High Court of Kerala 2003).

Rajagopal, B. (2005) 'The role of law in counter-hegemonic globalization and global legal pluralism: lessons from the Narmada Valley struggle in India', *Leiden Journal of International Law*, vol 18, no 3, pp 345–387.

Rajagopal, B. (2007) 'Pro-human rights but anti-poor? A critical evaluation of the Indian Supreme Court from a social movement perspective', *Human Rights Review*, vol 8, no 3, pp 157–186.

Salman, S.M.A., and McInerney-Lankford, S. (2004) *The Human Right to Water: Legal and Policy Dimensions*. Washington, DC: World Bank.

Schmidt, J.J. and Mitchell, K.R. (2014) 'Property and the right to water toward a non-liberal commons', *Review of Radical Political Economics*, vol 46, no 1, pp 54–69.

Singh, N. (2016) 'Climate Change and human right to water: problems and prospects', in *The Human Right to Water: From Concept to Reality*. New York: Springer, pp 83–103.

State of West Bengal v Kesoram Industries, 10 SCC 201 (Supreme Court of India 2004).

Subhash Kumar v State of Bihar & Ors, 1 SCC 598 (1991).

Sultana, F. and Loftus, A. (eds) (2012) *The Right to Water: Politics, Governance and Social Struggles*. Oxford: Earthscan.

Susetha v State of Tamil Nadu, 3418 Appeal (civil) (Supreme Court of India 2006).

Swyngedouw, E. (2009) 'The Political economy and political ecology of the hydro-social cycle', *Journal of Contemporary Water Research and Education*, no 142, pp 56–60.

Takacs, D. (2008) 'The public trust doctrine, environmental human rights, and the future of private property', *New York University Environmental Law Journal*, vol 16, p 711.

Takacs, D. (2016) 'South Africa and the human right to water: equity, ecology and the public trust doctrine', *Berkeley Journal of International Law*, vol 34, no 2, pp 55–108.

Taylor, M. (2015) *The Political Ecology of Climate Change Adaptation: Livelihoods, Agrarian Change and the Conflicts of Development*. Oxford: Routledge.

UN Economic and Social Council (2002) *General Comment No. 15: The Right to Water (Arts. 11 and 12 of the Covenant) (No. E/C.12/2002/11)*. UN Committee on Economic, Social and Cultural Rights (CESCR).

UN High Commissioner for Human Rights (2010) *Climate Change and the Human Rights to Water and Sanitation*, 1–6.

UNDP (ed.) (2006) *Beyond Scarcity: Power, Poverty and the Global Water Crisis*. New York: UNDP.

Water Aid (2016) *Water: At What Cost? The State of the World's Water 2016*.

Wazirpur Bartan Nirmata Sangh v Union of India (High Court of Delhi September 29, 2006).

Wazirpur Bartan Nirmata Sangh v Union of India (High Court of Delhi June 1, 2006).

Weston, B. and Bollier, D. (2014) *Green Governance: Ecological Survival, Human Rights and the Law of the Commons*. Cambridge: Cambridge University Press.

Woodhouse, M. and Langford, M. (2009) 'There is no human right to water for livelihoods', *Waterlines*, vol 28, no 1, pp 5–12.

Index

For Product Safety Concerns and Information please contact our EU representative GPSR@taylorandfrancis.com Taylor & Francis Verlag GmbH, Kaufingerstraße 24, 80331 München, Germany